# Conflicts in Environmental Regulation and the Internationalisation of the State



This book examines the global regulation of biodiversity politics through the UN Convention on Biological Diversity (CBD), the WTO and other international treaties. Using historical-materialist state and regulation theory, it assesses how the discourse and politics of sustainable development have contributed to the internationalisation of the state.

The authors argue that sustainable development, far from being a fixed concept, is a conceptual terrain on which different and conflicting symbolisations of and responses to the ecological crisis struggle for hegemony. Furthermore, they show that the multilateral environmental agreements are not at all a means to counteract neoliberal globalisation but, on the contrary, form an integral part of the ongoing transformation process. Focusing on the CBD, the FAO International Treaty on Plant Genetic Resources for Food and Agriculture and the Agreement on Trade-related Aspects of Intellectual Property Rights (TRIPS) in the World Trade Organisation, this co-authored volume addresses the following issues:

- state theory, regulation theory and International Political Economy
- biodiversity protection and valorisation of genetic resources
- access to genetic resources and sharing of benefits which arise out of its use
- enforcement of intellectual property rights and their impact on biodiversity.

This book will be of interest to students and scholars of international politics, International Political Economy, Environmental Studies, Development Studies and Political Ecology.

**Ulrich Brand** is Professor of International Politics at Vienna University, Austria.
**Christoph Görg** is Professor of Environmental Governance at Kassel University and Senior Researcher at the Helmholtz Centre for Environmental Research-UFZ in Leipzig, Germany.
**Joachim Hirsch** is Emeritus Professor in the Department of Social Sciences, University of Frankfurt, Germany.
**Markus Wissen** is Assistant Professor of the Institute for Political Science at Vienna University, Austria.

## RIPE Series in Global Political Economy



This series, published in association with the *Review of International Political Economy*, provides a forum for current debates in international political economy. The series aims to cover all the central topics in IPE and to present innovative analyses of emerging topics. The titles in the series seek to transcend a state-centred discourse and focus on three broad themes:

- the nature of the forces driving globalisation forward
- resistance to globalisation
- the transformation of the world order.

The series comprises two strands:
The *RIPE Series in Global Political Economy* aims to address the needs of students and teachers, and the titles will be published in hardback and paperback. Titles include:

**Transnational Classes and International Relations**
*Kees van der Pijl*

**Global Political Economy**
Contemporary theories
*Edited by Ronen Palan*

**Gender and Global Restructuring**
Sightings, sites and resistances
*Edited by Marianne H. Marchand and Anne Sisson Runyan*

**Ideologies of Globalization**
Contending visions of a new world order
*Mark Rupert*

**The Clash within Civilisations**
Coming to terms with cultural conflicts
*Dieter Senghaas*

**Global Unions?**
Theory and strategies of organized labour in the global political economy
*Edited by Jeffrey Harrod and Robert O'Brien*

**Political Economy of a Plural World**
Critical reflections on power, morals and civilizations
*Robert Cox with Michael Schechter*

**A Critical Rewriting of Global Political Economy**
Integrating reproductive, productive and virtual economies
*V. Spike Peterson*

**Contesting Globalization**
Space and place in the world economy
*André C. Drainville*

**Global Institutions and Development**
Framing the world?
*Edited by Morten Bøås and Desmond McNeill*

**Global Institutions, Marginalization, and Development**
*Craig N. Murphy*

**Critical Theories, International Relations and 'the Anti-Globalisation Movement'**
The politics of global resistance
*Edited by Catherine Eschle and Bice Maiguashca*

**Globalization, Governmentality, and Global Politics**
Regulation for the rest of us?
*Ronnie D. Lipschutz, with James K. Rowe*

**Critical Perspectives on Global Governance**
Rights and regulation in governing regimes
*Jean Grugel and Nicola Piper*

*Routledge/RIPE Studies in Global Political Economy* is a forum for innovative new research intended for a high-level specialist readership, and the titles will be available in hardback only. Titles include:

1 **Globalization and Governance***
*Edited by Aseem Prakash and Jeffrey A. Hart*

2 **Nation-States and Money**
The past, present and future of national currencies
*Edited by Emily Gilbert and Eric Helleiner*

3 **The Global Political Economy of Intellectual Property Rights**
The new enclosures?
*Christopher May*

4 **Integrating Central Europe**
EU expansion and Poland, Hungary and the Czech Republic
*Otto Holman*

5 **Capitalist Restructuring, Globalisation and the Third Way**
Lessons from the Swedish model
*J. Magnus Ryner*

6 **Transnational Capitalism and the Struggle over European Integration**
*Bastiaan van Apeldoorn*

7 **World Financial Orders**
An historical International Political Economy
*Paul Langley*

8 **The Changing Politics of Finance in Korea and Thailand**
From deregulation to debacle
*Xiaoke Zhang*

9 **Anti-Immigrantism in Western Democracies**
Statecraft, desire and the politics of exclusion
*Roxanne Lynn Doty*

10 **The Political Economy of European Employment**
European integration and the transnationalization of the (un)employment question
*Edited by Henk Overbeek*

11 **Rethinking Global Political Economy**
Emerging issues, unfolding odysseys
*Edited by Mary Ann Tétreault, Robert A. Denemark, Kenneth P. Thomas and Kurt Burch*

12 **Rediscovering International Relations Theory**
*Matthew Davies and Michael Niemann*

13 **International Trade and Developing Countries***
Bargaining coalitions in the GATT and WTO
*Amrita Narlikar*

14 **The Southern Cone Model**
The political economy of regional capitalist development in Latin America
*Nicola Phillips*

15 **The Idea of Global Civil Society**
Politics and ethics of a globalizing era
*Edited by Randall D. Germain and Michael Kenny*

16 **Governing Financial Globalization**
International Political Economy and multi-level governance
*Edited by Andrew Baker, David Hudson and Richard Woodward*

17 **Resisting Intellectual Property**
*Debora J. Halbert*

18 **Neoliberal Hegemony**
A global critique
*Edited by Dieter Plehwe, Bernhard Walpen and Gisela Neunhöffer*

19 **Global Standards of Market Civilization**
*Edited by Brett Bowden and Leonard Seabrooke*

20 **Beyond Globalization**
Capitalism, territoriality and the international relations of modernity
*Hannes Lacher*

21 **Images of Gramsci**
Connections and contentions in political theory and international relations
*Edited by Andreas Bieler and Adam David Morton*

22 **Global Public Policy**
Business and the countervailing powers of civil society
*Edited by Karsten Ronit*

23 **The Transnational Politics of Corporate Governance Regulation**
*Edited by Henk Overbeek, Bastiaan van Apeldoorn and Andreas Nölke*

24 **National Currencies and Globalization**
Endangered specie?
*Paul Bowles*

25 **Conflicts in Environmental Regulation and the Internationalisation of the State**
Contested terrains
*Ulrich Brand, Christoph Görg, Joachim Hirsch and Markus Wissen*

* *Also available in paperback*

# Conflicts in Environmental Regulation and the Internationalisation of the State

## Contested terrains

Ulrich Brand, Christoph Görg,
Joachim Hirsch and Markus Wissen

Routledge
Taylor & Francis Group
LONDON AND NEW YORK

First published 2008
by Routledge
2 Park Square, Milton Park, Abingdon, Oxon OX14 4RN

Simultaneously published in the USA and Canada
by Routledge
270 Madison Avenue, New York, NY 10016, USA

*Routledge is an imprint of the Taylor & Francis Group, an informa business.*

Transferred to Digital Printing 2010



Typeset in Garamond by
Taylor & Francis Books



*British Library Cataloguing in Publication Data*
A catalogue record for this book is available from the British Library

*Library of Congress Cataloging in Publication Data*
Conflicts in environmental regulation and the
internationalization of the state : contested terrains /
Ulrich Brand ... [et al.].
p. cm. – (Routledge/RIPE series in global political economy ; 25)
"Simultaneously published in the USA and Canada by Routledge."
Includes bibliographical references and index.
ISBN 978-0-415-45513-8 (hardback : alk. paper) – ISBN 978-0-203-92850-9
(e-book : alk. paper) 1. Environmental policy. 2. Environmental protection. 3.
Biodiversity–
Political aspects. 4. International relations–Environmental aspect. 5.
Globalization–Environmental aspects. 6. Biodiversity conservation–Law and
legislation. 7. World Trade
Organization. 8. Sustainable development. I. Brand, Ulrich, 1967-
GE170.C643 2008

ISBN10: 0-415-45513-8 (hbk)
ISBN10: 0-415-58679-8 (pbk)
ISBN10: 0-203-92850-4 (ebk)

ISBN13: 978-0-415-45513-8 (hbk)
ISBN13: 978-0-415-58679-5 (pbk)
ISBN13: 978-0-203-92850-9 (ebk)

# Contents

# Abbreviations

| | |
|---|---|
| ABS | access and benefit sharing |
| AIA | Advanced Informed Agreement |
| ALCA/FTAA | Free Trade Agreement for the Americas (Spanish/English) |
| ARIC-I | Asociación Rural de Interés Colectivo Independiente (Independent Rural Association of Collective Interests) |
| BMU | Bundesumweltministerium (Ministry of the Environment, Germany) |
| BS | benefit sharing |
| CBD | Convention on Biological Diversity |
| CGIAR | Consultative Group on International Agricultural Research |
| CGRFA | Commission on Genetic Resources for Food and Agriculture (FAO) |
| CI | Conservation International (NGO) |
| CIBIOGEM | Comisión Intersecretarial de Bioseguridad y Organismos Genéticamente Modificados (Mexico) (Intersecretarial Commission on Biosafety and Genetically Modified Organisms) |
| CIEPAC | Centro de Investigaciones Económicas y Políticas de Acción Comunitaria (Centre of Economic and Policital Research for Communitarian Action) |
| CIMMYT | Centro Internacional para el Mejoramiento del Maíz y del Trigo (International Agricultural Research Centre in Mexico) |
| CITES | Convention for International Trade in Endangered Species of Wild Fauna and Flora |
| CNI | Congreso Nacional Indígena (National Indigenous Congress) |
| COCOPA | Comisión de Concordia y Pacificación (Commission for Concord and Pacification) |
| COMPITCH | Consejo de Organizaciones de Médicos y Parteras Indígenas Tradicionales de Chiapas (Council of Organizations of Traditional Physicians and Midwifes in Chiapas) |
| CONABIO | Comisión Nacional para el Conocimiento y el Uso de la Biodiversidad (National Commission for the Knowledge and Use of Biodiversity) |

| | |
|---|---|
| CONAI | Comisión Nacional de Intermediación (National Commission for Intermediation) |
| CONANP | Comisión Nacional de Áreas Naturales Protegidas (National Commission of Nature Protection Areas) |
| COP | Conference of the Parties |
| CPB | Carthagena Protocol on Biosafety |
| ECOSUR | El Colegio de la Frontera Sur (The College of the Southern Border) |
| EPA | Environment Protection Agency (USA) |
| ETC Group | Group on Erosion, Technology and Concentration (NGO, formerly RAFI) |
| EU | European Union |
| EZLN | Ejército Zapatista de Liberación Nacional (Zapatista Army of National Liberation) |
| FAO | Food and Agricultural Organization |
| FR | Farmers' Rights |
| GATT | General Agreement on Tariffs and Trade |
| GBF | Global Biodiversity Forum |
| GEF | Global Environmental Facility |
| GMO | Genetically Modified Organism |
| GPA | Global Plan of Action (on PGR) |
| GR | Genetic Resources |
| GRAIN | Genetic Resources Action International (NGO) |
| IARC | International Agricultural Research Centers |
| ICBG | International Cooperative Biodiversity Group |
| IGC | Intergovernmental Committee on Intellectual Property and Genetic Resources, Traditional Knowledge and Folklore |
| IIFB | International Indigenous Forum on Biodiversity |
| IMF | International Monetary Fund |
| INE | Instituto de Ecología (Institute of Ecology) |
| INPI | Instituto Nacional de Propiedad Intelectual (National Institute of Intellectual Property) |
| IP | Intellectual Property |
| IPGRI | International Plant Genetic Resources Institute |
| IPR | Intellectual Property Right |
| ITPGR | International Treaty on Plant Genetic Resources for Food and Agriculture |
| IUCN | The World Conservation Union – International Union for the Conservation of Nature (NGO and intergovernmental organisation) |
| IU-PGR | International Undertaking on Plant Genetic Resources |
| IMF | International Monetery Fund |
| LEGEEPA | Ley General para el Equilibrio y la Protección del Medio Ambiente (Mexican Environment Law) |
| MAT | Mutually Agreed Terms |

| | |
|---|---|
| MEA | Multilateral Environmental Agreement |
| MS | Multilateral System (for the exchange of PGRFA) |
| NATO | North Atlantic Treaty Organization |
| NC | The Nature Conservancy (NGO) |
| NGO | non-governmental organisation |
| NIH | National Institutes of Health (USA) |
| OAU | Organization of African Unity |
| OECD | Organization for Economic Cooperation and Development |
| PAN | Partido de Acción Nacional (Party of National Action) |
| PCT | Patent Cooperation Treaty |
| PGR | Plant Genetic Resources |
| PGRFA | Plant Genetic Resources for Food and Agriculture |
| PIC | Prior Informed Consent |
| PLT | Patent Law Treaty |
| PPP | Plan Puebla Panamá |
| PRD | Partido de la Revolución Democrática (Party of Democratic Revolution) |
| PRI | Partido Revolucionario Institucional (Party of Institutionalised Revolution) |
| RAFI | Rural Advancement Foundation International |
| RIBMA | Reserva Integral de la Biósfera Montes Azules (Integral Biosphere Reservation of Montes Azules) |
| SAGARPA | Secretaría de Agricultura, Ganadería Desarrollo Rural y Pesca (National Ministry of Agriculture, Stock Farming, Rural Development and Fishery) |
| SBSTTA | Subsidiary Body for Scientific, Technical and Technological Advice |
| SEDUE | Secretaría de Desarrollo Urbano y Ecología (National Ministry of Urban Development and Environment) |
| SEMARNAP | Secretaría de Medio Ambiente y Recursos Naturales (Ministry of Environment, Mexico; today SEMARNAT) |
| SPLT | Substantive Patent Law Treaty |
| TK | Traditional Knowledge |
| TLC/NAFTA | North American Free Trade Agreement (Spanish/English) |
| TNC | Transnational Corporation |
| TRIPS | Trade Related Aspects of Intellectual Property Rights |
| TT | Technology Transfer |
| TWN | Third World Network (NGO) |
| UNAM | Universidad Nacional Autónoma de México (National Autonomous University of Mexico in Mexico City) |
| UNCED | United Nations Conference on Environment and Development (1992) |
| UNCTAD | United Nations Conference on Trade and Development |
| UNEP | United Nations Environment Programme |
| UNO | United Nations Organization |

| | |
|---|---|
| UPOV | Union Internationale pour la Protection des Obtentions Végétales (International Union for the Protection of New Varieties of Plants) |
| WB | World Bank |
| WIPO | World Intellectual Property Organization |
| WRI | World Resources Institute (NGO) |
| WTO | World Trade Organization |
| WWF | World Wildlife Fund; today Worldwide Fund for Nature (NGO) |

# Introduction

## Genetic resources and the internationalisation of the state

Since the beginning of the 1990s there have been numerous attempts to politicise and regulate crucial societal problems on an international scale. Environmental problems belong to the most important subject-matters of these efforts. This is visible not least in the fact that the series of UN conferences which took place in the 1990s was opened by the *UN Conference on Environment and Development* (UNCED) in 1992 in Rio de Janeiro, Brazil. An important result of UNCED was the *Framework Convention on Climate Change*, dealing with a subject of indisputable significance and urgency. A second and equally important result of UNCED was the *Convention on Biological Diversity* (CBD) which addresses the loss of biological diversity and its economic, social and political implications. The CBD came into force in 1993, but 15 years later the impact of this international agreement as well as of related political terrains still remains contested. After many years of further negotiations and programme formulation at the international level, designed to concretise and realise the commitments entailed in the convention, it is recognised that the implementation at the national and regional/local levels remains crucial and that many problems of an effective "new biodiversity order" (Le Prestre 2002) are linked to weak implementation. The relationship between the international and other policy levels is subject to further considerations. Therefore, this question has been placed at the top of the agenda at the ninth Conference of the Parties (i.e. member states) in Bonn, Germany, in May 2008.

A second dimension is the highly contested relationship of the CBD and other international agreements. The CBD itself is part of a contested governance system which embraces other environmental agreements dealing with species or ecosystem conservation – such as the UN *Food and Agricultural Organization* (FAO). In particular, the relationship between the CBD and the World Trade Organization (WTO) and its treaty on Trade Related Aspects of Intellectual Property Rights (TRIPS) is one of the most controversial issues in international politics. This is not the case merely by chance. The CBD itself is not purely an environmental agreement designed for the conservation of endangered species. It addresses issues that range from the protection of ecosystems over the rights of local people, the economic value of genetic resources and the handling of technologies related to their commercialisation all the way up to intellectual property rights and the fair sharing of benefits arising from the use of biodiversity. Thus, environmental concerns are strongly connected to the regulation of economic and technological processes, questions of justice as well as international and domestic power relations. Because of this broad scope and the integration of environmental and development concerns, the CBD is "perhaps the first true sustainable development convention" (Le Prestre 2002: 1). For the same reason, the fate of the CBD, its strengths and weaknesses and its impact on the new international order in general provides an ideal field of investigation to estimate the impact of the concept of sustainable development since the Rio Conference (Conca *et al.* 2007).

This book deals with one central aspect of this broad research field, that is with the conflicts concerning the appropriation of *genetic resources*. Genetic resources are one element of biological diversity, or biodiversity. This concept is actually a new one which encompasses not only the variety of species but also the diversity of ecosystems as well as the hereditary characteristics of plants and animals (cf. Chapter 2). The latter, that is the genetic properties, gain importance as a crucial resource in the context of emerging genetic technology for industrial production. Some observers call genetic resources the "oil of the information age" which indicates their enormous real and potential value. With that, a specific political dynamic has emerged around the conditions and forms of use of genetic resources.

The appropriation of biological and genetic resources is not at all a new phenomenon. Rather, for centuries it was an integral part of the colonial world order, and more recently it continues to be a key strategy in the neocolonial world order. However, during the phases of colonialism, classical imperialism and post-war neocolonialism there were rarely open conflicts around these issues, even if biological resources were central for colonial or imperial politics. This was largely the case up until the early 1980s when the issue became a subject of international concern. Three different developments are important here: the *material erosion* of genetic diversity – in particular seeds valuable for food and agriculture – which has become obvious ever since the late 1960s. Second, the *distributional* effects between Northern and Southern countries – connected with the use of genetic resources, especially in the agricultural sector – became a political issue in the early 1980s and initiated a process which was called by some observers "seed wars". This led to political struggles about the creation of a political, legal and institutional framework in order to regulate the emerging conflicts as well as rising environmental concerns regarding the erosion of biodiversity and distributional conflicts about their economic use. Third, *technological* developments, that is genetic engineering, brought a new dynamic to the conflict field. Therefore, in the 1980s, a global regulation system was created with the CBD as its central element. When analysing these regulation systems, it is important to understand that questions of how to deal with global environmental problems are inextricably linked to political and economic issues which arise from the regulation of the exchange of genetic resources as well as stemming from social conflicts regarding food supply and the potential development of "marginalised" areas.

Against the background of this historical dynamic we want to outline and analyse central elements of the *international political economy of genetic resources, that is its contested institutional and discursive terrains, in the context of the societal transformation towards post-Fordism and the internationalisation of the state.* For this, we take into account the global regulation system as an outcome, medium and presupposition of social struggles, power relations, ecological concerns and challenges for biodiversity governance.

At the centre of our inquiry are the mechanisms of the regulation of access to genetic resources and the sharing of benefits arising out of their use. *Access and benefit-sharing* (ABS), however, are not merely technical or juridical issues. Because of the specific nature of this conflict field and the strong relationships between environmental and distributional dimensions, the question of ABS constitutes the key issue of this global regulation system. Here, environmental issues such as the erosion of genetic diversity are strongly connected to the regulation of the global economy and of the most modern technologies. Moreover, ABS affects social dimensions such as the rights of marginalised local or indigenous people. Last, but not least, it encompasses a complex field of processes where the "enclosure" of formerly free available commons (Kloppenburg/Balick 1996: 177) is at stake.

This imposition is mostly vindicated via economic reasoning but is linked to the far-reaching transformation of societal practices and living conditions around the world. Decisive here is the imposition of *intellectual property rights* on genetic information which constitutes very specific and exclusive private property rights over crops and medicinal plants. Intellectual property gains importance through the extension of patents on genetic resources – including those practices which are criticised as *biopiracy*. Another important issue is the question of how those nation-states rich in biodiversity can "get their share" of the economic use of "their" resources but also of concern is how local people are involved in this benefit sharing. In this context, to some degree new forms of property titles are discussed and different models of benefit-sharing are tested.

In these highly contested processes a huge variety of different actors with diverging and sometimes opposing interests and very different power resources are involved. Indeed, the variety of actors involved includes: international and national corporations, governments, international organisations, non-governmental organisations, research institutes, epistemic communities as well as local actors such as indigenous communities and local farmers. Our investigation focuses, on the one hand, on the conflicts between these different actors and how they are inscribed institutionally and discursively at different levels and, on the other hand, on how the institutional selectivities of this complex multi-level system shape these conflicts, that is the political economy of contested terrains. Over time we became aware of the relevance of these multiple institutional levels and the complex relationships between them. A global management of genetic resources is emerging which connects supranational, international, regional (in our case NAFTA), national and local levels of policy formulation and implementation. What is at stake is, however, how these levels are produced and linked to each other: how international and supranational processes influence and are influenced by local and national structures, configurations of interest and power relations. Therefore, it is not only about the implementation of international politics as a top-down process but also how international conflict terrains and global strategies are influenced from the bottom up, that is by local problems and the interests of locally embedded actors. What is important, in other words, is the upscaling and downscaling of politics connected with the transformation of the nation-state. The global management of genetic resources is a good example, too, because the importance of the local is at least formally acknowledged. However, we need to enquire as to whether or not this formal acknowledgement is sufficient and which possibilities of shaping societal relations really exist for local actors. These questions are dealt with in a case study about Mexico.

Thus, this book focuses on the forms in which these complex conflicts are dealt with at the international, national, regional and local levels. It examines how the complex regulation system has emerged, how the conflicts are institutionalised at the various levels as well as what opportunities exist for the democratic shaping of these processes, particularly regarding the relationships between these levels.

This global regulating system is analysed as an example of the transformation of political forms of steering, participation and control, and for the shaping of global societal relations. The transformation of the state and the state system, connected with the new round of capitalist globalisation, that is the neoliberal and imperial restructuring of politics and economics, can be analysed productively and with many insights for other policy fields.

We propose an enhanced perspective on these transformations, calling this the *internationalisation of the state*. By this we mean that the nation-state does not disappear. On the contrary, our aim is to show that the formal sovereignty of states over genetic resources plays a crucial role in the creation of global markets. Nevertheless, this does not mean that nation-states – in particular those in the global South where most of these resources are located – have real control over their use and the distribution of benefits. On the contrary, it is our contention that global markets and international institutions such as the WTO impose pressure on and hence, undermine the national steering capabilities of these nation-states. Economic processes, however, are analysed in this book in the tradition of a critical (International) Political Economy which sees them as always embedded in political and socio-cultural processes. Thus, the accelerating internationalisation of the economy requires multi-level political-institutional processes and structures.

And this transformation towards the internationalisation of the state is not only about the changing capacities of control but also concerns the comprehensive transformation of relations of political and societal domination. They are linked with new forms of economic production and reproduction as well as the far-reaching transformations of societies and cultures. Our main assumption is that these developments need to be taken into account in order to adequately evaluate the possibilities of political shaping. This is the reason why the present inquiry not only analyses the possibilities and limits of political control of social and economic processes but also addresses the connected power relations and societal processes of domination. Thus, we analyse the conflicts about genetic resources against the background of broader societal changes; what we call – from the perspective of regulation theory – *the transformation from Fordism to post-Fordism*. We see the resource conflicts as an expression of new global conflict lines connected with a new phase in capitalist development. In this transformation process new technologies (such as information and communication technologies or new forms of biotechnology) are as important as *new societal relationships with nature*. This concept, developed in the tradition of the critical theory of the so-called Frankfurt School, highlights the way societies are interconnected with the biophysical conditions of their existence, while at the same time constructing nature practically (i.e. technically, economically) and discursively (e.g. culturally or scientifically). Finally, we want to analyse the opportunities that exist to realise a democratic shaping of societal relations under changing conditions.[1]

In the *first chapter* we sketch out the socio-theoretical framework and the recent transformation towards a "knowledge society" as well as the changes of the state and the state system. Moreover, we outline the concept of societal relationships with nature. In the *second chapter* we move to the concrete conflicts about genetic resources and analyse especially the Convention on Biological Diversity. We concentrate on the elaboration of a recently negotiated agreement which intends to regulate access and benefit-sharing, the so called *Bonn Guidelines on Access and Benefit Sharing*. We argue that these guidelines are not only selective towards different societal interests, they are not capable of delivering a stable compromise among the dominant groups, too. This became visible before and especially during the conference of the World Trade Organization (WTO) in Cancún in September 2003 with the first public actions of the *Like-Minded Group of Megadiverse Countries*. It is decisive that the overall tendency towards a growing commercialisation of biological diversity is lifted to a new level by this new actor.

In the *third chapter* we analyse another contested terrain: the use of biological diversity in agriculture. This area was shaped in recent years by the negotiation of a new agreement, the *International Treaty on Plant Genetic Resources for Food and Agriculture* (ITPGR). In contrast to the Bonn Guidelines this treaty deals with a partial segment of genetic resources, that is those which are decisive for the development of new agricultural products and therefore for worldwide nutrition. Unlike the Bonn Guidelines which were negotiated in a relatively short time, the ITPGR remained highly contested for many years and its adoption almost failed on several occasions. These processes clearly show the problems of the global management of genetic resources which have to do with the mentioned role of intellectual property rights (IPR) and especially the patenting of genetic materials.

Therefore, in the *fourth chapter* we focus on the international conflicts around the validity of IPR and especially the revision of the TRIPS Agreement (*Trade Related Aspects of Intellectual Property Rights*) which took place in the context of the WTO. The relationship of different agreements to each other plays an important role, especially the relationship between the CBD and the TRIPS Agreement. Moreover, what was at stake concerned how other forms of property and, more generally, other problems beyond and against "modern IPR" could be strengthened: ecological questions or those concerning worldwide nutrition. In recent times a change can be observed here: the more the neoliberal transformation of societies is criticised, the more the conflicts are shifted to other terrains such as the *World Intellectual Property Organization* (WIPO). Furthermore, bilateral pressure on weaker actors is growing.

In the *fifth chapter* we take the example of Mexico in order to better understand the relationship between different political levels or scales of conflict and to analyse the impact which international regulations have on the national and local levels. We will see that the tendency towards the commercialisation and valorisation of biological diversity which is inherent to neoliberal and imperial globalisation comes not only from "outside" Mexico but is produced by internal interest coalitions themselves. In Mexico a valorisation paradigm has become dominant and this affects those forces which are in opposition to this paradigm to perform societal relationships with nature differently. However, it would be a misleading perspective to assume that the "logic of globalisation" or global hegemony of "the" neoliberalism is dominating all other processes and actors. Indeed, certain local actors are able to influence processes on the international terrain. The indigenous rebellion of the Zapatistas in Chiapas in the Southeast of Mexico and its consequences for emerging global social movements and for a new internationalism are an example that a contested and highly contradictory shaping of societal relationships with nature does exist. This is the reason why the internationalisation of the state should be interpreted as a reconfiguration of global domination in which the resistance of social movements is incorporated, a factor which is constantly questioned by these movements. In the *final chapter* we will put these different lines of analysis together and draw some more theoretical conclusions about the International Political Economy of genetic resources and the internationalisation of the state.

The present volume is based on a research project conducted at Frankfurt University, Germany, from 2000 to 2003. The findings of this project were completely revised and actualised for the present publication in English.

Whereas originally a more general idea of the global regulation of biodiversity was the starting point of our research, very soon we became aware of the complexity of this research field. In particular, the issue of intellectual property rights and the different conflict areas connected with this highly contested and dynamic topic made it difficult to proceed as originally planned. Some very interesting aspects of the issue, for example the negotiations of the Cartagena Protocol on Biosafety (see Gupta 1999; Gupta/Falkner 2006), had to be ignored, while others, such as the upscaling and downscaling of politics and the politics of scale, gained more space in our research than we had previously planned. In particular, this latter topic became more paradigmatic for our search to develop an appropriate theoretical approach to deal with the complex subject matter and the global conflicts connected. Even if we started with the basic theoretical assumption of an internationalisation of the state (Hirsch 2003, 2005; Poulantzas 2001) as outlined below (see Chapter 1) we realised that we had to deal very carefully with the relationships between the different political and socio-economic levels. To address this challenge we referred to the discussions in critical Geography concerning a "politics of scale" (Brenner 2004; Brenner *et al.* 2003; Keil/Mahon 2008; Smith 1995; Swyngedouw 1997) and tried to synthesise our theoretical approach on the state with these discussions. The result is an enhanced and especially internationalised version of critical state and social theory or what we could call – in order to underline a core contribution – a *neo-Poulantzian approach to International Political Economy (IPE)* which takes into account the up-scaling and down-scaling of the state. Building on Nicos Poulantzas' understanding of the state as a "material condensation of societal relationships of forces" (cf. Poulantzas 1980) and combining it with more recent theoretical developments, we deal with institutions which try to regulate economic, political and social processes on different scales (for a more theoretical foundation cf. Brand 2005, 2006). We believe that this theoretical proposal, which needs more elaboration, is helpful for any kind of IPE dealing with the political regulation of neoliberal and neoimperial globalisation. Thus, beside the empirical findings this theoretical approach is the main outcome which is built upon the material collected in our research about the global regulation of genetic resources.

We would especially like to thank Ana Esther Ceceña and Alfonso López Ramírez from the National Autonomous University in Mexico (UNAM), Silvia Ribeiro from ETC Group and all our interviewees who gave us so much information and explained to us their perspectives on highly complicated and contested topics and terrains. Karin Blank was invaluable as a research assistant and made her own contribution in the field of the organisation of interests of indigenous peoples. In updating our findings for the English publication, we strongly benefited from Gregor Kaiser's critical comments and his comprehensive knowledge on the details of a dynamically developing field. Corinna Heinecke supported us with much advice about recent developments of the CBD. Miriam Heigl delivered useful information concerning the recent developments in Mexico. Special thanks go to Irene Wilson for the translation of the German version of the research project. The anonymous referees of the Routledge/RIPE Series in Global Political Economy as well as the Series Editors, and here especially Randall German, gave us enthusiastic and extremely useful feedback in order to improve the first draft of the book. Dan Hawkins helped us in editing the final text and Simone Buckel in administering the growing list of references. Last but not least we would like to express our gratitude to the German Volkswagen Foundation which made the extensive research possible and especially Alfred Schmidt for his valuable advice and encouragement during the preparation and elaboration of the research.

We hope that this study is a useful theoretical and empirical contribution to the emerging knowledge about the contested terrains of the international political economy and the changing societal relationships with nature.

Ulrich Brand, Christoph Görg, Joachim Hirsch and Markus Wissen
Berlin, Frankfurt/Main, Kassel, Leipzig, March 2008

# 1 The regulation of nature in post-Fordism

There is a strongly pronounced tendency in environmental sociology and politics to discuss questions concerning how to deal with ecological problems more or less in isolation from broader societal and politico-economic changes. Even global environmental problems such as climate change or – in our example – the erosion of biological diversity are generally discussed in relative separation from the structures and processes of societal production and reproduction and the problems and conflicts involved in them. And even where such a connection is in fact made, the dominant tendency is to discuss international environmental problems in the context of the formation of cooperative mechanisms and instruments to deal with common threats. The way in which environmental problems are intertwined with global and national distributional problems, as well as with connected power relations, usually remains underestimated or even ignored.

Such an approach, however, increases the threat that the problems involved become abbreviated. This is particularly obvious in the issue areas we deal with in this book. The problem area "regulation of biodiversity" is characterised by a high degree of overlap between global environmental and distributional conflicts (Göerg 1999b). In order to make clear the manner in which these problems are intertwined, the following section shall *first*, explain the concept of *societal relationships with nature*, focusing on the constitutive interdependence of the societal process with nature. Furthermore, working within and indeed expanding upon this objective, this study centres on the emerging political-institutional terrains of international and Mexican biodiversity politics. However, this process can only be understood if the national and international *societal* developments are taken into account. Political economy and questions of hegemony are therefore decisive in order to understand the complex nature of international and national regulatory projects as well as non-intended societal regulations. Due to this fact, the *second* part of this chapter shall present and discuss the most important elements of the regulation approach. The manner in which this approach perceives the new phase of capitalist development will be outlined in order to enable us to focus, *third*, on the special meaning of knowledge and, *fourth*, on central characteristics of the emerging post-Fordist relationships with nature. In the *fifth* section, the transformation of the state and the state system will be dealt with. The aim here is to describe and further develop certain concepts of materialist state theory which will be important for the following empirical analysis. In this sense, following some general theoretical remarks on the state and political institutions, we will introduce, with key emphasis, the concept of the "internationalisation of the state" and of "second order condensations of societal power relations". Chapter 1 will end with some remarks on the methodology applied in the empirical investigations.

## 1.1 Societal relationships with nature

Even when, in the 1970s and 1980s, the ecological crisis began to receive ever greater public attention, it still needed quite some time before its relevance for social science research and the development of a theory of society was fully appreciated. Indeed, it was a far from easy task to conceptualise how social and non-social, "natural" processes are linked with one another. The reasons for that are manifold and sometimes contradictory, depending on the theoretical traditions involved. The history of sociology was accused of having ignored the material-substantial implications of social processes and consequently, it was considered that a new environmental paradigm was required (Catton/Dunlap 1978; for a critical discussion see Buttel *et al.* 2002; Görg 1999a). Moreover, the need for a bio- or ecocentric perception of society and nature was highlighted. By contrast, constructivist approaches to nature became prominent, at least since the 1990s, focusing more on the (discursive) construction of nature (Haraway 1991; see also Castree/Braun 1998). The underlying debates between naturalistic/realistic and constructivist/productivist approaches have yet to be adequately resolved. While in the Marxist tradition the "production of nature" became a prominent and stimulating concept (Smith 1984; Harvey 1996), the question remains as to how other aspects, such as discursive/symbolic dimensions (i.e. regarding natural sciences) could be integrated into such a concept as well as how it could possibly grasp the limits to the production of nature (Benton 1989; Castree/Braun 1998; Swyngedouw 2004).

Besides the debate about naturalism and constructivism, now widely discussed in a broad variety of approaches such as the actor-network-theory (Latour 1999), there was a further reason which was partly responsible for this delay. Different theoretical traditions, regardless of their sometimes opposing approaches to societal development, generally agreed on one point regarding the direction of progress. For the dominant version of modernisation theory of the 1970s, based on the work of Talcott Parsons, the direction of societal development was predetermined: towards the greater control of nature and the decreasing relevance of nature for the development of society (Parsons 1975; for a criticism see Görg 1999a). Even for most of the theories of the post-industrial society (see Bell 1967; Touraine 1972) it was clear that in the course of societal development, the "game against nature" would be replaced by a purely social game. Nearly the same argument occurs in the Marxist tradition. At least in the official socialistic ideology, environmental problems did not exist and natural limits to social development were not acknowledged. Social development was considered as if nature did not matter. Ecological problems were considered to be a secondary contradiction, subordinate to the development of production relations, which were considered to be of primary importance. It took until the 1980s for the relevance of environmental problems in the development of society and for the formation of social theory to be recognised (see Smith 1984; Beck 1992; Altvater 1993; O'Connor 1988). And even then it was difficult to acknowledge that the relationships between society and nature have their own logic which must be dealt with carefully. It is correct that capitalist accumulation is one central cause for the destruction of nature. In recent years, however, capitalism has demonstrated its capability of reacting to environmental problems, therein trying to manage its own conditions of production, giving rise to a new, "post-modern" regime of accumulation. Therefore, in order to analyse the relationships between society and nature in more detail, it is helpful to periodise capitalist development. In line with this objective, in the next section we refer to the concept of post-Fordism in order to indicate a new phase of capitalist development.

Some time was needed until it was recognised that the problems which were put on the political agenda by the ecological crisis were not isolated environmental problems. When this was realised, there was room for the recognition of the fact that we are dealing with a comprehensive *crisis of societal relationships with nature* (Becker/Jahn 1987). The concept of societal relationships with nature places a quite different emphasis with regard to both aspects concerning the naturalistic-constructivist debate and the concept of progress and social development. Although this concept stems from within the Marxist tradition, it was also developed in the social theory and the historical diagnosis of the "dialectic of enlightenment" in the Critical Theory of Horkheimer and Adorno (see Horkheimer/Adorno 1987; Görg 1999a, 2003a). According to that concept, society cannot be discussed independently of nature, nor does the process of history aim towards an ever more comprehensive control of nature. First, the social process is constitutively mediated through nature. The difference between nature and society, however, should not be eliminated. While naturalistic and some constructivist approaches assume that nature and society are not distinctive entities, Critical Theory emphasises that we should make a distinction between both (without falling back into a dualistic approach). In particular, this is the case with respect to the aim of liberation and social emancipation. Moreover, nature is, in one aspect, a social construction, constructed discursively through language and materially, through human labour. Nevertheless, it remains a materiality of its own which displays qualities that societies cannot really control. That leads to the second point: certainly, the process of modernity is based on an increasing domination of nature, but this domination does not lead to an increasing control over nature; rather, it rebounds in the destruction of nature and in ever greater dependency on the results and secondary effects of the domination of nature.

Thus, basically, society cannot free itself from its dependencies in relation to nature because the social process always contains material-substantial elements (e.g. "natural" resources) and this process is therefore dependent on the metabolism of nature. The mutual interdependence of nature and society is central, however, not only to one side, namely society, but also to the other, nature. At the beginning of the twenty-first century, nature untouched by human activity is virtually non-existent. Marx and Engels were completely aware of this tendency to transform nature from the state in which it was found (Marx/Engels 1978: 44). But in spite of this transformation the material-substantial conditions of human existence retain their own meaning, which can be respected or ignored by human activity – the latter with potentially destructive results for society in the form of ecological risks. Following Adorno's critical theory, this complex understanding can be designated the *non-identity of nature* (Görg 2003b).[1]

The starting-point for a critical theory of societal relationships with nature, however, is the acknowledgement of a practical, that is economic/technical and cultural/discursive, construction of nature, and not the supposedly unchangeable laws of nature to which humanity or society has to adapt.[2] Today, scientific constructions, closely linked to the technical and economic strategies of their application, have come to the foreground, finding their expression in the diagnosis of the "knowledge society" (see below). It should be remembered, however, that there are also other practical constructions of nature, as we shall see more precisely in the following section. There is an unavoidable *pluralism of societal relationships with nature* which, however – particularly under bourgeois-capitalist conditions – is characterised by an *asymmetric relationship* and a dominance of capitalist forms, particularly in production and reproduction. However, even in the age of globalisation, cultural interpretations and, linked to them, forms of knowledge and practices, can be found which cultivate a completely different treatment of nature. These ("traditional") forms of knowledge and their agencies have been given an even greater value by the ecological crisis, for one can discover in certain practices dealing with the tropical rain forest or in certain forms of agricultural production, elements of a more respectful handling of natural resources. This poses the question as to whether these practices are today only marginal or if they have a chance of influencing the *shaping of global relationships with nature.*

After all, the dynamisms in this process are determined by actors with quite different strategies. As we have just been able to see in the example of biological diversity, the dominant strategies for the appropriation of nature increasingly aim at its commercialisation, that is at a purely economic appraisal of nature. Here, the *tendency towards the domination of nature* is continued in a new and even stronger form, even after the experience of the ecological crisis. In the tradition of the Critical Theory not every form of the appropriation of nature is defined as the domination of nature (for if it was, the development of society without the domination of nature would be inconceivable). The domination of nature tries to completely subject or subsume nature to particular objectives and ignores every "own" meaning, every non-identity of nature. This is without doubt the case in the tendencies towards the commercialisation of biological diversity, for nature is reduced here to its usability and all other elements which are not compatible with this are ignored.

As a result of more than 30 years of environmental struggles, it is now widely recognised that the increase in the domination of nature has led to a "domination of secondary effects" (Beck *et al.* 1994). Societies are confronted more and more with the consequences of the unhindered domination of nature, and that causes additional costs and a great deal of trouble. The idea of the complete control of nature has, therefore, at least partially been abandoned and the scarcely controllable risks involved in the appropriation of nature have increasingly been taken into account, that is in terms of uncertainty or the precautionary principle. The use of nature is, therefore, increasingly accompanied by attempts to mitigate its destructive effects prophylactically or to eliminate them reactively – to engage in the protection of nature and the environment. But the question is whether this means an *acknowledgement* of the non-identity of nature or whether it represents the attempt at an *accompanying mitigation* of the domination of nature because of the uncontrollable consequences. The question is, in other words, whether society acknowledges that nature cannot be totally subsumed under societally based targets, thereby allowing room within which nature's own logic can be respected; or alternatively, whether we are dealing with a *reflexive form of the domination of nature* which takes into account that although we sometimes face negative consequences, our targets will not be changed. This question can be answered via an examination of the fate of strategies which follow a different, less destructive form of the appropriation of nature, that is a handling of nature which attempts to respect its non-identity. Considerable attention will be paid to this in the following section.

This approach differs from the often analysed questions, which refer to the general chances of the protection of the environment and nature under capitalist conditions. Of course, it also deals with whether and to what extent aspects of the protection of nature can hold their own against commercialisation strategies. But precisely in the field of biological diversity, the problem cannot be grasped via a confrontation of economic and ecological dimensions, or of conservation and use strategies. On the contrary, conservation and use, as well as ecology and economy, are indissolubly intertwined. Biological diversity, or biodiversity, in the comprehensive scientific meaning of the concept, is a societal construction (which, e.g. does not exist in the same way in all cultures). In this construction, economic interests and geostrategic objectives have already been included, as we shall see more precisely in the process of the establishment of this new scientific concept (Flitner 1995a).

At the core of this are those parts of biological diversity which are of potential economic value: the *genetic resources*. By genetic resources, we mean those components which contain hereditary characteristics – that is not coffee once it has been roasted, but the seeds or its hereditary elements and generally, the genetic parts of animals and plants. These genetic resources are of particular interest because they represent an input for the newer biotechnologies and genetic technologies and the industries based on them, the so-called *life science industries*.

Because of these new scientific and technological developments there is a growing economic interest in nature. At the same time "humanity" cannot completely renounce the use of "biodiversity" as a whole. Since biological diversity also includes the diversity of the animals and plants used in agriculture and for food supply, it is of central importance in ensuring human nutrition. Thus, the conservation of nature cannot be played off abstractly against its use. The strategy of the "pure" protection of nature can only be a solution for, at most, parts of biodiverstiy, for example tropical rain forests or coral reefs. Even in tropical rain forests, however, nature supplies a habitat for many people and for greatly differing forms of human use (Hecht 1998). Thus, even there, the strategy of protection clashes with interests in the use of biodiversity – and even the setting up of a nature reservation is basically only a change in the form of use, namely use for ecotourism or as "natural capital". It is, simply put, always a question of the different forms of the socialisation of nature, of different cultures and practices in the handling of biodiversity. In this constellation the tendency towards their greater commercialisation determines the direction of the organisation of societal relationships with nature. In the process, other societal forms are marginalised and tend to be dissolved. The threat emerges that nature in the form of genetic resources is reduced to its economic value, that is subsumed under the practices of capitalist valorisation (in German: *Inwertsetzung*; see, for the specific meaning, Görg 2004). In the reflexive domination of nature, in addition, even conservation efforts are functionalised within broader strategies of valorisation. The question, however, is not use versus protection; rather, it concerns which strategies of use can the conservation of biological diversity aim for. And, as a further topic, which is central to this study: how can "the democratic content of socio-environmental construction" (Swyngedouw 2004: 24) be enhanced? This question can be translated into an inquiry about the specific meaning which is given to the strategies of the valorisation of nature, the strategies of nature protection and the practices of a different organisation of relationships with nature in the global management of nature.

The *starting-point* for our investigation, therefore, is the *specific historical conditions* in which these different strategies clash with one another. For, with regard to the objects of the conflicts – biodiversity or genetic resources – we have to ask about the new forms of their scientific description. Moreover, we can observe a new constitution of the object itself, which does not "exist" prior to its connection with new processes of its practical appropriation. The very definition of the problem – that is the numerous disputes over dominant or even hegemonic interpretation patterns – is closely connected with the economic, and beyond them the social and cultural, interests in biodiversity. Are we dealing with the loss of habitats which are supposedly *close to nature* – particularly, tropical rain forests and coral reefs? Is it a question of the "loss of the apothecary of nature" or the "treasure chest" of the tropical rain forest, that is its potential *economic* value? Or, is there a real threat of endangering culturally embedded *practices and systems of interpretation*, which have often already socialised certain elements of biological diversity for hundreds of years and have even created certain aspects of that diversity? Even the political structures and terrains are in a state of upheaval. The valorisation of genetic resources is an integral part of a new phase of capitalist development and of what we call *post-Fordist relationships with nature*. If the novelty of this phase – W. F. Haug (2001: 451) calls it "biocapitalism as a mode of production" – and its central elements are taken into account, the conflicts over the organisation of relationships with nature could be more appropriately grasped. Moreover, the chances of alternative forms can be estimated more precisely.

## 1.2 The formation of a post-Fordist mode of regulation

The background to the change in the economic importance of biological diversity and in particular of genetic resources is the transition to a post-Fordist accumulation regime and mode of regulation following the world economic crisis of the 1970s. We examine this development using the concepts of *regulation theory*, particularly those which have been developed in France, Britain and Germany since the 1980s (Aglietta 1979; Boyer 1986; Lipietz 1987; Robles 1994; Hübner 1989; Jessop 1990; Hirsch 1990; Esser *et al.* 1994; Grahl/Teague 2000; Brand/Raza 2003; Graz 2006).[3] It is thanks to this theory that there has been an opening up of the field for a differentiated analysis of capitalist formations which differ in space and time, and which have their own specific strategies for the valorisation of capital and their own political institutional forms and relationships of social forces. The historical development of capitalist societies does not follow an objective logic but is determined by the conflicting activities of social actors. It is characterised by a crisis-determined succession of relatively stable formations, each of which has its own forms of economic production and reproduction, its own specific configurations of institutional systems, but also, its own particular crisis dynamics. The central question dealt with by regulation theory is thus directed towards the problematic conditions of the maintenance and reproduction of capitalist society, which is characterised by structural conflicts and antagonisms, that is towards the question of why this social relationship "reproduces itself despite and because of its conflict-driven and contradictory character" (Lipietz 1985: 109, our translation). The institutionalist approach of regulation theory proceeds from the assumption that "the state" and "the market" as social regulation mechanisms presuppose behavioural patterns which correspond to the dominant societal structures (this is in contrast to the understanding of a dichotomy of the market versus the state; see the criticism of Underhill 2000 on international political economy). These are, as a rule, guaranteed by a complex of socio-political institutions and norms, in which the economic and political/administrative processes are embedded. The valorisation of capital – the production and accumulation of surplus value – is a fundamental component and the driving force of capitalist society, but it must not be understood simply as a mechanism which objectively prevails and to which social behaviour adapts unconditionally. In the terminology of regulation theory, a system of regulation is necessary for this to happen. The concept of a *mode of regulation* describes "the totality of institutional forms, networks, explicit or implicit norms, which ensure the compatibility of relationships within the framework of an accumulation regime, both corresponding to the state of societal relationships and over and above their conflict-driven characteristics" (Lipietz 1985: 121, our translation). It circumscribes a complex field of societal organisations and institutions: including the state apparatus, the ideological orientations and the subjectivities which are materialised in it, and thus, it also involves the hegemonic relationship in the sense of generalised ideas about the order as well as the development of society.

We must therefore differentiate between "regulation" and "regulating". This is central to the theoretical placement as well as the methodological design of this study. The concept of regulation relates to the fact that capitalist society – based on private production guided by the market, wage labour and competition – can have no controlling centre as a result of the contradictions and conflicts which are contained in it. Its reproduction takes place via permanent disputes and struggles between contending actors. These can lead to more permanent social compromises which manifest themselves in specific systems of institutions which thereby lend the overall capitalist system a certain durability, albeit only temporarily. Accordingly, the reproduction of capitalist society must be regarded as a "process without a steering subject". Every mode of *regulation* contains a complex of institutional *regulating* dispositives, which is, in itself, heterogeneous and contradictory. It follows from this that the concept of a state which guides society must also be rejected. The state – as an apparatus of force – is an important point of intersection of regulative processes, but in its specific shape and institutional configuration, it is at the same time their object and product.

The political-economic regulation approach also analyses the changes in economic structures, using for this the concept of the *accumulation regime*.

This characterises a mode of reproduction which, for a certain time, can guarantee a relatively correlating relationship between the material conditions of production and societal consumption (Lipietz 1985: 120). In order to be relatively durable, a historically developing accumulation regime requires the establishment of a compatible mode of regulation. The accumulation regime and the mode of regulation are not to be regarded as functionally correlated, however, but underlie their own specific conditions of creation and development. They are always the contested product of social conflicts and thus, are always historically alterable.

This leads us to another aspect, which is of utmost importance for this study: every phase of development contains its own specific politico-social *axes of conflict* in which the relationships of class, gender and nature (among others) and the social antagonisms connected with them are expressed. In our example, the struggle over the (non-)commodification of societal relationships with nature is institutionalised within the framework of a mode of regulation. At the same time, this conflict is also grounded in divisions of labour and inequalities based on class, gender and "race" (especially indigenous peoples).

The *relative consistency and durability* of a historical form of capitalism depends on the formation of a mode of regulation which enables social conflicts to remain relatively compatible with the valorisation process of capital which is materialised in the accumulation regime. For this to be the case, not only must the general orientations for societal development be shared (e.g. the welfare-state class compromise, the neoliberal market and competition model), but *terrains* must also be defined and accepted by the most important actors on which political and social struggles can be fought out without resulting in fundamental interruptions to the accumulation process. A state which is in the process of internationalising itself is decisive for the creation of this terrain.

"Secular" *crises* of capitalism occur when an existing connection between accumulation and regulation breaks down as a result of the dynamism of social struggles formed by its own structure. The cause of the crisis of "Atlantic Fordism" (Aglietta 1979), which became manifest in the 1970s, must be seen, above all, in the fact that the productivity reserves which existed in this mode of development no longer sufficed to guarantee capital profitability under the conditions of the Keynesian, welfare-state mode of regulation and the corporative class compromises which characterised it (see Hirsch/Roth 1986; Altvater/Mahnkopf 1999; Hirsch 1995). This caused the forms of the international division of labour and the relationships of dominance, which had emerged after the Second World War, to fall into crisis (Hirsch 1993). Even though this interpretation is widely accepted, in more recent discussions on the existence of a post-Fordist mode of development there is disagreement about the extent to which this has already taken shape, and if so, whether its existence is relatively stable or whether it is already in a state of crisis. In the more recent theoretical discussions of regulation there is some dispute over which criteria should be used in order to be able to identify a more or less stable (post-Fordist) mode of development (see Alnasseri *et al.* 2001; Candeias/Deppe 2001; Hirsch 2001; Brand/Raza 2003; Purcell 2002 on the economism of the – French – approach). This has led to a rather far-reaching critique of the regulationist methodology, at least with respect to the discursive connotations it has acquired. The critique refers especially to the concept of regulation. As Goodwin and Painter (1997: 23) point out, "the notion of regulation has become associated in the literature with stability, coherence, functionality, and structure." They consider it important (and possible) to rework this notion in favour of a view of regulation as a "process, unevenness, tendency, and practice" (ibid.). In this sense, but without wishing to clarify these issues here, it is of importance to the following examination that we assume the existence of a contested and still crisis-ridden, but nevertheless quite clearly outlined, post-Fordist mode of development and the corresponding relationships with nature.

The outlines which have taken shape, in a relatively clear fashion, are closely connected with the politico-economic strategies which are subsumed under the label of neoliberalism (Plehwe *et al.* 2006). The neoliberal strategy for solving the crisis – which has essentially been established by internationalised capital in cooperation with the governments of several dominant capitalist states, led by the USA – is essentially based on a far-reaching deregulation of capital and financial markets, together with the comprehensive privatisation of public enterprises, pension funds and social security systems. In this way, and based on new transport, information and communication technologies, the internationalisation of production has been given a qualitatively new impetus. Cross-border company networks have increasingly been created which make it possible to install global chains of value creation for the flexible use of socially and geographically differing processing conditions ("worldwide sourcing"). The development of transnational company networks led to the tendential dissolution of the borders of the "national economies", which under Fordist conditions of accumulation and regulation were more strongly oriented towards domestic markets. This increased the independence of multinational corporations from the regulative environments and social compromises of individual states. The Fordist connection between Taylorist mass production and mass consumption supported by the welfare state became looser, formalised wage relationships were eroded and the distribution of income generally shifted in favour of capital (Cox 1987; Marglin/Schor 1990; Altvater/Mahnkopf 1999; Jessop 2002: 80–91).

## 1.3 The knowledge-based economy and the "knowledge society"

This trend was accompanied by a surge in the subjugation of further sectors of society under the process of capital valorisation by way of "commodification", that is the transformation of the products of labour and natural resources into commodities. Based on new technologies – particularly information and communication technology as well as biotechnology and genetic engineering – this led to a greater capital penetration of the services sector and to the increasing importance of the so-called *life science industries* (Dolata 1999; Gibbs 2000). The key characteristic of these industries is that they are, in a particular fashion, "knowledge-based", that is founded on the systematic production and appropriation of the results of scientific research. With this development, comprehensive and systematic rationalisation processes became possible, especially in the field of immaterial labour (research and development, production preparation, production-related services). At the same time, new opportunities for investment and the creation of surplus value were opened up to capital, besides the services sector, primarily in the field of information and telecommunications, agricultural production and the pharmaceutical, cosmetic and health industries. As a whole, the post-Fordist accumulation regime combines an increase in the rate of exploitation (wage reductions, rationalisation and the intensification of labour) with a new thrust based on the "internal appropriation of land" via the subjugation of further spheres of society – not least in the extended dimensions of the human psyche and the human body – under the process of capital valorisation.

On the part of private capitalist companies, ever since the beginning of the 1990s, a dynamic process of reorganisation has taken place culminating in the development a life science industry within the central business areas of agriculture and pharmaceuticals (and in some cases veterinary medicine). This process has been accompanied by considerable fusions and take-overs. In order to show the concentration processes and the power of the large companies, we will outline some of the available statistical information.

The world market for *seeds* is about 21 billion US-Dollars (USD) and the top ten companies, after a major concentration process, today control half of the world's commercial seed sales. The world's largest seed company is, since 2005, US-based corporation Monsanto, which boasts seed sales of 2.8 billion USD per year; followed by DuPont/Pioneer, also based in the USA, with 2.6 billion; Syngenta (Switzerland) with 1.2 billion; and the French Limagrain Group (more than 1 billion.). These mega-companies are followed by: the German KWS; Land O'Lakes (USA); Sakata (Japan); Bayer Crop Science (Germany); and Taikii (Japan), all of which have seed sales ranging roughly between 360 to more than 600 million USD (ETC Group 2005a).

The top seed companies, namely Bayer, Syngenta, DuPont and Monsanto, are also leading *pesticide* producers, a market which is worth almost 30 billion USD per year; an amount which is greater than the market of commercial seed sales, and in which the concentration of companies is even stronger: ten companies control 84 per cent of the global market and further concentration processes have been predicted. Bayer and Syngenta each sell more than 6 billion USD worth of pesticides per year, followed by the German BASF (which sells more than 4 billion USD), and the three US-based companies, Dow (annual sales of almost 4.4 billion), Monsanto (almost 3.2 billion) and DuPont (2.2 billion) (ETC Group 2005b: 6, based on 2004 with data provided by Agrow World Crop Protection News).

The largest market which has to do with genetic resources is by far the one of *pharmateuticals*, with annual sales of more than 240 billion USD. The top ten companies control 59 per cent of the world market and in recent years a strong concentration process has taken place: Hoechst and Rhône-Poulenc formed Aventis in 1999 and five years later Aventis merged with Sanofi; Glaxo Wellcome and SmithKline Beecham merged in 2000 to form GlaxoSmithKline; in 2000 Pharmacia merged with Monsanto but Monsanto became an independent company again in 2002 and in 2005 bought the Mexican seed company Seminis (see Chapter 5). In 2002, too, Pfizer merged with Pharmacia. This brief list of mergers is far from exhaustive, but rather seeks to focus on the most prominent ones in the pharmaceutical industry. The US-based company Pfizer has sales of more than 46 billion USD, followed by GlaxoSmithKline (United Kingdom, 32.8 billion); Sanofi-Aventis (France–Germany, 32.2 billion); the two US firms Johnson & Johnson (22.1 billion) and Merck (21.5 billion); Astra Zeneca from Sweden and the United Kingdom (21.4 billion); and the two Swiss companies Hoffman-La Roche (19.1 billion) and Novartis, which has sales totalling 18.5 billion USD (ETC Group 2005b: 2, with data for 2004 obtained from PJB Publications and 2005 Fortune Global 500).

Compared to these figures, *biotechnological companies* which concentrate on genetically modified drugs or seeds are relatively small, with revenues ranging from 10.5 million USD per year (Amgen, which is the leader, mainly selling five genetically modified drugs) to 5.4 million (Monsanto) or 4.6 million (Genetech, majority owner Hoffman-La Roche). However, all these companies have specific interests in a great variety of genetic resources which are the basis for their research and product development and where market growth rates are considerable (ETC Group 2005b: 4, with data based on the journal *Nature Biotechnology*, without pharmaceutical companies). Nonetheless, to date most sales of genetically modified crops stem from corn, soybeans and rice (Wisner 2005: 13–14).

Despite the vast potential market that these companies hope to conquer, there has also been growing resistance to what we will term "green genetic technology" in agriculture and in the production of foodstuffs. Indeed, this technology has been confronted with considerable resistance not only for technical reasons and because of the ecological effects, but also because consumers, at least in Europe, are not (yet) willing to accept these products. As part of a survey known as Eurobarometer, it was found that in only four of the fifteen participating countries the public supported the use of biotechnology in creating food (Gaskell *et al.* 2003: 17). This also supports the argument that the establishment of new technologies should not be taken for granted but that it depends on societal conditions and that it must, therefore, be established politically. It can also be observed, however, that the life science industry did not at all regard the "wars of perception", in which it was, at first, on the losing side (see the study by Mitsch/Mitchell 1999; see also ETC Group 2005b), as being lost battles; rather, the industry had begun waging a major campaign to push for and promote the public acceptance of "green genetic technology".

Independently of how this process continues in detail, it is certain that an overlapping trend will be further strengthened. Due to the general difficulty of estimating market conditions, to the immensely high costs of research and the development of new products, and to the necessity of safeguarding the fundamental relationships of ownership during the development of new markets and branches of production, the *securing of property rights* with regard to the new products, technologies and their genetic material has meanwhile become a primary concern of industry and research – from the large corporations to small research institutes. Corresponding to this is the enormous increase in the number of patent applications, particularly in the northern industrial countries. According to data of the NGO RAFI (now the ETC Group)[4] in the year 2000 there were three million patent applications waiting to be processed by the US American Patent Office alone (the great majority of them from the sphere of biotechnology and genetic technology); this number representing half as many patents as had been granted in the past 200 years (RAFI 2000a). In Europe, patent applications, lodged at the European Patent Office, in the field of gene technology rose from approximately 1000 in 1985 to 30,000 in 2005 (Wullweber 2004). As trends in the sphere of research on the human genome or on human parent cells show, the issue at stake is increasingly that of the granting of property rights to genetic resources and thus to the "natural" basic materials themselves. The resources here are of a different type to those of the Fordist phase, however. The issue is not so much one of the availability of natural products as such, but increasingly, of the information contained within them, of the "genetic code" (see for the difficulties concerning the term "code", Kay 2000). These resources, thus, do not simply exist, but are constituted as resources (namely as economically useful and thus valuable material) via scientific descriptions and technological developments. At the same time, the basic material of the economic and technical valorisation processes is becoming an increasingly contested object, upon which exclusive and monopoly-like property rights, primarily concerning patents, are imposed.

### *1.3.1 Knowledge-based industrialisation and a new economic imaginary*

The concept of post-Fordism is useful to explain what is often described as the emergence of a "knowledge society" (Stehr 1994; Wingens 1998; Krohn

2001; for a critique, Hirsch 2002b). This conceptual term is somewhat fuzzy, however, since every human society is based on "knowledge" in one form or another. Furthermore, it could suggest a technological concept of society in which the economic and political relationships of domination and exploitation are ignored. What has changed over the last decades, obviously, is the relevance of science in society. To determine exactly what this increasing relevance means, however, is not an easy task. The question of a scientific explanation about how society is constituted is necessarily connected with the question of how we, as scientists, know about it. This strong interrelationship between *the way the world is* and *how we perceive it*, is highlighted by the notion of a "co-production of science and society" (Jasanoff 2004a). According to Science and Technology Studies we have to take a symmetrical stance towards scientific knowledge about nature and social institutions (such as states) that produce and guarantee social order (Latour 1993; Jasanoff *et al.* 1995). But to say that science and knowledge are central for our understanding of both the natural and the social world does not necessarily exclude power relations and domination from the scope of investigations. Herein, of central importance are the shifting and *contested boundaries* between science and society, which are by no means given and fixed. In particular, science policy and research funding strategies (both from private and state actors) contribute to the restructuring of boundaries, blurring some of them (e.g. between basic and applied research) while creating new ones (e.g. between society and specific, constructed versions of actors, called "users" by Jacob 2005). Because in the so-called knowledge society, scientific knowledge supposedly increases its relevance, the scientific diagnosis about the character of society is even more difficult and contested. Consequently, ambiguity reigns concerning the appropriate role of scientists. Moreover, scientists as intellectuals are themselves continuously engaged in struggles for a new diagnosis of society – and their interests and revenues (e.g. funding for research) are strongly connected with this diagnosis (see Fischer 2000).

In this debate the concept of post-Fordism highlights the importance of information and scientific knowledge connected with a new mode of accumulation and new forms of societal regulation. This does not mean that we seek to explain societal development by initially looking at only one constitutive element of society – modes of production – for, if this was the case, it would only be fair to label such an approach as representing a reductionist understanding of the social (Jasanoff 2004b: 20). Instead, we ask which structural elements of society can best represent a starting point for further explanation of social change. Against this background, the "knowledge society" is less an expression of a "post-industrial" phase of capitalism; rather, it describes a new, more systematically "knowledge-based" form of industrialisation. The phrase "the end of industrial society" does not entail that the form of industrialisation – from early manufacturing to Fordist Taylorism – has continually changed in the course of history or that the dissolution of traditional factory-based forms of production, which is becoming visible in various sectors, means the complete end of industrialisation as such (Becker 2001). While Fordism – characterised by the Taylorist mass production of consumer goods – was based essentially on the commodification of the material conditions of the reproduction of labour, today, the production and appropriation of information and knowledge is becoming of central importance.

Another aspect of the role of knowledge in current transformation processes is highlighted by Bob Jessop (2003). He argues that the concept of the knowledge-based economy became, step-by-step, a hegemonic paradigm and *economic imaginary*, that is the consolidation of an accumulation regime and mode of regulation which gives the different actors a social, material and spatial-temporal frame of reference. The economy and politics as social processes and everyday actions by myriad actors are – through struggles – framed and constructed through corridors of rational and legitimate action: investment strategies, the organisation of the labour-process, laws, financial priorities of state budgets, norms of production and consumption in a broad sense, and international politics. It articulates accumulation strategies, technological innovations, cultural norms such as the learning society, perceptions of nature and political processes. Therefore, the knowledge-based economy becomes a grand narrative and a strategic vision. Certain strategies, projects and visions emerge and gain power, legitimacy, and through that continuity. The post-Fordist forms of appropriation of nature are also part of this economic imaginary and an emerging "eco-governmentality" (Goldman 2001).

In this sense, post-Fordism is, for Jessop, an imaginary of a progressively more integrated world-market which is able to orient and give coherence to accumulation strategies and to realise political and hegemonic visions of social life which put past and present narratives and experiences, and actual problems and future prospects together in a coherent framework. Jessop's main argument is that the economic imaginary of the "knowledge-based economy" is becoming, ever since the 1990s, the dominant and hegemonic discourse because it creates the framework within which struggles for political, intellectual and moral leadership as well as concrete technological-economic reforms take place. This imaginary – which is still full of tensions – orients actions throughout society: from everyday practices, via the educational and scientific system to state institutions and private firms.

Within the framework of the post-Fordist mode of accumulation and regulation, three tendencies which are basic to capitalist society are accentuated in a specific way:

(1) The capitalist production of commodities is principally dependent on being able to avail itself of *general conditions of production* which cannot be produced in the form of commodities, for example natural resources or

human labour-power. To an increasing degree, these include the necessary knowledge for societal production and reproduction. In the course of capitalist development this is less and less available to capital in the form of a "natural" factor of production or of a "free productive power". Rather, it must be produced systematically. The production of knowledge under capitalist conditions, however, remains dependent on production contexts which are outside the direct creation of surplus value. Its infrastructural foundations – for example in the form of basic research or the qualification of workers – is inadequately developed under market conditions, and complementary financial and organisational inputs, primarily on the part of the state, are necessary: schools, universities, state-financed and state-organised research institutions, as well as government subsidies for industrial research and development. With the growing importance of scientific knowledge for production, the amount of knowledge produced by the firms themselves grows, but at the same time, this increasingly requires recourse to "intellectual commons" in the form of generally available qualifications, information and understanding (Jessop 2000). In other words, in order to be successful in the competition for innovations, firms require a suitable social and infrastructural environment, for example in the form of "technology poles" or "industrial districts". This is a factor which, particularly, accounts for the competition between different locations. The relationship between private capitalist production of knowledge and the creation of knowledge as a generally available "intellectual common" takes on new and more complex forms in this process.

(2) Capitalist development is characterised by the continual process of *primitive accumulation*, that is the transfer of previously – in the form of "commons" – freely available conditions of production to private property. Marx essentially dealt with this in the context of the private appropriation of land. Today, the process of primitive accumulation relates in a particular way to knowledge, which is now being "commodified" and "enclosed" in a more robust manner than previously (May 2000). The concept of primitive accumulation must not be used as a historical category here; instead, it describes a structural relationship between capitalist and non-capitalist modes of production and of living, which in the course of history – driven by the process of the creation of surplus value – is configured in ever new forms (see Alnasseri 2003). For David Harvey (2003), the process of "dispossession" is at the core of the actual phase of imperialism. A prime example of this is the more or less legal expropriation of indigenous knowledge by the corporations of the life science industry in the form of "biopiracy" (Görg/Brand 2001; Ribeiro 2002b). In this sense, the global system of biopiracy and the establishment of a of biopatents work closely together (Delgado 2002: 103–135, 179–191).

(3) Finally, the capitalist process of development is characterised by a continuous contradiction between the growing societalisation of production

> and private appropriation. The production of societal knowledge is, increasingly, less attributed to individual persons or firms, but takes place more and more in complex systems and networks. This makes it ever more difficult to identify particular owners and distribute to them the use extracted from such knowledge. Companies basically endeavour, on the one hand, to maintain as free an access to knowledge and information as possible, while simultaneously striving to reserve both elements for themselves as private property. Concurrently, the dominance of short-term profit interests which accompanies the liberalisation of the global financial markets ("shareholder value") is the reason why longer-term and difficult to calculate research and technology strategies tend not to be realised. It is left to the state to regulate the contradictions and conflicts arising from the double meaning of information and knowledge as both private and "common" property.

The thing which is special about knowledge or, to put it more precisely, about the information relevant to knowledge, is that it – in contrast to material goods and land – can be multiplied and utilised without great cost (Becker 2001; May 2000; Nuss 2006). The meaning of intellectual property rights therefore lies either in excluding others from using them, or in forcing them to pay licensing fees, either by means of technical safeguards (e.g. copy protection in the case of computer software) or by means of legal entitlements (patents, copyrights). In this way, information which is technically, freely available and utilisable is kept economically scarce. The establishment of intellectual property rights also presupposes specific relationships of power in society, for example not generally available technological competencies, organisational and financial potentials and above all, the existence of a state guarantee of the corresponding legal entitlements. The growing importance of such strategies of scarcity and exclusion leads Rifkin (2000) to claim that "property" is increasingly losing its meaning in favour of the right of "access". This argument only makes sense, however, if the concept of property is confined to material goods. It is evident that the concept of property is both being considerably extended and experiencing a considerable change in its meaning in society in the course of the growing transformation of information and knowledge into private property. At the same time, the appropriation of intellectual property rights, as we shall display in the following section, continues to be essentially tied to the disposition of material resources and the ownership of capital.

Despite the centrality of the disposition over resources and capital which is related to property, Jesse C. Ribot and Nancy Peluso introduce an important distinction between property and access. Their starting-point is the question of who actually benefits from or enjoys things, and what are the power-shaped processes and circumstances to do so (2003: 154). Property means socially acknowledged and supported enforceable claims through ownership or titles defined by law, custom or convention to use something and benefit from the use. This implies the exclusion or restriction of others from that use. Access (to natural resources), however, is a broader concept and considers the fact that some people and institutions "control resource access while others must maintain their access *through* those who have control" (ibid.: 154; italics in original). Access through those who have control does not mean only the right to benefit (like property) but the *ability* to benefit from things which can also be enabled without property rights. Struggles take place on the division of benefits between different social positions which intend to gain, control or maintain access to resources. The move from concepts of property or tenure to access "locates property as one set of factors (nuanced in many ways) in a larger array of institutions, social and political-economic relations, and discursive strategies that shape benefit flows" (ibid.: 157). Therefore, access analysis needs to identify and map the means, processes and relations by which access and the flows and distributions of particular benefits from particular resources are gained, maintained and controlled (and denied, one could add) for individuals, groups and institutions as well as the related underlying power relations within particular circumstances (ibid.: 161). The distinctions between property and access and between the right to benefit and the ability to benefit is important for this study. We intend to show how forms of access and property are specifically related in the conflicts around genetic resources.

### *1.3.2 Intellectual property*

The process of the commodification of knowledge takes on different forms (see Jessop 2000). It refers, first, to the transformation of generally available resources of knowledge into private property, for example, when pharmaceutical firms use indigenous knowledge on the effects of medicinal plants for bioprospecting and then have the results patented. A further aspect is the production of knowledge as a commodity, for example in the form of research results, consultations, analyses, designs, concepts and so on. This field is becoming an ever more important segment of the services sector in the course of the differentiation of complex enterprise networks. And finally, all the forms of the transformation of intellectual labour into wage labour, which produces products for the market, and thus, the real subordination of intellectual labour to capital, fall into this category. The commodification of information and knowledge has two aspects: on the one hand, they themselves become a saleable commodity, and on the other hand, exclusive access to them becomes an ever more important foundation for the appropriation of material, for example genetic resources. That is, the ability to appropriate knowledge presupposes the existence of already accumulated knowledge.

One particularity of this process is the tendency to separate knowledge and machinery. Whereas in the earlier phases of industrialisation, prior to the Fordist one, knowledge relevant to production was, in principle, incorporated into the machinery and a durable part of machine systems, it is now becoming more independent of these and is taking on the form of an independently produced and tradeable commodity. What is significant in this connection is the rise of the computer software industry in relation to the producers of hardware.

The dominant form of intellectual property rights, next to copyright, is the patent. This is of course an old form of property right, primarily used by technologically advanced states and industries in order to ward off competitors, but in recent times its content and its importance have changed considerably. The meaning of this can be illustrated with an example: from 1990 to 2000 demand for patent protection for the main biotechnology subclass (micro-organisms or enzymes) "reached approximately 188,213 patent publications rising to a preliminary total of 299,163 patent publications by the end of 2003" (Oldham 2004: 3).

An important extension of patent protection lies in the possibility of patenting not only inventions but also discoveries (plants with specific properties, micro-organisms, genetic resources) – as far as they are commercially viable. In 1980 the US Supreme Court decided that the patenting of living organisms was permissible, which opened up a new era in the field of intellectual property. This means that products, processes and even forms of life – plants, animals, human cells and human DNA – are potentially "intellectual" property. The quantitative dimensions of this development are, as we have already stated, considerable. These developments continue, however, to be the subject of worldwide protest – and opposition to the patenting of living material, in particular, is at the core of this struggle. In defence of the extension of patent protection, it is argued that without it the process of technological innovation as a whole would be hindered because private companies would only devote increasing expenditures to research and development if they could realise corresponding profits in the form of innovation rents. This is because the production of knowledge – for example in the sphere of biotechnology – is very complicated, but transferring and copying it is relatively cheap (Janssen 1999). This claim should be qualified on several points, however. With the increasing speed of innovation even patent protected knowledge loses its value ever more quickly. The difference between biotechnology and the software industry is interesting here. The latter places its bets less on the securing of innovation rents via formal legal protection than on permanent (apparent) innovation, that is it operates with a planned shortening of the economic life of its products. This is probably connected with the different innovation costs, but also with the different technical and legal possibilities of establishing intellectual property rights. It is of importance for the connection between patent protection and innovation that patents can always also be used to prevent innovations. The guaranteeing of intellectual property rights can therefore, in the extreme case, lead to making research completely impossible. In every case it causes the diffusion of innovations to become slower and in some cases the exclusion of entire regions or social groups from their use.

### *1.3.3 Lines of conflict*

While, with the appearance of a "knowledge society", among other things, the vision of an egalitarian and democratically formable "world society" as a "global village" is painted, new relationships of power and patterns of exclusion are establishing themselves precisely through this development. The post-Fordist "Infocom"-capitalism is characterised by new axes of social conflict, actors and contradictions, and at the same time, by new forms of politically regulating them. "What is to be explored, finally, is on which territory the struggles over the shaping of society and over the political, ethical and legal limits of the use of biotechnology will take place" (Haug 2001: 453, our translation).

This leads, first of all, to the paradox wherein the increasing amount of information available to the masses is accompanied by a concentration of specialised knowledge within national and international elite groups which have access to the necessary means – material resources and political power (Comor 1999: 121ff.). A universalised "culture industry" arises, centred around the internationalised media industry complexes, which commercialise information as a commodity and in so doing, proceed to market virtual realities and world views on an increasing scale. While information and knowledge are thus concentrated in an increasingly important stratum of information and communication managers, the experiences and the social relationships of the "infotainment" consumers are formed by commodities and the market (Rifkin 2000: 181ff.). Because the social conditions and the infrastructural requirements for the preparation and transformation of information into knowledge are unequally distributed, there is a tendency for society to be divided into "haves" and "have-nots" with regard to information or, rather, knowledge. The "haves" have access to the relevant knowledge complexes, and the "have-nots" are left with a mass of trite and irrelevant information. The extension of the information and communication possibilities, with their growing economisation and in view of the highly unequally distributed financial and material resources, tends to increase social inequality. The inhabitants of the "global village" have greatly varying technical and social opportunities with which to use the existing technologies. "Knowledge" thus becomes a decisive mechanism for social exclusion.

The fact that the spread of communication and information possibilities is accompanied by a clear tendency towards the privatisation of information and knowledge at the same time increases the socio-geographical inequalities on the global scale. World Bank figures show that developing countries (low and middle income countries), in 2002, had to pay 9.3 billion USD more than they received for royalties and licence fees (Bödeker *et al.* 2005).The establishment of unhindered informational freedom under these conditions becomes an instrument with the help of which the capitalist centres with their developed information and communication industry safeguard their economic, political and cultural dominance (Comor 1999: 14ff.; May 2000: 73ff.). Access to scientific and technological potentials becomes an ever more decisive factor for international relations of dominance and dependence. According to an estimate by Sachs, roughly 15 per cent of the world's population produce the essential technological innovations and about half of the world's population is able to take possession of these and use them (Sachs 2000). The rest is threatened with being cut off and marginalised – technologically, economically and culturally. This phenomenon, which within a particular society leads to new processes of exclusion, on the global level, adds a new dimension to the conflicts between "North" and "South".

These developments are the context for our empirical investigations: for the conflicts over the use and valorisation of biological diversity and especially genetic resources, it is essential that one analyses the development of the international regimes that act to safeguard intellectual property (see in more detail Correa 2000; May 2000; Drahos/Braithwaite 2003; May/Sell 2006; as well as numerous case studies on Latin America in Brand/Kalcsics 2002). At issue here is the determination of who profits from the advantages which arise from the use of genetic resources. There is a certain paradox in the fact that the most "modern" actors (research institutions and above all transnational, high-tech corporations) are dependent on access to these resources and, thus also to a certain degree, on "traditional" population groups in the global South. The core of the conflict over access, benefit-sharing and intellectual property rights basically stems from this paradox. After all, such resources are primarily to be found in Southern countries and, within them, often in the habitats of "marginalised" population groups, especially indigenous peoples. In addition, for the appropriation of genetic resources, the "traditional" knowledge of how to use these resources plays an important role, for it often serves as a "filter" in the search for economically valuable substances (Kuppe 2001: 147ff.). As a judgement of May 2000 in the case of the Indian Neem tree shows, it has not at all been finally decided how these regulations will be interpreted in the future. One of the patent claims of the US company, W.R. Grace, to the traditionally long-used genetic properties of this tree – as pesticides and medicine – was rejected by the European Patent Office. Nevertheless, it is still valid and 20 others still remain legal (www.genenet-info.org, 8 March 2005).[5] Patent law has come under discussion, which is partly due to international political pressure, but is also connected with the fact that overlapping regulations from different systems demand clarification.

## 1.4 Tendencies in the emergence of post-Fordist relationships with nature

The post-Fordist form of rationalisation and capitalisation implies a fundamentally altered societal meaning of nature. At the same time, the changed relevance of knowledge and science has consequences for societal relationships with nature (see in more detail, Hirsch 2002b). Simultaneously, the institutional treatment of the environmental crisis leads, together with other changes, to a new mode of regulation. One specific foundation of the Fordist mode of accumulation and regulation was the unhindered exploitation and destruction of natural resources. This formed one of its fundamental moments of crisis. The "ecological crisis", which became ever more obvious at the beginning of the 1970s, was closely bound to the crisis of Fordism. The disputes in the context of the "ecological crisis" have led, in recent decades, to a stronger consideration of ecological and socio-ecological aspects. This must not, however, be thought of as a societal learning process in which "society" as a whole begins to adjust itself to its ecological problems, but rather, as a conflict-driven process in which different actors with different interests in the conservation and use of nature conflict with each other. To some degree, even weaker actors succeed here in finding consideration within the framework of the asymmetric compromises of an emerging "ecological capitalism". Through the discussion of the "ecological crisis" the destruction of the material foundations of life has become an integral and contested component of post-Fordist restructuring (on the connection between the discussion on sustainability and that of globalisation on the fringes of the Rio+10 conference" see Conca *et al.* 2007; Brand/Görg 2007).

New issues such as socio-ecological problems, gender issues and also North–South relations have been placed onto the political agenda since the crisis of Fordism – transmitted through the criticism of social movements, critical academics and non-governmental organisations (NGOs). That does not mean that societal relations in those areas did not develop in a crisis-driven manner previously, but that they gain increased societal relevance in the crisis of Fordism and also that they are being transmitted through new social actors. The constitution of "new" problems is itself a political process which in turn is connected with comprehensive socio-economic changes such as the new valorisation of nature.

In these disputes, "corridors" of problem-solving are created. This relates first to the political and legal terrain of the interpretation and treatment of problems. Second, a dominant understanding of problems emerges which defines the legitimacy of certain viewpoints and demands, that is that which is regarded as "reasonable" and "feasible". Thus, the dominant discussion of the ecological crisis in the "Northern" countries treats this as an environmental problem with a strongly technical solution perspective, through which privileges are granted to strategies of "ecological modernisation" (Toke 2001; Jänicke/Jacob 2004). Concern is concentrated primarily on the rationalisation of the flows of energy and materials for economic activities. Social aspects are often overlooked or de-legitimised, and questions regarding resource problems *and* distributional conflicts are *simultaneously* underemphasised. Concurrently, the general tendency towards the control of nature and the subsumption of nature under the conditions of capitalist valorisation processes continues to be in force, although in a reflexively broken form. The appropriation of oil, with all its international economic and geopolitical implications and its consequences for the realm of nature, remains a central condition of "fossilistic" capitalism (Altvater 1993) – despite all the solemn assurances of the inevitable beginning of a "solar epoch". The disputes over "strategic resources" (Ceceña/Barreda 1995) such as oil, gas and, increasingly, fresh water, are a central component of international politics. Nonetheless, new resources are also "constituted" (Heins/Flitner 1998) by means of new technologies and socio-economic developments. This is true, particularly in regard to genetic resources in connection with genetic technology (and in the near future we can expect the reconstitution of parts of nature through nano-technology). The industries of the so-called life sciences are the most powerful driving force in the reorganisation of societal relationships with nature in this field, whereby much – from the economic and technical options available to the design of the political and legal environment – is still uncertain or disputed. This concerns important questions such as the acceptance of new patterns of production and consumption in the spheres of health and nutrition (e.g. the acceptance of genetically modified crops) and the legitimacy of specific and competing appropriation strategies of nature (e.g. private property on "life"). Alejandro Toledo speaks of a necessary "double mutation" in order to make possible the intensive capitalisation of nature: In addition to the systematic recording of potentially commercialisable resources there must be a "semiotic conquering of territory" (A.Toledo Ocampo 2000: 285). Signs and meanings are redefined, that is new forms and outputs are defined as rational and appropriate. Of decisive importance for post-Fordist relationships with nature is that nature must be newly defined in order to be able to serve as a "raw material" and as a resource for capitalist valorisation. In this context, there is a demand by capital, that individualised "private property rights" to nature should be privileged, which is a completely unacceptable demand in many cultures.

These processes, thus, are always contested and, to some degree, politically regulated. With regard to the political-institutional design of relationships with nature, an increase in international environmental politics can be observed. The UNCED conference in Rio de Janeiro in 1992 played an important role here, although national and international policies had already existed since the 1970s. The institutions of international environmental politics are, however, not at all problem solving instruments which support one another, as is often assumed, but rather, they find themselves in an asymmetric and often contradictory relationship to one another. The following study will also focus on this aspect.

Furthermore, we can observe an increased importance of "civil society actors" in environmental politics, connected with general changes of governance structures in politics. Moreover, the enormous *knowledge-dependency* and *uncertainty* under which politics takes place here needs broader consideration. In many decisive questions it is still unclear today exactly where the problems lie and how they can be dealt with. The political terrains of socio-ecological politics, especially at the international level, are only just being structured, and in this process, in view of the intricate conflicts even "weaker" actors succeed in bringing forward their perceptions and interests. One important characteristic of post-Fordist capitalism is, therefore, that the *struggle over information and knowledge* is becoming an increasingly central sphere of social and political conflict (Drahos/Braithwaite 2003; Delgado 2002). Particularly "knowledge" – in the sense of the ability to generate and define societal problems symbolically and to bring this into political processes in the form of "expert opinion" – is becoming an ever more important power resource. Not only has the process of social production and reproduction become "knowledge-driven" in this way, but politics has as well. This has effects on the political constitution, and on the institutional and discursive treatment of problems, as well as on political strategies and forms and on constellations of alliances.

## 1.5 The transformation of the state and the system of states

The formation of a post-Fordist mode of development was accompanied by major changes in the institutional mode of regulation. This included not only the establishment of "neoliberal" norms and behavioural styles, but also a restructuring – completed with the breakdown of the Soviet Union – of the international relations of dominance and dependence, which is characterised by a complex arrangement of cooperation and conflict between the state groups of the capitalist "triad" (North America, Europe, South-East Asia/Japan) under the leadership of the USA. It involves an economic and political differentiation among the peripheral countries but also their far-reaching subordination. One major and still existing characteristic of the North–South relationship is that many Southern countries have become laboratories for neoliberal policies. The debt crisis, apparent ever since the beginning of the 1980s – in addition to authoritarian regimes and military dictatorships, which were a part of internal class struggles – became the central lever for achieving this.

In spite of the – measured in monetary units – statistically declining economic importance of many regions of the capitalist periphery, they remain de facto essential sources of cheap raw materials, foodstuffs and labour for the "high-technological mode of production" (Haug 2001: 450). The Fordist as well as the – vastly more precarious – emerging post-Fordist compromises in the capitalist metropolis continue to be based, among other things, on cheap pre-products produced in the peripheral countries – whether through the relocation of sections of companies, through pre-products bought on markets or through lower environmental standards – and the cheaper reproduction of labour power – whether through unpaid work by women, foodstuffs, raw materials or cheap household employees. There is still a rising inequality in the worldwide distribution of income which is not clearly divided along North–South lines, but rather, differs among regions (Wade 2001; with respect to gender relations Wichterich 2000).

### *1.5.1 Global governance?*

There is a certain degree of consensus within the social sciences regarding the general diagnosis of a fundamental change in the state and the system of states in the course of "globalisation". There is dissent, however, on how this can be conceptualised theoretically and what it means for the various collective actors and persons affected. What role does the nation-state play and what is the role of international politics, particularly the system of institutions that is developing there? How do the different levels of politics – from the local to the international – behave towards one another? Which actors, with what interests, play a role? What is the relationship of political/institutional and socio-economic structures and processes? To what extent do socio-ecological problems affect economic and political processes? Do they play any role at all in the social scientific and socio-political discussions?

A striking shortcoming of many political science approaches which are concerned with the internationalisation of politics is the way in which they theorise about the state (this is also true of many historical-materialist approaches; see Shaw 2000: 87). It can generally be stated that their understanding of the state usually falls into an understanding of either organisational sociology or system theory. Usually, the "modern" state as a form of political domination is presupposed without explaining its existence and theoretical functioning. This can be seen particularly clearly in the contributions of social sciences to global governance or global public policy, which claim to grasp the current upheavals analytically (see e.g. Rosenau 1995; Reinicke 1998; Held/McGrew 2002; survey and criticism in Hewson/Sinclair 1999; Brand *et al.* 2000; Saxe-Fernández 2005; Brand 2005; Behrens 2005). "Reduced to the simplest common denominator, 'global governance' means accompanying, politically, the process of globalisation" (Enquete-Kommission 2002: 415, our translation). It is about dealing with the problems arising from globalisation by means of public, cooperative policies, that is the "principal modalities of global-rule making and implementation" (Held/McGrew 2002: 11). The state is understood as an independent, sovereign and rationally behaving actor, operating in the name of a more or less unified and definable "national interest".

In a similar fashion to that found in regime theory (Krasner 1983; Rittberger 1993), the question of international cooperation between states is the core issue. International institutions are created for this purpose. Herein, from the outset, the issues are questions of political regulation, which are combined with the assumption that adequate forms would contribute to the resolution of diverse problems. The creation and the growing importance of international institutions is, as a rule, explained by the increase in international problems, which are due, not least, to the weakening of nation-state institutional systems with regard to "good governance". Global governance has, as is the case with governance approaches in general, a bias towards problem-solving theory, while neglecting, or at least underestimating, power relations and relations of domination (Mayntz 2005). Most governance approaches suggest that there are new institutions or measures *because* there are problems; that is, they explain them in a functionalistic manner. However, this is increasingly criticised within regime theory and governance approaches themselves. Constructivist concepts which have recently gained importance within regime theory (Risse 2003; Wendt 1999; Wiener 2003) address precisely the processes of problem constitution which have been hitherto neglected by more traditional regime-based theoretical approaches. They explore the ways international regimes, once they have been established by nation-states, gain a momentum of their own and change the problem perceptions and politics of their member states.

International political economy goes beyond regime theory in many of the aspects which are important for the issues we are discussing here (Scherrer 2003). Underhill (2000: 806) identifies three common assumptions of the various IPE approaches:

> i) that the political and economic domains cannot be separated in any real sense, and even doing so for analytical purposes has its perils; ii) political interaction is one of the principal means through which the economic structures of the market are established and in turn transformed; and iii) that there is an intimate connection between the domestic and international levels of analysis, and that the two cannot meaningfully be separated from one another.

However, in IPE the state is often understood in a rather formal manner as an institution equipped with rule-setting power. It is not seen as a materialisation of contradictory social relations. Thus, IPE runs the risk of conceptualising the relationship between politics and economics as a dichotomy, thereby undermining its own claim of having developed an integrated approach (see Bieling 2003; Shaw 2000). Against this background, "global governance" is "more a site, one of many sites, in which struggles over wealth, power, and knowledge are taking place" (Murphy 2000: 799) To sum up, neither the relationships between politics and economy nor the wider societal context of political institutions, including the ecological aspects of socio-economic processes, are treated carefully.[6] To remedy the complementary deficits of regime theory and IPE, a reference to regulation theory and to materialist state theory may be helpful.

### *1.5.2 State and institutions: theoretical considerations*

The inadequate anchoring of IR and IPE approaches in a theory of society has unmistakable consequences for the ability of the analysis to offer robust explanations, especially as far as the current transformation processes of states and the system of states are concerned. We summarise both processes under the term "changes in statehood". Therefore, this term does not only mean the transformation of the international political system. We return here to concepts of historical-materialist state theory which were developed in the 1970s by reverting to the approaches of Marx and Gramsci (see among others, Poulantzas 1978/2002; Jessop 1990; Hirsch 1997, 2002a). The state and the system of states are understood here within the framework of social theory as a specific form of the political under the conditions of capitalist socialisation. Accordingly, a fundamental structural characteristic of capitalist society is the creation of an apparatus of physical force which is formally separated from the classes in society. However, the separation of "state" from "society" and of "politics" from the "economy" does not represent a dualism because they are simultaneously constituted to each other. The state form, which was already defined by Max Weber as the central characteristic of the modern state, is based on the fact that the capitalist mode of appropriation of surplus value, that is via private production, wage labour and the exchange of commodities, both requires and accelerates the independence and centralisation of compulsory physical force. The "particularity" or "relative autonomy" of the state is the foundation of a specific mode of regulation of class and other social relations. It is this which makes possible in the first place, the formulation of a "politics of capital" which goes beyond divided, competing interests, to be able to, at the same time, enable the exploited and dominated classes to be politically integrated in both a repressive and a consensual way. The state, thus, simultaneously becomes, in this specifically determined form, a material condensation of specific relationships between social classes and groups, that is one characterised by its own institutional structures and routines (Poulantzas 1978/2002: 154ff.). It is neither an independent subject, nor simply an organisation with a rational purpose, but the expression and a component of societal relationships characterised by fundamental contradictions and conflicts. It, therefore, necessarily forms a heterogeneous complex of apparatuses with divergent societal relations which are relatively independent of one another and often operate against each other. The formal separation of state and society is at the same time the form of the presence of the political in the economy. The state – as an integral part of the capitalist relationships of production – is not outwardly in opposition to the economy. The contrasting of "state" and "market" as spheres which are separate from each other, a view which is often found in academic literature, is therefore misleading.

Although "the state" and "(civil) society" are formally separate, at the same time they form a contradictory unity. Both spheres are to be understood as complex relations of domination based simultaneously on both force and consensus (Gramsci 1991: 783ff.). "Civil society", as a part of bourgeois society, is not a sphere in which there is no domination; rather, any domination is permeated by power relations and coercion. At the same time, it is the sphere of political self-organisation, of the articulation of interests and discussion, and thus a field of struggle on which hegemony, that is legitimate concepts of societal order and development, is fought over. This implies varying, and even contradictory, practices. Since civil society is permeated by conflicts but nevertheless remains an integral part of societal power relations, it can, in the terminology of Gramsci, be regarded as a part of the "extended state". At the same time, the existence of a developed civil society is a decisive precondition for democratic processes. The latter gain their political effectiveness, however, above all, from the fact that they are based on procedures for the formation of opinion and for decision-making which are institutionalised and formalised by the state. Decisive from the point of view of the dominant actors is the extent to which they are able to present their intentions as legitimate and socially necessary or in the general interest of society, to what extent they are able, via compromises, to partially change the viewpoints and interests of other actors – but also their own – and/or to exclude groups with fundamentally opposing interests. The *discursive* and the *institutional* levels complement each other here. One central subject of dispute is the institutional safeguarding of the actors' own interests through state policies or the hindering or even prevention of the institutionalisation of contrary interests. A major indicator for the establishment of specific positions is, therefore, the extent to which they succeed in becoming "the national interest".

The concept of civil society cannot simply be applied in an analogous way to the international scale: processes of societalisation that transcend state borders do exist – although in a highly fragmented fashion – but there is no world state. The concept can be used, however, to show that there are political processes on the international scale beyond traditional state diplomacy or force. Alex Demirovic (1997: 258) emphasises the character of transnational actors such as business associations, NGOs, elites and their networks, all of whom, despite pushing for divergent interests, come to represent a civil side-stage in which a global consensus is worked out. Nevertheless, this should not be interpreted separately from the processes which take place within nation-states (see also Görg/Brand 2003; on the risks of transferring Gramscian categories to the international scale see Germain/Kenny 1998).

These characteristics are founded in the structure of capitalist societies. Their basic components consist not only of the particular form of the state but also the pluralism of (national) states (see for more detail, Hirsch 2002a: 35ff.). The system of nation-states is an expression of the contradictions and rivalries which characterise global societal relations. The pluralism of the states is itself a mode of reproduction of the capitalist political form. It divides social groups and classes along state borders and thus creates the basis both for the formation of competing "national" class formations and for "corporate" and other societal compromises. The decisive basis for this is the fact that the individual states dispose over the (legitimate) monopoly of physical force, which makes possible both the control of cross-border movements and material redistributions as the basis of social compromises. It makes possible the formation of highly differentiated conditions of production and valorisation and thus creates the inequalities which represent an essential basis for the global accumulation process. The fragmentary structure of the system of states is, thus, a decisive precondition for the stabilisation and reproduction of capitalist society within the context of the world market. It is, as is the form of the state itself, a fundamental part of the capitalist mode of production. This explains why, on the global scale, neither a coherent, comprehensive "state" nor a relatively coherent "civil society" can arise. And at the same time, it becomes clear that the political organisation of capitalist society consists of a complex intertwining of different scales: local, national and international.

Before we go into the current process of the transformation of the state and the system of states in detail, we must first deal with the relationship between the capitalist political form and political institutionalisation (on this see Esser *et al.* 1994). Social forms are, in Marxist terminology, objectivations of societal relations – resulting from the general principles of socialisation, that is private property, wage labour, the exchange of commodities and competition – which confront individuals in a reified form. This is particularly true of the commodity and money forms and of the political (state) form. The capitalist form of the state – its "relative autonomy" and the formal separation of "state" and "society" – is an expression of a social structure in which the direct and conscious creation of society, and thus the existence of radical democracy in the form of an immediate political community, is impossible. General references for consciousness and behaviour constitute and reproduce themselves in the *social forms* of bourgeois society. These become concretised in *social institutions*, or to put it the other way around, the general societal form determinations are materialised in the institutions. Institutions can be understood as consolidated modes of behavioural orientation, routinisation and coordination, which both guide social activity and make it possible, thus lending social systems a relative continuity (Giddens 1988). It is essential, however, that societal institutions cannot simply be understood as direct manifestations of social forms, fully determined by them. Rather, social forms lay the foundation for, support and limit processes of institutionalisation. A complex intermediary context must be taken into account here between the societal structure (or the social forms), the institutions and the activity which reproduces them (Görg 1994). The regulationist category of "institutional form", as introduced above, is a concept which mediates between social forms and institutions. It stands for historically and spatially specific concretisations of social forms as certain modes of state intervention, inter-state relations, monetary systems and wage relations which materialise themselves in national and international social and political institutions.

Since social activity, which is expressed in institutions, is neither "purely" objectively determined nor without conflict, it is possible for institutions and processes of institutionalisation to generate contradictions to the social forms. Societal crises and conflicts express themselves in such a process. The capitalist social forms are a fundamental and structural component of the mode of production, whereas institutional forms and institutions, as a product of social activity and societal struggles, are changeable within this mode. This means that the capitalist social forms do not determine a specific system of institutions and the political form therefore does not determine a particular form of the state. Each concrete form of the state is a historically determined expression of the capitalist political form. The "particularity" or the "relative autonomy" of the state, the specific characteristics of the monopoly of physical force and the institutions and institutional forms which are taken by the material condensation of class and other societal relations are historically and spatially variable, according to the specific economic conditions of reproduction and societal power relations. The strong condensation of political processes at the level of the nation-state, with all its "extensions" in civil society, is only one possible form of the institutionalisation of the political. It was expressed in the north-western part of the world under the conditions of class compromise which characterised "Atlantic Fordism" until the second third of the twentieth century. It follows from this that the question as to the possible "end" of the "(nation-) state" is wrongly formulated. Instead, it must be examined whether, and if so how, the capitalist political form is institutionalising itself in a new way, and what conflicts and what possible societal processes of crisis are implied in this.

However, there are some limits to this form of theorising when we consider the state in countries of the capitalist periphery. Here we can find a variety of state forms that differ in many aspects from those in the advanced capitalist countries, ranging from South-East Asian developmentalist states to the "failing states" in large parts of Africa. This variety results from the specific articulation of capitalist and non-capitalist sectors within these countries and the respective modes of integration into the world market. Because of the very different socio-economic conditions and colonial heritages in the capitalist periphery, it is not possible to formulate a coherent theory of "the" peripherial state.

Materialist state theory as outlined above refers to developed capitalist societies with their specific economic and class structures. In the capitalist periphery, these conditions exist, at best, only partially (Castro Escudero *et al.* 2005; Olivier Costilla 2005; Burchardt 2004: 131–146; Schedler *et al.*

1999; Heigl 2007; see the current transformations in Latin America, Velinga 1998). In the capitalist periphery, we often find highly segmented class structures, important sectors of non-capitalist production, economies with a high degree of structural heterogeneity, a minor degree of wage labour and, therefore, a civil society which is "less" developed. Independent social actors, self organised societal associations and a free public often exist only in a rudimentary form, non-governmental organisations very often are organised and financed from abroad. In many cases, the "modern", centralised state apparatus is – due to the colonial heritage – set upon societies which do not have corresponding economic and social structures. For these reasons, the relative autonomy of the state with respect to the dominant classes or class fractions often does not really exist. That means that a central characteristic of the capitalist state is missing. Likewise, there generally does not exist an "extended state" in the Gramscian sense, as a complex unity of state and civil society. One decisive factor for state structures in the periphery is the presence of land reform, that is the abolishment of quasi feudal or oligarchic agrarian structures, which is important for industrial development with its consequences for class structures, the establishment of comprehensive market relations and the existence of a state apparatus equipped with relative autonomy. This, for example, makes an important difference between Latin American and South-East Asian states. In general, peripheral states are, therefore, more repressive ones. The regulation of class relations and economic exploitation is, to a higher degree, based on physical force. Compared with more developed capitalist countries, the relation of "consensus" and "repression" appears to be significantly one-sided. However, some features have been changing, particularly in the last two decades, with democratisation processes and neoliberal transformations in many countries (Schedler *et al.* 1999).

Especially in African countries, we can observe a special weakness or even absence of a centralised state apparatus and the increasing importance of economic activities based on the violent exploitation of large parts of the population by warlords. As a result, governments lose control over parts of their territory; the monopoly of force erodes in favour of clan-based warlords fighting each other; and new and extremely violent modes of societalisation, beyond the capitalist state and market, emerge, which are nevertheless integrated in the capitalist world market (see Seibert 2003). But also in Latin American and South-East Asian countries, whose state forms are more similar to those in the capitalist core, we have to introduce some differentiations. The state in these countries must be conceptualised against the background of their subaltern status in the world economy. Especially due to the debt crisis, supranational and Northern dominated institutions such as the International Monetary Fund (IMF) exert a much greater influence on the societal power relations and the state politics than in Northern countries, often preventing a national or even an internal bourgeoisie from developing itself (Poulantzas 2001; see Alnasseri 2004).

### *1.5.3 The internationalisation of the state*

The neoliberal globalisation strategy, which has been established since the 1980s as a reaction to the crisis of Fordism, has aimed at a comprehensive restructuring of global class relations. Central to it was the destruction of the class compromises which characterised the Fordist state of the post-war era. The structure of the states and of the state system was profoundly altered – a process which was further accelerated by the collapse of the Soviet Union and the accompanying unilateral political dominance of the capitalist triad centres under the leadership of the USA. The background for this was the growth of the internationalisation of capital during the Fordist epoch (Poulantzas 2001), which had led to a fundamentally altered relationship between capital and the states of the now relatively unified capitalist world system. The new thrust of capitalist globalisation is based on the increasing internationalisation of production which, at the same time, works to accelerate the it. This means that state–society relations as well as the relations of states to one another have been considerably modified; a process which we describe as the *internationalisation of the state* (Jessop 1997; Hirsch *et al.* 2001; Robinson 2001; Hirsch 2002a). This process has several dimensions which are important for understanding our empirical research. Although an internationalisation of the economy already took place during Fordism, the processes of deregulation and privatisation since the 1980s have led to an increased dependency of the nation-state apparatuses on the international capital and financial markets. The actors of these processes – particularly multinational corporations and international financial institutions, but also international networks of managers, politicians and members of international institutions – determine in a quasi "de-politicised" way, that is relatively independently of formalised democratic political institutions and decision-making processes, essential contents of individual state politics. This is a contested process and implies the shifting of relationships of social forces. This development is expressed in the type of the *competition state*, which is centrally aligned towards ensuring the optimal conditions for the valorisation of capital (Hirsch 1995). It involves a considerable shift in the state's capacities for economic and socio-political intervention. The transformation towards a competition state has also involved changes in both the configuration of the state apparatuses and in the modes of political decision-making and decision enforcement. The position of the finance ministries and of the central banks, which are largely autonomous in relation to the democratic political institutions, was strengthened and the political parties transformed themselves in this process from "mass integrating" organisations for the mediation of interests to medial agents for the procuring of legitimacy. This is not the result of a historical law in whatever form – whether that of the logic of capital or of the modernisation and differentiation of societies – but of a political strategy which has been partially enforced through hard struggles. We can also call a state whose apparatuses are increasingly dependent on international capital and finance markets and which interiorise these dependencies and the related relationships of forces, "internationalised competition states" (Hirsch 2005: 145–51).

As a consequence, the opportunities of the states for the coherent and integrative regulation of society within national boundaries have been diminished. The result is increasing economic and social fragmentation. As we have shown above, the geographically fragmented system of nation-states, which both politically and economically integrates its component parts in various ways and which is mediated by competition, is a general structural characteristic of the capitalist world system. Nevertheless, the competition among different capitals, between socio-political areas ("locations") and wage-earners (migration plays an important role here) has a greater importance now than during Fordism. Material and ideological integration paths and macro-economic coherencies are less stable than under Fordism and are less tied to the nation-state. The apparatuses of the competition states, particularly those of the impoverished parts of the periphery, are given the role of handling the consequences of the neoliberal transformation with their restricted regulatory possibilities and of creating a minimal amount of social cohesion. In the following section, we will sketch out some central features of the transformation of statehood. The process in which "civil society" actors gain in importance beyond that of the Fordist corporate institutions can be described as the *relative privatisation of politics*. Decision-making and decision enforcement by the central bureaucracy changes to "governance" in the form of the "cooperative" or "negotiating" state, that is the state integrated into a complex network of state and private actors. The private actors, who are increasingly gaining in political importance, are not only multinational corporations but also the so-called non-governmental organisations (NGOs). At the international level, too, structures of the "negotiating state" are developing. This creates room for activities involving the definition of problems and the representation of interests which are conducted by NGOs (see Görg/Brand 2003; Demirovic 2003; Hirsch 2003; McCormick 2005). One change is decisive here: in periods of neoliberal hegemony, conditions to which political activity has to adapt – or is perceived to require adaptation – are claimed, ever more, to be determined by the "market" and the political processes are, to a certain extent, economised, that is relatively, directly subordinated to economic rationality. NGOs are thus, particularly as organisations which carry out, for instance, development projects, among other things "a consequence of neoliberal marketization" (Murphy 2000: 795; on private financial regulation Palan 2003; Malcher 2006).

A further process can be described as the *denationalisation of state politics*. As a result of the transformation towards a competition state, of the limited possibilities for a coherent regulation of the national economy and the increasing cross-border economic interlinking, an *increased importance of subnational and local regulating contexts* can be observed (Brenner 2004; Wissen

2000). Regulating institutions and social compromises of forces manifest a move from the level of the central state to towns, communities, semi-states and – often crossing national borders – economic regions, which strive for their own paths to development and policies of locational competition.

The denationalisation of state politics comprises a further aspect. The process of globalising accumulation is accompanied by the *internationalisation of political regulating complexes*, which can comprise formal institutions, regimes, treaties or networks. In the discussion among political scientists, the process of the denationalisation of state politics and the newly emerging political institutions and their interrelationships are examined using the concept of multilevel analysis. This is done particularly with regard to the European Community or Union and the supranational political processes and institutions emerging from them (Scharpf 1999; Hooghe/Marks 1999; Bache/Flinders 2004). Attention is directed primarily towards the cooperation of the supranational and the national levels. The dominant perception in political science is that this is a process for dealing with the economic globalisation process, a process which generates crises and problems that need to be addressed by politics. But this explains only a part of the dynamism. First of all, it is a case of a "new constitutionalism" (Gill 2000), which is advanced by the capitalist triad centres in order to institutionally safeguard their global dominance, in general or with respect to specific actors – for example in the form of the North Atlantic Treaty Organization (NATO), which has been reshaped into an instrument of global military intervention, of the International Monetary Fund (IMF), the World Bank (WB), the Organization for Economic Cooperation and Development (OECD) or the World Trade Organization (WTO). Shaw understands international politics as the generalisation of the "global-Western-state-conglomerate" which grew out of the increasing international cooperation among the capitalist core after World War II (2000: 199ff.). At the same time, regional integration processes serve to strengthen the position of individual groups of states in the relationship between competition states (European Union, North American Free Trade Area). The growing weight of international organisations and regimes means that state apparatuses that are based on a large number of national societies gain in relevance. International political institutions are supported by the states and their autonomy is limited: "they usually remain the creatures of the most powerful of their state members" (Murphy 2000: 793). Nevertheless, they develop their own political dynamism, which even the most powerful actors cannot completely escape. John G. Ruggie (1982), with respect to the "embedded liberalism" of the post-war constellation, elaborated that although there were generally accepted compromises, their specific institutional, international and national implementation was contested. This is probably even truer with regards to the post-Fordist mode of development and its "disciplinary neoliberalism" (Gill 2000). The following study concentrates on the detailed analysis of this process.

One particular aspect of this development is the *internationalisaton of law*. The development and enforcement of law tend to separate themselves – although in very different ways – from the states. Examples of this are the setting up of international courts: the European Court, the International Criminal Court, but also the dispute settlement body within the WTO, which is concerned not only with jurisdiction but also, de facto, with the creation of law. The international enforcement of private property relationships – not least those concerning intellectual property rights, for example within the framework of the WTO-TRIPS Agreement – is especially gaining in relevance. Of particular importance is the development known as *lex mercatoria* of a private civil law, supported chiefly by international companies and lawyers' firms. This is not necessarily new: state legislation was always strongly based on the formalisation and codification of socially developed legal norms. But the generation of law relatively independently of the states has increased in importance and has begun to, ever more clearly, transcend individual state borders (Günther/Randeria 2001; for a critical voice, see Buckel 2003). Here the difference between the creation of law and the enforcement of law is important, however – a difference which is not always clearly made in political science. The enforcement of the law in the case of conflict remains tied to the monopoly of force in the individual states and is dependent on its relative effectiveness. "Strong" states can evade international jurisdiction more easily than "weak" ones, if indeed, they submit to it at all – as can be seen in the case of the behaviour of the USA towards the International Criminal Court. Generally speaking, however, it can be assumed that there is a stronger separation of the creation of law from the enforcement of law.

The internationalisation of capital, which was further advanced by the transition to post-Fordism, and the transformation of the system of states have changed the *conditions for the formation of hegemony at the international level*. The pattern of the present world order shows a contradictory picture in this regard, where a strong tendency towards transnationalisation is counterbalanced by a global world order still fragmented through nation-states. On the one hand, there is a system of international organisations and regimes – from the CBD to the WTO – in which the interests of international capital and of the dominating states in the safeguarding of the capitalist market economy and private property as well as the policy of neoliberal deregulation are solidly institutionalised – even if they remain contested. In the course of increasing cultural and communicative interaction, there are also indications of tendencies towards the creation of interstate "civil society" structures. And last but not least, in the wake of the internationalisation of the state and of capital, an "international managerial class" is growing, consisting of company and state functionaries, members of international organisations, scientists and think tanks, and representatives of NGOs, which is in a position to formulate concepts of world order and development, cutting through political conflicts, and over and beyond national borders. Kees van der Pijl (1999: 138) names this group a *cadre class* which is in a position to conduct "planning and the propagation and monitoring of social norms". This group tends to represent the long-term interests of the reproduction of capitalist societies and can therefore take a position which might be in contradiction to company managers. This is particularly clear in the current discussion on the negative implications of the transformation processes. Michael Goldman (1998) uses, for the specific sphere of environmental and resource politics, the term "global resources managers", who do not per se have an interest in the short-term valorisation of nature. On the other hand, not only is this "international civil society" characterised by considerable inequalities of power and asymmetries but it is still fragmented along state and regional borders.

The present international order basically possesses the *characteristic of hegemonic instability.* The post-Fordist accumulation regime is more prone to crisis than the Fordist one and we cannot speak of coherent modes of regulation. The terrains of international political conflict are still emerging. With regard to the dimension of international leadership, the USA finds itself in a dominant position, above all in a military sense but also with certain qualifications economically, but it is at the same time unable to develop the political leadership which is crucial for hegemony. This, in turn, is connected essentially with the enforcement of the neoliberal mode of politics, which places its stakes on economic privatisation, deregulation and the liberalisation of market forces, and which largely does so without a materially supported integrative organisation of societal relations, both in a national and in an international framework (Hirsch 2002a: 145ff.) This is the basis for the complex relations of cooperation and conflict among the "strong" states. They are forced to cooperate out of their common interest in the safeguarding of the existing economic and political order or its transformation in their interest, and this expresses itself in the development of comprehensive international organisations and regimes. Simultaneously, this relationship of cooperation is in tension with the diverging interests of specific states and groups of states, not least in the relationships between the USA/NAFTA and the European Union, and due to the increasing unilateralism of US politics in some important areas, which are expressed in repeated policies of refusal and blockade. Capitalist competition continues to be determined essentially by the existence of reproduction complexes which are regulated by individual states.

Following the collapse of the Soviet Union and the end of system confrontation, the capitalist periphery is – despite dynamic socio-economic developments in some regions, especially in Asia – in a greater position of dependency, which expresses itself in reduced political and economic room for manoeuvre. In view of failed "development" strategies and uncertain societal and economic perspectives, we can hardly speak of a stable hegemonic relationship. Instead, we could speak of a *fragmented hegemony,* in which worldwide, the compulsory elements of a permanent crisis management move, ever more strongly, into the foreground (Brand 2004).

The issue is to grant these processes within the metropolitan countries legitimacy and to safeguard them with political institutions. This constitutes a large part of present-day international politics. Regarding North–South relations we could, modifying Gramsci, speak today of hegemony in the sense of "coercion with selective elements of consensus".

The process of the "internationalisation of the state" leads, as a whole, to the fact that the *liberal democratic institutions* at the level of the individual state *threaten to run idle*. Relevant political processes of discussion, consensus formation and decision-making shift to levels on which those affected have only very limited influence. The more political alternatives and decisions are formulated at the international level, the more strongly they will, in view of the heterogeneous complex of regulating systems which is to be found there, become the subject of non-transparent and hardly democratically controllable negotiation processes between powerful state and private actors. It is, therefore, not completely wrong to speak in this context of a process of the "refeudalising" of politics (Maus 1991; Anderson/Goodman 1995). It is not least these problems which feed the debates on an "international civil society". The hoped-for democratic quality of the latter breaks down, however, due to the complex fragmentation and power structures which characterise the international political system just as much as global societal relations.

Nevertheless, *nation-states retain a central importance* (see e.g. Panitch 1994; Murphy 2000: 797; Jessop 2003) and, as we will see, even gain importance because in international biodiversity politics *national sovereignty* over biological diversity and especially genetic resources is granted as a precondition for conservation and use. This is based on the fact that their "monopoly of legitimate physical force" despite some differences and qualifications – and despite developments since September 11, 2001 – remains, at the core, unchallenged. Moreover, the core nation-states, in particular, still have, within their complex apparatus, an enormous amount of material resources as well as knowledge and experiences. The states thus remain the decisive institutional configurations in which the capitalist form of the political is manifested. They continue to be the decisive guarantors of the social order – particularly of property relations – and of societal cohesion. The states, therefore, continue to be the centre of the regulation of class and other societal relations and it is essentially up to them to provide the general conditions of production (infrastructure, research, technology) and reproduction. Even though the "state–capital" relations are being formed anew under the conditions of globalisation and internationalisation, this does not mean that corporations become "stateless" or independent of the state. Although they can relate more flexibly to the states – by establishing and utilising different locational conditions – and can, up to a certain point, play them off against one another, they remain dependent on them for the enforcement of their interests, whether through their reliance on the potential force of the state or for reasons of political legitimacy. It is, therefore, no coincidence that the most important multinational corporations, as a rule, have their headquarters or their operative centre in the dominating states of the capitalist triad.

### *1.5.4 Internationalisation and the problem of "scale"*

As we have seen, the internationalisation of the state means that state–society relations are being newly configured on a global scale. The institutional shaping of the state apparatus, thus, takes on a more complex spatial form. Political-administrative regulating and controlling functions increasingly shift from the nation-state to the subnational, regional and international levels. A system of more or less strongly institutionalised bodies of regulations arises which differ, but also intersect and overlap, in their range of activity and which are related to materials and problem areas which are different from, although not independent of, one another, and in which different societal relationships of power are concentrated. This is essentially a problem of scale.[7] Scale is not simply given; rather, it is a social construction resulting from conflict-driven processes of socialisation and dispute. Scale should therefore, in principle, also be regarded as a dispositive of power, which both symbolises and defines processes of politico-social inclusion and exclusion. It can be regarded as the result of institutionalised social compromises and power relations, with which – not least via the externalising of costs and the imposition of relationships of exploitation – the cohesion between different conflicting actors is produced. As such it is "the *spatial* resolution of contradictory social forces" (Smith 1995: 61). For example, the historical constitution of the nation-state level required sustained (class) struggles over the centralisation of political power in the course of the transition to capitalism (Poulantzas 1978/2002: 123ff.).

It follows that politico-social struggles are always, among other things, about the dominance of a certain scale within the geographical-institutional asymmetric setting or about the reconfiguration of that constellation ("re-scaling"). The transition from Fordism to post-Fordism can be understood as a process in which the global capital and financial markets, with their dominant actors, gained in importance over the nationally institutionalised social compromises and relationships of power. At the same time, there was a growth in the importance of subnational and regional levels, together with the social relationships that were institutionalised there. In view of the lack of a centralised apparatus of force at the global level, the different regulating complexes form, in a certain sense, a relatively incoherent *network of international regulation*. However, this network is characterised by the fact that in its individual components highly unequal ranges of activity and relationships of power and dominance are manifested. It is marked by asymmetries and, thus, highly selective possibilities for the definition of problems and the consideration of interests. In connection with the present study, we are concerned, above all, with the conflict-driven relationship between the Convention on Biological Diversity (CBD), the Food and

Agriculture Organization of the United Nations (FAO) and the World Trade Organization (WTO) with its Trade-Related Aspects of Intellectual Property Rights (TRIPS) Agreement. We will also examine how the scales of regulation, constituted by these agreements, behave towards the national, local and regional scales. The incoherence and contradictory nature of the international system of regulating and the connected complex asymmetry of scales is reflected in the fragmentation of a "world society".

### *1.5.5 Hegemony and the second order condensation of power relations*

In recent years there has been an ever stronger awareness that the regulating potential of international regimes can only really be estimated when the complex relations of tension between overlapping areas of regulation and their interplay are taken into account (Young 2002; Rosendal 2001; Thomas 2002; Oberthür/Gehring 2006; Raustiala/Victor 2004). This is particularly true of the relationship between environmental regimes and regimes for the liberalisation of world trade.

In order to understand this conflict-driven interplay of international regulating processes, we revert to the concept of *international hegemony*. In order to explain the way in which we use this term, we must first discuss the different meanings of the concept of hegemony, which is usually not very precisely defined in academic literature. Hegemony should not be equated, as it often is, with the dominance of one or several states. Such dominance is important, but the historical-materialist concept of hegemony goes beyond this (see Cox 1993; Scherrer 1998; Gill 2000, 2003; Borg 2001; Bieler/Morton 2004). International hegemony, in the sense of a transnational and consensus-based societalisation, has two dimensions: namely, (1) the establishment of certain living conditions and concepts of value in societies, and (2) the ability, through governments and other societal groups, to formulate projects, follow strategies and work out compromises on the international scale.

(1) International hegemony first comprises norms of material production and consumption which are established in the core societies or regions and penetrate dependent societies in a "capillary" and broken fashion, that is they are translated into a multitude of strategies and activities. This is accompanied by the generalisation of certain concepts of social order and development and, thus, the consolidation of expectations and planning horizons. It is not about the relationships of dominance between states in the narrow sense (i.e. the opportunity to wield relational power) but about complex processes in which greatly differing economic, political and cultural actors are involved. Of particular importance is the "interiorisation" of production and consumption norms and of societal concepts of development by groups *in* the dependent societies or regions. An integral part of international hegemony is political leadership in the sense of the ability to ensure that subordinate societies recognise the existing

international relations of dominance by means of economic and political compromises.

(2) International hegemony is expressed, politically and institutionally, either in the clear dominance of one state or one region (such as England in the nineteenth century) and/or in a system of international institutions and "regimes" within which the definition of problems and suggestions for dealing with them are struggled over and in which generally effective decisions can be taken and binding compromises negotiated. The system of international institutions can, under these conditions, be understood as the institutionalisation of hegemonic relations which, especially in the political terrain, define rules of negotiation, and participation and access rights. In contrast to the nation-state the power relations are, however, much more complex because you have the operation of, first, nation-states, in the form of governments, as well as, second, "private" international actors. Cox (1993: 62), therefore, speaks of relations between states and classes. Hegemonic relations and their institutionalisation always remain contested under the conditions of social antagonisms and competitive relationships and the complex condensation of power relations based on them.

The international system of institutions is not a coherent unit – not even in the sense of still-to-be-created. The apparatuses, which are, partially, in a contradictory relationship to one another, are terrains on which very specific and differing social power relations are concentrated. In addition, resources and the legal possibilities of sanction play a large role in the effectiveness of specific institutions (Shaw 2000: 215ff.). This perspective differs from regime theory, which usually simply diagnoses the existence of overlapping regimes or network-like institutions without explaining the phenomenon. In our approach, by contrast, international institutions are embedded theoretically in the more comprehensive process of the transformation of the state.

Our central argument is to understand the international state, its international political institutions and its discourses as "second order condensations of societal power relations". Drawing on and further developing the analysis of the nation-state as given by Poulantzas, the term "second order condensations" refers to the complex process of institutionalising societal power relations on various geographical scales in the course of the internationalisation of the state. This process is driven by "social strategies and struggles for control and empowerment" (Swyngedouw 1997: 141) and can result in new relations of dominance. These power relations are inscribed into the structure of the international state apparatuses. Those forces are, most of all, national-governmental actors which are themselves asymmetric material condensations at the level of the respective nation-states and which, through time, come to form a condensation of competing domestic forces that is given cohesion via formulating "a national interest". Besides the nation-states, there are class actors and other representatives of societal interests such as feminist, environmental or indigenous groups. This varies from field to field. Moreover, international political institutions are both material condensations – as in the case of the European Union – and at the same time actors on other terrains.

The concept of "second order condensation" does not mean a priori existing asymmetries between different scales. If there is any hierarchy, for example between the national and the international levels, it is a question of historical developments. Material condensations of second order might – not must, as crises are always possible – express hegemonic projects in the sense that the dominant forces are able to universalise their interests and that there are poorly contested orientations of the relevant political and social actors.

The materiality of the internationalised state exists besides the national (and even local) states of the international institutions. These spheres, although not being centralised, are organised, as well as being formally separated from economic power, stabilised over a certain time period, as well as disposing over expertise, knowledge, legal means and financial resources. The legal means usually need to be implemented by nation-states and the resources come from them. More generally speaking, the international institutions do not dispose over their own state power to stabilise capitalist relations and to give hegemonic projects continuity, but this power is "lent" by the (powerful) nation-states. However, there might evolve a certain density and resistance against the will of dominant states.

Our approach has one important implication for the present study: this concerns the question of coherence among different international political terrains. As we said, a major debate in international politics as well as in International Relations and especially in regime theory relates to the problems of creating coherence among international political institutions. Incoherence in international politics is, from the outlined perspective, less the inability of nation-states or international political institutions to control political processes than of contradictions and tensions among the dominant forces and the inability to create consensus among the dominated. From our perspective, the lack of coherence has to do both with the contradictions and tensions among different policy fields and with the material condensations of the balances of forces in specific political institutions and with the asymmetries among them (see Görg/Brand 2006).

Moreover, incoherence is a mode of international politics in the sense that among political institutions we can observe a strong asymmetry even if they were to regulate the same issues. The fact that in many cases several political institutions claim responsibility for the same issue area is not only a consequence of incoherence but can be interpreted as a power strategy of dominant forces. Because, if a terrain is too politicised and weaker actors become stronger the more powerful actors have the possibility to re-evaluate other terrains (or to act in a bilateral mode). We can borrow, from John Braithwaite and Peter Drahos (2000, ch. 24), the concept of *forum-shifting* which means that a political actor, especially the USA, can shift strategies from one organisation to another, leave an organisation or pursue strategies towards several ones. But beyond Braithwaite and Drahos we can understand forum-shifting not only as a strategy of actors towards existing terrains but also as a strategy to change power relations and to structure the terrain itself.

## 1.6 Questions of method

The theoretical framework presented and the outlining of the problem should have made clear the complexity of the subject at the centre of the study. Not all of the aspects mentioned here will be pursued to the same degree of detail in the following sections. In the foreground are two questions: first, the relationships between the different political negotiation terrains and international agreements; and second, the connections among the various political scales and their production. Two further things must also be taken into account. First, the field is characterised by an enormous historical dynamism. New actors, who had scarcely been in evidence previously, appear continually, and new terrains and fields of conflict receive new meanings which could not previously have been foreseen. The danger is, therefore, great that important factors can be either lost from view or over- or underestimated. We are not dealing with solid *ceteris paribus* conditions but with a developing conflict terrain. Second, the descriptions by the social and natural sciences in these conflicts are not neutral instances but themselves form an element of the developing conflicts. Whether or not tensions exist between the CBD and the TRIPS Agreement is not at all a neutral description but a part of the subject matter. Since, last but not least, actors participating in the conflict try to substantiate their positions and weaken those of their opponents with the assistance of these descriptions, there can be no neutral observer positions in the strict sense of the word, but only a permanent self-reflection of one's own experiences.

In view of the historical dynamism, being too strict regarding the subject-matter to be examined, and its temporal classification, always entails the risk of losing sight of important processes and new developments. An additional problem is the consideration of the materiality of the problem subject-matter. If we take our introductory remarks on the concept of relationships with nature seriously, we have to take into account the biological-technical characteristics of the subject-matter concerned, namely biodiversity or genetic resources. This is true both for the definition of the problem and the solution options referring to it. Where exactly the problem lies or what an adequate treatment of the problem would be, is not simply predetermined; rather, corresponding descriptions develop as components of the development of the conflict. An interdisciplinary approach which critically examines the corresponding bioscientific descriptions would, strictly speaking, be indispensable here. For example, for many years the economic interest in genetic resources was simply regarded as given in view of the options for applying genetic technology, which gave impetus to the talk of a "green gold of the genes". Recently, by contrast, the signs have been increasing that in industry, very varying economic/technical strategies are being applied and that the economic value of biodiversity cannot simply be assumed (see Macilwain 1998). The possible misunderstandings are also serious if certain attributions with regard to the ecological importance of biological diversity are simply used uncritically (e.g. regarding the relevance of biodiversity for the stability of ecosystems).

We have attempted to methodically limit the barriers and possible dead-ends in both directions. To start with the second problem: we do not present an interdisciplinary examination of the access and benefit-sharing problem here, which would include natural or technical sciences. However, since before beginning this investigation we already had at least a few years of familiarity with the particularities of this field (including the natural science and technical dimensions; see Görg *et al.* 1999), we believe that we have been able to avoid the larger traps.

On the other hand, the temporal dynamism presented us with some difficulties. In the course of the study, terrains (such as the WIPO; see Chapter 4) and actors (for example the Like-Minded Group of Megadiverse Countries; see Chapter 2) appeared, the importance of which was not to be foreseen or which, at the time of the planning of the project, did not even exist. Other developments, by contrast, have taken much longer than was to be expected (e.g. the revision of the TRIPS Agreement; see Chapter 4), while others in turn fortunately found themselves in an important phase precisely in the time-period of the study (the adoption of the Bonn Guidelines within the framework of the CBD; see Chapter 2; and the process within the framework of the FAO; see Chapter 3).

The kernel of this study is represented by open questionnaire interviews which were conducted on the different negotiation terrains between May 2000 and July 2001 (see the list of interviews at the end of the book). Our interview partners were experts and representatives of the various conflict parties or actor groups participating in the process. The objective was not to receive representative answers to the subject-matter but to gain a deeper insight into the negotiation processes and their difficulties from the viewpoints of the different parties to the conflict. The interviews were analysed in different phases and according to various criteria. In particular, we tried to figure out the interests of groups involved and their positions regarding the subject matter, as well as trying to understand the conflict dynamic and, particularly, changes in the perception of the problem and the formulation of interests. Last but not least, we tried to estimate the relevance of the different terrains, their relationships with each other and their impact on different political scales, in particular the national and the local scale.

Participant observation at conferences – at least one for every terrain – was also essential. Without direct observations it is very difficult to estimate the exact meaning of specific purposes and the environment, the mood and the political tensions where negotiations took place. We participated as observers at COP 3 in Buenos Aires (1996), COP 4 in Bratislava (1998), COP 5 of the CBD in May 2000 in Nairobi and COP 6 in The Hague in 2002; the negotiations of the Bonn Guidelines in October 2001 in Bonn; the Sixth Extraordinary Session of the CGRFA in June 2001 in Rome; and some other national and international conferences (expert and NGO meetings). The evaluation of the TRIPS revision process was more complicated. Access to the negotiation process in this case was extraordinarily difficult. While we attended a conference designed as an invitation to the global civil society at the WTO's headquarters in Geneva in July 2001, to discuss the topic of intellectual property rights, we were not allowed to participate in formal negotiations and relied upon the negotiation documents and published reports of the negotiations. A further terrain, which was not at the centre of the *access and benefit-sharing* problem but which comprised an important theme in this context, could also only be dealt with by us to a limited degree: the question of indigenous knowledge. Here, we observed some meetings of the International Indigenous Forum on Biodiversity (IIFB) which took place in connection with the formal meetings of the CBD in Bonn and The Hague.

# 2 On the value of nature

## The Convention on Biological Diversity and the commercialisation of genetic resources

At first sight, the Convention on Biological Diversity (CBD) appears to be an important contribution to the conservation of biological diversity. Moreover, it could be perceived as offering a new direction for conservation strategies because, compared to earlier agreements, a variety of ecological, social and economic aspects are taken into account, creating one of the most comprehensive environmental agreements of the past century. In recent years, however, the ambivalent character of the Convention has become ever more evident. Indeed, some of the most important challenges (i.e. the integration of different aims, national implementation and its relationship to other agreements) are still far from being resolved. Above all, it becomes obvious that the CBD, in general, expresses a tendency towards a view of biodiversity which is primarily concerned with its economic utility and (exchange) value. The frequent accusation is that the CBD has even accelerated the commercialisation of biological diversity and thus contributes less to its conservation and more to its marketing.

The problem cannot be grasped, however, with an isolated appraisal of the Convention, for it concerns the entire context of different international agreements and negotiation processes. Only if their interplay and the multitude of their mutual tensions are examined, can an assessment of the actual regulating effects be possible. Only then can it be understood in what way post-Fordist relationships with nature in the field of biological diversity are regulated by a political institutional framework and what this framework actually means. Furthermore, it must be clarified that post-Fordist relationships with nature are created and formed in a specific way with the support of just such a framework (and not with economic-technological strategies alone).

The CBD was signed, in a complex and difficult negotiation situation, in the run-up to the UN Conference on the Environment and Development (UNCED) in Rio de Janeiro in 1992 (on the negotiation process see Sánchez/Juma 1994; Swanson 1997; Arts 1998; McGraw 2002). It was originally conceived as an "umbrella convention" which was to align the many already existing agreements on the protection of species (e.g. the Convention on International Trade with Endangered Species of Wild Fauna and

Flora) and make them more efficient. But in the course of the negotiations up to 1992 and in its further development after the first conference of the convention parties, the CBD increasingly changed its appearance. While in the first drafts for a framework agreement, which had been submitted by the International Union for the Conservation of Nature (IUCN) in the late 1980s, distribution conflicts and their regulation (i.e. the regulation of technology transfer, of benefit-sharing etc.) did not yet play any great role (IUCN 1989; Arts 1998), this changed increasingly in the negotiations on the CBD in the run-up to the Rio conference in 1992. The text of the CBD, which was passed at that conference, can only be understood adequately if economic aspects and the distribution conflict over the use of genetic resources are taken into account. The reasons for this are complex. Certainly, of central importance is the CBD's specific character of compromise as well as the economic interest in genetic resources which had significantly increased during the late 1980s. The idea of the "reconciliation" of "North" and "South" and of "ecology" and "economy", which characterises the concept of sustainable development, shapes the CBD to a particularly high degree. It is precisely for this reason that the Convention makes evident the pitfalls contained in this concept. In analysing these shortcomings both the limits of this double "reconciliation" and the definite effects of the international institutions which try to translate this model into action are unveiled (see also the contributions in Görg/Brand 2002; Brand/Görg 2007 and Conca *et al.* 2007).

In the following the assumption advocated by many observers will be closely scrutinised, according to which the CBD is naturally opposed to processes of the economisation of social relationships and the institutions which support this – particularly the agreements within the framework of the World Trade Organization (WTO) and the neoliberal restructuring of social relationships as a whole (IIED 2001; Ling/Khor 2001; and the assessment in many of our interviews). According to this position, in the CBD and in the other agreements signed in the Rio process (particularly Agenda 21), interests in the protection of nature and of weaker actors are taken into consideration. This is without any doubt the case. However, questions remain as to how the different processes are articulated, with each other as well as regarding the actual relevance of these international agreements for local actors.

If the CBD is to be examined in the following as an example of the internationalisation of the state, it will be done by means of a detailed examination of interests and political strategies which are condensed in those aforementioned institutions. In other words, it is not only considered as an instrument for solving problems or for guaranteeing global relationships of power and dominance, but also as a terrain on which different actors struggle for their interests. By taking the double position of international institutions as regulating authorities and as an expression of global relationships of power into account, the role of the CBD can be appraised in a more precise manner. The central issue concerns the specific relationship between conservation and use. As we shall see, very specific ideas and practices exist, each representing different mixtures of protection and utilisation and each partially interconnecting – at times surprisingly – but at the same time, also standing in a relationship of tension to one another.

Our argument is that the CBD itself tends, under post-Fordist neoliberal conditions, to create a framework for the commercialisation of nature. Not even the CBD can escape from the valorisation paradigm which is central to post-Fordist relationships with nature. On the contrary, this paradigm is institutionalised in the CBD, although in a different way to that of the WTO agreements. At the same time, the commercialisation of nature is not a unilinear process but one which is characterised by social struggles and contradictions. This may allow weaker actors (e.g. Southern governments, NGOs or indigenous peoples) to bring their interests to bear in the negotiations and to be at least partially considered in the compromises. However, it must be examined exactly *how* this occurs as well as how this process allows a degree of manoeuvrability for certain interests. In doing so, it must be taken into consideration that in part, the interests of the different actors are first constituted, or further developed, during the negotiations and the implementation processes of the CBD as well as in light of the experiences in the different countries. This becomes especially clear in the questions as to how access and benefit-sharing are to be ensured and what strategies are used by different actors as a reaction to this.

At the centre of the study are the processes connected with the regulation of access to biological diversity and the fair sharing of the benefits from its utilisation. The negotiations on the *Bonn Guidelines on Access to Genetic Resources and Fair and Equitable Sharing of the Benefits Arising out of its Utilization*, which began in Bonn in October 2001 and were signed in April 2002 at the sixth conference of the signatory states in The Hague, as well as the following negotiations of an *International Regime on Access and Benefit-sharing* right after the approval of the Bonn Guidelines, are interesting examples of the process of the constitution of interests and their condensation in an international agreement. A further example is the Like-Minded Group of Megadiverse Countries, which was formed at the beginning of 2002 and whose appearance on the international stage was perceived as jeopardising the solutions achieved in the Bonn Guidelines. Closely connected with the questions of access and benefit-sharing are issues such as the safeguarding of intellectual property, the rights of indigenous peoples and local communities and finally, of course, the issue of nature protection. It is important to us, first, to follow and understand more precisely the international processes within the CBD: what interests are articulated there and how are they condensed? In addition, the CBD should be placed in relationship to other international agreements; particularly the *Food and Agriculture Organization of the United Nations* (FAO) and the Agreement on *Trade-Related Aspects of Intellectual Property Rights* (TRIPS) within the WTO. Both will be briefly discussed here and will be dealt with in their own right in the following chapters.

The following section will present a brief summary of the issue of biodiversity and the problems connected to it. This will be done by looking at the development of the CBD and the most important regulations associated with it. Following this, important interest groups and their positions will be outlined and some central lines of conflict presented. It is remarkable in the context of the CBD that central concepts and measures based upon them are still characterised by considerable uncertainty and are part of the political disputes. This is reflected in the major role of experts but also in the fact that the argument of uncertainty is itself repeatedly brought forward in order to strengthen certain policies. Neither the exact extent of the erosion of biological diversity nor its ecological, economic or social consequences can be determined precisely. Moreover, the character of the problem – in terms of what the issue actually is; what is central and what is arbitrary? – is anything but clear. This is directly connected with the central concept of biological diversity or biodiversity, which does not in any way represent a precisely defined scientific foundation for the entire process.

## 2.1 Biodiversity as a political concept

The concept of biological diversity was, for a long time, used synonymously with the concept of the diversity of species. Today, by contrast, it describes a much more comprehensive *synthesis of three elements*: the diversity of species, genetic diversity and the diversity of habitats and ecosystems. This new meaning of biodiversity is relatively recent – and it involves far-reaching political and economic implications. The new concept first established itself in the second half of the 1980s. However, even as early as the late 1970s, the report “Global 2000” to the US President outlined an interesting shifting of weights, as a prelude to later developments, in which *ecological* threat scenarios are connected with the strong emphasis on the *economic* value of biodiversity and its *geostrategic* importance (Flitner 1999). Of decisive significance for the presently established meaning of the concept was the *National Forum on BioDiversity*, a conference of the *National Academy of Sciences* and the *Smithsonian Institution* in the USA in 1986, at which the abbreviation was introduced (Wilson 1988). Nonetheless, another aspect can be found at this conference which characterises the new concept to this day: the partially metaphorical elevation of biodiversity to the “diversity of the living” – a metaphor which eludes scientific description because it means everything and therefore nothing in particular (see, as an example: Wilson 1992; for a critique: Hertler 1999). The synthesis of the three main elements of biodiversity – as problematic as their scientific foundation may be – led to the restructuring of the corresponding spheres of politics. This becomes evident, not least, in the negotiations on the CBD, which began in the late 1980s.

In contrast to popular notions, it is not obvious what exactly the central problem is and what the "loss of biological diversity" is actually all about. Both in the public discussion and in scientific research a whole number of arguments are brought forward which are not congruent but rather, accentuate the very different and partly contradictory aspects of this subject-matter. Most common is the listing of data on the *extinction of species*, that is estimates of the rate of extinction of the species which exist worldwide. In the red list of the species threatened with extinction in the year 2000, The World Conservation Union – IUCN, for example, points out that worldwide, of 4763 mammal species 24 per cent are threatened with extinction, as are 12 per cent of 9946 bird species (according to the first Global Biodiversity Outlook in 2001). But this does not really represent the whole threat associated with the diagnosis "loss of biodiversity". The estimates have a high uncertainty factor, since all the figures are imprecise and provisional. It is not without reason that the extensively studied and "famous" mammals, birds, fish, reptiles and amphibians form the core of the IUCN's "red lists", as little is known about other species (in particular insects and microbes) and therefore, specific figures can hardly be presented. All other data are good estimates at best, because not only the rate of extinction but also the total number of the species is unknown and therefore must be extrapolated. Indeed, the *Millennium Ecosystem Assessment* claims that there are 1.7 to 2 million identified species and realistically estimates the number of species existing today at between 5 and 30 million (BSR 2005: 19).

According to this interpretation of the problem, which is geared towards the diversity of species and its erosion, at present we are experiencing the sixth wave of extinction in the course of the Earth's history. In contrast to earlier phases of the mass extinction of species this phase is unequivocally caused by anthropogenic changes in the environment (e.g. WBGU 1999). Putting aside existing uncertainties, the threat is obvious. The "Global Biodiversity Outlook 2", published in 2006, points out: "a common message emerges: biodiversity is in decline at all levels and geographical scales" (Secretariat of the CBD 2006a: 5). This trend can be reversed only when immediate measures, such as the creation of protected areas or pollution prevention programmes, are taken seriously. The so-called 2010 Biodiversity Target – to reduce significantly the erosion of biodiversity as a contribution to poverty alleviation – may only be reached when "unprecedented additional efforts" are provided (BSR 2005: 14; Secretariat of the CBD 2006a: 5). In these notions, other descriptions of the problem arise, thereby pointing out the ecological or economic usefulness of "biodiversity" at very differing levels. From an *ecological* point of view, it is argued that biodiversity offers inestimably valuable functions or services for the conservation and stability of ecosystems; something which also impacts upon the existence and causes of other environmental problems (from climate change to soil degradation and desertification), and the possibilities of treatment. In particular, the Millennium Ecosystem Assessment, which up to now is the most comprehensive study about the connections between ecosystem services and human well-being, focuses on such services and highlights the important value of biodiversity (or "Life on Earth", see Figure A in MASR 2005: vi) for all the services nature provides. As a main message, the Millennium Ecosystem Assessment argues that 15 of 24 ecosystem services are in decline with unforeseeable consequences both for nature and global society (BSR 2005: 22–41).

Seen from the perspective of the Millennium Ecosystem Assessment, the value of biodiversity is not restricted purely to the wealth of species diversity; it also refers to the contribution biodiversity provides for social use and human well-being in a wide range of areas: from nutrition and shelter, moving on to clean water and flood control as well as to basic conditions for social relations (security, social cohesion etc., see Figure A in MASR 2005: vi). Nonetheless, this redirection of conservation issues is not at all unproblematic. While the Millennium Ecosystem Assessment itself is not framed within a purely utilitarian perspective – it focuses merely on trade-offs between different kinds of use and also on limits of human utilisation – it does offer ground for such a reductionist perspective (see McCauley 2006; and the comments in *Nature* vol. 443, 19, October 2006: 749–750). Nevertheless, in recent years, the *economic* value of biodiversity has become even more important, representing itself in a broad variety of socio-economic fields of application. The reason why this concern has grown cannot only be seen in this broad variety of ecosystem services that biodiversity provides. Of more importance are recent technological changes due to the development of new bio- and genetic technologies. The issue at stake is primarily that of the loss of "natural capital". The leading metaphor here is that of the tropical rain forest as "nature's pharmacy" or more simply as a "capital asset". The main interests in the use of genetic resources are represented by pharmaceutical and agricultural companies. Due to the fact that throughout the world, biodiversity provides a wide range of ecosystem services, in many places quite different forms of utilisation are affected, from nutrition and leisure activities in developed countries to farming communities in developing countries and the population of the tropical rain forest regions. Ultimately, the protection of biological diversity, thus, is interconnected with economic interests and global social relationships of power and dominance.

The different arguments illuminate not only reasons which complement one another, for instance, why the erosion of biodiversity represents a problem. They also give quite different and sometimes contradictory answers to the questions: *What* exactly is the problem? *How* and *for whom* did it become a problem? *Why* and *with what means* should this problem be tackled? In international biodiversity policy it is therefore not a question of the solution to a given, more or less objectively existing problem, which must only be more or less well grasped and treated. On the contrary, the problems connected with the scientific definition of concepts show that there exists a struggle, with global dimensions, fought out over the regulation of relationships with nature – with locally and nationally different characteristics. Despite the widely recognised agreement that erosion of biodiversity is a major problem, an important question remains: What exactly is the problem and how should it be described and defined?

In view of the indicated perspectives on different aspects of the problems involved, the assumption that the concept of biodiversity is based on an immanent progress of knowledge in biology or its individual subdisciplines is misleading. It is evident that in view of the problems concerning the subject-matter, an *external* reason must be sought based on the fact that very differing disciplines have agreed on a new definition of their field of study and on the overlapping and intersecting nature of, partly competing, disciplines such as taxonomy and molecular biology. The key to understanding the problem's constitution lies in external societal conditions and in the history of the establishment of the concept. This gives reason to argue that the unification of heterogeneous research fields is not based on reasons of content but is *strategically motivated* (Hertler 1999: 49).

The difficulties mentioned above are closely connected with the expansion of the meaning of the concept and particularly with the increased emphasis on the economic value of biodiversity. With genetic diversity, in addition to the diversity of species, growing attention was paid to the potentially economically valuable elements of biological diversity, that is: the genetic resources. This dimension was not completely new (on this cf. the following chapter on the conflicts at the fringes of the FAO). But as a new creation, the concept of biodiversity was particularly suitable for taking up the different interests which come together in this field of conflict and to reflect on the societal fronts of the conflict (for more detail see Flitner *et al.* 1998; Brand/Kalcsics 2002; Takacs 1996; Swanson 1997). Its width and its lack of focus predestined it to absorb these different conflicts, on the one hand, but also to conceal them again, on the other. Finally, in the subject-matter of biodiversity, traditional themes concerning the protection of nature are combined with newer attempts at environmental protection going under the title of *sustainable utilisation*. Here, questions of the *utilisation of resources* are related to *access* and more or less exclusive *rights of utilisation*, and thus with questions of the *sharing of benefits* and *intellectual property rights*. The latter is indicated through the widely accepted concept of "genetic resources" instead of "genetic goods" (in our study we use both concepts interchangeably, having in mind that the concept of genetic resources is biased towards utilitarianism, cf. Kaiser 2003). The transition from a pure "umbrella convention" on the protection of species and nature to a convention which attempts to regulate not only the ecological dimensions of the problem but also the economic and social dimensions, is neither trivial not accidental – and neither can it be simply reversed at will.

## 2.2 Functioning and provisions of the CBD

The conservation of biodiversity is quite a complex cross-sectional task, encompassing the classical protection of nature and the organisation of agriculture as much as related questions, connected with the regulation of global trade and the transfer of the modern biotechnologies. The text of the CBD was completed in Nairobi on 22 May 1992, and it was then laid out for governments to sign as of June at the UNCED after which it finally had to be ratified by national parliaments. On 29 December 1993, the CBD entered into force, exactly 90 days after the thirtieth member state had ratified the Convention. As of June 2006, there were 188 member states which had ratified the CBD (counting the states of the European Union individually, even though it appears as one signatory). One important exception is the USA, which signed the CBD, after having hesitated for some time, but has not yet ratified it (which gives it, as a signatory, the right to take part in proceedings without being bound by any obligations, see section 2.2.1). The CBD is a case of *soft law*, which is signed by governments and ratified by national parliaments but which does not possess any sanction mechanisms (as for example, does the WTO with its Dispute Settlement Mechanism) but which first has to be implemented nationally. This latter aspect often opens up dispute over interpretations. The same is obvious at the international level: By the time of the eighth Conference of the Parties – which took place in Curitiba, Brazil, in March 2006 – all governments had to have submitted a report about the national status of biodiversity, especially protected areas. But only around 25 per cent of the parties, that is governments, did so.

Even if it, in contrast to the Framework Convention on Climate Change (FCCC), is not so-named explicitly, the CBD is, in fact, a framework convention which not only has to be translated into national law but whose provisions are also the subject of further negotiation at the international level. This occurs at the *Conference of the Parties* (COP) which takes place every two years, in the *Subsidiary Body for Scientific, Technical and Technological Advice* (SBSTTA) and via a relatively small secretariat in Montreal (cf. Le Prestre 2002).

Broad participation of non-state actors in the process is permissible and it has been welcomed in quite an impressive manner.[1] However, it is important to note that the active personnel involved in the CBD is fairly limited in number. It is always the same important delegation members, experts and NGO activists who meet at the conferences. The permeability of delegations and NGO activists is quite large in some countries (which only seldom means that NGO activists have a great deal of influence). This is obviously sensible in view of the great complexity with regard to both content and procedures. But there seems to be a recent shift which might have to do with the growing importance of the CBD: while the participation of NGOs was discussed for the first time in such an open and critical way, at the same time the CBD seems to have opened it more to private companies and their associations (ENB 2006a: 23).

One important element of the material implementation of the CBD is the funding mechanism, the *Global Environmental Facility* (GEF), which – against the decreasing resistance of the Southern countries – is established at the World Bank and still retains only a temporary status. Funding mechanisms not only shape the implementation process of the CBD, through concrete projects, but they also influence the development of the CBD as a political terrain itself. For example, before the COP in 2006, held in Curitiba, 2 million USD (United States Dollars) were considered necessary by the CBD secretariat in order to allow a comprehensive participation of delegates from Southern countries. But only 25 per cent of this amount was provided by Northern countries, which eventually meant that there were not enough delegates from Southern countries present at the COP.

In general the CBD follows three equally important objectives:

> The *conservation* of biological diversity, the *sustainable use* of its components and the *fair and equitable sharing* of the *benefits* arising out of the utilisation of genetic resources, including by appropriate access to genetic resources and by appropriate transfer of relevant technologies, taking into account all rights over those resources and to technologies, and by appropriate funding.
>
> (CBD 1992, Art. 1; emphasis added)

In other words, the CBD already lays down in Article 1 that the expected (economic) benefits which are to be shared will be created by guaranteed *access* to biological diversity. However, the concept of access includes three very different forms: access to natural resources, access to knowledge and – closely connected with the latter – access to technology (ibid.: Art. 16). Access must, however, be accompanied by fair benefit-sharing, namely of those benefits which arise from economic utilisation. Seiler and Dutfield (2001: 15) call this "a cornerstone of the CBD philosophy of convergence between the interests of the North and the South". In addition, the *prior informed consent* (PIC, CBD: Art. 15.5) of the "providers" to access must be attained and access should take place on *mutually agreed terms* (MAT, ibid.: Art. 15.4). This should be the case not only for access to genetic resources but for all forms of access. A decisive innovation in international law is also the prescribing of *national sovereignty* over natural (not only genetic) resources in the CBD (ibid.: Preamble, Art. 3 and 15). It is described in Article 3 as the central "principle" of the CBD. The previously existing principle of the *common heritage of humankind* to plant genetic resources is thus replaced by the *common concern* named in the preamble. This does not mean that nation-states are therefore also the owners of genetic or biological goods (cf. the critique of Harry/Kanehe 2005). National sovereignty here means only that states have the right to lay down national rules and laws governing the treatment of biological diversity. Thus national sovereignty is a central element in the international system of property rights in dealing with genetic resources.

The CBD also encourages technology transfer, that is it is concerned with access to technologies as a part of benefit-sharing. Article 1 states, however, that the objectives of the CBD are to be achieved by "taking into account all rights over those resources and to technologies". The second half of this regulation is an important formulation and represents a limitation to the technology transfer Southern countries had hoped for because it restricted this transfer according to the existing intellectual property rights laws (see below).

The rights of the local population and particularly of the indigenous population are mentioned in Articles 8(j) and 10(c), but they are relatively vaguely formulated. The rights are not listed in the preamble or independently, but are subpoints of the regulations for *in situ* conservation and thus are subjected to an instrumental approach (Stoll 1999). The role played by indigenous peoples (although the concept of peoples is not used in the official political process but that of communities) and locally based practices in the conservation of biological diversity are thus recognised, but their rights are only explicitly strengthened as far as they serve conservation and sustainable utilisation. Furthermore, they are subjected to national laws. Article 8(j) lays down that each party to the contract (i.e. the national government) *shall*:

> subject to its national legislation, respect, preserve and maintain knowledge, innovations and practices of indigenous and local communities embodying traditional lifestyles relevant for the conservation and sustainable use of biological diversity and promote their wider application with the approval and involvement of the holders of such knowledge, innovations and practices and encourage the equitable sharing of the benefits arising from the utilization of such knowledge, innovations and practices.
>
> (CBD 1992, Art. 8(j))

The fact is also significant that the Convention only covers those *ex situ* genetic resources which fall under the regulations of the Convention *after* the coming into force of the Convention in December 1993 or after the individual countries have signed the Convention (ibid.: Art. 15.3). Thus, neither the *ex situ* collections which have been set up over the centuries, nor botanical and zoological gardens, nor scientific collections and genetic bases in the field of agriculture, fall under the regulations of the CBD. This not only means that collections of inestimable economic value are excluded. It also changes the field of conflict, since, particularly in the field of the biodiversity utilised in agriculture, it is unclear whether the research institutes and biotechnology firms in the Northern countries are actually dependent on access to *in situ* material or whether the *ex situ* collections in the North are adequate for their purposes (see the following chapter on this). Similarly, the botanical and zoological gardens are confronted with an increased demand from companies, as the latter believe they will be able, in this way, to achieve faster and more secure access to genetic material. It is also important that the regulations of the CBD are not of equal standing but are related to one another in a hierarchical fashion. National sovereignty is binding. The facilitating of access (ibid.: Art. 15.2), the principles of mutually agreed terms and prior informed consent are, by contrast, "shall be" clauses; this means that governments are urged to take appropriate measures. Altogether, the formula "subject to national legislation" is central, and regulations – particularly concerning intellectual property rights – are limited in the sense that they are not allowed to contradict other international norms.

At the various conferences of the parties to the Convention, the regulations of the CBD have been developed further by a number of decisions, whereby with regard to access and benefit-sharing, questions of national legislation, the precise definition of concepts and the assessing of various experiences are in the foreground (see Siebenhüner and Suplie 2005 who describe this development as a process of institutional learning).

## 2.3 Positions of the most important actors

Due to the complexity of interests, a clear classification is not easy to make. The argument is continually brought forward that all the countries of the world are in one way or another providers and users of genetic resources. Nevertheless, this fact is of little importance for political momentum. As the home of "wild" and *in situ* biodiversity is primarily in the countries of the geographical South, but their commercial valorisation until now has primarily taken place in the countries of the North, this leads to a North–South conflict, which is largely the reason for the existence of the CBD in its basic structure (Svarstad 1994). The actual front lines of the conflict are much more complex, however, for neither "the South" nor "the North" are uniform entities, but are divided in themselves by a multitude of differences of interest between individual countries – not all Southern countries are rich in biodiversity; only a few Northern countries are capable of commercial valorisation – as well as within both Southern and Northern countries (between governments, the rural population and different sectors of industry).

In the following – with no claim to completeness: for example botanical gardens are not considered – the positions of the most important political actors and groups of actors in international negotiations are outlined. Essentially, this is an attempt to present a first picture of the most important interest groups and their view of the terrain of conflict, in order to be better able to appraise the individual conflicts and the approaches to the regulation of access and benefit-sharing.[2] International institutions, which also sometimes appear as actors in the questions which are at the centre here, particularly the FAO and the World Intellectual Property Organization (WIPO), are not listed separately. Their representatives are above all actively interested in the clarification of certain facts or in the thematic boundaries to other processes (the FAO wishes to remain responsible for agro-biological diversity and WIPO wishes to take over a guiding function for intellectual property rights). Therefore, as actors, they are more or less directly involved in the design of the CBD.

### *2.3.1 Governments of the Northern countries*

In countries where a strong domestic pharmaceutical, agricultural and cosmetics industry exists, the governments quite clearly represent the interests of "their" agricultural or pharmaceutical companies or research institutes in access, planning and legal security (see the interviews at CBD/19, CBD/22). Despite all their differences they are united on one point: there must be no softening of "hard" rules on the security of intellectual property. Especially, the TRIPS Agreement must not be questioned, which is why the CBD is not a suitable place for the legally binding regulation of intellectual property rights. In addition, the great uncertainty and the high costs of biotechnological developments are emphasised, as it takes several years before new products can be developed. Different forms of benefit-sharing are therefore proposed: from, for example, short-term sample fees and advance payments to participation in possible profits from marketable products (see below).

In the concrete access agreements, the Northern actors see the Southern governments in a moderating, supervisory position rather than as direct partners to the agreement (the stakeholders at this level are not the governments; for a typology of access agreements see Trommetter 2005: 758–781). The governments or national authorities are regarded as necessary, however, in order to create trust for the users in the direct providers (CBD/4). On the question of the rights of indigenous peoples and local communities, the Northern governments tend to point out that at the international level at best, guidelines and awareness of the problems could be created, but that these are "internal" concerns of the Southern countries. With regard to important concepts such as benefit-sharing, prior informed consent or the participation of indigenous peoples and the importance of "traditional knowledge", they point to the great conceptual uncertainties. In part, the work of the WIPO is registered positively, which has, in recent years, taken up the subject of traditional knowledge. The change in the attitude of those involved and the creation of trust are regarded as central concerns of the international negotiations (CBD/4). With regard to the national implementation of the CBD, the Northern governments see no need for action. This view clearly exists on the part of the provider countries. The Northern governments warn here of "over-regulation", that is the inflexible and time-consuming legal and administrative regulations for the preparation of bioprospecting. The Northern governments argue very ahistorically because the previous appropriation of genetic resources plays no role in their position. There are clear differences among the Northern countries, however. There are governments which are more ready to compromise, such as the Swiss (Novartis in particular plays an important role here) and the Norwegian (Novo Nordisk plays an independent role), which are concerned with establishing a stable legal framework in which the providers should share in the economic benefits. The hard-liners, on the other hand, are Japan, the USA, Canada, Australia and New Zealand, which have formed the *JUSCANZ Group* (the abbreviation is formed from the initial letters of the countries' names). As was previously mentioned: the US government plays a special role as a signatory state which did not ratify the CBD but acts as a "party" of it. It is one of the architects of the global rules for the appropriation and protection of biodiversity in other countries without the duties to implement the results in its own territory (and without the duty to finance the CBD secretariat and its activities). The particular interest of the JUSCANZ Group is that the developments within the CBD should not contradict WTO regulations. These governments tend to play the role of "brakesman", but they are also very influential in the formation of compromises due to their importance.

### *2.3.2 Governments of the Southern countries*

The most important difference between the individual countries is whether or not they in fact "possess" considerable biological diversity – and if so, to what degree? Thus, for example, Brazilian governments (liberal and leftist ones) tend to represent the interests of commercialisation with regard to "wild" diversity, while with regard to agro-biological diversity, it tends towards free exchange which should not be restricted by commercial interests (something which has been institutionally laid down in the multilateral exchange system in the FAO, see below). This is connected with the fact that in Brazil an immense "wild" diversity exists, while the country is dependent on other countries for seeds from genetic resources. However, the leftist government under President Lula da Silva – surprisingly, for many observers – has been an even stronger advocate of the interests of agrobusiness and especially the introduction of genetically modified soy bean seeds. Argentina, by contrast, which is not a country of great biological diversity in either area, but which is one of the most important cultivation areas for genetically modified plants, usually votes with the USA and Canada.

Since the end of 2001 there has been an interesting development among the provider countries, in which the position of the Megadiversity Countries has played a special role. The governments of Bolivia, Brazil, China, Colombia, Costa Rica, Ecuador, India, Indonesia, Kenya, Malaysia, Mexico, Peru, the Philippines, South Africa and Venezuela (later on Madagascar and the Democratic Republic of Congo joined the group), represented by environmental ministers or their delegates, signed the *Cancún Declaration* of the Like-Minded Group of Megadiverse Countries in mid-February 2002, following a meeting in Cancún, Mexico (where at the end of 2003 the WTO ministerial meeting was going to take place) and prior to the World Summit on Sustainable Development in Johannesburg in June 2002 where the group intervened in the debates. The aim of this group is, or was, to create a common agenda for sustainable development and to strengthen cooperation. Sixty to seventy per cent of the world's biodiversity and 45 per cent of the world's population live in these countries. The responsibilities for biological and cultural diversity and the great potentials for development are emphasised. The protection and the sustainable use of biological diversity, access to genetic resources, benefit-sharing and the protection of traditional knowledge are the core issues. A central point is national sovereignty over natural resources. Cooperation is to consist above all on a common position at international conferences and the exchange of experiences with regard to national implementation. The national legislation has a clear objective, namely "that the adoption of an appropriate legislation can become a highly useful tool in participating efficiently in international markets and in taking full advantage of their own comparative advantages" (SEMARNAT 2002a: 116). The fact that this group has coordinated itself alone can be understood as an answer to the situation of competition of Southern countries with regard to Northern actors, which is felt to be disadvantageous. Its formation is also part of a several-year process during which the interests of the megadiverse countries have been given a more distinct contour by the negotiations. Their perception of the present situation is: "The megadiverse countries have not yet obtained the considerable benefits that should come with their privileged position, in large part due to the absence of political will and long-term strategies" (Caillaux/Ruíz 2002: 161; cf. a critical account in Ribeiro 2005: 73–74). Responsible for this are also the asymmetrical international relationships of power. The perception of two structural problems in the exercise of national sovereignty is also interesting:

> On the one hand, it is difficult to enforce this right over genetic resources due to their reproduction capacity and natural dispersion, as well as the fact that these are scattered over several countries – none of which can claim exclusive rights over them. On the other hand, defining sovereign rights is complicated even further by the existence of vast ex-situ collections that often contain genetic resources originating in other countries.
>
> (SEMARNAT 2002a: 115)[3]

Until 2005 the group was very active. The last known document is dated January 2005, when experts and governments met in Delhi to discuss how an international regime on access and benefit-sharing might be arranged. The Delhi declaration stated that the proposed international regime on access and benefit-sharing should include

> mandatory disclosure of the country of origin of biological material and associated traditional knowledge in the IPR (Intellectual Property Rights) application, along with an undertaking that the prevalent laws and practices of the country of origin have been respected and mandatory specific consequences in the event of failure to disclose the country of origin in the IPR application.
>
> (New Delhi Ministerial Declaration of Like-Minded Megadiverse Countries on Access and Benefit-Sharing 2005)

This group's subsequent disappearance is at first glance quite surprising. On the one hand, at COP 8 in March 2006, the Like-Minded Group of Megadiverse Countries acted as such (see the final report: Secretariat of the CBD 2006b: 22) and on the other hand, the Group and other developing countries, especially the "Group of 77 and China", made joint statements and proposed a common position on access and benefit-sharing.

Two interpretations are possible. One could argue that the small group of megadiverse countries was able to build a consensus among the larger Group of 77 and China. In this case the so-called provider countries clarified their interests and strengthened their positions in negotiations which are now represented politically by the Group of 77 and China. The other interpretation is that the Southern countries were again divided by their heterogeneous interests and strategies as well as their situation of competition (see below) and their relative political weakness compared to Northern countries. One indicator of this could be the WTO ministerial conference in Hong Kong at the end of 2005 where the governments of India and Brazil accepted the invitation of special treatment from certain Northern governments. Future developments of this issue depend on negotiations, alliances, conflicts and compromises.

The governments of the so-called provider countries find themselves in a dilemma. On the one hand, they want their countries to share in the valorisation of genetic resources, and on the other hand, they fear that potential investors may be discouraged by regulations which may appear to be too strong. They are part of a *supply competition* which is exploited by the Northern governments in the CBD negotiations and by firms in the search for suitable locations for bioprospecting. Even if they sometimes still act as one in the Group of 77 and China – an NGO representative described that as a bargaining tool (CBD/15) – there are differences among the provider countries. These can be explained primarily by inner-societal relationships. In the Philippines and in India, for example, issues pertaining to biodiversity are highly politicised and interlinked with the disputes over the rights of indigenous peoples and small farmers. The African Group, led by government representatives of Ethiopia and Cameroon, plays an important role in the questioning of patents based on traditional knowledge.[4] Finally, concerning the delicate subject of the rights of indigenous peoples and local communities, it is not at all clear if Southern governments represent those interests internationally and promote strong international regulations.

### 2.3.3 *Biotechnological companies and business associations*

Heterogeneous positions are also to be found on the part of the biotechnological companies of the Northern countries. There are firms which continue to place their bets on a weak legal framework for access and benefit-sharing and a correspondingly stronger one for intellectual property rights, and which are present at the negotiations via government positions or as lobbyists (see the representative of the Japan Bioindustry Association, CBD/10). Basically, for the hard-liner position, the entire process is cumbersome, because what is at stake is the creation of good conditions for the companies (CBD/10). The representative of the already mentioned Japan Bioindustry Association made fun of the – from his point of view – not very helpful CBD: it was obvious that neither economists nor natural scientists had been asked for their advice (CBD/10).

Other companies place more emphasis on juridical security, particularly in questions of access and intellectual property rights. These companies, which take the CBD process seriously, also see, in the creation of a common understanding of the problem, in negotiations over guidelines and the building of trust, the central value of the international agreements (CBD/13). Explicitly, they are concerned with the lowering of transaction costs via a worldwide harmonisation of requirements, which would facilitate the exchange of genetic material (CBD/13; CBD/6). Despite their aim of legal certainty, they are sceptical with regard to legal rigidities and "over-regulation". The question of whether or not companies were present at the conferences was answered by a representative of Novartis, who said that the Swiss Government delegation was in permanent email communication with this company (CBD/13). By contrast, on the part of the Federal Government of Germany it was claimed that industry did not show much interest (CBD/27).

The activities of business associations, some of which describe themselves as NGOs or are described as such in the United Nations system, were less visible, for many years, in the CBD process (in contrast to the negotiations of the Biosafety Protocol where they were quite active). This was the case because they exerted their influence on national strategies in the run-up to the conferences or because representatives of the associations or of individual businesses were part of "Northern" delegations (on the classification and the different positions see Brand 2000: 203ff.; on the business associations see CEAS 2000: 27ff.; on business lobbying see Finston 2005). But since 2005, the visibility of the associations of biotechnological companies has increased; especially in relation to access and benefit-sharing as well as technological issues.

Indicators of the growing awareness are that the International Finance Corporation Group, a member of the World Bank Group, highlights in an internet based "Guide to Biodiversity for the Private Sector" the growing importance of biodiversity for business, especially in those countries with emerging markets (IFC 2006). The Organization for Economic Cooperation and Development (OECD) published, in 2004, a "Handbook of Market Creation for Biodiversity" and the International Chamber of Commerce (ICC) recently established a Task Force on the Convention (see Secretariat of the CBD 2006b: 25). But more important for the negotiation process is that in 2005, the American biotechnology industry founded the American BioIndustry Alliance (ABIA, www.abiallience.com). One of its aims is to ensure that its foot is already in the door when negotiations on access and benefit-sharing become more effective. Its members advocate the "full patentability of biotechnology inventions and the maintenance of the current minimum standards for the protection of intellectual property" (quoted in IP-Watch, Nos 10/11, 2005/6: 5). The president of ABIA, Jacques Gorlin, as well as the executive director, Susan Finston, are well-known experts in trade and intellectual property rights issues and have strong roots in governmental politics and experience as lobbyists for pharmaceutical and other industries. Members of the group include Bristol Myers-Sqibb, Eli Lilly, Pfizer, Procter & Gamble and Tethsy Research. At the last CBD meetings in Granada (Spain) and Curitiba (Brazil), Susan Finston played an influential role.[5] It seems that the CBD has become more important for business because it is here that progress is made (i.e. its interests in securing access and property rights) and the CBD might be used by Southern governments as a means of enhancing the legitimacy of its demands in the TRIPS negotiations. Therefore, biotech companies and business associations might consider the CBD as a more important agreement than they had previously thought.

One of the central obscurities in the further negotiation process is the question of to what extent the (transnationally acting) biotechnological companies do in fact have a strong interest in access to new resources. There is first, as we have already mentioned, the existing *ex situ* collections in the North (which might be adequate for agricultural aspects, for example; cf. the next chapter). In addition, varying technologies and production methods exist within the industry, which make themselves noticed in diverging appraisals. A "renaissance of the classical search for natural substances" (Heins 2001) confronts the approach of combinative chemistry, which believes that new products can be designed without inputs from nature. Again and again it is stated that the high expectations of the "green gold of the genes" have not yet been fulfilled – not *yet*, it must be emphasised.

### *2.3.4 Non-governmental organisations and social movements*

Precisely because of the fact that policies, in the highly complex field of biodiversity politics, are conducted "under conditions of uncertainty" and that considerable contradictions exist – particularly between the different institutions such as the CBD and the WTO – there is considerable room within which different (types of) NGOs can manoeuvre. Indeed, they can contribute, as experts or as part of epistemic communities, to the clarification of complex issues. With a huge number of informal meetings, conferences, consultations, reports and publications, and direct lobbying and campaigns, the NGOs and other non-state actors, and also individual scientists, have contributed to the definitions of the problems, even if these were not at all uniform. Individual NGOs often develop their political positions and effectiveness in cooperation with other NGOs. An important example for such networks and forums is the *Global Biodiversity Forum*, founded in 1993, which usually takes place before major international conferences such as the COPs or other CBD meetings or at the regional level (as well as in conjunction with the Climate Convention FCCC). The Forum – and alongside it, organisations such as the IUCN, WWF and WRI, see below – are important in developing policy proposals, supporting implementation and getting private companies and business associations involved in the political process. Specific campaigns such as "No patents on life" or "Ban Terminator" are also important.

NGOs and social movements – the latter play an important role particularly in Southern countries – have considerable differences among themselves (cf. the separate section on indigenous NGOs and peoples).[6] This is an expression of a complex conflict field in which very different interests organise and articulate themselves. There are not only differences with regard to the political proposals, the setting of priorities or specific attitudes. Even the problems to be regulated are perceived differently. The conservationist NGOs such as the World Wide Fund for Nature (WWF), the World Resources Institute (WRI) or the "semi-NGO" The World Conservation Union (IUCN) regard themselves primarily as representing the interest of the "protection of nature", although social questions are becoming more widely discussed (see Delgado Ramos 2004: 33–42). They play a major role in the formulation of expertise, particularly legal expertise, and in the implementation of specific projects. Their political style is cooperative and is concerned with integrating all the stakeholders, particularly companies. The clearest expression of this was the Global Biodiversity Forum which was organised in the run-up to the sixth COP, and which was dominated by these NGOs. The annoyance at the very industry-friendly position of the Global Biodiversity Forum (see for example, GBF 2002) apparently led to the fact that at the COPs an *NGO Caucus* has recently spoken out as an NGO alliance. The two major conservationist NGOs – beside the WWF which is the largest in the world – Conservation International (CI) and The Nature Conservancy (TNC) are very important in the implementation of protection policies. Despite the fact that they are not very present in the context of international biodiversity politics, they play a major role in the implementation process. One important example which shows the differences among NGOs is their consideration of indigenous and local peoples: the "Big Three" which receive most funds and dominate the world's conservation agenda – WWF, CI, TNC – are "increasingly excluding, from full involvement of their program, the indigenous and traditional peoples living in territories the conservationists were trying to protect" (Chapin 2004: 17).

Other NGOs, by contrast, set about agitating, particularly on the fringes of the FAO, in order to draw attention to the problem of the erosion of genetic resources, but also to the problem of distribution in the utilisation of genetic resources. They did not participate in the CBD process until the mid-1990s, and have far fewer resources than the protection of nature NGOs. Of particular importance are the relatively small but politically very efficient NGOs, Genetic Resources Action International (GRAIN), Erosion, Technology and Concentration Group (ETC Group, previously Rural Advancement Foundation International - RAFI) and the Intermediate Technology Development Group (ITDG). These NGOs perceive themselves as trying to articulate the social interests particularly of small farmers in Southern countries. The dominant process of biodiversity policy is seen as the privatisation of genetic resources and the expropriation of the previous owners, namely indigenous peoples and rural communities. Scepticism with regard to Southern governments has recently been expressed, for example as criticism of the Like-Minded Group of Megadiverse Countries, which is accused of forming a cartel and "a front for selling their biological diversity to the highest bidder" (Ribeiro 2002a: 40).

Following Rio and the coming into force of the CBD in the 1990s a large number of NGOs appeared in the biodiversity debate, such as the US American BioNet (Biodiversity Action Network; now dissolved), the African Centre for Technology Studies (ACTS, founded by the later General Secretary of the CBD, Calestous Juma) or the Third World Network (TWN). The latter represents similar positions to the above-mentioned GRAIN or ETC Group.[7]

Despite all their differences, the NGO types named have in common that from their point of view the CBD can only achieve effectiveness via the broad involvement of different groups of actors as well as ensuring that their involvement is bound by legal safeguards. The NGOs that are active on the fringes of the CBD (and the FAO) also have a critical position regarding the TRIPS Agreement and generally on the free trade logic institutionalised in the WTO. They believe that the CBD must be strengthened in opposition to these. Positions vary with regard to private industrial companies and research institutes directed towards commercialisation. As a whole, the participation of those companies which are open to the process is regarded as positive while their minimal participation is viewed as regrettable (CBD/6; CBD/18). Many NGOs see new opportunities for political manoeuvring being opened up by the CBD which allows them to formulate critical questions.

But quite different "NGOs" have also become active, although less publicly than the "political" NGOs, namely consulting firms and business associations (for the latter see above). Consulting firms offer their expertise as private-sector actors, for example in producing reports for national implementation. They use the international conferences as a market where they can offer their services. They do not involve themselves directly in the political disputes, but are important actors through their knowledge of technical or organisational questions. They play a major role in the implementation of the CBD at the national level.

Our own research (see also Arts 1998; Brühl 2001) has shown that the influence of NGOs is greater when their proposals move relatively closely to the dominant perception of the problem and to political interests and proposals. Specific expertise and information for suitable interventions, cooperation and contacts between NGOs, lobbying and negotiating capabilities are those aspects which are regarded by the NGOs themselves as central (Arts 1998: 258ff.). Particularly apparent is the idea of "political neutrality" which becomes a strong selection mechanism, integrating certain actors and subject-matters into this process and ignoring others, or at least diminishing the possibilities they have for clear articulation. NGOs and other actors compete with one another with their articulated interests and strategies, something which is often overlooked in the official view of the "NGO community".

### *2.3.5 Indigenous peoples*

The representatives of indigenous peoples repeatedly point out that the question of biological diversity is not about its economic value but about a complex connection between humans or peoples and their cultural practices regarding nature. They demand their rights in self-determination and rights over land as well as referring to human rights (Harry/Kanehe 2005). Therefore, they sometimes question benefit-sharing procedures (for example, in the Hoodia case, see below), because basic needs such as education, health care and so on must be guaranteed without being coupled with the exchange of genetic resources or traditional knowledge (ibid.: 100–101). The issue of access from their point of view is not one of effective and legal appropriation but an invasion of their societal practices (most strongly when it is a case of sacred plants or knowledge). The protection and use of biodiversity and other natural goods is put into a comprehensive framework by indigenous peoples, that is neoliberal globalisation, militarisation, development aggression, migration, political misrepresentation, as well as all other forms of discrimination (see for example, the Baguio Declaration 2004). A document written by indigenous women highlights that the central principles of the CBD should be expanded to other spheres of life and politics. "Free and prior informed consent should include the full and effective participation of indigenous women in the decision-making process" (ibid.).

Moreover, questions of indigenous knowledge cannot be separated from those of the control of these practices, which includes the control of territories (which is more than just land) and resources. Resources are regarded from the perspective of the safeguarding of food. Especially important is their fundamentally different understanding of questions of property. Property for indigenous peoples is often collective property, and private property of elements of nature is usually ruled out completely (Posey 1999). Their ideas are therefore incompatible with "modern" forms of property (least of all with patents). The indigenous representatives draw attention to the same demands again and again: their rights should be recognised internationally, including the rights to their territories and resources as well as their effective control. At the bottom of this is the demand for self-determination (CBD/24, CBD/26). They have an attitude of scepticism towards their own governments (see Chapter 5). In the countries in which there is national legislation, indigenous rights usually find no consideration (CBD/25). One important problem here is that the existing legal systems in these countries are dominated by modern, Western-based concepts. The idea that different forms of development should be given a political voice within the framework of the CBD is of central importance to those groups that see themselves as representatives of indigenous peoples.

The international level is regarded as a possible supportive sphere, particularly in very specific disputes at the national and local levels (relating to Article 8(j), protection, sustainable utilisation, benefit-sharing and the role of knowledge). Nevertheless, the main criticism of the CBD process is that the Convention itself – and its further development – were, and are being, developed without the participation of indigenous peoples. Especially in the case of the regulation which affirms a national government's sovereignty over natural resources, it is clear that the rights of indigenous peoples do not play any significant role: "The CBD is an agreement to take our resources from us." And the WTO is considered as having more influence on the CBD than do the indigenous peoples (CBD/26). Furthermore, their rudimentary and belated participation primarily serves to legitimise government policies. The momentum of development embedded in neoliberal policies – the biotechnologically motivated interest in genetic resources and in the spreading of genetically modified organisms (GMOs) and correspondingly in biosafety regulations – is emphasised (CBD/25): "The system does not allow any discussion and overrides all those who wish to question it" (CBD/26). With regard to the relationships of power existing at the international level, which are to the disadvantage of indigenous peoples, there are no illusions; it "is a foreign terrain" (CBD/26) on which they must operate diplomatically. Room for manoeuvring is seen more in the process of implementation and concretely in the individual regions where in some cases alliances can be made with environmental ministers, for example on the administration of national parks. Tactical references to the CBD are made here. The question of a moratorium on bioprospecting has been in discussion ever since the mid-1990s. This concerns the recognition of "the right to say no" to bioprospecting – not in order to refuse the utilisation of resources in general, but in order to criticise its dominant form (CBD/26). At the same time the ambivalent role played by the governments of the Southern countries is questioned (CBD/25).

A major debate on the part of the indigenous peoples or their representatives at the international level is whether they should get involved in the international processes at all (CBD/24). However, despite many calls to boycott participation in the international forums, the need to participate in the CBD process, in order to exploit the existing space for contestation, remains the dominant position.

The indigenous peoples have developed networks in recent years as a way of giving voice to their concerns at international negotiations. The official forum in international negotiations is the *International Indigenous Forum on Biodiversity* (IIFB).[8] In recent years the Indigenous Forum has successfully developed a platform for entering into international politics. The COP and the signatory states can no longer ignore the representatives of the indigenous peoples and the IIFB, although appropriate recognition of their concerns is still lacking. The problem of expertise is central, for example in order to be able to adequately appraise juridical issues. Benefit-sharing could start here, for instead of monetary compensation benefit-sharing could take place in the form of capacity building of technical knowledge in the different areas. However, in the opinion of the IIFB, the precondition for access and benefit-sharing is the mutual respect of the sovereignty of all the negotiation partners – and in its opinion, this is still missing (Blank 2002). The IIFB demands from the signatory states of the CBD the recognition of its existence as a forum of legitimate self-representation and of the rights of indigenous peoples (in the sense of the ILO 169), as well as the chance of real prior-informed consent or of the granting of an effective right of veto in access and benefit-sharing questions ("the right to say no": CBD/24; CBD/25; CBD/26).

There are also other networks. Prominent is the Indigenous Peoples' Biodiversity Network (IPBN), which is especially oriented towards the CBD. Since 1998 the Indigenous Women's Biodiversity Network (IWBN) has been formed, which emphasises the specific way in which women handle biological diversity (particularly in nutrition), the poverty of women and their under-representation in the negotiations, and which demands greater participation and education for women (IWBN 2006). An important strategy is also to get female indigenous representatives into the government delegations.

### 2.3.6 *The importance of gender relations*

Inequalities based on gender and gender specific conflicts over the use of genetic resources, that is the gendered division of labour, are central in the appropriation of biological diversity. Recent studies show that mainly women gather wild plants and manage them, that they are homegardeners and plant domesticators as well as herbalists and healers and also, that women generally keep the seeds. And it has also been discovered that in many regions women act as the principal farmers, often breeding plants informally (cf. the case studies in Howard 2003). However, these contributions are seldom recognised in research or in political processes.

> In those areas where women have traditionally held control, because of modern technologies and perceptions, women have lost substantial influence and control over production and access to genetic resources to men who benefit from extension services and have the ability to buy seeds, fertilizers and the required technologies. In this way women have also lost their status and self-determination and are not compensated in any way.
>
> (FAO Focus 2006: 2)

This has to do with dominant gender relations which make the contributions of women invisible, thereby obscuring gender relations. Patricia Howard argues that, first, it is necessary to open up the "black box" of the household which includes the societal relationships with nature in the domestic sphere, that is gendered knowledge and the transmission of knowledge, the social division of labour, the relationship between subsistence and market-oriented production, concepts of masculinity and femininity and the systems of indigenous rights to and knowledge about plants (Howard 2003: xviii, 2; GTZ 2002: 10–11). This is an expression of the "cult of domesticity" which is the dominant and mainly Western idea that the (domestic) sphere of reproduction and subsistence is less important than the sphere of production. This, in fact, privileges men and patriarchal norms and obscures the fact that the domestic realm is enormously productive and crucial for the development, conservation and sustainable use of plant genetic resources.

Therefore, it is of utmost importance that women participate beyond concrete conservation and sustainable use activities as well as in political processes. The CBD does not develop a legal framework for this but, at least, states in its Preamble that women play a "vital role ... in the conservation and sustainable use of biological diversity" and considers the "need for the full participation of women at all levels of policy-making and implementation for biodiversity conservation" (CBD 1992, Preamble) as important. Nonetheless, this remains a rather instrumental understanding of the role of women in biodiversity politics.

In international biodiversity policy, however, women continue to play hardly any role. The German GTZ (2002) did organise a lunch-time workshop at COP 5 in Nairobi, and at COP 6 in The Hague, isolated voices were raised that the role of women needed to be given more importance in the discussion and policy process. There are also recommendations to integrate gender issues into the CBD process through capacity building and the raising of public awareness, gender-sensitive programmes, the creation of gender focal points and a gender monitoring and evaluation system, as well as gender balance in expert panels, among other proposals (GTZ 2002: 29–32). But in the political institutions and on the part of the NGOs the subject is non-existent. And despite the formation of the IWBN, even at the meetings of the indigenous peoples, it is not at all easy to get the subject of gender relationships onto the agenda. This indicates a very hard structural selectivity which makes the discussion of gender-specific aspects extremely difficult. There appears to be a consensus among the actors – including the critical NGOs and indigenous peoples – that discussion at the international level should be about so-called hard issues, namely about politico-economic questions, whose gender implications are neglected.

However, the provisions of the CBD have a strong gendered impact. Because of the gendered structure of decision-making processes in state politics and at the local level they tend to strengthen a patriarchal system when mechanisms of benefit-sharing, prior-informed consent or mutually agreed terms are implemented or when protection areas are created. Who is going to benefit and who obtains which rights? Whose food-security is in danger? These are not gender-neutral questions since socio-economic processes as well as those of political decisions are notoriously patriarchal (see BUKO-Kampagne 2005: 117–124).

Not by chance, women and their collective organisations tended, until now, to articulate themselves outside the established political terrains. Thus, internationally there are a number of social movements which expound the problem of control over genetic resources from the perspective of women (see e.g. Akhter 2001; Ayales *et al.* 2002; Baguio Declaration 2004).

### *2.3.7* ***Preliminary summary***

We can see that the conflict lines are much more complex than suggested by the notion of a North–South divide between biodiversity-rich countries in the South and Northern high-tech countries. Even if a broad variety of actors with complex interests are involved, central characteristics of the conflict field can be figured out. The dominant groups of actors – Northern and Southern governments, private companies and research institutes – assume a potential win–win constellation because benefits which can be shared arise in the first place from the access to biological diversity and its valorisation. It is only the forms of appropriation and the distribution of the potential benefits which remain controversial. The governments of most countries – and thus the protagonists of the CBD – also agree in their opinion that insisting on national sovereignty over biological or genetic resources by the CBD is important. The considerations behind this vary, nevertheless. It was through the criticism by the Southern countries of the centuries-old practice of utilisation by Northern actors without benefit-sharing that this gene robbery came under pressure to legitimise itself, and this led to the formulation of national sovereignty regarding biogenetic resources. The interest of Southern governments, companies and research institutes was strengthened by the commencing "biotechnological revolution". With the provision of national sovereignty, an important condition for the valorisation of biological diversity was created.

The debate on environmental politics, which placed the erosion of biological diversity at the centre of attention, contributed considerably to the politicising of North–South relations and to creating a new terrain for compromises. Nonetheless, during the CBD negotiations in 1992 and the continued shaping of the CBD as of 1994, questions concerning the structural inequalities in North–South relations disappeared into the background. Within the framework of the ecological perspective it is usually argued that the national sovereignty of states should be limited and a global management programme set up in order to take action against the exploitation of natural resources. Countries such as Brazil, nevertheless, have always resisted such an external influence. However, the situation today seems to have changed, to the extent that biodiversity-rich countries now recognise the value of their natural resources and attempt to use but also to conserve them. Therefore, the protection of nature becomes interesting for these countries. Discussions on military and ecological security can even overlap and mix here (for Brazil, see Acselrad 2002). External experts on nature protection are in some cases consulted for the corresponding supervision and control.

Disposal of resources concerns not only their protection, however, but has more to do with their valorisation and with planning security. This is the case because on the one hand, the Northern businesses and research institutes require clear legal relationships for the case that marketable products result from bioprospecting. This legal security can be granted best by the nation-state with its monopoly of legitimate force. National sovereignty is the necessary condition for the distribution of rights of disposal and the state's guarantee of these rights: state-political activity is not opposed to the interests of the economic actors here, but it is the precondition for valorisation. On the other hand, the issues at stake are much broader than suggested by the conventional view on ABS (access and benefit sharing). As the Millennium Ecosystem Assessment (MASR 2005) pointed out, the decline of ecosystem services is a major obstacle to reaching the Millennium Development Goals of the United Nations. In this context, the commercialisation of biodiversity and its monetary evaluation, in general, is considered as putting at risk, or even ignoring, important contributions of biodiversity for development purposes.

In the context of the CBD, however, only a few actors, particularly the representatives of indigenous peoples and some NGOs, focus on the issue of the erosion of biological diversity from a perspective which does not primarily centre on its potential economic value. Protection and utilisation are closely connected here, too, but from the perspective of local living conditions. This leads to a critical appraisal of international biodiversity policy and its dominant development.

## 2.4 Problems and fields of conflict

The CBD is recognised by most actors as a fundamental compromise. More was not to be expected in the negotiations up until 1992, a fact admitted by some NGO activists who are well acquainted with the process (CBD/11, CBD/15). There is a far-reaching consensus among the participants that the design of the Convention is still in the making. The further shaping of the contents of the CBD and their effects are also highly controversial and even a representative of the CBD secretariat has come to the conclusion that the immense inequalities of power between the countries and other actors is a major problem (CBD/18). The conflicts are condensed in a certain way, in that the central concepts of the CBD are unclear and in the opinion of various actors still require concretisation. In this section, therefore, the controversies surrounding the central aspects will be examined more precisely.

### *2.4.1 Access to genetic resources and benefit-sharing as the central conflict*

As we have already stated, by "access" the CBD means different things: access to natural resources, access to knowledge and, closely connected with the latter, access to technologies (see below). With regard to access to genetic resources, the governments of the individual signatory states have the right to regulate access (CBD 1992: Art. 15.1). According to the CBD access should generally be "facilitated" (ibid.: Art. 15.2). The idea behind this is that the potential utility of genetic resources should be made available to everybody. In fact, however, the economic value of biodiversity is strongly pushed into the foreground. This leads to a different appraisal of natural goods as resources and of knowledge. Certain forms of knowledge are, however, not market-compatible per se (traditional healers and forms of exchange in agriculture; cf. the next chapter on this). In order to facilitate access they first have to be made into market-compatible forms of knowledge and thus they must be comprehensively modified in their social foundations. Not only are questions of property rights, in the narrow sense of the word, at stake here, but also of key importance are comprehensive societal practices.

There is a great potential for conflict with regard to the concrete application of the prior-informed consent (PIC) and mutually agreed terms (MAT) regulations. While the Northern governments are clearly in favour of flexible and voluntary conditions, some Southern governments (i.e. the members of the Like-Minded Group of Megadiverse Countries) push for binding regulations. They consider the CBD to be inadequate on access and benefit-sharing because it simply formulates general principles but does not lay down a uniform terminology or procedures (CBD/23). But among the (so-called) provider countries, the concretising of the regulations is controversial. The CBD, in Article 15.5, lays down that prior-informed consent is to be granted by the providing signatory state, if the latter does not decide otherwise.

In the negotiations on the formulation of the access and benefit-sharing mechanisms it has, furthermore, become clear that prior-informed consent and mutually agreed terms require comprehensive knowledge on the part of the national and local actors if there is a real desire to reach an "informed consent". Since this requires technical, economic and juridical knowledge in order to comprehend the complex regulations and to be able to weigh their consequences, and since this knowledge does not simply exist, these mechanisms require the setting up of corresponding structures. One important instrument which is recognised by all actors as such is the construction of corresponding competencies: *capacity building*. However, it is still an open issue as to what is specifically understood by this term (see below). In addition, although the need for necessary flexibility is often emphasised, many Southern representatives see a danger in this in that it tends to favour the stronger actors.

Indigenous peoples and critical NGOs increasingly politicise the problem of access by making accusations of *biopiracy*. This stance is intended to criticise the historical and current process of the appropriation of genetic resources, which is regarded as being illegitimate. This does not become more legitimate through the formal agreement of state authorities, as indigenous peoples and local communities are not involved in the process in which the legal framework is developed (see Ribeiro 2005; Egziabher 2005; Shiva 2005; cf. the detailed study of Central American countries in Delgado Ramos 2004: 107–182).

> For ETC Group, biopiracy is the privatization through intellectual property systems of biological resources and/or related knowledge, regardless if this is legalized by a national law or a bioprospecting agreement, and even if a so called "benefit-sharing" agreement is included. As Alejandro Argumedo, a Quechua activist describes it, benefit-sharing in bioprospecting contracts is like waking up while your house is being robbed, and the thieves try to calm you down by offering to share the benefits derived from the commercialization of your stolen goods.
>
> (Ribeiro 2002a: 37; cf. BUKO-Kampagne 2005, http://en.wikipedia.org/wiki/Biopiracy)

One consequence of this for the above position is that potential local beneficiaries of benefit-sharing must also have the right to refuse access. However, Southern governments have also claimed the term as their own, although they have a rather different understanding of it (SEMARNAT 2002a: 124). By biopiracy they do not mean a fundamentally illegitimate process of the privatisation of nature; rather, they use the term when the legal conditions (particularly prior-informed consent and mutually agreed terms and thus their participation in benefit-sharing) are not met (cf. section 2.2). And even companies use the term when their intellectual property rights are abused.

Altogether, the question of access illustrates that the decisive actors – Northern governments and biotechnology firms – determine the momentum of the CBD. Since it is unclear, however, how strong the interest in genetic resources for pharmaceutical research actually is, and the fact that often these firms prefer bilateral access via bioprospecting agreements, these dominant actors are not necessarily interested in internationally binding rules on access. This is why the negotiations on rules for access and benefit-sharing started relatively late, knowing the controversial character of the regulations. However, the lines of conflict have become clearer due to the negotiations of the past years. The question concerning *how* appropriation takes place in a specific case is determined less by formal legal regulations and more by their interpretation. And that is ultimately dependent on the corresponding power resources.

Just like access, benefit-sharing can also take very different forms. Within the CBD and its further development, it remains closely interlinked to access to genetic resources. However, this ignores a perspective, which for ecological reasons would certainly be worth considering, namely to fairly distribute the utility which arises from access *forborne*, that is from the fact that certain parts of nature are not put to any economic use. This is indeed often demanded of the rural population near to nature conservation areas without any compensation taking place, particularly in the Southern countries. Furthermore, access also includes non-monetary benefit-sharing in addition to monetary benefit-sharing – which arises primarily from the commercialisation of genetic resources. In addition to participation in research results and capacity building, non-monetary benefit-sharing can include technology transfer, although this can lead to conflicts with regard to intellectual property rights and patents (Seiler/Dutfield 2001: 24f.) Benefit-sharing is therefore not only a question of monetary payment. The provider countries hope for technology and knowledge transfer from benefit-sharing, while the so-called user countries or the private biotechnology companies hope for an improvement in the conditions for further bioprospecting and the further processing of the samples – for technology transfer can also be used for that purpose. Benefit-sharing thus plays a central role in the Convention but, surprisingly, it plays a very subordinate one in the further negotiations. On the part of the receivers and valorisers of genetic resources, it is repeatedly emphasised that in the negotiations on the CBD high expectations have been awakened with regard to the potentially sharable economic benefits. This could lead to disappointments and conflicts (e.g. CBD/2; CBD/4; CBD/9; CBD/10; CBD/15). A representative of a research institute active in the area of seeds described benefit-sharing linked to the actual development of products as a "lottery" (CBD/1).

Benefit-sharing involves a double distributional conflict. At the *international* level it is highly politicised because the motivation of the Southern governments is essentially determined by their interest in sharing in the biotechnological developments which are driven forth by companies or research institutes located in other countries. At the same time, it remains a decisive accusation on the part of the provider countries that the benefits are monopolised by the receivers of genetic resources (e.g. CBD/11). The other level of conflict, namely that of the *national* distribution of the potential benefits within the provider countries themselves, is discussed much less in the international negotiations. A representative of Novartis openly speaks of an inner-societal conflict with regard to benefit-sharing: the indigenous peoples in most cases prefer to achieve respect and an improvement in their living conditions (schools, roads, hospitals), while money as a medium of benefit-sharing is as a rule demanded by the governments (CBD/13). The interests of the Northern actors, however, go beyond simply paying money to the governments and in return receiving legal access and the safeguarding of intellectual property rights. Their practices and proposals show that they are primarily concerned via benefit-sharing to improve the conditions for valorisation, for example by equipping local research institutes with technical apparatus and know-how. The interest in non-monetary benefit-sharing is supported by NGOs and indigenous organisations. The governments also increasingly agree to non-monetary benefit-sharing because in their estimation this comes closer to the interests of the users as well as favouring technology transfer. This continues to be accompanied by the demand for monetary participation in the development of marketable products, however.

Finally, it is important that, as mentioned before, the genetic resources acquired before the coming into force of the CBD do not fall under the Convention (CBD 1992: Art. 15.3). This means that the material collected until 1993 now continues to be freely available for commercial purposes. This fact has hardly been politicised until now, however.

To summarise, benefit-sharing has a rather weak legal status in the CBD. This is an expression of the international relationships of power regarding this question. The Northern actors wish to secure legal access for themselves and cannot therefore completely withdraw from the discussion on benefit-sharing. On the other hand, in this area no initiative on their part is to be expected. In addition, the large quantities of genetic resources with immense economic potential which were already available in the Northern countries before December 1993 are excluded. Initiatives, for example by botanical gardens to introduce a voluntary retrospective benefit-sharing, have been greeted benevolently, but they only cover a small fraction of the material which flowed from the South to the North over the last centuries.

One danger for weaker interest groups is that the specific meaning of "fair and equal benefit-sharing" is left to bilateral negotiations. Power resources play a decisive role here, but so does knowledge on the part of the providers about potential benefits and their sharing. Furthermore, benefit-sharing is seen here exclusively in the context of the valorisation of genetic resources, so that other perspectives – such as benefit-sharing for the conservation of biodiversity without at the same time having the perspective of its marketing – are devalued. This commodifying perspective is strengthened by the competition of the "suppliers" among themselves. A study by the NGOs GAIA and GRAIN (2000) points out that in spite of an immeasurable quantity of literature on benefit-sharing,

> the scope of the benefit-sharing debate is actually very narrow. Almost without exception, benefit-sharing examples focus on bilateral and contractual agreements, generally between some company or institute from an industrialised country interested in a resource or knowledge, and some country or community from the South that can provide it. *Benefit-sharing is defined from the perspective of the bioprospector: how much money is paid, and whether other non-monetary benefits flow to the provider.*
>
> (GAIA/GRAIN 2000: 2, our emphasis)

A very limited and commercial interpretation of the use arising from genetic resources is dominant, mainly because the use for local communities, for the direct securing of their means of living, for health and food security, is ignored, as is the protection of nature and public research. Only marketable exchange values are of interest. In future it will have to be shown whether benefit-sharing is a clear right with which certain standards are set or, rather, whether it is a symbolic act in order to do justice to the regulations of the CBD and to legalise the appropriation of biological diversity. The starting-point should be the interests of the local communities and farmers. It will be decisive to what extent access and benefit-sharing are seen as inseparable and in what way this inseparability is specifically created. Second, with regard to access and benefit-sharing, the question of control over knowledge and resources, but also over land and social processes, is central. Thus, *ex situ* collections have largely been removed from the control of the Southern countries, which is why they are not a major issue in the international negotiations. In contrast to this, intellectual property rights are the central lever for the control of genetic resources and their further development.

### *2.4.2 Indigenous rights*

There is no doubt that the CBD is an international institution within which the concerns of indigenous peoples are heard relatively strongly. Article 8(j) of the CBD lays down regulations for national implementation, but without clearly defining the demands and legal consequences. A special role is played here by the different legal opinions on the relationship of indigenous knowledge and the scientific knowledge used in the industrial context (see Agrawal 1995; Brush 1993; Posey 1999; on the role of indigenous peoples in the CBD see Potvin *et al.* 2002). But even if the indigenous peoples have succeeded, in recent years, in procuring more weight for their demands, their interests nevertheless continue to play a subordinate role in the specific momentum of the negotiations.

At the international level, the area most fraught with conflict is that of intellectual property rights. The pressure placed on the governments by private companies with regard to the strict safeguarding of intellectual property rights is diametrically opposed to the interests of indigenous peoples. The degree of manoeuvrability is regarded as being very minute here: "its (patents are) the only game in town" (CBD/24). The recognition of already existing and inalienable indigenous rights has not happened, never mind the notion of respecting the fact that "traditional" knowledge is not only knowledge of different things but also a "different institutional treatment of knowledge by human societies" (Kuppe 2002: 117, our translation). Nor do the attempts to strengthen alternative concepts such as community rights or traditional resource rights find any support. Even regarding benefit-sharing, ambiguity predominates.

> That's gonna be a difficult thing. Where do the benefits come from? What kinds of benefits are they? Are they monetary, capacity building or what? What exactly is the range of benefits? How are they identified and made broadly known to the community?
>
> (CBD/8)

In general the governments of the provider countries fear that by granting the indigenous peoples a legal position that is too strong, they would be effectively discouraging potential bioprospectors (CBD/15). The strengthening of the rights of indigenous peoples is avoided in the CBD in two ways: their voluntary character is emphasised, and they are made "subject to national legislation and policy" (e.g. the governments of Australia and Malaysia at COP 6 in 2002; ENB 2002: 2). Furthermore, the broadly interpretable regulations of Article 8(j) involve the problem that indigenous peoples would have to be educated for effective participation. Nevertheless, indigenous peoples, supported for example by the governments of Norway and India, were able to enforce one important aspect. At COP 5 (2000 in Nairobi) in Decision V/16, the provision of prior-informed consent was extended to indigenous peoples. Since then the representatives of the indigenous peoples have the status of official consultants and no longer simply that of observers.

In order to show some of the complex dimensions of bioprospecting, access and benefit-sharing, it is pertinent to offer a brief outline of the case of the San people and the Hoodia cactus in Southern Africa (see Harry/Kanehe 2005: 97–103; Hoering 2004; Frein/Meyer 2005: 140–153). To use this cactus in modern medicine a benefit-sharing agreement was negotiated which is often perceived as being a success story as an instrument to alleviate poverty. But it still remains quite unclear whether this objective can really be reached, as well as, generally, how this agreement should be judged.

The plant was used for hundreds of years by the most ancient people of Southern Africa, who live mainly in the Kalahari, in order to suppress hunger while hunting. In 2001 it became known – by chance – that the Council for Scientific and Industrial Research in South Africa (CSIR) had undertaken bioprospecting and found that the substances of the Hoodia plant can be used for a drug against obesity, that is to suppress appetite and aid slimming, which is, in the Northern countries, a billion dollar market. CSIR obtained a patent and gave the pharmaceutical company Phytopharm a license. (For a time, the transnational company Pfizer owned the licence, after having bought it for 21 million USD, but later this company withdrew from this branch and the licence went back to Phytopharm.) Among other things, it was agreed that CSIR had the right to cultivate and produce Hoodia for the South African market while Phytopharm had market rights for the rest of the world. When Phytofarm's Richard Dixey said in 2001 that the San people had died out and, because of this, could not be part of a benefit-sharing agreement, a lot of concern rose up and San from four countries raised their voices. The criticism of Phytofarm and CSIR by the San people and international NGOs led to the negotiation of a benefit-sharing agreement which was signed between the San people and CSIR in March 2003. As part of this agreement, the San people were to receive 8 per cent of the royalties that CSIR receives from Phytofarm; Phytopharm is not part of the agreement. At the same time, the San people are not permitted to use their knowledge for commercial ends. A Hoodia Trust was created (also with members of CSIR and the South African Government) and benefit flow to the San people was expected to reach one million USD. A small amount of money has been paid already, but held back until the San have established a legal entity. The Working Group in Indigenous Minorities in Southern Africa (WIMSA) played an important role here in promoting the interests of the San people in the negotiations.[9] The ambiguity lies in the fact that the benefit-sharing agreement was negotiated after the appropriation of the Hoodia substances through CSIR and that "prior informed consent" – a crucial requirement of the CBD, see below – was not given.

> It is important to note that the San were compensated for their traditional knowledge and not for any right they might have in the genetic resource itself. It was the CSIR, and not the San who consented to access to the genetic resource itself.
>
> (Harry/Kanehe 2005: 99)

Some say that at best, the agreement is "very lenient punishment. Imagine if a thief could get away with his crime as long as he gave his victim a few percent of whatever the receiver paid him for stolen goods" (Hoering 2004: 18). It is questionable whether the patenting was in line with the commitments of the CBD. Moreover, there is criticism that not only prior-informed consent was not given but that the San people could not decide about the patent registration. With the agreement, Phytopharm has the right to monopolise the knowledge of the San people concerning Hoodia for commercial ends (Harry/Kanehe 2005: 99). Even in the case of a prior-informed consent, the San people could only wait if the pharmaceutical company was able to develop and market a product with Hoodia which might be the "first African blockbuster drug" (Rachel Wynberg, quoted in Hoering 2004: 19). The Hoodia case is also important in demonstrating the fact that indigenous peoples "are seen only as traditional knowledge holders and not as territorial rights holders whose consent must be sought before accessing resources within their territories" (Harry/Kanehe 2005: 97). Another criticism is that the flow of benefits is not related to another provision of the CBD, that is the acknowledgement and protection of traditional knowledge and peoples. And finally, the major interests for CSIR and especially the South African Government are less related to the San people than they are to other lucrative ideas related to the seed production, cultivation, processing and eventual exportation of Hoodia and many other plants (Hoering 2004: 23).

However, WIMSA and many San people are glad that at least today they have the agreement and that they did not file a suit against the patent, which they saw as having little prospect of success. And throughout the process of the last years, there has been a recognition of their knowledge which has helped to raise their consciousness. In many other cases the traditional developers of the plants have not received anything. For example, the cancer drug Vincristine embodies the knowledge of indigenous peoples from Madagascar and genetic resources from a periwinkle and is marketed by the US-based corporation, Eli Lilly, to the tune of 100 million USD annually (see other examples in Sharma 2005).

There is another ambiguity in the San case which can reveal some challenges of the existing ABS regulations. Even if there are doubts as to whether the patenting was legal and whether some other provisions were respected or not, in pure legal terms the negotiation of the CSIR was not in contradiction of the CBD. The consent of indigenous peoples is only required when it is subject to domestic law.

> Therefore, what transpired with the Hoodia was totally consistent with the Bonn Guidelines [see below, the authors]; that is, the CSIR acted as the South African national authority granting access to the Hoodia. The indigenous San peoples had no established legal right vis-à-vis Hoodia under South African law and so their consent was not required.
>
> (Harry/Kanehe 2005: 99–100).

The lesser meaning of indigenous rights in the context of access and benefit-sharing can be clearly seen in the fact that the negotiations on access and benefit-sharing and on Article 8(j) are conducted separately and the indigenous perspective is very seldom integrated into other processes. The major institutionalised forum in the CBD is without any doubt the *Ad Hoc Open-ended Inter-Sessional Working Group on Article 8(j) and Related Provisions* of the CBD (established in 1998). It works on elements of a *sui generis* system for "the protection of the traditional knowledge, innovations and practices of indigenous and local communities" (Secretariat of the CBD 2006b: 38, 154) and intends to collaborate with the Working Group on Access and Benefit-sharing. But the latter seems to be hesitant. The Working Group on 8(j) is an important but weak form of institutionalised interests.

As a whole, indigenous peoples are threatened with being made into a "negotiation mass" by the Southern governments. At the international level, the governments speak prominently of the protection of their "own" indigenous peoples, but they are criticised that this takes place only from the perspective of valorisation. One proof of this is that the much quoted indigenous peoples were not consulted in the initiative for the Like-Minded Group of Megadiverse Countries (Ribeiro 2002a). An indication of the rather symbolic meaning of indigenous rights could also be that the governments often do not speak of indigenous peoples but of indigenous communities or groups. This implies a weaker status in the sense that "communities" cannot claim rights, or traditional and indigenous knowledge (CBD/24), which more or less suggests a resource.

For Article 8(j) the issue will be decisive whether the rights of indigenous peoples (and of local farmers) are instrumentalised for facilitated access, or whether in the case of conflict the rights of these weaker actors can be protected. Here, the question of information is certainly a central one, as is the concrete design of prior-informed consent and mutually agreed terms. Even though conflicts over the rights of the indigenous peoples, like those of other local users of biodiversity, have been repeatedly discussed at international conferences, ultimately, they are fought out in the countries themselves as well as at specific locations. Concrete strategies and the successful organisation of interests play an important role here. Competition between subordinate actors is, by contrast, supportive of the exercise of rule. For one thing, it legitimises regulating intervention by states; and at the same time it places the users in a position of advantage, because they have a weaker "negotiation partner". In the case study on Mexico (see Chapter 5) this will become even clearer.

### 2.4.3 *Intellectual property rights*

The safeguarding of intellectual property rights is the central concern for the industrial utilisation of genetic resources and in particular for gene technical utilisation, and it is taken into consideration in various places in the CBD. Taken from within the question as to whether and to what extent this is permissible or what consequences it has from an ecological, economic and social point of view, one of the most important and controversial fields of conflict between different international regulations systems is unearthed. Intellectual property rights are discussed primarily in connection with the rights of indigenous peoples and in access and benefit-sharing questions. The representatives of the Southern governments (CBD/23) and most NGO representatives (CBD/7; CBD/11) see a central weakness of the CBD with regard to the low value of intellectual property rights in the sense of the protection of indigenous peoples and the provider countries. Nevertheless, the CBD is regarded as the only international forum in which weaker actors can assert their interests with regard to intellectual property at all. An adequate legislation on access must contain regulations on intellectual property rights. "It's going to be the make or break situation" (CBD/11). However, this poses the question as to what power the Southern governments and NGOs have to cause a break. Hardly anyone is thinking of withdrawing from existing international agreements; instead, of concern is the question of to what extent other forms of intellectual property will be accepted. These discussions are taking place on the fringes of the World Trade Organization's TRIPS Agreement and have more recently been strongly reinforced in the World Intellectual Property Organization (WIPO). The debate is centred on whether, in addition to "hard" intellectual property rights protection (particularly by patents), other protection standards should be accepted.

### *2.4.4 Protection and conservation*

With the CBD, the economic value of biodiversity has shifted to the centre of the problem, while ecological and socio-ecological problems threaten to be displaced to the background. Thus, representatives of the German Federal Ministry of the Environment, for example, complained that protection interests were neglected in the CBD (CBD/20, CBD/22). This is connected with the fact that the economic value of biodiversity was, and is to some degree, largely regarded as an unproblematical lever or incentive for the greater protection of biological diversity. The dominant line of argument concerning the treatment of biological diversity on the fringes of the CBD is that the economic value of genetic resources is suitable to open up new sources of finance for nature protection and to increase the public appeal of biological diversity. The consequence of this perspective is that the aims of the CBD are to be achieved *via facilitated access*. This argumentation and practice is surprising at least in that the original point of nature protection areas was *to limit access*. Whether it was a case of pure protection areas or biosphere reservations with graduated forms of use, the practice was always aimed at conserving threatened species or ecosystems by regulating and limiting human access. The classical concept of protection, however, which aimed at the waiving of utilisation altogether, had already come under criticism at the beginning of the 1980s and was replaced by the concept of sustainable use.

Two arguments are used here, each of which is based on the other. First, the setting up of protection areas should not take place against the will of the affected population, which must be allowed to continue to follow its sustainable forms of use, or an acceptable compensatory income must be created. In addition, a second argument says that an incentive to ensure that protection areas are respected, for example in the form of income sources from ecotourism or even bioprospecting, must be promised. If the first argument places the economic use in the service of conservation, the second argument places conservation in the service of economic use – for only if nature is conserved can it be profited from, whether this is through tourism or in the form of bioprospecting. The two demands thus not only complement each other but there is a threat that they turn the relationship between protection and use upside down. To put it more pointedly, the tendency towards the commercialisation of biodiversity itself works to promote the conservation of nature (see e.g. Escobar 1996; Heineke/Wolff 2004: 30). This means in fact that the protection of nature itself becomes profitable, and protection interests are no longer, per se, in contradiction to commercial interests in the exploitation of nature. It cannot be ignored, however, that in many parts of the world the exploitation of minerals or biological resources such as tropical wood by far exceeds the use and conservation of genetic resources. For this reason alone, it is doubtful whether the economic value of biodiversity does in fact represent a central lever for its conservation. Moreover, there are many doubts that economic benefits for the valorisation of genetic resources will fund conservation (Siebenhüner *et al.* 2005: 40–41).

Doubts are also appropriate because the described practice of bioprospecting does not at all make the long-term conservation of the ecosystems necessary. Once the information has been gained, that is the genetic characteristics discovered, interest in the conservation of the organism studied is exhausted, at least from an economic perspective (cf. articles in Burrows 2005).

The conservation issue reveals once again the missing enforcement mechanism of the CBD. Because there is no sanction mechanism, the parties, that is the governments, cannot be forced to evaluate the impact of their policies and to initiate actions that protect and conserve biodiversity. The aim of creating a funding mechanism for a global protection area system by 2008, and then establishing it as a system by 2010 and thereafter working towards the creation of a network of biotopes by 2015, seems at present to be a completely unfathomable task. Moreover, at COP 8 in March 2006 the activities of the large conservation NGOs – especially Conservation International, The Nature Conservancy and the WWF – with respect to protection and conservation were praised, despite the fact that they involve hardly any indigenous peoples and local communities within their organisations or campaigns.

More recently the relationship between the conservation and utilisation of biodiversity was newly framed by emphasising its relevance for the functioning of ecosystems and for ecosystem services. In particular, the Millennium

Ecosystem Assessment highlights the interconnections between ecosystem services and human well-being in a broad variety of societal sectors and fields of social actions: from nutrition and shelter up to recreation and tourism (MASR 2005). Compared with earlier assessments, the Millennium Ecosystem Assessment focuses more on the trade-offs between different ecosystem services and reveals that some – in particular regulating and cultural services – are degraded because of substantial net gains in provisioning services (in particular agriculture and fishing). Thus, the Millennium Ecosystem Assessment requires that gains and losses regarding the use of ecosystems must be assessed more carefully and taken into account in decision-making processes. Moreover, it states that some ecosystem services could be under-valued and therefore they could risk being destroyed if the focus is only on market values (MASR 2005: 98). Even if in the Millennium Ecosystem Assessment economic and other flexible governance instruments are recommended, it also identifies the limits of pure economic approaches and requires at least the incorporation of non-economic considerations.

### *2.4.5 Participation and capacity building*

The imposition of national sovereignty over biological diversity raises the question of to what extent local population groups can articulate themselves under the principle of national sovereignty and what rights they have – particularly if their interests are contradictory to those of nation-state actors or dominant interest groups. There seems to be at least an implicit consensus between the governments here that the rights of local communities are internationally recognised but that this recognition is of a primarily symbolic nature. One catchword in the CBD is *participation*. Four levels play a role here: participation in the international negotiations, in the formulation of national legislation, in the negotiating of access agreements and finally, in the process of the appropriation of resources and knowledge. Participation in the appropriation of resources and knowledge is the most important for the prospectors because this can increase effectiveness. At the other levels, however, there is a danger that participation may serve to cultivate legitimacy for a process which is basically without an alternative. It is therefore essential to determine what exactly is meant by participation, not only at the level of the consultations but also with regard to the decision-making processes; above all in order to be able to influence decisions in the case of deviating appraisals. This is made considerably more difficult by the fact that it is a case of political processes with many preconditions in which the implications of certain developments and decisions are not at all obvious.

At first glance, there is no politically explosive material behind the formula of capacity building. However, different forms of application of the CBD are expressed in the ideas of what capacity building should look like concretely, that is which strategies and material resources should be applied. Capacity building not only means educating people or making the CBD publicly known, it also means the development of comprehensive organisational, institutional and knowledge capacities. It is exactly this subject which has been placed at the centre of examination in the so-called Bonn Guidelines on Access and Benefit-sharing, which are outlined below. A critical examination of their suggestions illustrates that the corporate and governmental side is principally concerned with improving the conditions for valorisation. By contrast, NGOs and indigenous peoples see, in capacity building, the opportunity to strengthen the rights of weaker actors.

The current proposals for capacity building at the local level have been criticised for leading in the wrong direction. Instead of training even more access and benefit-sharing experts who can negotiate bilaterally with corporations, the aim should be to create strong, legally binding, multilaterally consensual and internationally sanctioned rules of the game in order to protect communities and to define the extent of corporate manoeuvrability (GAIA/GRAIN 2000).

### *2.4.6 **Problems of national implementation***

The implementation of the CBD regulations at the national and local levels should take place on two planes: first, through national policies (national reports, legislation, nature conservation projects etc.) and second, through specific access agreements. The interests of the Southern countries, and especially the megadiverse countries, are relatively clear here, namely "that the adoption of an appropriate legislation can become a highly useful tool in participating efficiently in international markets and in taking full advantage of their own comparative advantages" (SEMARNAT 2002a: 116).

In the estimation of many participants, the national implementation and legislation processes have been accorded the highest priority at present. Nevertheless, they are tackled, if at all, only on the part of the so-called provider countries – and even here it tends to be slow.

The implementation of the CBD is considered to be part of a third phase of the convention, after the negotiation of the convention text and, after 1993, the policy development within the CBD (Zedan 2005: 2–4). However, in recent years, national implementation has been considered one of the central weaknesses of international biodiversity politics (see Executive Secretary of the CBD 2005). The implementation process is considered too slow (Secretariat of the CBD 2006b: 184, i.e. Decision VIII/8). Therefore, national biodiversity strategies and action plans should be discussed in detail at COP 9 in 2008.

Although it is seen as problematic that no access legislation for appropriate user measures has been enacted in the user countries, this issue has not been brought forward in the negotiations. A few years ago, a survey of the legislation in more than ten Southern countries, which at the time had either been passed or was under review, clearly showed that in the majority of cases there was no independent legislation; the state had almost exclusive property rights over genetic resources (there were exceptions in the countries of the Andes Pact: *Comunidad Andina de Naciones*, CAN); and finally, that PIC had to be negotiated and approved by the state (Caillaux/Ruiz 2002: 162–166). The statement by a member of the Brazilian government is probably expressive of the general opinion held by those countries, namely that the provider countries "should provide a very robust legal enabling environment" (CBD/23). The inadequacy of the legislation is connected with the fact that it is not clear what content the national regulations should have and to what extent they should be binding. Different demands are formulated by different actors.

Another aspect of the protracted implementation of the CBD is based in the competition among the national state apparatuses. The leading environmental ministries usually carry little weight with regard to such issues, something that will become even clearer in the case study of Mexico (see Chapter 5). A further aspect named by an interviewee is the lack of sanction mechanisms in the CBD in the case of non-implementation (CBD/12). The national implementation of the CBD not only means the provision of an adequate legal framework but also requires that the nation-state is able to finance and to apply it (as a part of public policies). Here too, there is a great deal of potential for conflicts to develop, as different interests are expressed in individual policies. At the international level, the monitoring of the implementation, via reports among other things, is prescribed, but this obligation does not seem to be taken particularly seriously.

The fact that room for manoeuvre does exist in the process of the national implementation of the CBD is demonstrated, for instance, by the Philippine legislation or by the corresponding regulations of the Andes Pact. They both go beyond the CBD and insist that existing local communities or indigenous peoples must agree before national implementation can go ahead (ten Kate/Laird 1999: 28f.; on the Philippines see also Liebig *et al.* 2002). From the perspective of the Northern governments, the Philippine legislation is considered to be "over-regulated" (CBD/1; CBD/4; CBD/9; CBD/10), while from the point of view of NGOs, it tends to be seen as a positive example (CBD/3; CBD/7; CBD/24). The representative of a Philippine NGO which contributed to the formulation also defends it (CBD/2). The problem here is seen in the fact that the legislation is not implemented. Moreover, the problem remains very obvious that the national regulations also have to be implemented institutionally.

### *2.4.7 A multilateral system of rules for bilateral policies?*

As shown above, with regard to the regulation of access and benefit-sharing, the CBD has created legal security in that, with national sovereignty, the regulating authority of the nation-state became internationally recognised. This recognition was demanded and enforced by the governments of the Southern countries in the CBD negotiations with reference to the centuries-old practice of the unregulated (and indeed, free-of-charge) appropriation of genetic goods. Behind such actions was the practice which today is condemned as *biopiracy*. There are, however, different understandings of what biopiracy exactly is. Two positions are important. The first is advocated primarily by governments of the South and most NGOs, which understand biopiracy as the gratuitous and legally unregulated appropriation of genetic resources. According to this view, the practice of biopiracy would be ended if appropriate legal regulations existed, particularly the fulfilment of prior-informed consent and benefit-sharing in access agreements (legalistic view). A second, alternative position can be outlined, which places stronger emphasis on the inequalities of power among the different actors and problematises the function of legal regulations in the practice of biopiracy itself (particularly intellectual property rights and patents, but also rules in the CBD). According to this point of view, legal regulations as such are not at all adequate as long as they do not remove the inequalities of power. Rather, it is decisive whether or not the weaker actors (above all rural and indigenous actors) are really able to defend their rights against stronger interests (this could therefore be described as the "realist" view).

In a strictly legalistic sense the CBD and the further negotiations on access and benefit-sharing can be seen as attempts to remove the practice of biopiracy by creating a legally binding framework for the appropriation of genetic material. However, in doing so the CBD at first favours a bilateral approach. Agreements between providers and users are to take the place of unregulated and gratuitous appropriation. Access agreements are not bindingly required in the CBD but it is highly likely that they will become the central instrument for regulating concrete access and benefit-sharing in the existing international and national framework. It is remarkable that, until now, access and benefit-sharing agreements have predominantly been made, above all, in the field of pharmaceuticals in contrast to their general omission in the fields of agricultural and cosmetic products. The agreements should regulate, among other things: who the partners to the agreement are (state agencies, research institutes, companies etc.); to whom the samples of genetic resources belong and whether the samples are to remain in the provider countries; which authorities are responsible; whether and how trust is to be safeguarded; how, for whom and by whom intellectual property rights are to be safeguarded (patents, licenses etc.); whether there is to be joint research; and finally, whether research results are to be returned and to what extent prior-informed consent, mutually agreed terms and benefit-sharing are to be safeguarded. Although it is not explicitly foreseen in the CBD, research institutes and companies such as Novo Nordisk and Glaxo Wellcome, as well as botanical gardens and research institutions such as the US National Cancer Institute or the CGIAR (Consultative Group on International Agricultural Research), have been developing their own guidelines for the conclusion of access and benefit-sharing agreements (Caillaux/Ruiz 2002: 167–170). Until now the accusation has been raised with regard to the access agreements that these not only focus on bilateral agreements, which already favour stronger actors, but also that the specific models for benefit-sharing represent an "exclusively economic approach" (GAIA/ GRAIN 2000: 1).

It is not enough simply to contrast bilateral and multilateral approaches to the regulation of access and benefit-sharing. Although the CBD clearly favours bilateral access models and is therefore considerably different from the multilateral system for the exchange of genetic resources as now established within the framework of the FAO (cf. the next chapter on this), it nonetheless is also, in fact, a multilaterally recognised framework in which these agreements and thus global markets for genetic resources are to be safeguarded. The differences thus refer primarily to what exactly is to be regulated multilaterally, what strength these regulations have over bilateral agreements and in what form they organise global exchange: what interests are taken into consideration and in what form? It is decisive as to whether or not, beyond national sovereignty, recognised minimum standards and enforceable rules continue to exist (see below).

## 2.5 The Bonn Guidelines on Access and Benefit-sharing

After a long consultation and a short negotiation process, at COP 6 in The Hague in April 2002, the Bonn Guidelines on Access and Benefit-sharing were officially adopted.[10] The Guidelines express the great importance which has been attached in recent years to the development of national strategies and policies. In different parts it is outlined, in very great detail, how national strategies should be developed, what roles they should attribute to providers and users, what prior-informed consent and mutually agreed terms and concrete material transfer agreements could look like (the latter in Appendix I), and finally, what benefits are possible (Appendix II). In a formal legal sense these are not a part of the Bonn Guidelines but they are closely connected with them. The Guidelines are regarded as "a useful step of an evolutionary process in the implementation of relevant provisions" (Decision VI/24, point 6, www.biodiv.org/decisions/), and it is regarded that there should also be a review process. Even in the Guidelines themselves, careful and less binding formulations dominate. They should be voluntary, simple, acceptable, complementary to other international instruments, flexible, transparent and subject to changes. Furthermore, the subject of capacity building is given an important position through the detailed listing of what is understood by it, and through the decision of the COP (VI/24 B) to set up an *Open-Ended Expert Workshop on Capacity-building* within the framework of access and benefit-sharing (which met for the first time in December 2002; report under UNEP/CBD/ABS/EW-CB/1/3). With this, this subject has gained an important position in the CBD process. At the end of the Bonn Guidelines there are regulations on intellectual property rights, in which on the one hand, countries are encouraged to integrate designations of origin for genetic resources with the registration of intellectual property rights, and on the other hand, the WIPO is granted an important function in the generation of proposals as to how the former can be implemented.

How should the Bonn Guidelines and the process of their creation be judged? At first glance, it is surprising that, in view of the difficult problems involved, the negotiations within the framework of the CBD could be brought to a close in an astoundingly short time. This quick success was not expected, particularly because of the dispute over the importance of intellectual property rights. Whereas this dispute dominated the first expert panel, the arguments in the second panel, in the working group and at the COP itself were more "objective" (CBD/27), that is interests were increasingly defined as objective issues. What is of decisive importance is that within the Bonn Guidelines a specific compromise was found and institutionalised which is strongly directed towards the commercialisation of biodiversity and which also has the special characteristic of being voluntary. The real meaning of the Guidelines probably lies in the way that they provide an orientation. They contain an enormous collection of knowledge which can be referred to in concrete legislation, strategy-building and negotiation processes.

At the same time the compromise which was found with the voluntary Guidelines did not lead to a real clarification of the points of dispute behind it. By contrast, in the same year in which they were adopted, the demand for legally binding regulations was once again put on the agenda. The group of megadiverse countries, in particular, had obviously come to the conclusion that under the Guidelines, benefit-sharing had not been dealt with in a satisfactory manner. In the final instance the compromise was not strong enough to be able to include the interests of all the leading actors. Even in this weak form it shows a very selective consideration of the different interest groups, as it essentially serves the need for clear rules for the commercialisation of genetic resources, while the interest groups that are less interested in commercialisation (indigenous peoples, nature protection organisations) do not expect much from the Guidelines. In addition, in the Guidelines, there is no questioning of the fact that the implementation of the CBD is only to take place within those countries that contain most of the world's biodiversity ("provider countries").

Nonetheless, the problem of the voluntary character of the Guidelines should not be misconstrued as meaning that the global market for genetic resources could indeed regulate itself. Rather, at this point it becomes apparent that binding state agencies are of central importance in constituting markets. In line with this notion, the most pertinent aspect concerns the explicit manner in which the interests of the different actors are included in the regulating institutions. Herein, the interests of the companies are expressed in the Guidelines with the statement that international law is to have a guiding rather than a binding effect (CBD/13). At the same time, regarding capacity building, the issue is whether and to what extent local actors have the opportunity to help design the conditions for the marketing of genetic resources. The proposal to set up a working group on this and to take into account, in a stronger manner, the role of intellectual property rights in the implementation of access and benefit-sharing agreements can be interpreted as a type of concession to critical voices by the supporters of the Guidelines.

Basically, the initiative of the countries of megadiversity does not go beyond the horizon of valorisation; rather, it simply refers to the conditions of benefit-sharing and the role of intellectual property rights in this. Indeed, the essential demands of the Bonn Guidelines from the Like-Minded Group of Megadiverse Countries' perspective refer to their unbinding character and the question of intellectual property rights – and precisely concerning the question of intellectual property rights, it is not to be expected that there will be any great concessions. In addition to the Philippines, Ethiopia, which does not belong to the like-minded group, is also pressing for a legally binding international access and benefit-sharing instrument (Caillaux/Ruiz 2002: 170–174). The initiative of the megadiverse countries is thus first of all directed towards an improvement of the marketing conditions in the sense of the "providers". However, in this respect it is also about building up political pressure on this terrain in the direction of a revision of the TRIPS Agreement. With this, the concrete appraisal of the access and benefit-sharing regulation within the CBD shifts towards the question of the relationship between the CBD and other international agreements. We shall deal with this question in more detail in Chapter 4.

Finally, it can be stated that the problems of the concrete appropriation of biodiversity are condensed most strongly in the concept of capacity building. Capacity building can therefore be understood as the terrain on which the concrete form of the implementation of the Bonn Guidelines is negotiated. Thus, it is not a neutral, technical problem, but a deeply politically determined terrain. Ultimately, here it is decisive whether it will be designed as technical, scientific knowledge for better valorisation, or whether capacity building offers a chance for weaker actors to be able to receive training in accordance with their interests and thereafter, to follow these interests. The "key areas requiring capacity building" (Appendix to Decision VI/24 B, www.biodiv.org) clearly place the emphasis on an improvement of the valorisation of biological diversity. For example, traditional knowledge appears when the issue – apart from the protection of that knowledge – is that of the stock-taking of biological resources in the framework of the Global Taxonomy Initiative. Indigenous peoples and local communities play a role only in the points concerning "development and strengthening capacities ... for participation in decision making and implementation" and "contract negotiation skills" (Appendix to Decision VI/24).

## 2.6 The emerging International Regime on Access and Benefit-sharing

Since COP 7 (2004 in Kuala Lumpur, Malaysia), an International Regime on Access and Benefit-sharing has been an official part of the work programme of the CBD (Decision VII/19). However, ever since the deliberations regarding the terms of reference for the negotiation of a regime, the main object of contention has remained the same: there is no agreement as to whether the agreement should be legally binding or non-binding. It is particularly the JUSCANZ Group, that is the governments of Japan, USA, Canada, Australia and New Zealand, that opposes a legally binding regime. This was one of the reasons for the relative failure of the Third Meeting of the Ad Hoc Open-ended Working Group on Access and Benefit-Sharing that was held in Bangkok in 2005. The JUSCANZ Group and the European Union insisted on further analysis of the gaps in already existing legal instruments dealing with access and benefit-sharing – resulting in an open list of legal elements to be considered at the next meeting. However, given the enormous expectations, the chair at the Fourth Meeting of the Working Group, held in Granada, Spain, at the beginning of 2006, surprisingly presented a draft version for an international regime (the so-called Granada Paper, CBD 2006). The Northern governments did not succeed in rejecting it but, as a compromise, most provisions were put into brackets. This can be seen as a success for the Group of 77 and China. Northern governments consider the Granada Paper more as a useful overview of positions and intend to slow down the process with the argument that more knowledge is needed about what should be done.

After long deliberations at COP 8 in Curitiba in 2006, it was agreed that negotiations for a regime should be completed by 2010. By the time the Working Group convenes its next meeting, it is expected that an international group of experts will have evaluated the feasibility of a "certificate of origin" which means that a patent can be applied for only when it is documented that the material was acquired legally. The negotiated compromise of the Bonn Guidelines should serve as a central orientation. However, the draft version for an international regime indicates that negotiations have hardly begun and that at least until 2010 there will be considerable conflict (regarding its final make-up?) (Secretariat of the CBD 2006b: 129–136).

At least three different positions became obvious: first, those wishing to complete negotiations as soon as possible; second, those keen on negotiations but still undecided as to their key components; and finally, those more interested in exchanging information and experience than in developing an international regime (ENB 2006: 22).

Of course, we are not able to predict either the results of the future negotiations or the concrete bioprospecting practices. But our analyses of the CBD process over the last 15 years, of the characteristics and importance of environmental and biodiversity politics at the international level, of the actual state of the CBD, and of power relations and political dynamics, suggest some conclusions.

Like the Bonn Guidelines, the regime will perform a number of duties: it will reiterate the importance of access and benefit-sharing issues; enhance the knowledge of the enormous complexity of the issues at stake; and help to understand and articulate the concrete interests in the negotiation process, particularly regarding the protection and use of biodiversity. Furthermore, at the concrete level, capacity building might be improved and there may be more money put into subjects related to it. However, as we argued above, capacity building is not a technical issue; rather, its concrete forms shape the way biodiversity is dealt with.

The major conflicts arising during the formulation of a regime are:

- Comprehensive rules in order to secure intellectual property rights for the so-called provider countries. Strict intellectual property rights are only foreseen for the marketable products of biotechnological "innovations". There will be a conflict if the "disclosure of origin" of genetic goods, that is a required certificate of origin when a patent is applied for, is negotiated, either in the CBD or in WIPO and TRIPS (an expert group on that has been established: cf. IP-Watch, 4 April 2006; Heineke/Wolff 2004: 29–30).
- In line with this, the effective forms of fair and equitable benefit-sharing will be contested as well. Here, derivatives of developments of genetic goods attain importance since the so-called provider countries want to see them included in the benefit-sharing mechanisms. "The issue concerning these materials is to what degree they are based on the original natural genetic material, and to what degree scientific knowledge has added value to the development of theses materials" (Chambers 2004: 63). But this is decisive for the question of benefit-sharing.
- The modalities of a regime: this regards the concrete meaning of the "binding" character of an international agreement as Northern actors will invariably favour a voluntary one in the sense that the multilateral voluntary principles form a guideline for bilateral agreements.

  > [I]f the nations want the regime to regulate the trade of genetic material with customs regulations, control the use of foreign genetic material under domestic jurisdictions, create a certification scheme, require disclosure of origin in patent applications, and create a system of legal redress and dispute settlement, then the regime will need to impose changes on domestic laws and therefore will need to be legally binding and ratified by national legislators.
  >
  > (Chambers 2004: 63f.; Frein/Meyer 2005: 154)

- An exemplary illustration of the strong position of the US biotech industry is given by the following quotation: "A mandatory patent disclosure obligation of any kind would radically increase uncertainty for industry and make it very difficult to invest in bioprospecting needed to develop and commercialize genetic resources inventions" (Finston 2005: 4). The draft version of an international regime is still full of brackets, something which does not mean that all issues are really controversial but rather, that some will serve as bargaining tokens. Among other things, the draft version still leaves open a number of issues: whether or not a regime should "facilitate" or "regulate" access to genetic resources; if alongside genetic resources their "derivatives and products" should also be included; and which provisions remain subject to national legislation (Secretariat of the CBD 2006b: 129–136).

- Additionally, the participation of indigenous peoples and local communities seems to be a problem for some actors in the negotiation of the international regime. The question of whether or not indigenous participation in the future negotiations on the regime should be enhanced was discussed exclusively by governmental representatives in closed groups, something which does not indicate an open attitude of the negotiators (see report of COP 8 in Secretariat of the CBD 2006b: 35).
- And finally, it is not clear how fast the regime will be negotiated. The negotiations about the regime should be finished in 2010. But, it is still not clear if the parallel "gap analysis" will slow down the process or promote it.

Despite the fact that through the Like-Minded Group of Megadiverse Countries the Group of 77 and China managed to create an internal compromise – the "G-77/China, speaking with a single voice for the first time in the ABS process" (ENB 2006: 22) – thereby gaining some power in the CBD process, this is not likely to shift the overall tendency of the CBD towards the valorisation of biodiversity in general and genetic goods in particular. The conflicts within the CBD remain most of all distributional among the dominant actors, including the governments of the provider countries. All of them are interested in global rules for the commercialisation of genetic goods and therefore they all generally accept the overall setting.

## 2.7 Summary: the importance of the CBD in the international process

Without doubt the most important achievement of the CBD process has been "to have developed an integral approach in the treatment of biodiversity" (Krebs *et al.* 2002: 25, our translation). The CBD is neither a pure nature protection convention nor one exclusively oriented towards valorisation. This is also the reason why the CBD is confronted with such multifarious claims (which is often seen as a problem, e.g. CBD/17). In contrast to other international conventions – such as the WTO – the CBD allows widely different actors with divergent views to bring in their perspectives and interests. In addition, the CBD – in contrast to the ILO Convention 169 on indigenous rights, for example – has weight in international law and "new" political questions can apparently be more easily raised and dealt with on a terrain which is structuring itself anew – which is one difference to the FAO. Its strong compromise character obviously represents the great attractiveness of the CBD.

The CBD has this compromise character primarily, however, because major Northern and Southern actors have an interest in the valorisation of genetic resources. The CBD has become a central terrain for negotiating modes of valorisation (particularly the conditions of access and the accompanying benefit-sharing). The compromises achieved here are not at all symmetrical even between the governments, and in addition the Northern and more powerful actors attempt to secure specific interests, namely those which safeguard their intellectual property rights, on quite another terrain: the TRIPS Agreement. The concrete forms of compromise formation in the individual fields of negotiation are therefore decisive, as is the relationship of the agreements to one another. Furthermore, it can be observed that the over-riding importance of access and intellectual property rights is emphasised by the Northern governments.

In many interviews it was pointed out that the CBD is not primarily or exclusively concerned with legal, institutional regulations, but above all with moral, symbolic and political questions in the broader sense. International regulations "can focus desire and attention" and can stimulate reflection (CBD/3). The Guidelines should supply a framework which acts more as an orientation for the concrete negotiation of access. This guiding and problem-focusing effect of the CBD is without doubt an important element of the entire field of conflict, because it includes the chance of a politicisation of new problem dimensions (traditional knowledge and indigenous peoples, biopiracy etc.) It only represents the *chance* of a politicisation, however, because the concrete institutionalisation of these problem perspectives is closely connected to the interests of the different actors and the chances of compromise formation. If regulations of the CBD and the Guidelines gain in importance in individual disputes, they become part of a *power bargaining*, that is of a negotiation process which depends on concrete power relations. This is true both for the Bonn Guidelines and the actual negotiations of an international regime as a whole (e.g. for the compromise with regard to voluntary guidelines) and for individual elements (e.g. what are recognised benefit-sharing mechanisms?) and of course, to a particular extent, for the working out of individual access agreements. In the specific negotiation situations over a bioprospecting agreement questions of power will remain central against the background of the Guidelines or a regime – whereby here, power ranges from discursive power to the establishment of an "own" view of the problem via the institutional capacities for the enforcement of own interests to become a sanctioning force in the nation-state. The compromise recognised by the parties to the conflict thus represents a type of *guiding corridor*, in the designing and further concretisation of which power potentials play a decisive role.

The recognition of power relations and the search for room for manoeuvre on the part of Southern governments can be seen particularly clearly in one thing: the CBD, which came into force at the end of 1993, imposes national sovereignty over genetic resources. The discussion which followed focused on the *in situ* resources to be found in the Southern countries. As we have discussed above, this ignores the immense resources which are already to be found in Northern institutions (private companies, research institutes, botanical gardens).

One important aspect of the global conflicts over genetic resources is the *structural supply competition* of the countries in which biological diversity exists. Because the majority of biodiversity-rich countries are confronted by a relatively small number of biotechnology firms, all of which have the technical know-how and the financial capacity for research using genetic resources, these countries find themselves in a difficult situation. Thus, the biodiversity-rich nation-states of the "South" find themselves trapped in a competitive relationship, one against the other, over the scarce demand on the part of the life sciences industries. This strengthens, above all, "Northern" interests in the appropriation of biological diversity because they can insist on "good conditions". By founding the Like-Minded Group of Megadiverse Countries an effort was made to achieve a uniform position on this question. However, even despite this cooperation, the competition mechanism has not been annulled. The situation is aggravated by the fact that genetic resources – in contrast to oil and therein, the strategy of OPEC – are not resources which are required in large quantities, but rather, as a rule, all that is required is the genetic information they contain.[11] The supply competition also contains an uncertainty factor in that governments may possibly invest in good conditions for valorisation (e.g. by establishing research capacities) and then perhaps go short in the bioprospecting agreements.

With regard to the prescribing of the rights of weaker actors, their representatives or allied NGOs have been able to see to it that these remain on the agenda in the negotiation process. This is connected with practical problems regarding the legitimate appropriation of genetic resources. The importance of local actors lies in the fact that they may have useful knowledge for the valorisation of resources. In addition, NGOs are an important element of the negotiation process at the international level with their knowledge and experience. Against this double background, critical NGOs and indigenous organisations have succeeded in politicising the process. They were helped by the fact that on the part of firms and governments there is great uncertainty as to how the appropriation of genetic resources can take place effectively and legitimately. This process remains controversial.

For without denying completely the interests of weaker actors, governments use three mechanisms to weaken them. For one thing, the voluntary character of regulations is emphasised (using the argument of the danger of "over-regulation"); this is followed, second, by reference to the national level; and third, reference to the complicated and to a degree, completely new situation. This is the case for questions of benefit-sharing, for prior-informed consent or for indigenous and local rights. Altogether, it has become clear up to this point that criticism of the valorisation paradigm is more difficult at the international level than at the national or local one.

Finally, there is one thing which should not be forgotten. The appropriation of biological diversity and in particular of genetic resources takes place permanently with or without a legal framework. It can take on legal forms if a national legal framework based on the CBD has been created. The individual access agreements can also be guided by international law, without the latter already being transposed into national law. In many cases the appropriation takes place largely without a legal framework or with one which does not satisfy the CBD. This is the reason why in addition to some NGOs and experts, increasingly, Southern governments also speak of biopiracy. It remains an open question here as to whether or not – at the material level of the ecological crisis, which consists in the continued and dramatic loss of biological diversity – the CBD can be an effective instrument, or whether perhaps the work of stock-taking will be "the scientific accompanying music to the burial of biological diversity" (CBD/17, our translation).

# 3 Limits to commercialisation?

## Genetic resources in agriculture and the conflict over a multilateral exchange system

The field of agriculturally utilised biodiversity has some particular characteristics, each of which can both help to illustrate the breadth and complexity of the problems involved and bring to light some of the central lines of conflict in a compromised and explicit form. These particular characteristics are manifold. On the one hand, the structure of the problems in the field of agro-biodiversity is characterised by the fact that genetic diversity is the product of human activities over centuries. Therefore, this genetic diversity must be examined from the perspective of the utilisation of nature. Moreover, genetic resources are – apart from their economic importance for capitalist industries – of central relevance for worldwide nutrition and even more for food security in certain regions. On the other hand, the quest for the conservation of Plant Genetic Resources for Food and Agriculture (PGRFA) is no less pressing due to the dramatic erosion of this genetic diversity. But this search becomes acute under different general conditions and throws up different lines of conflict and actor constellations.

The conflicts over the appropriation of genetic resources in this field already have a much longer history and their principal features can be traced back to the beginning of European expansion and colonialism (Crosby 1972, 1986). However, these conflicts have been given a new momentum by the technological and political shifts in the 1980s. The institutional attempts to deal with these conflicts also go back further and they had already produced important regulating procedures before the CBD. At the same time, the changes which are connected with the coming into force of the CBD, and later with the TRIPS Agreement, are also central for later developments and allow insights into some central dimensions of the problem. Finally, the characteristics of this problem field are reflected in an agreement which has some distinct features compared to the CBD and is characterised above all by an alternative approach to access and benefit-sharing: a Multilateral System for the exchange of genetic resources.

At the centre of the following presentation is the genesis and the enhancements of the *International Treaty on Plant Genetic Resources for Food and Agriculture* (ITPGR). This agreement, which is based within the framework of the FAO, was concluded in November 2001 following lengthy negotiations and thus belongs to the more recent agreements in this field. It regulates international exchange for the subsection of genetic resources which play a particular role for world nutrition. In this part of the PGRFA a multilateral exchange system was developed in the International Treaty, mentioned above, which is intended to take into account the specific characteristics of these resources. The negotiations, however, were extremely difficult and were repeatedly threatened with failure. Even after their conclusion in 2001, some important issues had not been appropriately resolved and it took another five years until the specific regulations concerning the multilateral system of exchange could be finished – and their actual effectiveness is still difficult to judge. Thus, in these negotiations, the question of the relationships of the different international agreements to one another as well as that of their relationship to the sovereignty of nation-states became highly divisive. The options of regulating the use of genetic technology remained controversial, as did the applicability of intellectual property rights and patents and the chances local actors and farming communities had of having their interests in these international agreements adequately considered (here, primarily in reference to the embodiment of the principle of *Farmers' Rights*, see below).

These controversial issues, however, are not expressions of the contradictions between ecology and economy or between conservation and utilisation. Rather, they represent a *conflict between different forms of human utilisation of nature*. These conflicts can be described as conflicts between traditional and industrial forms of agriculture (whereby the traditional forms can vary quite considerably). At first glance, this produces a North–South conflict. But this line is shaped in various ways, since in Southern countries the conflicts of interest reproduce themselves internally in face of a variety of agricultural production systems. In the field of PGRFA in general, it is particularly difficult for nation-states both in the South and in the North to formulate a policy which is consistent within itself. At the level of nation-states a whole number of different ministries and authorities are involved (from the usually responsible agricultural ministries through environmental, trade and justice ministries to the foreign ministry) and the complexity of the problems requires particular expert knowledge (Petit *et al.* 2000). Even the interests of the most important actors, that is the governments of the nation-states, can be internally clarified and harmonised only with difficulty and in lengthy negotiations as part of a state project. In addition, the interests of the biotechnology industry and those of traditional, non-capitalist forms of agriculture conflict with each other here, although usually not by way of direct confrontation between the actors but rather in the form of different interpretations of the problem and the corresponding strategies. Representatives of indigenous or rural groups are personally present at the negotiations only in exceptional cases, for example at the conference "In Safe Hands" on the occasion of the FAO Conference in June 1996 in Leipzig. In the course of the negotiations, their interests were often represented by certain NGOs (primarily seeds NGOs; see Brand 2000: 209ff.). The NGOs in this sector usually proceed from a decidedly political standpoint and understand themselves explicitly as the representatives of these interests. They are concerned above all with the respect of particular characteristics of agriculturally utilised biodiversity and, related to that, with the propagation of Farmers' Rights and the limiting and prevention of intellectual property rights. The conflict is also partly interpreted as one between the private and the public spheres, between genetic resources as private property or as a public or common good (e.g. Petit *et al.* 2000). It remains quite unclear here what exactly is meant by a public good and in what relationship it stands both to private property and to Farmers' Rights. In any case, the strategy of capitalist valorisation, including its ecological effects, is at the centre of the conflict. On closer examination, however, this strategy is supported by a broad spectrum of actors with quite varying interests: from middle-class seed growers to biotechnological TNCs. Since in this area a considerable process of concentration and displacement is taking place, an identity of capital interests cannot be assumed, particularly between small, specialised firms and the large monopolies. Nevertheless, they ultimately work together in the privatisation of genetic resources. The industry is permanently represented at international negotiations by relatively powerful associations – such as, for example, the lobby association Bio or the international association of plant producers, ASSINEL. However, it is not clear whether they in fact have an interest in obtaining access to genetic resources in the Southern countries or whether the material which already exists in the *ex situ* collections or in the large gene data bases in the North is sufficient for them (RAFI 2001a; see Le Buanec 2005).

The cooperation of the actors and their interests in this field is therefore complex. In the following section we shall elaborate on how a valorisation paradigm was established in this area even though it met with considerable opposition from a number of actors, much of which stemmed from the highly explosive conflict over the exclusion of the patentability of PGRFA. The different interpretations of the problem and their interplay in the genesis of the ITPGR play a decisive role here. We shall deal in particular with the important function of experts and the constitution of a dominant interpretation of the problem and the options relating to its resolution. In addition, we shall examine the fundamental question of the coherence of state and inter-state politics. Are the tensions and contradictions between different fields of politics (agriculture, environment, trade, property rights, to name only the most important) decisive for the fact that "governments can't make a policy" (Petit *et al.* 2000)? Or are these contradictions between different parts of the state apparatus (and the connected international state apparatuses) the expression of antagonistic social relationships at the national and international level – and is the search for coherence in fact an attempt to establish a power-shaped stabilisation of these relationships? What does the search for a coherent policy in this area actually mean? To what extent can suitable answers to the problem structure in fact be formulated? And how do national and international lines of conflict overlap here? We shall have to deal here above all with the questions of whether and to what extent the "multilateral system of exchange" with regard to the regulation of access and benefit-sharing and compared to the CBD is in fact as differently designed and constituted as sometimes claimed. What is the function of the ITPGR and the Multilateral System with regard to the problem and to the interests of locally situated actors? Can it in fact consolidate their interests and does it increase their chances of gaining influence? In order to be able to answer these questions it is first necessary to take a closer look at the complex problem, the history of this conflict and the genesis of the ITPGR before the agreement itself can be examined in more detail.

## 3.1 Structures of problem and conflicts

Although critical voices have pointed out that "wild biodiversity" in many – and often in the really important – cases is not really so "wild", instead being interlinked to greatly differing forms of human intervention (Hecht 1998) – this is particularly pertinent with regard to the very beginnings of the field of agricultural biodiversity. The diversity of the hereditary characteristics of useful plants, in which we are interested here, is closely connected with the practices of humans. Without human cultivation of nature this diversity would not exist in its present form and in no way should this practice be abandoned. The waiving of the agricultural use of PGRFA by concentrating solely on the protection of these varieties is neither meaningful nor possible. Rather, and decisively, the conservation of this threatened diversity can only be achieved in the long term by the further use of such varieties (Oetmann-Mennen 1999). The conflict in this field is therefore interlinked with specific societal practices related to the production and reproduction of genetic diversity. This has in fact been recognised in the international conflicts and negotiation processes – at least in principle – in the form of Farmers' Rights (see below).

With the provision of Farmers' Rights it is recognised that the present diversity of PGRFA has arisen over the centuries, and is conserved today, by means of farming practices. What is controversial, however, is whether and if so which rights actually derive from this and how the tension between collective rights and private intellectual property rights should be dealt with. The particular characteristics of this field are condensed in the distinctions between the bilateral model of the CBD and the multilateral approach of the International Treaty, as well as going far beyond them. At the same time, it appears to be completely unclear whether the present politico-legal developments will be of benefit to the socio-economic practices and the living conditions of social groups which have created this diversity – or if they will result in their weakening. This in turn would not only have far-reaching consequences with regard to the direct living conditions of these people but also with regard to food security and the development opportunities of entire regions (Egziabher 2002). In order to be able to estimate the relevance of international regulation processes for the local and regional levels more precisely, first the problem as well as the interests and action capacity of the different actors need to be examined more closely.

### *3.1.1 Plant genetic resources as a cultural product: the concept of Farmers' Rights*

The introduction of the concept of biodiversity was accompanied in the public debate by a considerable mystification of the subject to the all-encompassing unity of every living thing (typical of this: Wilson 1992). In this process, not only were the bio-scientific problems forgotten which were linked to the introduction of the new concept (Hertler 1999), but what was also ignored was that biodiversity comprises not only tropical rain forests and coral reefs – and beyond that perhaps specific varieties of threatened mammals and bird species – but also the species of cereals and vegetables used by humans. Pat Mooney (1998: 14) remarked derisively of this short-sighted view of the problem, in which rice, potatoes and wheat do not exist: "In general, nothing requiring spoons or chopsticks is biodiversity." In spite of the fact, however, that the introduction of the concept was linked to economic motives from the very beginning (Flitner 1999; see Chapter 2), the bracketing of the biodiversity used and produced by humans seems to be very helpful. It contributes to the obscuring of the real interests in genetic resources and therefore to the reframing of the field.

In contrast to the conservation of species diversity, in this case the diversity within species, in particular the diversity of genetic characteristics which is embodied in different varieties and subspecies, plays a decisive role.[1] This genetic diversity is not something naturally given, but rather, is the product of cultivating practices of seed selection over thousands of years. This selection created the varieties with the desired characteristics – yield, length of stalk, resistance to dryness or dampness and so on. Due to non-intended mixture with related wild varieties it also produced a broad range of certain varieties that were not directly used (so-called traditional varieties and land races). The technique of selection, which is not dependent on knowledge about the laws of heredity, determined farming and breeding practice for a long period of time beginning with the Neolithic "revolution" – which itself was not so much a historical rupture, but a long passage of transition. It was not replaced until the end of the nineteenth century by the technique of cross-breeding according to Mendel's laws of heredity. However, in large parts of the earth it still determines the practice of rural agriculture today (de Boef *et al.* 1993). In addition, it is accompanied by certain forms of a gender-specific division of labour, culturally based patterns of interpretation and traditional norms and rules for the treatment of plants and animals. Of central importance here is the practice of the *free exchange* of seeds. This form of use can only continue to function within the farming communities if suitable varieties are freely available and knowledge freely circulates as to which varieties fare best under which conditions or which varieties are particularly resistant to certain types of damage. Without these social practices the diversity of useful plants and animals would not have developed and it is very much a question whether these practices are in the process of disappearing along with the disappearance of biodiversity. One further specific feature of agricultural biodiversity is the fact that due to the free exchange of the last 12,000 to 15,000 years it has spread over almost the entire earth and is seldom restricted to individual areas. At the same time, there continue to be centres of diversity in which even today we can find a particularly large number of varieties of wheat, corn or potatoes, with all the inherent conflicts arising from this.

The fact that the agriculturally used genetic diversity of today is the product of previous generations was explicitly recognised on the international stage in the 1980s with the introduction of the concept of Farmers' Rights. "Farmers' Rights means rights arising from the past, present and future contributions of farmers in conserving, improving and making available plant genetic resources, particularly in the centres of origin/diversity" (Resolution 5/89 on the International Undertaking of the FAO). This recognition was preceded by fierce political struggles, however, and until the coming into force of the International Treaty in 2001 it was not a legally binding commitment. Even today it is still not clear what this recognition really means and whether it is of any practical relevance.

In order to be able to more fully understand these conflicts a brief glance at their history is necessary. The interest in genetic resources was formed at the latest with European expansion and the conquering of America, which in view of the more strongly noted interest in gold and precious metals is often forgotten. As Alfred W. Crosby (1972, 1986) has shown, the conquering of the "New World" was a dividing-line in the history of evolution, since continents which had been almost completely separated for millions of years were now brought into cultural contact (the "Columbian exchange" according to Crosby 1972). Even the conquest itself was, from the beginning, accompanied by interest in genetic resources: it was intended to ensure control of the trade in spices and specialities with East Asia.[2] New varieties were introduced to Europe which led to a decisive change in living conditions there. The planting and exploitation of certain biological resources such as sugar cane, tobacco and bananas have, in the longer run, had perhaps an even more decisive and at least as fateful a role in colonisation as the exploitation of mineral resources. The influence on living conditions is due not only to the introduction of luxury goods such as tobacco or chocolate, but even more so to the potato, without availability of which, as a cheap staple food, the development of European capitalism would have taken a different course. Therefore, globalisation is also in this respect of the same origin as the establishment of capitalist relationships on a world scale.

Already, in this first phase of capitalist globalisation the availability of biological and genetic resources played an important role in the division and subjugation of the world. As in a "botanical chess game" (Brockway 1988), the European colonial powers handled entire countries and regions as stakes in their geopolitical strategies and thus constituted the globe as the object of their imperial interests.

The practice of collecting genetic material for aesthetic or research purposes and the establishment of botanical or zoological gardens is also grounded in this early phase. As we have seen in the previous chapter, these collections and the genetic resources contained in them are of increasing importance today. However, grave changes in the treatment of genetic diversity and of PGRFA in particular can be observed in the last 100 years (on the phases see Heins/Flitner 1998). The first change concerns the application, already mentioned, of Mendel's laws of heredity and the accompanying emergence of a breeding practice based on scientific methods. With the technique of cross-breeding the possibility arose of deliberately introducing genetic characteristics into a new product in the course of breeding. This technique was introduced long before the introduction of the concept of DNA as a theoretical basis for the understanding of genes was developed. With this, the commercial and political interests in genetic resources received a decisive impetus. The concept itself was defined at that time by the Soviet geneticist Nicolai Ivanovic Vavilov. He was the first to point out the *geographically unequal distribution* of the diversity of useful plants and combined this with the reference to their *economic importance* – bringing into discursive existence the concept of *genetic resources* (Flitner 1995b).

Since the beginning of the twentieth century the collection of seeds has become a goal of national politics. Driven by scientific advances in breeding and accompanied by commercial interest in their marketing, the central goal is at first still food security, that is national autarchy in the supply of food (for a more comprehensive outline of this, see Flitner 1995a). Already at that time lengthy collection journeys were put together in order to provide more supplies of genetic resources for breeding. At the Fifth International Congress for Hereditary Sciences in Berlin in 1927, Vavilov presented for the first time a map with the centres of diversity, accompanied by the theory that the present-day centres of diversity are also the areas of origin in which certain plants were first domesticated (see Flitner 1995a: 53ff.). Even if this theory has since been refuted, at the time, the search for genetic resources was given an additional impetus after this presentation. Looking back, this phase has been described as the phase of "the primitive accumulation of genetic resources" (Kloppenburg 1988: 153–156). However, this – just like the primitive accumulation of capital itself – was not a purely technical or economic process involving the collection of material but was instead more expansive, being accompanied by far-reaching socio-economic changes in production and the social division of labour. And it was accompanied by the state-secured establishment and enforcement of property rights which were necessary to enable capitalist valorisation of this diversity as an industrially utilisable resource (Kloppenburg 1988; Fowler 1995).

This process is also not a historically completed phase, instead being a tendency that permanently accompanies the development of capitalism, just as the primitive accumulation of capital itself continually subjects itself to new social relationships. That means, capitalism imposes its relationships of production and property on other social relations and secures itself against these (Alnasseri 2003). It is only the way in which this is done which changes in the course of time. During the colonial epoch it was based more on the military and security through violent forms (despite being complemented by legal measures such as, for example, prohibitions of exports). In the late and post-colonial eras, security by means of legal and political measures is of increasing importance. The enforcing and securing of capitalist relationships of production, however, the valorisation and primitive accumulation in a Marxist sense (the separation of the producers from the means of production, see Görg 1998), is accompanied by violent social transformation. This remains evident even today. Indeed, in contemporary times, this violent form of capitalist accumulation and its social and ecological consequences stand out ever more clearly.

### *3.1.2 The "green revolution" and the erosion of diversity*

Following the Second World War (WWII) the appropriation of PGRFA entered a new phase. Overlapped by bloc confrontation and accompanied by a strong tendency towards the industrialisation of agriculture, plant genetic resources became an element in the Fordist appropriation of biodiversity (Brand 2000: 179ff.). Starting out from the industrialised agriculture of the countries of the North, the model of the intensification of agriculture was exported to the global South. Methods of production directed towards increasing yields and the planting of large-scale monocultures were introduced there during the phase of the "green revolution". Although not all regions of the South were affected by this to the same degree – while it has been put into practice massively in some countries in South East Asia, such as India and the Philippines, the African Continent has largely been spared – the effects were and remain serious. The introduction of high-yield varieties, so-called hybrid varieties, at the beginning increases the yield per acre in many areas, but soon the far-reaching social and ecological effects become visible. Since the expensive high yield varieties can only meaningfully be introduced if they are planted on a large scale and with a considerable amount of capital deployment (the deployment of labour and chemical products) there is a strong tendency towards cash-crop products for export, which results in a considerable process of concentration and social polarisation and contradicts the declared goal of increasing food security. In the process, the gender-specific division of labour also changes and there is a displacement of the practices which were responsible for the genesis of the diversity of useful plants.

Above all, however, the green revolution was accompanied by a considerable shift in international power relations. Thus, countries such as the Philippines became strongly dependent on the seed companies of the North, primarily from the USA. This, too, contradicted the goal of securing the food supply (Pelegrina 2001). But this shift was based on the strategic decision to use the "green" preventatively against the threatened "red revolution" – not only against the growing influence of the Soviet Union and China but also against a social transformation and a change in property relationships. Therefore, the green revolution was introduced without social reforms, especially without land reform for a fair distribution of the ownership of land. During this phase, in addition, certain actors appeared for the first time which, to this day, remain of great importance in the area of plant genetic resources: the International Agricultural Research Centres (IARC) such as the International Rice Research Institute (IRRI) in Manila or the International Maize and Wheat Improvement Centre (CIMMYT) in Mexico. Originally founded with the support of the Ford and Rockefeller Foundations as central institutes of the green revolution in which the high-yield varieties were developed (see Fowler 1995: 182ff.) their function has now changed considerably. With the fusion of these institutes into the Consultative Group on International Agricultural Research (CGIAR) with its seat at the World Bank, one of the most important international organisations has been created which today still retains influence in this sphere.[3]

Yet another development is important for our discussion. Already in the early 1960s it became clear that the green revolution had considerable ecological effects: first, because of the increased use of fertilizers and pesticides with regard to the ecological context (soil, waters etc.), and second, from the erosion of genetic diversity itself. It is no longer a contested point that the intensification of agriculture and the introduction of high-yield varieties is the main cause of the loss of agricultural diversity. Experts from the FAO such as Otto Frankel and Erna Bennet pointed out this development very early on and alarmed the expert public (Frankel/Bennet 1970). As a result, *gene erosion* became an important topic at the FAO, although at the beginning in a rather conventional way. At first, the International Agricultural Research Centres were given the task, together with already existing and yet to be founded national gene databases, of conserving the agriculturally used diversity in the form of *ex situ* collections. This is problematic not only from an ecological point of view, because *ex situ* cannot secure the further evolution of seed and its co-evolution with possible pests. Even more important are the socio-economic effects, because the setting up of large *ex situ* collections favours commercial breeding and strengthens the trends towards concentration and monopolisation as well as international dependencies and the unequal exchange of genetic material (Mooney 1981).

### *3.1.3 The "seed wars" and the International Undertaking*

The last point led to the so-called "seed wars" of the 1980s. While the high-yield varieties and the fertilizers, herbicides and pesticides necessary for their cultivation had to be bought for money from private companies of the North, the *germplasma*, which was stored in the seed banks and made available from there to the commercial breeders and to the state or semi-state breeding programmes, flowed gratuitously into the North. This unequal exchange, even at that time secured by intellectual property rights, was denounced by the countries of the South at a conference of the FAO in Mexico in 1981, and a threat was made to stop the gene transfers. Heated disputes followed which inspired the *Washington Post* to make a comparison with the "Star Wars" programme of the then US President Reagan. These conflicts ended, although only temporarily, with the founding of the Commission on Plant Genetic Resources (CPGR, now CGRFA) and the conclusion of the International Undertaking on Plant Genetic Resources of 1983. The former was, and still is, the place at which further official intergovernmental negotiations take place, accompanied however by several informal negotiation circles. The International Undertaking, however, as a set of agreements which are not binding in international law, forms only the starting-point for the regulation of access and benefit-sharing in the treatment of genetic resources. It was in fact not at all able to resolve the conflicts via mutual agreement. Although it contains (in the amendment of 1989) a recognition of Farmers' Rights, the question of benefit-sharing could not be resolved. A corresponding international fund remained de facto without finances. Nevertheless, with the International Undertaking, Farmers' Rights were formally recognised as a counterweight to the rights of commercial plant breeders.

The reason for the very limited regulation capacity of the International Undertaking goes beyond its non-binding character in international law. Ultimately in the International Undertaking the central question of access was not really clarified. In contrast to the CBD, the International Undertaking is based on the principle of the *common heritage of humankind*. This principle, which was developed in the UN Convention on the Law of the Sea, is referred to in order to subject certain regions (the sea bed, the Antarctic, outer space) and the resources to be found there to international management. Thus, these regions are removed from the states' desire to exploit the resources or at least they must subject this exploitation to international standards (among other things in the consideration of ecological perspectives). At first glance this principle also appears appropriate for subjecting the diversity in the sphere of PGRFA, on which further breeding and thus the development of ecologically adjusted seed depends, to an international regulation mechanism. There is a trap involved, however. As a "common heritage of humankind" not all plant genetic resources are freely available to the same extent. The products of commercial breeding are subject to specific property rights which couple access to the payment of fees or licenses, while the traditional varieties are passed on freely and gratuitously. The area of application of the common heritage principle therefore only extends to a part of the PGRFA – and therefore this principle cannot function here either.

As Jack R. Kloppenburg and Daniel L. Kleinman (1988) analysed very early on, the contradictory interests of Northern agriculture and the Southern countries were not adequately taken into consideration in the International Undertaking. Therefore, it was not really suitable as a regulating authority for access and benefit-sharing. Present attempts to establish a "global commons" mirror the same problem, as does the Multilateral System within the framework of the International Treaty. How should they deal with intellectual property rights and the interests of commercial plant breeders and the biotech industry? Since the commercial breeders do not agree to subject their varieties, which are protected by special forms of property rights, to free exchange, the countries of the South were on their part also no longer prepared to do so. Instead, Kloppenburg and Kleinman suggested another mechanism which in their opinion could prove more suitable for conciliating contradictory interests: the principle of *national sovereignty* over genetic resources. In a certain way, they anticipated the principle which was in fact to become the foundation of the CBD several years later. They did *not* do this, however, in the attempt to establish a bilateral regulation mechanism aimed at individual contracts. On the contrary, their idea was that national sovereignty should do justice to the international interdependencies which decisively characterised the sphere of PGRFA (in contrast to "wild biodiversity"). Their proposal shows that the nation-state level of regulation offers itself as a compromise model for the global conflicts over plant genetic resources. In this model, however, the interests of the different groups of actors are taken into account very selectively.

The main part of their central paper deals with the derivation of the international interdependencies. They attempt to prove that no region and no country on this earth is autarchic in respect to the question of the availability of plant genetic resources. All regions are more or less dependent on the flow of genetic resources in order to be able to adjust crops and industrially used biological resources to changed ecological conditions or to be able to incorporate new characteristics into seeds. The thesis of the *international interdependencies of plant genetic resources* forms the theoretical background for regulation models such as the multilateral exchange system within the negotiations on the International Treaty. Its details are, however, far from being uncontroversial. This model does not in any way take into account the dependencies between modern, industrialised and traditional agriculture. These dependencies concerning plant genetic resources, however, are absolutely central, since one sector requires (and uses or destroys) the diversity while the other produces it (Flitner 1995a: 207). Between these groups of actors the model of national sovereignty does not create any conciliation, and it does not even attempt to do so. Also, in regard to different regions of the world and different varieties of plants, the differences are in part considerable, and this has also influenced the strategy of the actors in the negotiation process (RAFI 2001a). Nevertheless, the model of international interdependence has remained the dominant description of the problem to which the different actors refer, although with very differing accentuation: some, in order to define plant genetic resources as "public goods" and to propagate the free exchange of genetic material, and others to legitimate free access to genetic material for commercial purposes and thus implicitly or explicitly play into the hands of the valorisation paradigm.

### *3.1.4 The situation following the conclusion of the CBD: "environmentalisation"?*

The International Undertaking was therefore not a suitable model for the regulation of access and benefit-sharing for a number of reasons. It was neither legally binding nor capable of being enforced, nor could one speak of an acceptable compromise between the different groups of actors. Altogether, therefore, appraisals of it have varied quite considerably. While its weaknesses were obvious, on the other hand the anchoring of Farmers' Rights was positively emphasised. In addition, as a whole, it was regarded as being the expression of a worldwide distributional conflict. Considering the relationships of power not much more was to be expected. Thus, the countries of the South had achieved some smaller victories and even the existence of the International Undertaking was, according to Pat Mooney (1998: 22), regarded by the countries of the North as an attack on intellectual property rights for plant genetic resources. Just as with regard to the Commission on Genetic Resources for Food and Agriculture (CGRFA), the very existence of an international regulation system was already judged to be a success in view of the conflict structure and the relationships of forces involved.

At the same time, there were a number of other processes in which it was attempted to deepen the international consensus with regard to the treatment of plant genetic resources. Of particular importance here are the *Keystone Dialogue* (Keystone 1991) and the *Crucible Group* (1994, 2000, 2001). Both are quite different. The Keystone Dialogue was oriented towards the model of consensus-oriented negotiation systems (see Fowler 1994: 196ff.; for a critique: Görg 2003a: 175ff.). Representatives of the different conflict parties (governments both from North and South, international organisations, industry, universities and NGOs) attempted, in a long-winded negotiation, to find consensus on the central issues of dispute.

Although such a consensus – or at least a compromise – was finally found, its practical meaning was rather trivial. Nonetheless, Pat Mooney (1998: 24f.), one of the participants and a central figure in the NGO camp, considers it successful in clarifying the central conflict lines. Moreover, governments and experts from the North tempered their previous view on plant genetic resources. But even if in this sense NGO activists and the countries of the South had achieved partial success here, this dialogue forum had practically no influence on the political regulation process, and particularly not on the GATT negotiations (Fowler 1994: 203f.; on the Crucible Group see below).

The conflict structure changed considerably at the beginning of the 1990s with the conclusion of the CBD and the negotiations on the GATT. In spite of all the remaining contradictory interests within the framework of the FAO, slow progress, or at least a serious negotiation process, could be observed. With the conclusion of the CBD, however, it was feared that in view of its more ecological orientation, the distributional conflicts and the accompanying politicising of the issue could once again be overshadowed. Frederic Buttel (1992) coined the expression "environmentalisation" for this threat. And there were good reasons for such fear. During the negotiations on the CBD, national sovereignty was anchored as the central principle of access. Moreover, in the appendix, in Resolution 3 of the Nairobi Final Act, the final document of the CBD negotiations before being passed to the UNCED Conference in 1992 in Rio de Janeiro, the particular role of the plant genetic resources was recognised. However, at least two dimensions had been excluded which were of central importance for the genesis of the seed wars in the 1980s: the issue of Farmers' Rights and the status of the collections which had been set up before the coming into force of the CBD. Both topics could not be clarified and had to be left to further negotiations (CBD Handbook 2001: 287f.). While the first was judged to be an enormous step backwards in view of the conflicts and negotiations of the 1980s, the latter is highly explosive in view of the economic importance of the collected material. It is clear in international law that contractual regulations cannot be applied retrospectively. But at the same time, it was precisely these collections that had provided a decisive impulse for the seed wars.

Particularly in the countries of the North and among the public there, the CBD had contributed considerably to mystifying the socio-economic and political explosiveness of the subject by focusing more on the conservation of species and nature. Nevertheless, the fear that the subject would be depoliticised was not justified. With the CBD and the recognition of the principle of national sovereignty, the conflicts were lifted onto a new terrain and in turn had an effect in this form on the continuing negotiations within the framework of the FAO. Thus, the second negotiation body, the Crucible Group, which had been assembled after the coming into force of the CBD, attempted to sound out the new situation. This situation was overshadowed in addition by the preliminary negotiations on the GATT and the TRIPS

Agreement. Thus, the first report of the Crucible Group (1994) was particularly important because the conflict lines which characterised the international discussion in the 1990s were already contained in it in a condensed form. In contrast to the Keystone Dialogue, the Crucible Group was not geared towards consensus. In its final reports (both in its first round in 1994 and in its second round in 2000/01) the different positions on which no agreement could be reached were placed side by side. With regard to the CBD there was agreement that the role of agro-biodiversity had not been well understood in the CBD and that the unsolved problem of the collections could grow to become the greatest gene robbery since Columbus (Crucible Group 1994: 31f.).

Ultimately, the topics intellectual property rights and Farmers' Rights/traditional knowledge became central within the Crucible Group. At that time, the discussions still stood in the shadow of the GATT negotiations and the formulation of the TRIPS Agreement. Although at this point the countries of the South had not yet taken a decidedly antagonistic position to the TRIPS Agreement (see Chapter 4 on this), for a part of the Crucible Group the consequences were foreseeable. The expansion of intellectual property rights would not only put pressure on the Southern countries; it would particularly undermine Farmers' Rights and the practice of the free exchange of seeds: "Farmers have the absolute right to save seed, to experiment with exotic germplasm, and to exchange seed with neighbouring communities. To deny these rights is to cut the heart out of conservation and enhancement of plant biodiversity" (ibid.: 87). As this only expresses one of the positions represented in the Crucible Group, the core of the conflict was clearly named. While industry insisted on the necessity of intellectual property rights for the protection of its innovations, the opponents see in intellectual property rights the main danger with regard to farming practice. In between there is a third, more mixed position, which emphasises the national room for manoeuvre (above all with a view to the *sui generis* clause in the TRIPS Agreement) and rejects the resistance to intellectual property rights, patents and the TRIPS Agreement as an unrealistic maximalist position.

In these debates ecological, socio-cultural and socio-economic dimensions are closely interlinked. Ecological questions concerning the conservation of the diversity of plant genetic resources are related to the different material (agricultural) practices and connected life-styles, which in turn are overlapped by global distribution problems and economic valorisation strategies. This complex set of problems is expressed in the tensions between different international agreements and the problems of their national implementation – and also in their local effects on specific places (see Chapter 5). In the International Undertaking, local practices are recognised as important with the concept of Farmers' Rights and it is also attempted in other ways to consider the particular material characteristics of this field (which to this extent represents a complement to Article 8(j) of the CBD, see above). Seen from the perspective of the International Undertaking, both the CBD and the TRIPS Agreement bear the threat of neglecting these particular characteristics by way of commercialisation: the former by means of privileging bilateral access agreements, the latter in addition by the international securing of minimum standards for intellectual property rights. In this dispute no agreement could be achieved, either within the framework of the Crucible Group – or even later – or on the question of an appropriate regulation of plant genetic resources or in the understanding and the role of the different agreements. With the conclusion of the GATT/WTO Agreement and the TRIPS Agreement a new political momentum became apparent in the negotiations. To this extent the fear of a de-politicisation of the issue has not come to fruition.

## 3.2 The genesis of a multilateral exchange system

Before the conclusion of the CBD it was already becoming apparent that the previous foundations of the International Undertaking were in need of revision. Already in 1991 the principle of the common heritage of humankind had been replaced in Resolution 3/91 of the Commission on Genetic Resources by the principle of national sovereignty and thus the most important difference between the (then not yet existent) CBD and the International Undertaking was settled. Nevertheless, the problem of the harmonisation of the two treaties was far from being resolved. In 1993, the official negotiations on the revision of the International Undertaking began in order to bring this into harmony with the regulations of the CBD, according to the official line of reasoning. It took until November 2001 before the revised version, the International Treaty on Plant Genetic Resources for Food and Agriculture (ITPGR) could finally be concluded. Between those two dates lay more than seven years of extremely hard international negotiations, which were more than once on the brink of failure. The reasons for these extremely tough negotiations, which at the same time were the main barrier to an earlier conclusion of the agreement, were questions of intellectual property rights, the particular treatment of biological diversity in agriculture and the principle of Farmers' Rights, as well as the – closely connected – national interests of major states and blocs of states (see on the regulations of the International Treaty: Mekouar 2001; Seiler 2003). Therefore, the very existence of this agreement can be booked as a success, since with it an important international problem area – with consequences for the development strategies of the poorer countries and their food security as well as for the further development of "green gene technology" – is regulated by an international agreement. Yet even now, some of the most important issues of dispute have not been clarified, nor is it foreseeable what actual effects the International Treaty will have, especially on farming communities in poorer countries and the global structure of agriculture. In spite of the apparent success of the negotiations in the final round, a delegate made a sarcastic remark on the outcome, saying: "We do not only not know whether the glass is half full or half empty – we don't even know what liquid is in it" (personal communication to the authors).

### *3.2.1 Negotiations on the harmonisation of the CBD and the International Undertaking*

In the following section, we describe in detail a complicated and long-winded negotiation marathon as it encapsulates very well the difficulties of the entire process. The replacement of the principle of common heritage by that of national sovereignty before the conclusion of the CBD demonstrates two things. First, at the end of the 1980s and the beginning of the 1990s there was generally a strong tendency to give preference to national interests at the expense of global commons. This took place in parallel to the often evoked upswing of multilateralism following the end of bloc confrontation, which reached its zenith in the "spirit of Rio". Therefore, it illustrates clearly the differences that ruled at that time between rhetoric and reality. However, we should not simply complain about this incongruity. Instead, what is needed is an attempt to provide a closer examination of the effects the multilateral institutions actually have and in what relationship they stand to national interests. If the safe-guarding of national interests took priority even at the zenith of multilateralism, then this is a clear indication of the increasing economic appreciation of genetic resources, because for their commercial use, national sovereignty and state-political regulation are both necessary. Second, it becomes apparent, in contrast to the official statements, that the negotiations on the harmonisation of the two agreements were not necessarily hampered by the two different central principles – here common heritage, there national sovereignty. Rather, the question of intellectual property rights (FAO/1; FAO/4) and, closely connected with them, hopes of bilateral benefit-sharing and the dimension of the multilateral system of exchange (FAO/1; FAO/2; FAO/5; FAO/7) must first be named, along with the special treatment of agro-biodiversity, including the question of Farmers' Rights (FAO/2; FAO/5). Up until the end it was not clear whether, in view of the opposing points of view on these issues, it would be possible to find a compromise or whether pure power politics would prevail (FAO/1; FAO/7). A further point of continuing controversy, which must be resolved in the ongoing process of the International Treaty negotiations, is the treatment of the *ex situ* collections. Thus, some of the most important issues in the whole process have not yet been entirely resolved. This gives rise to the question of what meaning such vague provisions actually have in the eyes of the actors which enter into them as well as in their regulatory content.

The official negotiations within the framework of the Commission on Genetic Resources for Food and Agriculture (including the extraordinary sessions of the CGRFA), an Inter-session Contact Group and the FAO

Council, comprised almost 20 meetings (see the overview in the ENB 2001). In addition, there were both smaller and larger conferences which dealt with related subjects, such as the FAO Conference in Leipzig in 1996, at which with the acceptance of the Report on the State of the World's Plant Genetic Resources and the passing of the Global Plan of Action (GPA), the two most important elements of the Global System on Plant Genetic Resources, next to the International Undertaking, were established. The World Food Conference of the FAO in 1997 and the follow-up conference in 2002 should also be mentioned. The International Undertaking is only one part of a global system for the conservation and utilisation of plant genetic resources. To be sure, even the negotiations in this sub-area of biodiversity are extremely complex and in their multitude of scientific, technical, economic and legal dimensions, are only really understandable by a small circle of experts. In this area, which is concerned with the special role of plant genetic resources for international politics, the global economy and the interests of national governments and other actors (TNCs, local communities etc.), expert and specialist knowledge plays a very special role (Görg/Brand 2003). Interests of different actor groups do not exist per se; instead, they become interpreted and formulated by this expert knowledge. Nevertheless, the often contradictory or even antagonistic structural conditions of the major actors are not even remotely transcended by this knowledge, as the protagonists of this knowledge or post-industrial societies like to claim. The history to date of the conflicts over plant genetic resources shows that in spite of, and even because of, dominant and generally accepted interpretations of the problems and interests involved (e.g. the international interdependencies in the plant genetic resources), the conflicts have not at all been brought closer to a solution, and it has not even become easier to attain compromises. The genesis of a multilateral exchange system can illustrate that something different occurred: interests are reinterpreted with reference to expert knowledge and this knowledge therefore becomes part of the power struggle.

The model of international interdependencies in the area of plant genetic resources has in principle established itself as the description of the problem since the 1980s. However, despite this, there was still a long way to go from that description to the development of a multilateral system of exchange. One decisive step in this direction was taken by the International Plant Genetic Resources Institute (IPGRI, a research centre based in Rome in the vicinity of the FAO). Via various preparatory documents since 1994, the idea of a Multilateral System was gradually established (Cooper *et al.* 1994; IPGRI 1996), until in about 1997 the core idea was accepted by the majority of the parties to the conflict (see ENB 1997). Up to that time, the negotiations had been repeatedly described as preliminary. Only with this idea of a multilateral system of exchange, developed by scientific and legal experts, was it possible for the parties in conflict to aim for a compromise regarding their interests. This demonstrates the great importance of scientific expertise in such a complex negotiation process. Its role should not be misunderstood as neutral, however, nor should it be perceived as being synonymous with a transitory mode of negotiation, beginning with hard bargaining and argument and ending with constructive discussion and dialogue. The Rome-based IPGRI suggested the need for a compromise between partially opposing interests. But these interests were rooted in structural conditions which were not in any way undermined by this suggestion and consequently, the distributional conflict was not overcome, but simply reproduced at a new level. Moreover, when the idea of a multilateral system of exchange was established in the negotiations, the distribution conflict really began. Therefore, the far-reaching consensus with regard to the interpretation of the problem (international interdependencies) and the approach to a solution (Multilateral System), did not at all lead to a real consensus; instead, it brought about the reformulation of the opposing interests at this new level. Experts thus played an important role in the formulation of a "general interest" to which all the conflict parties could agree. But, they were included precisely as a part of the more far-reaching bargaining processes. It can be contended that the supposed general interest ultimately remains a "generalised" particular interest. This remains true despite the fact that it exists within the framework of a compromise between several interest groups, as we will see in the following.

From the perspective of one of our interview partners, in the negotiations following the formal agreement on a Multilateral System, at first nothing really changed. Rather, the different positions were upheld and ensured that they would mutually block one another until in 1999 the chairman of the Commission on Genetic Resources, Ambassador Gerbasi from Venezuela, presented a new draft. This draft was the starting-point for further negotiations, which took place primarily in a smaller contact group of representatives from 41 countries (FAO/7; see ENB 2001: 2). However, not only was there a dire need that the relevance of the Multilateral System for the global regulation of plant genetic resources be made clear, but also, the national delegates had to understand the system to the extent that they were able to formulate the interests of their countries, or what they perceived these to be, with regard to this proposal. In the context of these clarifications it may well be concluded that in the opinion of observers, the international negotiations almost resembled a university seminar – although this was only the case for a brief period – before the conflict over the specific regulation of access and benefit-sharing, and with it hard bargaining, exerted itself once more.

Despite the fact that ever since 1999 the most important core ideas of the agreement had been on the table, the negotiations were repeatedly threatened with failure. The core part of the revised International Undertaking was to be the Multilateral System, although both the question of benefit-sharing (and closely connected with it the problem of intellectual property rights) and the dimension of the Multilateral System were controversial. The regulation of benefit-sharing proved to be particularly difficult. Here, most of the actors recognised that the main benefit of the International Treaty was the Multilateral System itself. Facilitated access to plant genetic resources, because of the international interdependencies, constitutes a benefit for all participants – but how can this benefit be "fairly and equitably" distributed as the International Treaty demands (ITPGR 2001: Art. 13.1)? In the International Treaty itself, first, non-monetary forms of benefit-sharing are dealt with such as the exchange of information and technology (Art. 13.2(a) and (b)), as well as capacity building, as mentioned in the last chapter (Art. 13.2(c)). With all these provisions, the question arises as to whether and to what extent the different groups of actors really benefit and whether the whole process can, in fact, be described as "fair". These questions are ultimately not clarified within the framework of the CBD. Furthermore, in the International Treaty there are two additional problems: do the practices which are responsible for the production and conservation of the genetic diversity of plant genetic resources actually benefit from these regulations? This question points above all to the problem of whether Farmers' Rights are strengthened or not. The second problem is directly linked to the form in which benefits are granted: what are the expectations, first, in the special case of *monetary* benefit-sharing and, related to this, with regards to the commercialisation of genetic resources?

In reference to this second question, during the negotiations an interesting turn took place. While the Southern governments had long insisted on monetary compensation, by the end of the negotiations even the lobby associations of industry had agreed to the payment of such compensation. In line with this, a corresponding model was formulated by the breeders association, ASSINEL, and a corresponding proposal was then brought into the negotiations at the third meeting of the contact group in August 2000 by the Norwegian Government (FAO/2; FAO/7; ENB 2000a). However, the formulation proved to be a stumbling block in a double, contradictory fashion. It said that a payment to an international fund would be due if intellectual property rights were claimed based on material from the Multilateral System, and it was therefore branded a door-opener for intellectual property rights. Nevertheless, on the basis of this proposal, at the third meeting of the contact group, it was possible to formulate a compromise which at the beginning all participants (although some with reservations) could agree to. At the fourth meeting of the contact group in Neuchatel, in November 2000, a severe set-back took place, which in the opinion of observers from NGO circles brought the negotiations to the brink of failure (RAFI 2000b; Mulvany/Redding 2000; see ENB 2000a, 2000e). However, it was not the reservations of the opponents of intellectual property rights which took effect here, but very surprisingly, those of the USA and other industrial countries.

Although the proposal for a monetary benefit-sharing, compensatory system had already been accepted by the commercial plant breeders and the biotechnological industry (see FAO/2; FAO/4), it was brought into question once again at Neuchatel by a group of four countries which delegated their group's leadership to the USA. According to observers (Mulvany/Redding 2000) it was more or less the same group which had previously almost blocked the negotiations on the biosafety protocol, namely the USA, Canada, Australia and New Zealand (Gupta 1999). These "foolish four" (RAFI 2000b) not only overlooked the fact that industry was prepared to pay monetary benefit-sharing, but they also took the position that without a principle willingness to pay a monetary compensation it would hardly be possible to achieve a compromise with the Southern countries. This event is of general importance beyond the individual case. At first it may seem surprising that the USA, which is usually seen as an outspoken representative of the interests of industry in such negotiations, ignored their agreement here and in doing so risked being declared the bogey-man of any further negotiations.[4]

The reasons responsible for such a seemingly contradictory position had to do with the supposed fears this country had with regard to the general importance of this regulation on monetary benefit-sharing. The USA and the other four countries were alarmed, as they had been during many other negotiations, that certain regulations of the TRIPS Agreement would be infringed (ENB 2000e: 5). Although the US delegation had at first agreed to the compromise on this issue in Teheran, it had placed this agreement in reserve, awaiting final assent by the government. But the delegation was apparently unable to definitively impose its position in the face of the opposing interests of the trade policy-makers. Or, as other observers who were in contact with one delegate expressed it: The "Biocrats" (meaning the bureaucrats in the field of biodiversity) could not convince the "high priests of patents" (from the trade ministries) of the meaningfulness of the regulations on plant genetic resources for food and agriculture. "They couldn't get their trade people in Washington to pay attention. Those people think a plant is some kind of industrial manufacturing facility" (quoted according to RAFI 2000b). This means that the international negotiations were endangered by the internal contradictions of a central actor: the divergent strategies of the agricultural and the trade ministries were an expression of the contradictions within the US government.

We can assume that behind these contradictions are diverging appraisals of the problem and the conflict situation, above all with regard to international agreements and the representation of national interests by these agreements. While it appeared obvious that the agricultural ministries would reluctantly agree to the proposed compromise on monetary benefit-sharing if a compromise could not be reached any other way, the trade ministries feared that with a commitment to benefit-sharing, the TRIPS Agreement would be violated and a dangerous precedent would appear, threatening the case for patent laws. What we can observe in this case, therefore, are complex relationships in this multilevel system, characterised by a multitude of overlapping concerns between the national and international levels. The national level does not disappear, but is present even in a fragmented and contradictory way at the international level. This is what we mean when we refer to the concept of a "second order condensation of power relations". Moreover, at least implicitly, most actors have the expectation that the government of a country should advocate a coherent policy and behave as an undivided, single actor. Even critical NGOs often refer to this fiction, even if they know that in different ministries very different interests and relationships of power are condensed. Accordingly these power relations, condensed in national ministries, often cause tensions in international negotiations.

The uncertainties of the USA (and a few other states) were dealt with at Neuchatel when a representative of the WTO explained the regulations of the TRIPS Agreement and the dispute settlement body (DSB) of the WTO. This explanation made explicit that the on-going negotiations on an agreement between sovereign states would depend on the expertise of another international organisation. It is interesting that the latter's representative had to disappoint the expectations of her, however. She pointed out that she could not supply a legal interpretation of the International Undertaking regulations in question and that fundamentally, the competence to examine violations of the TRIPS Agreement lay with the member countries of the WTO (ENB 2000c). With regard to the relationship between the International Undertaking and WTO/TRIPS this means that it can only be explained in advance by legal expertise to a limited degree. Since here different views of the law clash, it ultimately depends on whether a member country ascertains that there has been a violation of the TRIPS Agreement and whether its view of things is confirmed by the dispute settlement body. This cannot be clarified completely in advance and the question of consistency between the two agreements continues to be unanswered (ENB 2000e: 11). Apart from this procedural problem, the WTO and its representatives are put in a political position in which they accrue a high degree of interpretational competence. There is at least a threat that in the end, the WTO's dispute settlement body will become the decisive authority in which decisions on conflicts are made.

The above description of this negotiation process illustrates that even behind the commonly shared interpretations of the problems, the distributional conflicts remained firmly entrenched. Moreover, we can observe the manner in which national and international actor constellations overlap each other, both horizontally (e.g. the relationship between the International Treaty and the dispute settlement body or TRIPS) and vertically (e.g. different US ministries and their impact on the International Treaty negotiations). One further interesting aspect is that the expectations regarding monetary benefit-sharing have always accompanied the formulation of the

Multilateral System and have been used as a door-opener for intellectual property rights. So, it appears clear that the functionality of the international fund which is to administer the monetary benefit-sharing and to transfer the money should be coupled to the granting of intellectual property rights. Even if this coupling ultimately remained without enforcement in the International Treaty,[5] the close connection between monetary payments and private property rights as an element of the valorisation of genetic resources becomes quite lucid. By contrast, the omission of intellectual property rights as part of the material to come under the Multilateral System was one of the central elements of the negotiations, as we will see in the following. They were condensed in the last round of negotiations in spring 2001 to the four words "in the form received". In order to understand their meaning, it is helpful to take a look at the Multilateral System which was finally passed at the International Treaty. It must be examined whether and to what extent this in fact represents – as is often claimed – a counterweight to the commercialisation and monopolisation of seeds.

### *3.2.2 The international seed treaty: strengths and shortcomings*

*The Multilateral System and the problem of intellectual property rights*

The Multilateral System of access and benefit-sharing to plant genetic resources as the central element of the International Treaty articulates the particular characteristics of the area of regulation and its subject-matter. The necessity of a system of its own on plant genetic resources for food and agriculture is formulated in the preamble to the International Treaty with a view to the particular nature of those resources, the erosion of genes, international dependencies and the general importance of these resources for world food security (ITPGR 2001: Preamble). With regard to these particular characteristics, in Part IV of the Agreement, the constitution of a Multilateral System for access and benefit-sharing is approached. However, the Multilateral System covers only certain plants, which are listed in Appendix I. This contains only 35 food crops particularly important for worldwide nutrition and 29 forages. In addition, the range is limited even further. It automatically includes only those varieties which are under the control of the signatory states (i.e. the states which have joined the International Treaty) and which are to be found in the public sphere. All others (e.g. private collections and commercial breeders) are only "invited" to include their varieties too (Art. 11.2); or governments should "encourage" them to do so (Art. 11.3). Within the framework of the Multilateral System a facilitated access is granted, which is explained in more detail in Article 12 and particularly in point 12.3. Facilitated access here means that access, free of charge or for minimal cost (12.3(b)), is to be granted exclusively for the purposes of research, breeding and training and explicitly not for other industrial purposes (12.3(a)). In contrast to Article 15 of the CBD, facilitated access here does not include marketing for profit but is, on the contrary, directed towards the establishment of a *public* sphere in which certain material is not to be subjected to commercial aims. The exchange in this public sphere is to be regulated by Material Transfer Agreements (MTA), finalised during the first meeting of the Governing Body of the Treaty in June 2006 (ENB 2006b).

The aim of the Multilateral System is to provide access to genetic material for public and private breeders under particularly favourable conditions. In the case of the commercialisation of plant genetic resources the International Treaty requires a monetary benefit-sharing regulated by an international fund, from which the measures of the FAO Global Plan of Action from 1996, that is for the conservation of plant genetic resources, are to be financed (Arts 13.3 and 13.4). For material from the Multilateral System the monetary benefit-sharing scheme does not serve the parties but the system itself. A monetary benefit-sharing falls due above all when the material is later no longer available without limitations, and in all other cases the recipients are only to be "encouraged" to pay. Nevertheless, the principle drawback of the Multilateral System is apparent here, namely, that such a multilateral system of exchange can only function if the most important plant varieties are included in it and if the material also continues to be freely available, that is when it is not removed from the public sphere and privatised. In both points, the Multilateral System shows considerable deficits. For one thing, the list of plants is kept to a bare minimum; indeed, even such important varieties for food security as soy beans cannot be found listed. Behind this tactic rest the interests of some developing countries, which wished to keep out of the Multilateral System those species of which they possess a considerable diversity of rural types and traditional varieties, as they hoped to attain considerably more advantages from bilateral trade. For example, in the case of the soy bean, which is among the most genetically modified plants, this strategy was adopted by China, which prevented the inclusion of soy beans in the Multilateral System due to the diversity of this species in China and the commercial interest in soy beans in the USA.[6] Brazil has repeatedly been accused of similar strategies (FAO/6; FAO/7). Despite the general criticism of developing countries against such a trend, even NGO representatives admit that the Southern countries are thus reacting in their own way to the general tendency towards commercialisation.

Another problem is connected with the already discussed question of intellectual property rights. If the commercially important varieties were to be placed under strong property protection and thus removed from the public sphere of the Multilateral System, this would in the long term undermine the substance of the Multilateral System. This danger was articulated publicly early on by NGOs with the term "grain drain". At its core this danger is also recognised in the International Treaty and subsequently, there were attempts to outline some precautions against it. The four words "in the form received" play the decisive role here. In Article 12.3(d) it is stated, "Recipients shall not claim any intellectual property rights that limit the facilitated access to the plant genetic resources for food and agriculture, or their genetic parts and components, *in the form received* from the Multilateral System" (emphasis added). What sounds good at the beginning – namely the prohibition of claiming intellectual property rights over plant genetic resources from the Multilateral System – is in the end decisively weakened. For "in the form" in which genetic resources are taken from the Multilateral System, namely as seeds, they are not capable of being patented in any case. Patentability requires as a minimal criterion novelty and invention – and both are clearly not given in the case of seeds.[7] For this reason the insertion was also controversial. It causes the clear exclusion of intellectual property rights, that is patents, to be turned unnoticed into an inclusion. As seed, in the form in which it is received, the material cannot be patented, but its genetic characteristics can. In order to counter this threat, a second insertion is made, which at first glance also excludes "their genetic parts and components" from patentability, in addition to the seeds themselves. In the process of negotiation, the two passages were highly controversial. After the US government, at the sixth meeting of the contact group in Spoleto in April 2001, insisted that the passage "or their genetic parts and components" should be placed in brackets (i.e. it is not consensual and must be further negotiated), the Group of 77 and China reacted by putting the other passage "in the form received" in brackets (see the NGO report on this at www.ukabc.org/iu4htm#3, of 3 May 2001).

In this example we can observe very clearly the contradictory interests of the developing and major industrial countries on the subject of patents on plant genetic resources. The formulation which was finally passed represents a compromise between the two positions, which ultimately really clarifies nothing and whose effectiveness is controversial. Both perspectives were incorporated within the compromise, at least formally, but the real challenge was avoided. From the point of view of the biotechnology industry it is not parts or components of plants which should be protected by patents anyway, but a technical procedure for isolating individual genetic characteristics and transferring these characteristics to other organisms (FAO/2; see also the statements of the lobby association BIO). In the opinion of representatives from this industry the new insertion to Article 12.3(d) does not change this. Perhaps they are right and it does not help in any way to combat "grain drain", nor does it work against the privatisation of genetic material from the Multilateral System.

### *The relation to other agreements*

As a result of the final compromise, the International Treaty does not present any clear conditions on this issue. Therefore, it will be finally interpreted in patent law. Additionally, of even more importance is the fact that the International Treaty is decisively weakened in relation to the TRIPS Agreement. The International Treaty would only have had a chance as an instrument against the patenting of plants if it had contained clear regulations on the exclusion of intellectual property rights. In the opinion of those actors who pushed for the exclusion of intellectual property rights, however, this ultimately political statement of the International Treaty is not as clear as they would have wished (FAO/6). At least the final adjustment has been postponed to the further negotiations and the implementation of the International Treaty (FAO/1). The relationship between the International Treaty and other agreements, in addition to the TRIPS Agreement and above all its relationship to the CBD, is not only a question of legal interpretation but also of the political weight of each agreement. Furthermore, these relationships are subject to continual change and re-interpretation. Whereas at the beginning of the negotiation process on the harmonisation of the International Undertaking the relationship to the CBD was the central focus, in the course of the following negotiations the problem of the relationship to the TRIPS Agreement moved more and more into the foreground.

Tensions in the relationship between the CBD and the International Undertaking and the International Treaty, respectively, continue to exist. This has to do, primarily, with the contradiction between bilateral and multilateral access and benefit-sharing (FAO/1; FAO/5; FAO/6), even if this tension is questioned by the industry (FAO/2). But now in the foreground of this relationship are the material differences between "wild biodiversity", on the one hand, and plant genetic resources and the particular characteristics of agriculture, on the other. Here, almost all our interview partners pointed out that the CBD – despite all the progress of recent years – does not have the necessary competences. In the CBD, among other things, the environmental ministries, which often regard agriculture in general as an enemy (FAO/5), lead the negotiations and not the agricultural ministries. This appraisal of the peculiarities of this field combines a judgement of international institutions with the responsibilities and competencies of nation-state apparatuses. With regard to the international level, this scepticism, which appears to be the expression of a general tendency, has an effect on the legal relationship of the two international agreements. Indeed, by the end of the negotiations, two alternative models for the international regulation on plant genetic resources for food and agriculture stood facing one another: a protocol to the CBD, similar to the biosafety protocol, and an agreement and an institution within the framework of the FAO. Concerning the variation of making the International Undertaking into a protocol to the CBD, the apprehension is articulated: "If it is a protocol of the CBD environment would take over agriculture" (FAO/7). Most of our interview partners therefore preferred the annexation to the FAO, which was in fact chosen (via Art. 14 of the FAO statute; see FAO/2; FAO/5; FAO/7). Nevertheless, almost all of the interviewees expressed a desire that *both* international institutions be closely related. This includes the obligation for the International Treaty to report to the Conference of the Parties of the CBD. This reference to the CBD is explained by the NGO side with the argument that the CBD is *more political* and less bureaucratic than the FAO (FAO/1; FAO/7). Interestingly enough, this viewpoint is confirmed by industry, although it comes to the opposite conclusion: the International Undertaking or the International Treaty should remain at the FAO precisely for this reason (FAO/4).

*Problems of bilateralism and the use value of genetic resources*

While at the beginning of the 1990s the fear of depoliticisation via "environmentalisation" was prominent, ten years later the tide has turned: the conflicts within the framework of the CBD are now regarded as more political and more explosive by both sides (NGOs and industry). The dangers of bilateral mechanisms in the exchange of genetic resources, privileged by the CBD, continue to be pointed out, however (FAO/1; FAO/6). Particularly, threats towards commercialisation and privatisation and the completely different power resources of those TNCs and indigenous communities connected to bilateral negotiations are mentioned. Altogether, it was clearly recognisable that in the sphere of plant genetic resources, a monetary benefit-sharing scheme was regarded with considerably more scepticism than in the context of the CBD. Just as on the NGO side the "hype about benefit-sharing" has been publicly denounced (GAIA/GRAIN 2000), in the interviews, too, there was clear criticism of the fixation on monetary benefit-sharing schemes and the accompanying commercialisation of genetic resources (FAO/1; FAO/5; FAO/7). Unrealistic expectations with regard to bilateral benefit-sharing are, according to this view, in addition to intellectual property rights, one of the main obstacles in the negotiation process (FAO/1). One representative of an international organisation declared the confusion of "usefulness" and "value", that is of use value and exchange value, to be one of the main problems (FAO/5; similarly FAO/7; a literal differentiation between *use value* and *market value* is to be found in FAO/1). While plant genetic resources have a high use value for agriculture, for breeding and for food, it is argued that their economic exchange value is often overestimated, particularly with regard to bilateral access agreements. This has negatively influenced the tactics of the Northern countries as well as some of the Southern ones, for the latter now expect more advantages from selling their resources than from implementing the Global Plan of Action (FAO/5; FAO/6). In contrast to this, representatives of industry have no problems with a monetary benefit-sharing scheme. As we have seen, this is also true of payments to an international fund in the case of the commercialisation of genetic resources. Representatives of the biotechnological industry propagate a win–win situation if intellectual property rights are also accepted (FAO/4).

From the point of view of the actors in the seeds area, the relationship to the CBD is therefore not free of tensions. They refer to the fact that the

CBD favours a particular model, namely that of bilateral access agreements with their orientation towards a monetary benefit-sharing scheme. This model has also gained importance within the framework of the plant genetic resources and the International Undertaking with consequences for the negotiation procedures and for the agreement itself (among other things with regard to the limited range of the Multilateral System). By contrast, with regard to national sovereignty, in principle no problem is seen, since it can be used by states in different ways. National sovereignty can be a door-opener for privatisation (as in the CBD), but it is not necessarily so, because it could also be used in order to create a Multilateral System (FAO/1; FAO/5). Therefore, their effects are judged as being very ambivalent, particularly on the part of the seed NGOs. A legal framework at the national level could protect the actors who are most vulnerable, that is the rural communities (FAO/1). Nevertheless, no plea is made for all controls to be left in the hands of national governments. A "privatisation" in favour of local communities is even demanded in order not to endanger the use value of biodiversity (FAO/1). In this case, privatisation is not understood as commercialisation or in the sense of private property rights, but in the sense of a local common. In addition, the difference between sovereignty and property is pointed out (FAO/5) and it is emphasised that national sovereignty does not mean real independence. A national agricultural system today is simply seen as nonsense (FAO/7), since no-one can escape the international interdependencies.

It is therefore too simple just to contrast the bilateral CBD model with the multilateral model of the International Treaty. There is no clear dichotomy between the two because the latter is permeated completely by the bilateral model in its genesis, its structure and its problems. The extent of the Multilateral System and some of the most important regulations of the International Treaty can only be understood if the practice of bilateral access agreements, privileged by the CBD, is kept in mind. The hope of bilateral benefit-sharing was the reason why only a small share of plant genetic resources was in fact covered by the Multilateral System. Furthermore, it is also held responsible for the fact that the real benefits of a Multilateral System and the responsibility regarding the conservation of the diversity of plant genetic resources (which is laid down primarily in the Global Plan of Action) is not really recognised by the governments (FAO/5). The functioning of the Multilateral System, established by the International Treaty, is not at all so clearly defined that it is really able to fulfil even this limited task of securing facilitated access and the free exchange of plant genetic resources in a public sector. Representatives from the NGO sector see the main problem as being whether or not it will be able to establish itself as an alternative model or whether the tendency towards commercialisation will in fact win through. Here – because of the central role of intellectual property rights – ultimately, the relationship to the TRIPS Agreement and to the WIPO will be decisive.

*Negotiations in the "shadow of hierarchy"*

As we have seen, the relation of the International Treaty to the TRIPS/WTO Agreement was an immanent problem in the negotiation process. We can state, therefore, that the relations between environmental and trade agreements are not only external in character, following their coming into force. At least in these cases the relationship between the new agreements and the TRIPS/WTO formed part of the discussions during the negotiation process, thus becoming part of the formulation of the agreements. Corresponding regulations even belonged to the most controversial elements in the genesis of the International Treaty. As we have seen, this concerned the question whether a binding monetary benefit-sharing scheme could be linked to the granting of intellectual property rights or whether this was an infringement of TRIPS. The answer in this case was made dependent on the interpretation of the parties to the agreement and here, it was the dispute settlement body of the WTO that was granted autonomy in making a final decision. The second point of controversy remained unsolved, literally until the last minute, and concerned the questions as to whether patents on plant genetic resources should be prevented, and if so, whether they could be prevented, with the aid of the International Treaty.

The interpretation of the TRIPS Agreement plays a central role here, particularly the *sui generis* clause of Article 27.3(b). It allows exceptions to the obligations imposed by the TRIPS Agreement on the national formulation of strict intellectual property rights standards, if a system of IPR protection "of its own kind" exists. Nonetheless, the exact nature of a *sui generis* system of intellectual property rights remains to be clarified, as does the still prevalent controversy surrounding whether or not existing or yet to be developed legal systems could actually do such a system justice. Particularly relevant here is the International Union for the Protection of New Plant Varieties (UPOV), which is recognised as a *sui generis* law but is rejected by its critics as being too similar to patent law (Leskien/Flitner 1997). By contrast, a draft law of the Organisation for African Unity (the OAU model law; see Ekpere 2000) is intended to do justice to the particular characteristics of agriculture in Southern countries (Egziabher 2002). However, from the point of view of the World Intellectual Property Organization (WIPO) it contradicts the requirements of the protection of intellectual property rights precisely for this reason. In the next chapter we shall examine the question in more detail as to the latitude which follows from the TRIPS Agreement and the developments within the framework of the WIPO, and what momentum can be observed there.

There is another dimension concerning the relationship of different agreements. Usually it is stated in the legal text that the agreement should be "mutually supportive" of other agreements, that they should not contradict any rights or obligations from other agreements, and that neither should they establish any hierarchy to other agreements – in other words, they should not be subordinate to the WTO Agreement (ITPGR, Preamble). What this means in practice is something which remains to be seen.[8] At the moment there are diverging appraisals of the relationship between the International Treaty, on the one hand, and the TRIPS Agreement, on the other. There are optimistic expectations that the International Treaty could establish itself as a technical explanation of the concrete problem area, in this case the treatment of plant genetic resources in food and agriculture (as is hoped for the Biosafety Protocol with regard to the treatment of GMO; see ENB 2000a: 12). In the final analysis, this expresses the hope that the International Treaty could be referred to as an authority for a binding definition in controversial cases. However, there are fears that the International Treaty will be subordinated to the TRIPS Agreement in the important issue of intellectual property rights: because of the link between a monetary benefit-sharing scheme and intellectual property rights, it is seen as being a door-opener for patents on seeds and plants in general (FAO/6).

Ultimately, much will depend on how the Governing Body, the controlling organ of the International Treaty, works and what decisions it will take after its constitution. The issue will above all concern what stocks are in fact designated to the Multilateral System and how the system treats those private actors, in other words, the commercial breeders and the biotech industry, which do not bring their stocks into the Multilateral System (Seiler 2003). Even if NGOs welcome the existence of this agreement as a whole and have put considerable effort into lobbying for its realisation, most of them remain sceptical with regard to the question of whether their main objective – to counteract the commercialisation and privatisation of plant genetic resources – could in fact be achieved. An important point in this context is the question: to what extent will it be able to exert itself against the TRIPS Agreement? The danger exists that the weak and imprecise text could result in an insidious transformation of the agreement (unnoticed) into a "crop genetic trade fair" (GRAIN 2001b).

### *3.2.3 The strategies of actors and the relevance of the international regulation system*

The central conflict during the negotiations on the International Treaty concerned the limitation of commercialisation in this partial area of biodiversity. Whether this agreement, in view of the bilateral framework of the CBD which is directed towards commercialisation and of the strengthening of the valorisation of genetic resources by the TRIPS Agreement, can in fact set such limits is highly questionable. A further weakness of the International Treaty stems from a rather sceptical viewpoint: the very weak anchoring of Farmers' Rights. The legal dimensions of Farmers' Rights have never been clearly stated or, indeed, defined before. Therefore, the International Treaty represents the first time in which Farmers' Rights have been formally endorsed by a legally binding instrument at the global level.

Moreover, with international recognition and the national anchoring at least an important step towards legal recognition has been made (Mekouar 2001; Correa 2003). According to these provisions, the rights of informal innovators (traditional farmers), should be equated with the rights already granted to formal innovators (plant breeders, Mekouar 2001: 21). Article 9 recognises the contributions of local and indigenous communities to the conservation and development of the diversity of plant genetic resources (Art. 9.1) and advocates the protection of traditional knowledge, fair benefit-sharing and participation in the political decision-making process (Art. 9.2 (a)–(c)). Nonetheless, these regulations do not go beyond general principles. The responsibility for Farmers' Rights is generally seen as lying with the national governments (Art. 9.2) and in their practical meaning they are explicitly subordinated to national legislation (Art. 9.3). This limits the relevance of Farmers' Rights in a double fashion: they can scarcely be used as a lever in conflicts with national governments; and they are not an adequate counterweight to the advancement of intellectual property rights, particularly in the area of the International Union for the Protection of New Plant Varieties (UPOV).

The weakening of the concept of Farmers' Rights has been appearing in the negotiation process for a long time. Even critical NGOs neglected to take up the issue again in order not to further endanger the negotiations (FAO/1; FAO/7). This is the expression of a fundamental problem: because of the general importance of plant genetic resources for food and agriculture, both NGOs and some of the contracting parties were primarily interested in successfully negotiating an agreement in this field. When the negotiations were threatened with failure at a certain point, on several occasions, the internationally active NGOs began campaigns in order to prevent this failure (see Mulvany/Meienberg 2000; Buntzel-Cano 2000; Berne Declaration 2001). Therefore, a compromise even with the hardest opponents seemed to be the top priority. Without the JUSCANZ Group an agreement would not have made much sense, because this group contains the most important producers and the largest exporters of agricultural products. In view of their massive strategies of obstruction the question arose whether the USA, Australia and several other countries were prepared at all to join the International Treaty (FAO/1). After the USA at first did in fact abstain from the International Treaty, in November 2002 it signed the treaty, much to everybody's surprise. It is less clear what the motives for this change of mind were. The US ambassador to the FAO declared that the USA had meanwhile recognised "the wisdom of the treaty" (according to Associated Press, 6 November 2002). Apparently, the US government considers the International Treaty to be less dangerous than it had at first thought, or it preferred to be on board, as in the case of the CBD, in order to be able to pursue a more effective strategy of hindrance from that standpoint.

A further problem is closely connected with this: the divergent appraisal of vague and unequivocal phraseology, which is also criticised by industry

(FAO/2; FAO/4). The issue here is primarily that of binding and clear, that is calculable, regulations concerning benefit-sharing, in which differing interests within industry also play a role (FAO/2). Thus, the conventionally working, smaller breeders appear to assume that they will not have to fall back on new material to a great extent, but will be able to work with the existing collections. For them benefit-sharing is therefore relatively unimportant. For some industrial countries the wish for more specification is, by contrast, an important instrument to raise the negotiations to a more technical level at which only the consistency of legal details is negotiated and no longer the differences of interests which are behind them. This strategy was applied by the industrial countries particularly in the negotiations on intellectual property rights (see the next chapter on this).

For the representatives of the NGOs, by contrast, it is, in principle, clear that the lack of clarity in the regulations of the agreement shifts the problem onto other terrains and can thus be used as a tactical element – sometimes to their advantage – if provisions, which they consider better, could not be integrated into the Treaty (FAO/1; FAO/7). Individual regulations and even the International Treaty as a whole are regarded as important in an instrumental way: in order to slow down the process of privatisation and to strengthen the interests of small farmers. In this sense, obligations should of course be as specific as possible in order to open up clear negotiation options against these dominant processes. But, since in the final analysis all regulations represent compromises which evolve according to the respective relationships of power and situational constellations of strength, many actors have no illusions about the results which can be achieved or their practical consequences. To a certain degree, the individual regulations themselves are not necessarily the real goal. They are, rather, more or less partial results which can be used as instruments in further negotiations, and even on other terrains and in other processes.

At least some NGO representatives show an awareness of the international regulating system and its momentum, which can be described as exceptionally realistic, realistic in terms of Poulantzas' theory of politics and law. According to this, international regulations are important for opening-up the strategic opportunities of different actors and in particular, the rights of weaker interest groups. They are not, however, an expression of a real general interest, because the contradictory individual interests and the corresponding relationships of power are not abolished. Other NGOs, particularly from the field of the environment, and interestingly, even representatives of industry, fabricate a contrasting hope of gratifying a real general interest, although with differing strategic intentions. Thus, a representative of industry articulated the continuing hope of a win–win solution for the regulation on benefit-sharing which would do justice to all interest groups – if both intellectual property rights and patents were accepted (FAO/4). In contrast to the position of many NGOs in the field of the environment, this position is not at all naïve, as the representative of industry is quite clear about the fact that intellectual property rights represent the most controversial and still most contested aspect of the entire negotiation process. But he assumes that in the struggle for hegemony, in spite of the individual interest involved, his position has a good chance of establishing itself as the general interest. The representatives of environmental NGOs, by contrast, are in many cases integrated into such a compromise as subordinate actors.

## 3.3 Structural selectivities and the limits to the multilateral exchange system

How little international treaties really have to do with the establishment of a global general interest can be seen in the negotiations on the International Treaty. As some observers confirmed, the negotiations have, for a long time, been moving away from a point of common orientation to problem solutions (conservation of genetic diversity on plant genetic resources, food security) towards the direction of a distributional conflict over valuable economic resources (ENB 2000e: 12). As figured out, a widely shared view of the problems has crystallised as the background to the negotiations. The further design of the institutional system and the development of the strategies referring to it are more the outcome of hard bargaining than of the argumentative clarification of the problems involved. Against the common background, diverging interests emerge all the more clearly. Such continuing contradictions of interest are usually accepted in the negotiations if they are about the interpretation of legal questions and it emerges "that there will be as many interpretations as there are lawyers" (ibid.: 11). But the fact that they touch the interpretation of the problem as a whole and affect all the individual problems in one way or another is quickly forgotten. Closely connected to this is the fiction of a general interest created by international institutions in a similar way in which it is supposed to be created by the nation-state. Through this fiction, or what could be termed the state fetish, both the continuing existence of contradictory, partly antagonistic interests and the structural selectivities of the institutional system are obscured. This is precisely what forms the character of domination of political institutions in the process of the internationalisation of the state. Political institutions secure the reproduction of contradictory interests and the connected relationships of power. In two closely connected fields it can be demonstrated which structural selectivities are in fact included in international treaties and how the International Treaty functions in the sense of the valorisation paradigm: in the fate of the existing *ex situ* collections and in the question of how and to what extent plant genetic resources are to be treated as a global common.

Since the conclusion of the negotiations on the CBD, it has become clear that the fate of the existing collections, above all those stored in the International Agricultural Research Centres, represents one of the most difficult problems for the further treatment of plant genetic resources. In 1994 these collections were placed under the auspices of the FAO and given to the Research Centres in the form of a trusteeship. A decisive issue is whether and to what extent these collections will in fact remain in the public sector. In addition, there is the question of compensation or of benefit-sharing with the original countries of origin. In the opinion of critics, these questions are not satisfactorily regulated in the new International Treaty (ITPGR 2001: Art. 15). With regard to the first point, there is the same problem as with the International Treaty as a whole: protection against intellectual property rights is only offered for material "in the form received", even if these four words are not explicitly mentioned in the article on the International Research Centres collections. What makes the issue even more burning is the suspicion that some International Research Centres are themselves prepared, in the future, to claim intellectual property rights over their own research results (GRAIN 2002a). This is probably understandable as a reaction to the advance of patents on genetic resources – why should public research institutes not attempt to protect their own results? But in this case it could cause the entire system to collapse, or to make a 180 degree about-face on its original objectives. The erosion of the Multilateral System would then only be a question of time, and in particular, the most economically valuable material would gradually disappear into the private sector. In addition, the International Treaty would even cushion this process by lending it legitimacy.

This becomes even clearer when we ask the question to what extent the International Research Centres belong to the public domain and what are the actual characteristics of this domain. To merely promise that they will exclusively serve general welfare, the conservation and further development of plant genetic resources and therein, the securing of world food, is not enough, particularly when the rights of the areas of origin and the problem of Farmers' Rights are simply excluded. As the NGO GRAIN has stated: "By unilaterally declaring all 500 000 designated accessions as 'public domain', the CGIAR [Consultative Group on International Agricultural Research, the authors] in particular has severed whatever relationship farmers and indigenous peoples may consider that they have towards those materials" (ibid.).

Here a certain interpretation of the problem – we need the *ex situ* collections of the International Research Centres in order to be able to confront the problem of food security – is subtly re-functionalised into the construction of a general interest. If this is examined more closely, however, it is plain to see that doubts are growing with regard to the generality of this interest. Moreover, the claim to generality even works in the opposite sense by privileging commercialisation and private interests, that is it represents a structural selectivity towards specific interests. While the official interpretation of the public domain emphasises that no exclusive property rights to the material may be claimed (Correa 2003), the contradictions between different legal systems and in particular the difference to collective property rights is not taken into consideration (GRAIN 2002a). In this way, under the problematic definition, "We have to protect the public domain for the security of world food", in a de facto process, a reverse tendency gains the upper hand, leading to the replacement of collective property rights systems by a public domain and the construction of a global commons, which in fact favours the privatisation of research results with the aid of intellectual property rights. Even if the International Agricultural Research Centres do not claim these private property rights themselves, this tendency could win through, because the Research Centres, according to the International Treaty, are not obliged to control the further use of material from their collections. What the receiver does with it remains outside their authority. In the final analysis, we cannot therefore speak of a satisfactory solution. Rather, GRAIN appears to hit the nail on the head with its question: is the FAO-CGIAR system "biopiracy by another name?" (ibid.)

Beyond the treatment of the collections, the question arises of what in fact is to be understood by the public domain and to what extent this represents an answer to the problems in the treatment of plant genetic resources for food and agriculture. In the run-up to the Rio+10 summit in Johannesburg in 2002 an initiative for a "treaty to share the genetic commons" was formed (RAFI 2001b). This initiative, in contrast to others, is completely awake to the fact that global common goods can only be meaningfully established if intellectual property rights are explicitly excluded to parts of the commons. That is basically a tautology since it only repeats the definition of a common good, according to which no-one can be excluded from the use of the good. Indeed, precisely for this reason it would seem that this demand has a contra-factual status in view of the present situation: the demand for plant genetic resources as a global common is identical to the demand for the prohibition of intellectual property rights on plant genetic resources. In view of the real relationships of forces, however, this demand will not be realisable, at least not in the foreseeable future. Therefore, the threat exists that via the de facto definition of a general interest, the powers for the privatisation of genetic resources will be legitimised or even be strengthened, because the underlying processes are concealed and the necessary demand for a prohibition of intellectual property rights is ignored due to power relations.

The actual effects of the International Treaty cannot yet be completely estimated. The Treaty is still too recent. Many important elements will, in addition, depend on the further design of the Governing Body. Therefore, in conclusion, only the fundamental problem of nation-state and inter-state regulation processes will be elucidated. On the one hand, there is official recognition of the fundamental problem that the conservation of agro-biodiversity is of decisive importance for the security of world nutrition. On the other hand, we can observe a tendency away from the common heritage principle and towards national sovereignty and intellectual property rights.

The solution to the problem is defined as the general interest and delegated to state authorities. It is precisely this solution, however, which plays into the hands of the valorisation paradigm. It is therefore not at all adequate to contrast the Multilateral System of the international treaty with the bilateral system of the CBD. The precise formulation of the Treaty cannot be understood without the system of the CBD – including its central elements: nation-state sovereignty and bilateral access agreements. The provisions of access and benefit-sharing of the CBD, therefore, form the framework in which not only by far the largest part of plant genetic resources are traded (only a relatively small share are subject to the Multilateral System), but which also contains the decisive conditions for the Multilateral System, that is the tendency towards the commercialisation and privatisation of genetic resources, connected with intellectual property rights. Even the apparently global general interest which is anchored in the Multilateral System of the International Treaty does not represent, as we have seen, a really binding instrument against the privatisation and commercialisation of plant genetic resources, but can even support them. This logic can easily be recognised in the latest publications of the Crucible Group (Crucible Group 2000 and 2001). Despite the scepticism of several members of the group (and against the judgements of mainstream political science) they accord the national laws and the latitude for action of nation-states a decisive role in the shaping of this field. But expectations regarding the creative power of the state are not at all free from contention. Doubts are articulated as to whether the complex socio-political questions can be dealt with by purely legislative measures (Crucible Group 2000: ix). Moreover, the problem is mentioned of whether or not the nation-state options for the regulation of bilateral negotiations, which are in the foreground here, are actually the problem: "In fact, many members of the Crucible Group argue that bilaterally oriented national access and intellectual property laws will ultimately have a negative impact on the resolution of these very issues" (Crucible Group 2001: xiii).

Apparently, even in this field bilateral negotiations are not at all marginal or even excluded, but form the centre of the problem. In addition, recourse to the (nation-)state as a regulating instance prevails, even where its inadequacy has long been obvious. This can be taken as a clear indication that even in complex international problem areas the state form continues to be relevant. First, the nation-state remains in power as a fetish from which a regulating effect and the construction of a general interest is expected. Furthermore, national and international regulations are not only a condensation of relationships of forces. Rather, these regulations can only develop their real strength vis-à-vis the contradictory interests which continue to reproduce themselves if they take on a specific *form*, namely the state form. Only when they take on this specific materiality or a relatively independent political form vis-à-vis the individual actors, can they develop their effectiveness in the regulation and reproduction of antagonistic relationships.

# 4 Politicising intellectual property rights

## The conflicts around the TRIPS Agreement and the World Intellectual Property Organization

In the previous chapters, the importance of the agreement on *Trade-Related Aspects of Intellectual Property Rights* (TRIPS) was frequently mentioned. Without exaggeration, this sub-agreement of the WTO can be described as the most controversial international agreement of the second half of the 1990s. While it was hardly known when it was signed within the framework of the GATT negotiations and its effects could only be estimated by a few experts, it quickly advanced to become one of the main points of criticism of the mechanisms and the effects of neoliberal globalisation. The TRIPS Agreement is so important not only because of the subject-matter which it regulates or because of its own objectives, but also because of its effects on other international treaties. This chapter explores the contested character of intellectual property rights (IPRs) and their institutionalisation on the international scale. It begins with an outline of the creation of the TRIPS Agreement and of the tensions and contradictions which are inherent in it (4.1). Second, we analyse the ways in which Southern governments, NGOs, social movements and indigenous organisations have politicised these contradictions since the late 1990s (4.2). This is followed by an examination of the growing importance of "scale jumping" strategies, which for example take on the form of bilateral trade agreements (4.3). Fourth, we explore the "forum-shifting" from the TRIPS Agreement to the World Intellectual Property Organization (WIPO) (4.4), before, in a fifth step, turning to the conflicts which have recently emerged within the WIPO itself (4.5). The chapter ends with some more general observations on multilateralism and hegemony (4.6).

### 4.1 Shaping an international regime of intellectual property protection: evolution and contradictions of the TRIPS Agreement

The TRIPS Agreement was negotiated during the Uruguay Round of the GATT which started in 1986 in Punta del Este (Uruguay), and concluded, along with the latter, in 1994. It contains internationally valid minimum standards for the protection of intellectual property. Besides the GATT and the GATS, which regulate international trade in goods or services respectively, it is one of the three central multilateral agreements that fall under the umbrella of the WTO. The socio-economic background of the TRIPS Agreement stems from the fact that, during the last decades of the twentieth century, the production and appropriation of immaterial goods has significantly gained importance in the process of the valorisation of capital. This increasing "knowledge intensity" of the economy has strong implications for the issue of property rights: whereas the generation and initial industrial application of new knowledge is a complex and costly process, the imitation of knowledge-intensive goods can be quite easy. Examples include the production of generics in the pharmaceutical sector or the copying of computer software. Hence, the rise of the "knowledge society" has also given rise to demands for a stronger protection of intellectual property:

> Due to the tendential de-materialisation of production and the increasing share of immaterial contributions to new products, from the point of view of industry and the political actors representing its interests, there is an imperative necessity to protect the knowledge and technological parts of these products via intellectual property rights, in order, in this way to once again lengthen the time-period necessary for product imitation, which is permanently becoming shorter, and to amortise the increasing expenditure on R&D.
>
> (Seiler 1999, our translation; cf. Chapter 1; Jessop 2003; Nuss 2002).

In most industrialised countries, intellectual property had already been extensively protected before the coming into force of the TRIPS Agreement in 1995. Patents, copyrights and rights on trademarks, industrial designs or geographical indications (e.g. "Champagne") guaranteed a high level of intellectual property protection. On the international scale, the protection of intellectual property was organised by the WIPO, an organisation which had been founded in 1970 and which acted under the umbrella of the UN since 1974. However, the agreements administered by the WIPO, among others, the Paris Convention on the protection of industrial property and the Berne Convention on the protection of literary and artistic works, were of a sectoral nature. Furthermore, they were not generally binding, but covered different groups of participants.[1] Above all, the WIPO had no sanctioning mechanism which would have made it possible to prosecute a claim for intellectual property protection internationally. From the point of view of Northern business, this became a problem the more the knowledge-intensity of production, as well as its internationalisation, progressed. In the course of these developments, the markets and genetic resources of developing and threshold countries gained importance, where the protection of intellectual property was far less extensive than in northern countries:

> In most developing countries, IP regimes were geared towards supporting national economic development, and therefore offered low levels

> of IP protection with weak enforcement mechanisms. Western-based MNEs [Multinational Enterprises, the authors] began complaining about the loss of hundreds of billions of dollars every year to piracy in developing countries.
>
> (El-Said/El-Said 2005)

It was against this background that the interest of powerful economic forces in establishing internationally enforceable IPRs increased.

However, the institutionalisation of strong IPRs on the international scale did not simply follow the demands of Northern business. On the contrary, it was a contested process, whose result was in no way predetermined. Even in the 1970s, it had been an open question who would gain the power of definition over the subject-matter of IPRs: the Group of 77 had succeeded in placing intellectual property rights on the international political agenda and discussing them in the context of the problems of development. The existing IP protection was criticised as a tool of protectionism which prevented the developing countries from closing the technological gap between themselves and the industrial countries, and which, therefore, reinforced the unequal development of North and South. With the support of the UNCTAD (United Nations Conference on Trade and Development), the Southern countries initiated a process for the revision of the Paris Convention on the protection of industrial property. This failed, however, due to the irreconcilability of the positions of the developing and the industrial countries (see May 2000: 83ff.). Against the background of this failure and the rising demands for stronger international IPRs in the course of the post-Fordist transformation, the perception of the problem shifted: the development policy dimensions of the protection of intellectual property moved into the background behind the question of legal security for knowledge-intensive enterprises, and the protection of intellectual property was increasingly regarded as a decisive condition for the valorisation of capital. In other words, the discursive claims were staked anew: a "universalising discourse of knowledge as property" evolved, which delegitimised the competing forms of the conceptualisation of knowledge as they had been pushed forward, for example by the governments of developing countries in the 1970s (ibid.: 85). The issue of intellectual property protection which Southern governments had tried to subordinate under development goals was (again) framed by "mercantilist policy ends" (May/Sell 2006: 204).

During the first half of the 1980s, Northern governments succeeded in establishing the protection of intellectual property as a subject of the GATT negotiations. In doing so, they acted in the interest of their influential information technology, pharmaceutical and seed industries. If there was resistance on the part of the developing countries, this was broken with bilateral negotiations and the threat of imposing penal duties or of withdrawing trade facilities. From the beginning of the 1990s then, the issue was no longer that of whether an agreement on the protection of intellectual property would be reached in the negotiations under the Uruguay Round of the GATT, but only, as Drahos puts it, of

> how far an agreement on intellectual property would deviate from the blueprint that had been provided to negotiators in 1988 by Pfizer, IBM, DuPont and other members of the international business community in the form of a draft proposal entitled *Basic Framework of GATT Provisions on Intellectual Property: Statement of Views of the European, Japanese and United States Business Communities.*
> (Drahos 2001: 9; cf. Drahos/Braithwaite 2003; May/Sell 2006: 161ff.)

The institutionalisation of the protection of intellectual property within the WTO can be judged as being quite paradoxical: "It is an aberration that TRIPS is located in a trade organisation whose main functions are supposed to be the promotion of trade liberalisation and conditions of market competition, whilst TRIPS is protectionist and curbs competition" (Khor 2001: 8).[2] However – and this was decisive for the "forum-shifting" (Drahos 2001: 5) from the WIPO to the WTO – the latter has the advantage over the former from the point of view of the dominant forces in that it not only widens the geographical scope of IPRs but also possesses the means to control their implementation and, in the form of the dispute settlement body, to enforce their compliance. In addition, it offers the opportunity to use trade issues as leverage to establish high IPR standards where they have not existed before. May and Sell summarise the features of the TRIPS Agreement compared with the weaker regime of the WIPO as follows:

> (1) to bring all members of the WTO under the same set of principles and minimum standards for the recognition and protection of IPRs; (2) to give this governance regime "teeth" by applying the WTO's dispute settlement mechanism to any international disputes regarding the undertakings within TRIPs; and (3) by linking IPRs to the wider issues of international trade at the WTO, to make significant inroads into the hitherto sovereign ability of countries to establish, govern, and regulate intellectual property in response to perceived national political economic priorities.
> (May/Sell 2006: 175; cf. Byström/Einarsson 2001; Correa 2000; Seiler 2000)

The TRIPS Agreement was a victory of dominant economic and political forces from Northern countries. However, with its establishment, conflicts over the international standards of the protection of intellectual property did not come to an end. This is largely due to four tensions and contradictions which are inherent in the concept of "intellectual property" in general and in the TRIPS Agreement in particular. The *first* contradiction affects the producers of technological knowledge themselves. The more rigid the protection of intellectual property, the more difficult it is for research to build on patented discoveries: private or public research institutions which attempt to develop new knowledge or products are forced to first try to obtain licenses from firms which possess the patents for the products and processes needed for research. This is not infrequently beyond the financial capacities of the institutions involved and therefore can block the research plans or cause high transaction costs. Of course, the disadvantages which result from this contradiction do not affect all producers of technological knowledge equally, but are divided unequally along several conflict lines: South vs North, public vs private, small enterprises vs big corporations and so on. Thus, IPRs may support knowledge monopolies and actually act to hinder precisely that which they are said to promote: socially useful innovations.

> Patents on genes, for instance, restrict their use as tools to identify possible therapeutic molecules and therapies. Access to these tools, and hence progress of science, may be slowed down, particularly in developing countries and in public research institutions, due to the need to obtain multiple licences and the escalation of research costs from license fees.
> (Correa/Musungu 2002: 19f.)[3]

A *second* contradiction lies in the fact that, through the strengthening of intellectual property rights, the advantages of achievements which more than in other spheres are the result of social processes are privatised. The private appropriation of the means and results of a socialised production is one of the constitutive characteristics of capitalism, but in the latter's "knowledge society" phase it becomes of particular importance. As we saw in Chapter 1, the production of knowledge is a highly socialised process, which makes it difficult to allocate the parts of an intellectual achievement to particular actors. "The increasing socialisation of knowledge production in networked economies makes it hard to distinguish legally between the intellectual property of different firms – let alone the individual knowledge workers – as a basis for allocating the returns to innovation" (Jessop 2003: 105). Within the context of biodiversity, this problem is illustrated particularly in the issue of "traditional knowledge": for generations, local and indigenous communities have contributed to the conservation and further development of biological diversity. The knowledge they have accumulated in this process is essentially collective and thus cannot simply be subordinated to a regime of private IPRs. The TRIPS Agreement ignores this fact. It strengthens the protection of private property which is "scientifically" achieved by individual actors and does not have a concept of "traditional" collective forms of generating and using knowledge or, to put it more generally, of the social character of knowledge generation. Consequently, it does not recognise, let alone protect these forms. There is an "inherent mismatch between the protection that was created for finite, inanimate objects coming out of industrial activity, and the flowing, mutable and variable properties of biological materials and associated indigenous knowledge" (Sahai, cited in May/Sell 2006: 196f.; cf. Hoering 2002). The universalisation of individualised private IPRs under the TRIPS Agreement increasingly jeopardises collective practices of knowledge generation. It strengthens the rights of private firms which use traditional knowledge as a key with which to access genetic resources and their commercial valorisation. As a result, indigenous communities may find themselves in a situation where their own practices of using genetic resources are restricted because private firms are granted property titles for an "invention" which is essentially based on the very knowledge and practices of these communities (see May/Sell 2006: 194ff.). Thus, the global regime of individualised private IPRs, as it is institutionalised in the TRIPS Agreement, largely builds upon collective practices and at the same time threatens them in their existence. In this way, it is in danger of undermining, in the long term, the conservation of biological diversity and thus its own foundations (see Delgado Ramos 2001: 482).

*Third*, besides strengthening the asymmetric relationships between different groups of *producers*, a rigid protection of IPRs as institutionalised in the TRIPS Agreement also disadvantages certain groups of *users* of knowledge and technology. On a global scale, these are the developing countries. Because of their often less-developed infrastructure for research and development compared to the industrial countries – all developing countries together account for only 4 per cent of world expenditure for research and development (Correa 2001) – they are dependent on technology transfer from the North. Strong IPRs make this transfer difficult and expensive, especially from the point of view of the least-developed countries (May/Sell 2006: 170). Furthermore, they threaten to undermine their access to medicine, their opportunities for the production of generica or their food security.

A *fourth* area of tension is marked by the relationship between the TRIPS Agreement and other institutions which regulate access to genetic resources, above all between the TRIPS Agreement and the CBD. The latter propagates the protection of biodiversity through its commercial use[4] and proves, in this respect, its compatibility to the former. However, there are a number of regulations in both agreements which, at least in their implementation, could come into conflict with one another: whereas the TRIPS provisions on patents, as well as on plant and animal variety protection, strengthen the ownership rights of individuals, the CBD provides for the protection of (collective) traditional knowledge and practices, for the protection and the sustainable use of genetic resources, for the "national sovereignty" over these, as well as for a sharing of benefits which arise from their commercialisation and for a technology transfer to Southern countries (cf. Chapter 2; CIEL/WWF 2001; Plahe/Nyland 2003: 33f.; Thomas 2002).

These inherent tensions and contradictions of the intellectual property issue in general and the TRIPS Agreement in particular have become more manifest since the late 1990s when Southern governments as well as NGOs, social movements and indigenous communities began to articulate their interests more emphatically. The result is a politicisation of the TRIPS Agreement. This is what we shall turn to now.[5]

## 4.2 Intellectual property rights and development: politicising the TRIPS Agreement

There are several factors which have contributed to the politicisation of the TRIPS Agreement. The *first* one is the review process of the controversial paragraph 3(b) of Article 27 of the Agreement which began in 1999. Article 27 requires countries to make patents available for any product or process which meets the criteria of novelty, inventiveness and industrial applicability. Exceptions to this rule which concern living material are stipulated in 27.3(b). They can be made for "plants and animals other than micro-organisms, and essentially biological processes for the production of plants or animals other than non-biological and microbiological processes". According to this, micro-organisms and non-biological and microbiological processes must be patented. Furthermore, plant varieties must be protected "either by patents or by an effective *sui generis* system or by any combination thereof".

At first sight, the provisions of Article 27.3(b) offer national governments significant leeway. They lack, for example, a clear definition of the term "micro-organism" (see Dutfield 2001) and thus seem to allow more than just one interpretation of it. They also leave open what is to be understood by an "effective" *sui generis* system for the protection of plant varieties so that national governments seem to have the possibility of shaping such a system according to their own needs. This has led even critical academics like Carlos Correa to argue that "How WTO members use the flexibility they have to implement the Agreement's obligations is a matter of policy decision" (Correa 2000: 97). With reference to the patenting of living material, Correa writes, "Nothing in the Agreement obliges Members to consider that substances existing in nature, biological or not, are patentable, even if isolated and claimed in a purified form" (ibid.: 228). This interpretation is in danger, however, of detracting from the really existing power asymmetries which counteract the formally existing room to manoeuvre. It cannot be denied that the TRIPS Agreement does not oblige its members to patent living material, but as long as it also does not oblige them to *exempt* living material from patenting, weaker actors from developing countries lack an effective means of resisting the demand for patents on life forms in bilateral negotiations with more powerful actors from Northern countries. And as far as the *sui generis* protection of plant varieties is concerned, representatives of the industrial countries have emphasised that they see such a system in the Convention for the Protection of New Varieties of Plants (UPOV) in the version of 1991.[6] This strengthens the rights of plant breeders – that is in fact, the Northern seed corporations – vis-à-vis those of the farmers. The latter are not allowed to exchange among themselves seed, retained at the time of harvest, which belongs to a protected plant variety, for the purpose of breeding and sowing. That this is a significant restriction becomes clear if one takes into account that the exchange of seed in many developing countries is an essential basis for food security and for the conservation and further development of biological diversity.

Its economic, social and environmental implications have made Article 27.3(b) "arguably the single most contentious element of the WTO agreements" (Thomas 2002: 187). Therefore, in the course of its review, government representatives from the South have insisted on clarifications and amendments which would enable them to, in actual fact, take advantage of the exceptions offered by it. They consider such substantial revisions also necessary in order to harmonise the TRIPS Agreement with other international systems of regulations, above all with the CBD. Thus, they demand, for example, the inclusion in the TRIPS Agreement regulations of the disclosure of the origin of genetic resources used in an invention, of prior informed consent and of benefit-sharing. Such regulations would enable them to monitor whether or not a patent was granted in a legally correct manner or if the invention the patent was granted for is based on biopiracy. The furthest-reaching and most substantial proposal stems from the "African Group", a group of African developing countries, which favours a general prohibition on the patenting of life.[7] Even if such a demand has no chance of being anchored in the TRIPS Agreement, it does not remain without effect. According to Hepburn, it not only influenced the debate in the TRIPS Council, "but also resonated with the concerns of campaign groups and the media, and helped raise awareness amongst a wider public" (Hepburn 2002; cf. GRAIN 2000: 8).

The *second* reason for the politicisation of the TRIPS Agreement is the fact that the period within which many developing countries should have implemented the TRIPS Agreement ended on 31 December 1999.[8] Implementation can be accompanied by domestic political conflicts which give rise to processes of interest formation and articulation. When the TRIPS Agreement was negotiated internationally, the national constellation of interests in many developing countries was less clear than in the advanced capitalist countries, where strong scientific and economic forces, stemming mainly from the seed, software and life sciences industries, were in favour of a tightening of the protection of intellectual property on an international scale (even if there were also economic groups and sectors from whose point of view this was less beneficial or even detrimental). By contrast, many developing countries lacked experience in the protection of private intellectual property rights which the industrial countries had had for a long time, a fact which was not least due to "very different traditions of thought about the possibility of property in knowledge" (May/Sell 2006: 209; cf. El-Said/El-Said 2005). Furthermore, knowledge on the complex IPR issue and on the possible consequences of an intensified protection of intellectual property was often missing. Finally, there were no societal forces that possessed anything like the power resources of the corporations from the North and that could correspondingly have influenced the negotiation positions of their governments. The necessity of enforcing the TRIPS Agreement domestically might have changed this situation. Now became more obvious the ways in which the Agreement and its dominant interpretation actually strengthened Northern capital's freedom to act and at the same time restricted the practices not only of indigenous communities but also of Southern research institutions, state apparatuses and enterprises which were interested in developing a national knowledge-based industry. Thus, together with the review of Article 27.3(b), the approaching end of the implementation period might have had the effect of strengthening protest against tight IPRs and developing alternative knowledge within Southern countries and, consequently, speeding up the formation of interests and the positioning of the Southern governments in the international negotiations.

*Third*, international NGOs, such as the Third World Network, the Canadian ETC Group (Action Group on Erosion, Technology and Concentration) or the Spanish group GRAIN (Genetic Resources Action International), as well as critical political intellectuals such as Vandana Shiva, persistently pointed out the threats contained in the TRIPS Agreement in relation to food security and health provision, especially for the people of the South. Furthermore, they contributed to developing and disseminating alternative knowledge on biological diversity and the protection of traditional practices. As far as the Third World Network is concerned, it also participated in the elaboration of the "African model law" of the Organisation for African Unity (OAU) of 1998, together with African scientific institutes under the leadership of the Scientific, Technical and Research Commission of the OAU (Ekpere 2000; Oh 2000; Kongolo 2001; Egziabher 2002; see also Chapter 3). This law contains provisions on the regulation of access to genetic resources, which are intended to put an end to biopiracy. It takes account of the decisive role of farmers (including the gender-specific aspects of their role) for food security and for the conservation and further development of biological diversity, and it protects their knowledge and strengthens their rights vis-à-vis the breeders' rights. Furthermore, it provides for benefit-sharing and prior informed consent, which must be obtained both from the state and from indigenous and local groups. The model law stays within the framework of the TRIPS Agreement. At the same time, however, it confronts it with a different accentuation: it protects local and indigenous groups against unlawful access to their resources, their knowledge and their technologies – a danger to which they are increasingly subjected due to the internationalisation of Northern IPR standards through the TRIPS Agreement. The law can therefore be used as a basis for national and supranational *sui generis* systems for the protection of plant varieties which in contrast to the UPOV also take into account the interests of subaltern actors. As such, it is an important Southern contribution to the Northern-dominated debate on the protection of intellectual property.[9]

*Fourth*, the criticism of and resistance to crucial stipulations of the TRIPS Agreement are (implicitly) supported by the developments in other international forums which also deal with conservation and the use of biodiversity. Among these are the adoption of the International Treaty on Plant Genetic Resources by the FAO in November 2001 (cf. Chapter 3), the passing of the Bonn Guidelines within the framework of the CBD in April 2002 and, most importantly, the negotiations about an international regime on access and benefit-sharing, to be finalised before 2010 (cf. Chapter 2). These developments lend additional legitimacy to many of the claims articulated by Southern governments in the TRIPS Council. Furthermore, Southern countries try to strengthen their position vis-à-vis the North by intensifying their cooperation. This is expressed, for example, in the constitution of the "Like-Minded Group of Megadiverse Countries" (cf. Chapter 2).

However, one has to avoid conceptualising the politicisation of the TRIPS Agreement only as a North–South issue. The constellation of conflicts is far more complex than that. Thus, the cooperation between Southern countries is always superimposed by a structural supply competition: every megadiverse country is principally interested in the exploitation of its biological diversity, provided that an appropriate share of the profits is guaranteed via benefit-sharing arrangements. Therefore, the megadiverse countries always compete over the issue of who offers the most favourable supply conditions to (Northern) corporations. Besides this contradictory constellation of interests *between* Southern countries, there are conflicts *within* every country, with the result that the "national interests" government representatives articulate in international negotiations are essentially an expression of domestic relationships of power. For example, indigenous communities and small farmers, whose rights are often disregarded by the states they live in, are represented in a very selective manner by their governments at the international level. Their traditional knowledge, which is referred to in nearly every Southern submission to the TRIPS Council in order to undermine the necessity of a revision of Article 27.3(b), is often perceived under the aspect of "national" development.[10] Furthermore, it seems to be used in an instrumental manner:

> Indigenous peoples' organisations and TK [traditional knowledge, the authors] holders sometimes suspect that the subject is being used by their governments to secure trade advantages that will be of no benefit for them. They see particular irony in a situation in which governments of countries where indigenous peoples are victims of human rights abuses are among those calling for protection of TK as a matter of justice in international trade.
>
> (Bridges Trade BioRes 2002)

The politicisation of the tensions and contradictions which are inherent in the TRIPS Agreement or mark the relationship between the latter and other international conventions, such as the CBD, have had a strong impact on the negotiations in the TRIPS Council. At the Ministerial Conference of the WTO in Doha in 2001, the Council was given the task, "besides other issues", of "examining" the relationship between the CBD and the TRIPS Agreement as well as the problem of traditional knowledge. This extended the mandate of the Council and for the first time a connection between these issues and the revision of Article 27.3(b) of the TRIPS Agreement was made on the part of the WTO. At the Ministerial Conference in Hong Kong, which took place in December 2005, governments acknowledged the need to continue discussions on the CBD–TRIPS relationship. Furthermore, the TRIPS Council was urged "to intensify" the consultation process. Negotiations will include "the possible requirement for the disclosure of origin, a mandated review of Article 27.3(b) on the patenting of plants and animals, and the protection of traditional knowledge" (CBD 2006). Thus, one can observe a shift in the perception of problems in the conflicts over genetic resources: besides the legal security of firms from the agro-business and the life sciences sectors, increasingly, developmental issues are symbolically represented. Actors such as NGOs, indigenous organisations and governments of the megadiverse countries have questioned the definitions of problems of Northern actors and expanded the range of the politically negotiable issues.

Up to now, this has not resulted in a substantial revision of the Agreements' most critical stipulations. However, neither have the Northern governments reached their declared objectives: to limit the review of Article 27.3(b) to questions of implementation, which would have primarily concerned developing countries, and to bring the process to a quick end. The agenda of the TRIPS Council remains controversial. The debate as to what should in fact be negotiated runs through the review of Article 27.3(b) like a red thread. Years after the beginning of the review process and after the Doha Declaration had mandated the TRIPS Council to examine the CBD–TRIPS relationship it is "clear that the differing levels of political ambition have yet to be resolved, with one official saying that 'some of us want to play soccer while the others want to play rugby'" (Bridges Trade BioRes 2006a).

## 4.3 Scale jumping: towards national and bilateral solutions of the IPR issue?

The politicisation of the TRIPS Agreement and the resulting polarisation of positions in the TRIPS Council have led the governments of industrial countries to consider how to reach their objective of a stronger protection of intellectual property in different ways. One result of this is a certain strategic shift which could be described as double "scale jumping" (Smith

1995). On the one hand, efforts are made to downscale to the national level proposals which government representatives of Southern countries have made with respect to the TRIPS Agreement. On the other hand, there are attempts to negotiate standards of intellectual property protection which are even higher than those of the TRIPS Agreement ("TRIPS-plus standards") in bilateral or regional trade agreements.

The first form of scale jumping can be observed in connection with the controversial issue of the disclosure of origin of the genetic resources and traditional knowledge used in an invention. From the point of view of indigenous communities and government representatives from many developing countries, a disclosure of origin is a means to prevent biopiracy. It would oblige a patent applicant to provide information on the country of origin and on the compliance with the legal requirements of the latter regarding prior informed consent and benefit-sharing. If the applicant did not comply with the disclosure requirement, the patent-granting authority would have to deny or revoke a patent. A disclosure requirement made in this way would not only enhance the transparency of patent granting, it would also make it superfluous for a government to verify, in every individual case, whether or not patents which are granted in other countries on the basis of its own genetic resources, comply with its national legislation. Such individual verifications and possibly following legal proceedings taken to revoke a "bad patent" require a long and costly process which is beyond the financial and juridical capacities of many developing countries. In recent years, therefore, there have been several proposals from Southern governments to search for an international solution to biopiracy, that is to amend the TRIPS Agreement, for example in Article 27.3(b), to include a disclosure requirement.[11]

However, these attempts have largely been blocked by the dominant actors, who argue that preventing the misappropriation or theft of genetic resources is not an issue to be regulated under patent law. Thus, following the US government, "those acts fall outside the purview of the TRIPS Agreement and are appropriately the domain of a separate regulatory system".[12] Similarly, according to the position of the European Union, a disclosure requirement is neither an additional nor a substantial criterion for patentability. The disregarding of the requirement, therefore, could not be punished under patent law and, consequently, could not lead to the refusal or withdrawal of a patent. Instead, a disclosure requirement should be "self-standing", that is a system of its own, outside the patent law.[13] Another argument refers to the mechanisms which will guarantee the disclosure of origin and the compliance with national regulations on access and benefit-sharing. As seen above, Southern governments have proposed an international solution within the TRIPS Agreement, which would oblige a patent applicant to provide information on the origin of the genetic resource used in an invention. A national patent authority would be obliged to verify this information. For the US government, such a system would be "a legal and administrative nightmare for all involved" (WTO 2000b: 5). From its point of view, the problem is not a "lack of safeguards in the TRIPS context", as stated by the governments of Brazil, India and other countries, but "the lack of clearly defined national systems directly regulating the use of genetic resources, particularly in the context of access and benefit-sharing (ABS) systems" (WTO 2006e: 2). The already cited position of the European Union, in principle, supports this view: a "sound and effective *national legislation* in the countries providing genetic resources" (emphasis in the original) is the "main legal instrument" of benefit-sharing (WTO 2001: 5). The European Union leaves it to the countries to verify compliance with their national regulations as well as the legality of patents individually.

Thus, whereas far-reaching minimum standards for IPRs have been institutionalised internationally and have strengthened the position of knowledge-based economic sectors, the upscaling of stipulations which would enhance the possibility of Southern actors to control what happens to the genetic resources of their countries has been, up to now, successfully prevented by dominant forces from the North. By defining them as matters of national legislation, governments from the industrialised countries try to keep the claims of Southern governments away from the international level and to downscale them to the national terrain of each WTO member. This affects not only the transparency of patent granting and the possibility of preventing or revoking "bad patents", but also the national regulations themselves, insofar as they may be developed in the shadow of bilateral agreements and the power asymmetries inscribed therein. This is the second form of scale jumping with which Northern governments attempt to circumvent the impasse the negotiations on the TRIPS terrain have got into.

Bilateral agreements are increasingly used, especially by the USA and the European Union, in order to force developing countries to do away with the exemptions from patentability allowed by Article 27.3(b) and to implement substantial stipulations for the protection of intellectual property which go beyond those laid down in the TRIPS Agreement. Even more so than in the negotiations on multilateral terrains, they offer the Northern governments the opportunity to exploit bilateral power asymmetries and the structural supply competition of the biodiversity countries. The US government states this quite openly: "Where genetic resources or traditional knowledge can be obtained from a number of sources, of course, the party seeking access likely will seek the resources or knowledge from the territory that provides the most favourable conditions" (WTO 1999b: 5). Bilateral agreements on the protection of intellectual property are often made via the loophole of aid agreements or trade and investment contracts (GRAIN 2001c; cf. CIPR 2002b: 180 f.; Correa 2004; Hepburn 2002; Lasén Díaz 2005). These then include, for example, the obligation not to exempt plants and animals from patentability, to guarantee the "highest international standards" for the protection of intellectual property or to implement the 1991 version of the UPOV, which strengthens the breeders' rights vis-à-vis the farmers', as *sui generis* protection for plant varieties. They also often provide for their own arbitration mechanisms. GRAIN (2001a: 6) therefore speaks of a "tunnel vision toward one global patent standard".

The increasing importance of bilateral agreements seems to be a major trend in international trade politics (WTO 2003: 26ff.). As such, it is a reflection of the resistance which Northern governments and business interests have been confronted with on the WTO terrain. "Having failed to advance their agenda in the post-1995 multilateral system, industrial countries are now seeking to reverse the centre of gravity away from the WTO" (El-Said/El-Said 2005). The failure of the Ministerial Conferences in Seattle 1999 and Cancún 2003, as well as the suspension of the Doha Round negotiations in the summer of 2006, underline this tendency. However, one has to be careful not to see a zero-sum-game between bilateralism and multilateralism at work here (Görg/Wissen 2003). At least two factors contradict this line of reasoning: first, even if it is contested, the TRIPS Agreement has in no way become meaningless. By institutionalising strong minimum standards of intellectual property protection on an international scale, it rather provides a solid basis for negotiating higher standards bilaterally. Second, the conflicts on one multilateral terrain, the TRIPS Agreement, have raised the status of another, the World Intellectual Property Organization, that is there is a certain forum-shifting back to the WIPO, which, however, has also resulted in the politicisation of this rather "technical" organisation. This can be seen in recent developments within the WIPO: the launching of a "Patent Agenda", which aims at harmonising the regulations of the protection of intellectual property internationally, that is internationalising the standards that are valid in the leading industrial countries; the negotiations within the "Intergovernmental Committee on Intellectual Property and Genetic Resources, Traditional Knowledge and Folklore" (IGC), where subaltern actors struggle for an institutionalisation of their rights; and finally, the initiative to establish a "Development Agenda" which was launched by the governments of Argentina and Brazil. The "resurgence" of the WIPO, which is indicated by these developments, is the subject of the following sections.

## 4.4 Forum-shifting: imposing higher IPR standards through WIPO's Patent Agenda?

The Patent Agenda was announced in August 2001 by the WIPO Secretary General and led, one year later, to a memorandum on the further development of the international patent system (WIPO 2002). Basically, it is concerned with the international harmonisation of the diverging national regulations on the protection of intellectual property. The Patent Agenda is based on three pillars: the Patent Law Treaty, the reform of the Patent Cooperation Treaty and the creation of a Substantive Patent Law Treaty (on the following see Correa/Musungu (2002) and GRAIN (2002b)). The *Patent*

*Law Treaty* was signed in June 2000. It came into force in April 2005 after it had been ratified by ten states. Its purpose is to harmonise internationally the procedure of applying for, and granting, patents. Accordingly, it contains a number of formal stipulations concerning, for example, the form, the contents or the point of time of a patent application. The applicant is required to provide only "an express or implicit indication to the effect that the elements [received by the patent office] are *intended* to be an application" and "a part which on the face of it *appears* to be a description" (Art. 5(1) (ii) and (iii) of the Patent Law Treaty, quoted and emphasised by Correa/Musungu 2002: 6). This regulation allows, according to Correa and Musungu, that the rights to an invention be secured at a very early stage, for example on the basis of a hypothesis which has not yet been verified.

The *Patent Cooperation Treaty* is an instrument which makes it easier to have an invention protected internationally. It creates the possibility of an international patent application. This is checked, first of all, in an international search for prior art – conducted by larger national or regional patent offices and at the cost of the applicant. The results of the search are documented in an international search report. The applicant can also apply for an international preliminary examination report, which in contrast to the search report contains a preliminary and non-binding appraisal of the patentability of the invention. The final decision on the granting of a patent is taken by the patent offices in those states in which the invention is to be protected. The Patent Cooperation Treaty has advantages for both the applicants and the national and regional patent offices: for the latter, their work is reduced because they can base their decision on the results of the international search or examination. The former can, on the basis of the search or examination results, better estimate how good their chances are of receiving a patent for their invention. But above all, they have gained time over potential competitors: "A patent application becomes 'claimed territory' before filing at the national level" (GRAIN 2002b: 2). The Patent Cooperation Treaty was finalised in 1970, extended in 1979 and amended in 1984. Since 2000, a new reform process has been underway which is intended primarily to simplify the procedure and to adjust the Patent Cooperation Treaty to the new standards of the Patent Law Treaty. The US government and the governments of other industrial countries are pursuing plans that are even farther reaching: they regard the Patent Cooperation Treaty reform and other reform efforts in the framework of the WIPO, primarily from the perspective of creating, in the long term, possibilities for the granting of a global patent.

The negotiations on the *Substantive Patent Law Treaty* (SPLT) are about establishing substantive patentability standards internationally and thereby going beyond the stipulations determined by the TRIPS Agreement. While the latter lays down minimum standards, the former, according to the governments of leading industrial countries, should establish best practice at the international level (Correa/Musungu 2002; Hepburn 2002). Correa and Musungu, therefore, describe the Substantive Patent Law Treaty as potentially the "most troublesome building-block of the proposed international patent system from the perspective of developing and least developed countries" (Correa/Musungu 2002: 16) because it would remove all the presently existing leeway for national regulations on the protection of intellectual property.[14]

The Patent Agenda can be interpreted as an expression of three problem constellations. The first is the "workload crisis" of the national and regional patent offices. Due to the rising importance of knowledge in the process of the valorisation of capital, the number of patent applications for the protection of intellectual property, and thus the workload of patent offices, have increased significantly (see Chapter 1 and Oldham 2004). The development of an international patent system, it is assumed, would make the examination of patent applications more transparent and avoid the duplication of work tasks, which would lead to an easing of the workload of national and regional patent offices.

Second, the developing countries have emphatically articulated their criticism of the existing system for the protection of intellectual property within the review process of Article 27.3(b) of the TRIPS Agreement. As analysed above, concepts such as benefit-sharing have gained in legitimacy. Against this background, the launching of the Patent Agenda in 2001 can be classified as a type of "counteroffensive" in the interest of dominant actors from the North in order to restore the blemished legitimacy of the patent system as a "public policy tool using the creation and exercise of private rights as a means of promoting the public good" (WIPO 2002: Annex I, p. 4). Because of its expertise, its "technical" character and its narrow mandate, which is limited to providing support for the protection of intellectual property, the WIPO appears to be, like no other, in a position wherein it is authorised to make the protection of intellectual property, both materially and symbolically, a priority once again.

Third, the Patent Agenda is an expression of the efforts of transnational corporations to simplify the procedure for the international recognition of patents, in speeding it up and making it cheaper and thus reducing transaction costs as far as possible, but also – in a substantial sense – in removing exemptions from patentability (see ICC 2002). If the Patent Agenda were successful, the national and regional patent offices would move closer together. A system would come into being which spanned different spatial scales and would guarantee an effective patent protection from the point of view of transnational enterprises. The WIPO, in this case, would get a national and regional "foundation" for the implementation of an international patent law, an advantage which could compensate for the lack of an international sanctioning authority. In addition, the SPLT would exceed the minimum standards for intellectual property protection laid down in the TRIPS Agreement, which would also correspond to the interests of enterprises from knowledge-based economic sectors. The latter, therefore, appear to be paying very close attention to the negotiations within WIPO. They articulate themselves not only indirectly, via their governments, but also directly. In Correa and Musungu (2002), the following description can be found of the behaviour of business representatives and patent lawyers in the SPLT negotiations:

> Curiously, despite their "observer capacity", the referred NGOs participated in the debates on the same footing as governments, making proposals or "reservations", supporting the positions of some governments or being opposed to government positions. There were no NGOs representing the views of consumers or developing countries.
>
> (Correa/Musungu 2002: 17)

It is, above all, the orientation of the SPLT negotiations to the interests of transnational corporations and the weak representation of subaltern interests which leads the critics of the Patent Agenda to make gloomy prophecies. For example, in a paper by GRAIN (2002b: 5), it is stated: "We may even see critics turn around and defend TRIPS, as it may suddenly appear a lesser threat compared to what WIPO comes up with."

## 4.5 Politicising a "technical" terrain: conflicts about intellectual property, genetic resources and development in the WIPO

The efforts of Northern governments and enterprises to internationally harmonise the standards of intellectual property protection according to their interests, reflect, however, only one side of the resurgence of the WIPO. Recent developments and debates in other WIPO forums indicate that governments of developing countries, NGOs and indigenous communities have managed to put their matters on the agenda of WIPO negotiations, too. One example of this is the setting up of the "Intergovernmental Committee on Intellectual Property and Genetic Resources, Traditional Knowledge and Folklore" (IGC) in 2001. The way in which IPR issues have been addressed here contradicts, to a certain degree, the way this has happened within the framework of the Patent Agenda. For example, a document which was prepared by the WIPO secretariat for the first meeting of the IGC, in order to provide an overview of matters concerning genetic resources, traditional knowledge and folklore, gives an account of discussions which challenge the dominant concept of individualised private IPRs and emphasises the social character of knowledge production:

> The equity of intellectual property rights is discussed not only in the balance between the rights of the creator and society as the user of his creation, but also in the balance of rights between the creator and society as the provider of heritage resources which he utilizes in his

> creation. This is the case especially where the provider has conserved the common heritage for generations under *in-situ* conditions, i.e. in the surroundings where the resource developed its distinctive properties.
>
> (WIPO 2001: 4)

Conserving and developing local technologies and products, including plant genetic resources for generations, is referred to as an *informal innovation* because it is not formally recognised as an innovative activity and, therefore, it is not protected by IPRs. It is pointed out that informal innovators now claim rights to intellectual property protection and that they criticise the unequal treatment of formal and informal innovative activity with regard to the protection of intellectual property, which is not objectively justifiable: "The creation of new intellectual property rights for formal innovations in a certain subject matter is seen as contingent upon the creation of cognate rights for informal innovations relating to the same or similar subject matter" (WIPO 2001: 5).

In spite of this quite reflexive outline of the "cultural bias" (Dutfield 2002) of the IPR system, which stood at the beginning of the IGC, the latter's work has not yet produced significantly substantive results. The claim for a legally binding international instrument for the protection of traditional knowledge, as raised by the governments of many developing countries, has been rejected by the representatives of developed countries (see Bridges Trade BioRes 2006c). As the South Centre has pointed out, "historically, the areas of interest to developing countries continue to be characterised by high levels of dialogue with no concrete outcomes" (South Bulletin 2002). As far as the IGC is concerned, this may also be due to the fact that many Southern governments, in contrast to the indigenous communities of their own countries, seem to question the significance of the IGC as a negotiation forum. The influence of the IGC on other WIPO discussions is considered to be quite low. Furthermore, there are concerns

> that a debate in the IGC would distract from or pre-empt a decision by the TRIPs Council on a proposal by a group of developing countries, calling for disclosure requirements and evidence of prior informed consent and benefit sharing related to genetic resources and traditional knowledge in patent applications.
>
> (Bridges Trade BioRes 2004)

Thus, the reflexivity of discussions seems to be closely related to the marginalisation of the forum where the discussions take place.

Another and seemingly more powerful attempt to politicise WIPO is the proposal to establish a "Development Agenda" (WIPO 2004). It was submitted to the WIPO in August 2004 by the governments of Argentina and Brazil, but rapidly found the support of the governments of other developing countries (WIPO 2005: Annex, p. 2). Starting from the crucial significance of technological innovation, science and creative activity for social progress and welfare, the proposal complains about a "knowledge gap" between wealthy and poor nations and reflects on the role which IPRs play in this respect. It criticises WIPO's present efforts at raising the patent protection standards and making them mandatory for all countries, irrespective of their levels of development. Instead it emphasises the "need to integrate the 'development dimension' into policy making on intellectual property protection" in order to create a balance between users and producers of knowledge. The paper proposes a "high-level declaration on intellectual property and development" to be adopted, for example, by the General Assembly of WIPO. In the face of "the inability of existing IP agreements and treaties to promote a real transfer of technology to developing countries and LDCs" it calls for a "Treaty on Access to Knowledge and Technology" as well as an inclusion of clear provisions on the transfer of technology in the treaties which are currently under negotiation in WIPO. Not least, it suggests a strengthening of civil society's participation in WIPO's activities.

The WIPO General Assembly welcomed the proposal in autumn 2004 and decided to examine it in "inter-sessional intergovernmental meetings". One year later the General Assembly established a "Provisional Committee" to continue discussions on the Development Agenda. Up to now, the debates have produced only limited results. Powerful forces deny the necessity of a Development Agenda for WIPO at all. "Simply put", a paper of the US government puts it, "WIPO already has had a robust 'development agenda' in all of its work for a long time" (WIPO 2005: Annex, p. 2). The statement of a US delegate goes even further: "We support WIPO. We would not want to change WIPO in a direction that would diminish that support" (IP-Watch 2005: 2). However, the critics of the proposal did not manage to suppress the discussion on it or to pass the development issue on to a subsidiary WIPO body such as the "Permanent Committee on Cooperation for Development Related to Intellectual Property". Southern governments have successfully resisted such attempts and demanded an integration of development aspects into all WIPO activities (ICTSD 2005a).

Within the NGO community, the idea of a Development Agenda has met with strong and positive resonance. In the "Geneva Declaration on the Future of the World Intellectual Property Organization" (TWN 2004), which was launched in September 2004, 25 NGOs welcomed the paper as "the first real opportunity to debate the future of WIPO". According to the signatories of the declaration, the WIPO had, for a long time, been responding primarily to narrow business concerns. Against this background, the proposal of a Development Agenda is seen as "a long overdue and much needed first step toward a new WIPO mission and work program". It seems to have driven the interest of developmental and consumer-oriented NGOs in WIPO and their efforts to engage themselves in pressing for more transparency and participation within the organisation's activities (see, for example de Paranaguá Moniz *et al.* 2005 and ICTSD 2005b).

The strong and controversial reactions may be due to the fundamental character of the proposal, the condensation of criticism of the WIPO it contains and the translation of this criticism into a reasonable demand: the establishment of a Development Agenda. The proposal is not a contribution to ongoing "technical" discussions on intellectual property protection within WIPO, but tries to reframe the entire debate. The term "Development Agenda" alone implies a criticism of the ongoing discussions in WIPO. It refers to and at the same time dissociates itself from the "Patent Agenda". Furthermore, it creates a link to the "Doha Development Agenda", which was adopted by the WTO Ministerial Conference in 2001 and strengthened the notion of development within the WTO, at least symbolically. Until recently, the task and the character of the WIPO had been considered to be primarily "non-political". The organisation itself had "consistently sought to reinforce the perception that it is a highly technical body to keep away most of the political and economic debate about IPRs" (Correa/Musungu 2002: 4). As such, the WIPO gained in importance from the point of view of Northern business and governments as soon as the tensions and contradictions of intellectual property protection became politicised within the WTO by Southern governments, indigenous communities and NGOs. Pursuing the further tightening of IPR standards in WIPO seemed to be a means for Northern actors to escape the impasse the TRIPS process was caught in. For WIPO itself, this forum-shifting offered an opportunity to gain lost ground vis-à-vis its institutional competitor, the TRIPS Agreement. However, the strong reactions which the proposal of a Development Agenda has provoked, both in favour and against, show that the politicisation of the IPR issue has reached the WIPO, too.

## 4.6 Contested terrains: summary and conclusions

The examination of the WIPO and the TRIPS Agreement illustrates the momentum which characterises the conflicts over intellectual property rights in the appropriation and valorisation of genetic resources. Both institutions are increasingly politicised by Southern governments, indigenous groups, social movements and NGOs, with the result that the positions of those groups have gained "representational strength" (Jenson 1989) in an environment which has traditionally been dominated by the perceptions and interests of Northern governments and business. Where in the past, the latter had been able to ignore questions of traditional knowledge or benefit-sharing, they must now at least take a position on these issues. The time-factor plays an important role here: while the industrial countries at the conclusion of the TRIPS Agreement had already been familiar with the complex subject matter of IP protection for a long time, many developing countries lacked the necessary experience and expertise. A process of position-finding and organising – partly supported by the expertise provided by NGOs – was therefore necessary before the governments of Southern countries began to articulate themselves perceivably. Further factors which support the politicisation process are the outcomes of the negotiations on other international terrains such as the CBD and the FAO, the attempts of Southern governments to strengthen their negotiation position by intensifying their cooperation and the fact that developmental and consumer-interest NGOs emphatically articulate their demands, especially for the prohibition of patents on living materials. Events such as the protests in Seattle at the end of 1999 or in Cancún in 2003 also play an important role: they have placed the developmental, social and environmental consequences of WTO policies in the public light and strengthened the legitimacy of "Southern" concerns. Even if this has not changed anything in the fundamental positions in the TRIPS negotiations, the *negotiation context* has changed: the review process of Article 27.3(b) of the TRIPS Agreement has increasingly been set under legitimation pressure.

Until now, the politicising of the TRIPS Agreement and the WIPO has primarily led to symbolic shifts in the debate on genetic resources: the definitional sovereignty of powerful Northern forces over what is negotiable and what is not has been questioned. In the frame of the TRIPS Agreement, they were not able to restrict the review of Article 27 to mere implementation issues. Instead, they saw themselves confronted with the demand of Southern governments for substantial revisions. As a consequence, the subject of the review process remains continually contested. Similarly, in the WIPO, governments of developing countries are increasingly reluctant to negotiate about an international high-level harmonisation of IPRs. By launching the proposal of a Development Agenda, they have called for a redirection of the entire system of intellectual property protection. Thus, what are at stake in both organisations are not only controversial issues to be negotiated within largely accepted institutional frameworks, but also the shape of the institutional terrains themselves.

In the conflicts over what should actually be negotiated in the review of Article 27.3(b), the ambivalence of the TRIPS Agreement, even from the perspective of the industrial countries is revealed: on the one hand, it has contributed to the establishment of minimum standards for intellectual property protection and has improved their stability. On the other hand, this was only possible by shifting the IPR issue onto the contested terrain of the WTO. The improved protection of intellectual property was thus paid for with the politicising of the IPR issue, which proved increasingly difficult to deal with in the framework of the TRIPS Agreement. In this context, the WIPO, which had been institutionally weakened by the establishment of an agreement on intellectual property protection within the WTO, gained importance again. A forum-shifting took place in the opposite direction to that of the 1980s and 1990s: the establishment of important negotiations on the protection of intellectual property in WIPO, the most obvious expression of which is the Patent Agenda. One can interpret this process as an attempt to "re-technicalise" and, at the same time, further develop the IPR issue in the direction of a substantial international patent law which abolishes the exemptions from patentability still formally allowed by the TRIPS Agreement. Driving forces are the governments and knowledge-based industries from the industrial countries which seek to free themselves from the dilemma in which the review of Article 27.3(b) finds itself. However, as the recent conflicts over a Development Agenda for the WIPO have shown, the forum-shifting did not simply work in the interest of its driving forces. By contrast, Southern governments and NGOs, which have been sensitised to the consequences of a high-level intellectual property protection by the TRIPS Agreement, seem to have succeeded in politicising the "technical" terrain of WIPO, too.

The repeated forum-shifting shows how far away the international IPR regime is from hegemonic forms of conflict regulation. There is no international institution which is accepted to a similar degree by both dominant and subaltern actors as a terrain for the settlement of conflicts. The conflict terrains themselves are the subject of the disputes. The result of this is that, above all, the dominant forces repeatedly attempt to shift the IPR issue onto that terrain on which the chances of institutionalising their interests currently appear to be best.

From a development perspective, a major threat lies in the effort to establish TRIPS-plus standards of intellectual property protection via bilateral trade agreements. The latter have significantly increased in number recently. For those forces which are interested in tightening intellectual property protection internationally, they offer a way out of the impasse which the TRIPS negotiations have got into and which the Patent Agenda of the WIPO seems to get into, too. In contrast to a forum-shifting from one international organisation to the other, bilateral agreements are much less spectacular. They do not offer a stage to (collectively) politicise the contradictions of North–South relations. Instead, they constitute a realm where bilateral power asymmetries can be fully played out. However, it would be too simple to argue for a "disassembly of multilateral economic order" based on the increasing number of bilateral trade agreements, as for example Dieter (2003) does. More probable is an overlapping of multilateralism and bilateralism (see Görg/Wissen 2003). This is especially due to two specific elements of the TRIPS Agreement: first, the conceptual uncertainties and the exceptions from patentability in Article 27.3(b), and second, the fact that the TRIPS Agreement formulates minimum standards but provides no upper limits for the protection of intellectual property. The first element involves the necessity of bilateral processes of clarification when conflicts arise as a result of different interpretations of Article 27.3(b) or of different national regulations which, nevertheless, all comply with the TRIPS Agreement. The clarification can take place in bilateral negotiations or within the framework of the WTO's dispute settlement body. In any case, power asymmetries come into play.[15] The second element comprises the possibility of bilaterally negotiating higher IPR standards without violating the regulations of the TRIPS Agreement. "TRIPS standards will be a floor from which further bilateral, regional and multilateral standard-setting exercise will proceed, a floor with no ceiling above" (Drahos 2001: 10).

The politicisation of the TRIPS Agreement and the WIPO are likely to lead to the further gaining in importance of bilateral strategies for the establishment of TRIPS-plus standards. A danger for multilateralism results from this if the latter – as in regime theory – is regarded primarily under the aspect of cooperative conflict solution. However, proceeding from the assumption that power relationships and bilateral power asymmetries are always inscribed in international agreements leads to a different conclusion: multilateralism is not simply a counterpart to bilateralism, which is seen to suffer if and when the latter gains strength. There is no zero-sum-game between the two. By contrast, multilateralism may form a solid foundation on which bilateralism can unfold its influence. This can be learnt from the TRIPS Agreement: by allowing exceptions to patentability but not proscribing them, and by setting minimum standards for intellectual property protection but setting no upper limit to them, it prepares the field for bilateral *power bargaining*. The TRIPS Agreement is not only the result, but also the precondition and medium for the unfolding of bilateral power asymmetries. Multilateralism, in other words, forms a powerful vehicle which helps bilateralism to establish itself.

The question remains to what extent indigenous communities, small farmers and other less powerful social groups can take advantage of the politicisation of the intellectual property issue analysed in this chapter, that is: does the contradictory constellation offer opportunities for local groups to challenge the unequal relationships of power which shape their living conditions and to maintain or recover control over the latter? This is especially important if one takes into account that struggling internationally against the illegitimate appropriation of genetic resources and traditional knowledge for many Southern governments is not necessarily an end in itself but may also be a means of strengthening their bargaining position in areas such as agriculture and services, which are backed by stronger coalitions of social forces domestically. This seems to apply, for example, to Brazil and India and constitutes an important difference between these countries, on the one hand, and many African countries, on the other (Kaiser 2006). For assessing the opportunities of local control over genetic resources, there is a need to not only analyse the unequal relationships between North and South and among Southern countries, but also analyse those that exist within each country.[16] This is what we shall turn to know, when we exemplarily examine the domestic relationships of power with respect to genetic resources in the megadiverse country Mexico.

# 5 The relevance of the national and the local

## The disputes over a valorisation paradigm in Mexico and Chiapas

To what extent do international political processes represent a political and institutional framework for the treatment of biological diversity in specific national and local contexts? Using a case study conducted in Mexico, and within it at the local level in the state of Chiapas, this chapter explores this question, and three implied subquestions. *First*, are international policies translated relatively directly into national regulations or do they encounter specific interests, relationships of forces, political and institutional conditions and socio-economic dynamics at the national level and if so, are they modified by these? The question of implementation is, as we have already seen, of central importance, especially for a framework agreement such as the CBD (Convention on Biological Diversity). The forms of intervention of different levels are manifold. We shall differentiate here between four such forms: (a) legal regulations, which in a certain sense create "corridors" in which certain confrontations take place, (b) the forms of the constitution of problems and proposals for dealing with them, (c) the ability of certain actors to assert their authority and (d) structural conditions.

*Second*, it must be examined empirically to what extent in Mexico a "coherent" national biodiversity policy is being formulated, as has been demanded repeatedly by national and international actors. If this is not the case, what are the main reasons for this?

*Third*, what role is played by the specific place, which is bound into spaces in a multitude of ways? There exist competing ideas and strategies on how to deal with the places, that is the natural living conditions and societal relationships to be found there. Here, too, there are specific relationships of forces which cannot simply be "derived" from the national ones.

The choice of Mexico, and Chiapas, for the case study, was arrived at for several reasons. Mexico is, for one thing, a so-called megadiverse country, and it is also geographically and political-institutionally closely tied to the major "demand country", the USA, namely within the North American Free Trade Agreement (NAFTA). Moreover, questions concerning the appropriation of biological diversity have enjoyed a high level of importance in both expert and public discussion since the end of the 1990s. For the purposes of research this latter fact means that the perspectives and interests of individual actors and also the lines of conflict have been clearly articulated. Mexico is also interesting with regard to the question of the "democratic formation of biodiversity policy". The "question of democracy" has been emphatically discussed since the 1994 uprising of the Zapatista liberation army (*Ejército Zapatista de Liberación Nacional*; EZLN) and its social basis. Particularly in the state of Chiapas in the south-east of Mexico, which is a hot spot as far as biodiversity is concerned and at the same time one of the country's poorest regions, the relationship of forces has shifted as a result of the uprising. How does this change the conditions of democratic politics and thus also of the appropriation of nature? Finally, Mexico was also chosen because the project group already had good contacts there, which meant better access to the subject of the research.[1]

From the outset, this case study was chosen not as a representative model but rather, due to its particularities compared to other countries. In other countries the conditions look partially quite different. In addition, it is only recently that experience with the formulation of a national biodiversity policy has been gained, that is with the development of a political-institutional framework for the appropriation and conservation of biological diversity, and this process has mainly been initiated by international politics. The study also represents a very topical extract from an extremely dynamic political field in which as yet no clear structures have formed, either in Mexico or in other countries and regions. Even if its results can therefore hardly be generalised, the case study nevertheless imparts central insights into post-Fordist relationships with nature and the internationalisation of the state.

First, the most recent development in Mexico and Chiapas will be outlined briefly. Second, different positions will be identified which play a role in the formulation and implementation of a Mexican biodiversity policy. This is a snapshot, since the problems, interests, alliances and political proposals shift continually. Finally, the controversial development of a Mexican biodiversity policy will be analysed and "concretised" for the state of Chiapas.

## 5.1 A megadiverse country in the process of transformation into a competition state

Mexico is a so-called megadiverse country with ten vegetation zones and two centres of biological diversity: one in the arid North and one in the neotropical South. Of the 250,000 known plant species worldwide, 22,000 to 26,000 are to be found in Mexico of which 40 per cent are endemic, that is they only exist in Mexico (Mittermeier/Goettsch Mittermeier 1992: 69; Delgadillo Macías 1994). Therefore Mexico, following Brazil, Colombia and China, lays claim to the fourth largest share worldwide. Mexico also belongs to the group of countries with the greatest number of animals worldwide (amphibians, reptiles and mammals). The number of known plants only gives an initial indication of Mexico's "natural" *genetic* diversity. The latter has, as yet, been only marginally researched (CONABIO 1997: 6). In addition, in Mexico there is a great diversity of cultivated plants. There are, for example, 41 types of maize with several thousand varieties (Taba 1995 according to Wullweber 2004: 107). This is connected with the fact that in Mexico, as in most of the megadiverse countries, there is great *cultural diversity*. Roughly 230 different languages and dialects are spoken (Toledo Ocampo 1998). About 10–12 per cent of the population describe themselves as indigenous, are largely engaged in agriculture and have developed their own medical systems. The latter play a major role, for in bioprospecting this knowledge serves as a type of "filter" for finding potentially economically valuable substances. With regard to agro-biological diversity, the circumstance is important that the culturally diverse areas are at the same time the socio-economically most "marginalised". The modernisation of the agricultural sector within the framework of the Green Revolution hardly took place here, which is why a high degree of agro-biological diversity has been conserved. In the regions with a large and diverse indigenous population a great plurality of relationships with nature have continued to exist up to the present day.

However, the peripheral Fordist development model (cf. the overall analysis of Aboites 1989) has led to a great erosion of biological diversity, both in agriculture and also with regard to non-domesticated diversity. Driving this erosion are: deforestation, the degradation of the soil, the industrialisation of agriculture, urbanisation, the externalisation of ecological costs in economic processes, population growth, poverty and inequality, water pollution, the lack of incentives for the protection and sustainable use of nature and inadequate state policies. One example of the latter is the existence of a state authority in the 1970s (*Comisión Nacional de Desmontes*) which was responsible for deforestation for the settlement of farmers. The erosion has increased further in the last 20 years (CONABIO 1998; SEMARNAT 2002b). A recent study of the Economic Commission for Latin American (ECLA/CEPAL 2004: 166) underlines the high environmental costs of the currently dominant development model in Latin America with its export-orientation and environmentally unregulated features.

### *5.1.1 Democratisation, neoliberal politics and NAFTA*

The context of the current conflicts over the treatment of biological diversity is the present peripheral post-Fordist transformation of society. This context is, first, not intransigent but its production and stabilisation are the subjects of permanent social confrontations. Second, it does not "influence" biodiversity policy "from the outside". The specific forms of the appropriation of nature are – as we shall show – themselves an integral part of the neoliberal post-Fordist transformation of society. The 1990s in Mexico were characterised by two contrary processes: first, neoliberal strategies were established, and second, a complex struggle over the democratisation of the political system took place. Both processes have, as we shall see, affected the societal relationships with nature.

Many observers place the year 1985 as marking the clearest turning-point in the history of the struggles for more democracy, when the system based on the rule of the state party PRI (*Partido Revolucionario Institucional*) showed itself to be incapable of reacting more or less efficiently in the dramatic situation caused by a devastating earthquake in Mexico City. In the presidential elections in 1988 the leader of the opposition and later first non-PRI mayor of Mexico City, Cauthémoc Cárdenas, only lost, in the opinion of many, because of irregularities in the counting of the votes (Laurell 1992). During the 1990s there was a struggle over democracy in Mexico (particularly aimed at the question of the reform of voting rights), which was conducted by many social movements and organisations such as the PRD (*Partido de la Revolución Democrática*).[2] Democratisation was "the" subject of that period. The PRI took up the challenge and presented its candidate in 1994, Ernesto Zedillo, with a victory in "clean" elections. This could not stop the erosion of the PRI, however, which is why in 1997 it lost the first direct elections for the mayor of Mexico City and the parliamentary elections. The change of government at the end of 2000 represents the preliminary end of the democratisation process in Mexico. Due to the confrontations of the 1990s, the state party PRI, which had ruled for over 70 years, was pushed out of the highest executive and continually lost parliamentary elections and elections of governors. Authoritative corporatism was thus clearly weakened and made room for a more formally democratic constellation. This development resulted in more business-friendly politics. The new President Fox Quesada (Popular Action Party) was not only the general manager of Coca-Cola Mexico, but half of his cabinet were businessmen. Never before in Mexican history have so many businessmen been appointed to the heads of government ministries (Camp 2003: 125; Teichman 2002: 505). However, at the beginning of his presidency Fox tried to establish broader social welfare and health policies as well as a tax on wealth, but under economic pressures and resistance from the private sector he soon withdrew his projects (ibid.). In the presidential elections of 2006 – after a probable election fraud and heavy conflicts – again the candidate for the Popular Action Party became president. Felipe Calderón comes, in contrast to Fox, from the political establishment and is much more experienced in combating political opponents. However, he started with much less legitimacy than Fox and has the options of creating it through political and economic compromises with his opponents and their social base and/or increasing repression of them.

Parallel to the struggles over the democratisation of the political system, neoliberal strategies were enforced. Since the 1980s state policies have been directed centrally towards international competition and the allied societal interest groups (Olivier Costilla 2005; Morton 2003; Heigl 2006). For a country such as Mexico this means an increased (subordinate) integration into the international division of labour. A central role is played here by low labour costs, the industrialisation of the agricultural sector, the opening up of the economy to foreign trade and investment, access to and the valorisation of resources, and security for investments and planning. The establishment of neoliberal policies began in the 1980s when the debt crisis became obvious. In 1982 Mexico was the first country which issued a moratorium, which in the ensuing period led to hectic activity on the part of creditors and debtors in the management of the debts (Imbusch 1988: 103–156; Boris 1996: 20–29). The "creditor-driven event" (Rainer Falk, a German NGO intellectual) strengthened the position of the banks and Northern governments, and not only with regard to the maintenance of the flow of payments. The governments of the debtor countries, which were dependent on loans, had to carry out so-called structural adjustment measures: state budgetary discipline, the opening up of the economy to foreign goods and investments, the liberalisation of the financial markets, the privatisation of public enterprises, the deregulation of certain sectors and fiscal reform (Guillén Romo 1997: 97ff.; Morton 2003: 641–646).

However, the shift towards a neoliberal economic and social policy cannot simply be attributed to external constraints. It started as a domestic strategy with an *apertura unilateral* (unilateral opening) from 1985 to 1993, especially through the accession of the GATT Agreement in 1986. Sectors such as mining, transport, real estate, telephones and parts of the oil industry were opened up to foreign capital. Other sectors such as petrochemicals, banks and agriculture had already been opened up step by step within the framework of the debt negotiations in the 1980s. A ministerial order limited foreign investment to 49 per cent and allowed the purchase of Mexican companies by foreign investors without official permit (Teichman 1995: 92). In 1993, the national treatment of foreign investment was legally codified (Ornelas Bernal 1994: 259–266). Privatisation policies – between 1983 and 2002 the number of (predominantly) state-owned enterprises was reduced from 1155 to 202, MacLeod 2004: 71) – also served internal economic groups. However, the currency crisis of late 1994 showed that macroeconomic stabilisation was not assured through the existence of NAFTA. During this time, and especially during the negotiations of NAFTA, Mexico became a major destiny for foreign direct investment (Luna Martínez/González Nolasco 2004: 37; on the recent development of NAFTA, cf. Aguilar/Graizbord 2006). At the same time, the Salinas government tried to create social consensus for neoliberal politics in the 1980s, especially through the introduction of the National Solidarity Programme (*Programa Nacional de Solidaridad,* PRONASOL) which comprised welfare policies and the promotion of productive projects and regional development (Morton 2003: 643–644).

The "interiorising" of external constraints (Nicos Poulantzas) in Mexico encountered a political system which was largely dominated by a complex socio-political structure where the state and the ruling party were decisive for access to political influence and wealth. Within that party, in view of the crisis of the old model of development, the forces oriented towards the world market and the USA asserted themselves. This had already begun under President Miguel de la Madrid (1982–1988) and was intensified under his successors Carlos Salinas de Gortari (1988–1994), Ernesto Zedillo (1994–2000), the first non-PRI president, Vicente Fox (2000–2006), and his successor Felipe Calderón (since 2006), who maintained a high degree of continuity in economic and social policies. The political earthquake – the defeat of PRI – led to a "modernization of neoliberalism" (Morton 2003: 647). The market reform policy process

> was a highly concentrated activity, invariably resting in the hands of a small policy elite from whom highly personalised policy networks spanned outward, incorporating domestic actors who closely collaborated in the market reform project – invariably powerful members of the private sector.
>
> (Teichman 2002: 501, also on the conflicts among different capital groups)

Market liberalism led to a concentration of assets in a small number of conglomerates who gained enormous political power, leading to corruption and cronyism (ibid.).

A further step was the creation of the North American free trade zone. The North American Free Trade Agreement NAFTA, which had been negotiated from the end of the 1980s, was concluded in 1992 and came into force in 1994, provides a central institutional framework for the safeguarding of the neoliberal model. Through it, Mexico became a particularly prominent "partner" of the USA. The Agreement reflects the economic policy paradigm of supply-side orientation and comparative cost advantages: it is above all about rules for investment (as the highest aim and including the securing of intellectual property rights), but also about financing, production and distribution. By contrast, regulations concerning demand or questions of the distribution of social wealth are not mentioned. The regional bloc-building in North America, as elsewhere, is intended to safeguard and advance the processes of economic integration in a certain region by means of a political-institutional regulative system. NAFTA represents a process of the "locking in" of neoliberal strategies which cannot simply be reversed by nation-state policies after joining the Agreement; and especially not by a clearly weaker member such as Mexico (Shaiken 1994). According to David Gordon (1994: 295) its purpose is to secure "institutional harbors promising the greatest protection" for capital.

NAFTA was also the model for a much more ambitious project. Starting in 1998 a Free Trade Agreement for the Americas (FTAA, *Acuerdo de Libre Comercio para las Américas*, ALCA) was negotiated in order to create a free trade zone for goods and services and investment security, and secure "modern" intellectual property rights until 2005 (the NAFTA chapter 11 on investment security was to be copied verbatim). Originally, the project was proposed by George Bush Senior in 1990 with his "Enterprise for the Americas Initiative". In November 2005 at the fourth "Summit of the Americas" in Mar del Plata (Argentina), it became clear that the project had failed because of the resistance of many of Latin America's presidents. The reasons are the enormous popular mobilisations against the FTAA since the year 2000 and the split in the Latin American bourgeoisie between those favouring capital oriented towards the US market, especially agro-business, and those preferring capital oriented towards the internal market, that is parts of the industry. A third reason is the victory of leftist governments in many countries such as Venezuela, Brazil, Argentina and later in Bolivia, Chile and Ecuador, all of which were under growing pressure to respond much more to popular demands than the former liberal and conservative governments had done (García 2006; on the resistance movement, see CLACSO 2004). Additionally, discontent had been brewing due to the fact that the US negotiators had taken advantage of the lack of experience of their Latin American counterparts which led to a de-legitimation of the process (Ruiz Caro 2005: 304). A central aspect of the critique of FTAA was the question of unrestricted access to natural resources for transnational corporations as the distribution of benefits is considered to be highly unequal, an overexploitation of resources is an integral part of free trade and the consequences for the environment are usually negative (Ruiz Caro 2005: 300–301). However, the US Government has continued to push for free trade agreements on a bilateral level with quite some success, especially in Central America, which implies even more favourable conditions for transnational capital from the USA and Canada. With respect to biodiversity issues, the free trade agreements promote intellectual property rights in the sense of the TRIPS Agreement and do not include any provisions on regulated access and benefit-sharing (ibid.: 301–305).

The importance of the USA for Mexican economic policy has in fact grown considerably in the 1990s, not least as a result of NAFTA. Exports more than tripled between 1994 and 2001, and imports more than doubled.[3] Half of the exports are however simply re-imports from the Maquiladora industry, which concentrates more than one million industrial jobs, but the value-added in Mexico remains low (Boris/Sterr 2002: 197; Oxfam 2003: 17; Casares/Sobarzo 2004).[4] This growth is above all due to the low wage costs in Mexico, which is why the Maquiladoras became increasingly important. Foreign direct investment, which is also strongly increasing, takes place, in addition to investment in the Maquiladora industry, in the form of the buying out of Mexican enterprises by international capital in the course of privatisations. In recent years there have been increased efforts to spread the Maquiladora model to other parts of the country, for example the south-east of Mexico.

As with other international agreements, the specific effect of NAFTA depends on its specific implementation. An agreement does not simply "have an effect" because there is a treaty text which only has to be implemented. The way in which the real implementation of an agreement takes place depends, rather, on the structural conditions and relationships of forces. However, "corridors" and terrains for future developments are laid down with the fundamental formulation. Therefore, we can also speak, in the case of Mexico, of a "disciplinary neoliberalism" (Gill 2003). It has been shown in recent years that the USA has repeatedly been able to assert itself in controversial questions. Environmental issues do not play any role in the NAFTA. However, US President Clinton had to stick to his promise following his election in 1992 to conclude so-called side agreements on work and the environment.[5]

Finally, it is important that the conflicts over democratisation and neoliberalism are characterised by increasing, overt state and societal violence. The state resorts ever more extensively to overtly repressive means in order to enforce specific policies. Particularly in rural areas, the interests of the agricultural oligarchy are defended in many cases by paramilitary groups (see www.ciepac.org, Ceceña 2006).

### *5.1.2 Resource and environmental policies in Mexico*

In Mexico, in the transition from the colonial to the neocolonial order during the nineteenth century, the exploitation of natural resources remained central. An important turning-point was the Mexican Revolution (1910–1917), at the end of which national sovereignty over natural resources was postulated and in the 1930s was put into effect in important sectors such as oil production and processing. During the peripheral Fordist phase of development the valorisation of natural resources was intensified and extended to other sectors. In Mexico these were the sectors which had already been important in the colonial period: mineral deposits and agricultural products such as maize, wheat, coffee, cocoa and tropical fruit. Following the Second World War, scientific plant breeding, that is primarily the development of hybrid varieties, was advanced with the support of the Rockefeller Foundation.[6] Since the 1950s cattle-ranching and logging have played a growing role, as has oil since the 1970s, large quantities of which were found in Veracruz and southern Mexico and which was produced on a large scale due to the high prices (Morton 2002: 33).[7] As a strategic resource, oil – which still exists in large quantities – and increasingly natural gas, is important not only for Mexico but also for the USA. In 2002 almost half of the oil produced was exported, the larger part of it to the USA (EIA 2003: 9). In the 1980s the production of vegetables intended for export was added to the list.

There have been repeated conflicts over the exploitation of resources, particularly in the form of conflicts over the disposition and cultivation of land (see the classic by García de León 1985). However, a belief in progress was dominant in which the growing exploitation of natural resources was hardly ever questioned due to the existing relationships of forces. State resource policy in the sense of the efficient valorisation of natural resources was central for the peripheral Fordist development model. The protection of resources was of virtually no importance.

Environmental problems came onto the political agenda in Mexico in the 1970s, however, primarily as a problem resulting from the peripheral Fordist development model. Important in the 1970s was the dreadful air pollution in Mexico City,[8] the precarious water supply of the capital and, later on, the toxic waste of the Maquiladoras located at the northern border with the USA. Less attention was paid, throughout the country, to the destruction of the environment by the massive extension of oil production or other regional conflicts such as the planned erection of an atomic power station in the state of Veracruz.

The first comprehensive environmental legislation was passed in 1971, which was intended to regulate air and water pollution, and waste and sanitary issues (Russell 1994: 231). The reasons for this were internal pressure because of the increasing air pollution in Mexico City, mentioned above, the environmental laws which had been passed in the USA and a growing international environmental consciousness, which was expressed in the *UN Conference on Human Environment* in Stockholm. The environmental law was closely based on that of the USA. In addition, national parks were established (see below). Particularly under the government of Miguel de la Madrid (1982–1988) there were attempts to take some of the edge off possible protests and criticism by the symbolic upgrading of environmental policy. Environmental aspects were taken up for the first time in the national development plan which was developed at the beginning of his presidency. In 1988 a Ministry for Urban Development and Ecology (SEDUE) was set up, thereby institutionalising environmental questions at a high level for the first time. As a whole, state environmental policy remained largely symbolic until the beginning of the 1990s, the financial endowment of the Ministry was modest and the control of the legal requirements weak. In 1988 the General Law on Environmental Equilibrium and Protection (*Ley General para el Equilibrio y la Protección del Medio Ambiente*; LEGEEPA) was passed (the version of 1988 with changes in January 2000 and 2005 is valid): public works and polluting industries were obliged to submit a proof of environmental tolerance, pesticides and the transport of toxic substances were regulated. At the present time there is no perceivable environmental movement to speak of.[9]

For a long time, genetic resources, with the exception of seed banks, did not play any role as a political problem or as a subject of state regulation. It was regarded as normal that the area under agricultural cultivation grew from year to year. It was only as a result of the Green Revolution that in the 1970s the erosion of agro-biological diversity was discussed, and the international discussion played an important role in this. Not only was the political-economic side of the appropriation of nature discussed (who profits from its valorisation?) but, as a part of international discussions, the ecological side also attained a voice. The dramatic erosion of agro-biological diversity in the course of the Green Revolution was politicised by epistemic communities close to the FAO. The Mexican government played an important role in the 1980s when the issue was to question gratuitous access to plant genetic resources. This led internationally to the setting up of the CGRFA (*Commission on Genetic Resources for Food and Agriculture*) within the framework of the FAO and in Mexico, with international financial support, to the upgrading of the International Maize and Wheat Research Institute (CIMMYT). This discussion remained limited to small groups of experts, however, and only took place in an international framework (cf. Chapter 3 on this).

During the administration under Salinas (1988–1994), SEDUE was incorporated into the newly established Social Development Ministry (*Secretaría de Desarrollo Social*; SEDESOL), and at the same time in 1992, around the Rio Conference, the norm-setting National Institute of Ecology (*Instituto Nacional de Ecología*; INE), the executive Federal Attorney for Environmental Protection (*Procuraduría Federal de Protección al Ambiente*; PROFEPA) and the National Commission for the Knowledge and Use of Biodiversity (*Comisión Nacional para el Conocimiento y el Uso de la Biodiversidad*; CONABIO) were established. This was the direct result of the more intensive international discussions on environmental policy since the beginning of the 1990s. Since 1994 there has been an independent Ministry of the Environment (today *Secretaría de Medio Ambiente y Recursos Naturales*, SEMARNAT). In 1996 as a part of the implementation of the CBD, the General Environmental Law was revised and utilisation aspects were added to those of protection. In 1996, in addition, a law on plant varieties was passed which is subject to the Agricultural Ministry and regulates the rights of plant breeders (CONABIO 1997: 30). One important regulation in it is that the property-holding of plants existing in nature is prohibited.

The NAFTA Agreement provides a unique regulation of natural resources, that is a special treatment of oil, electricity and petrochemicals which in fact means that US companies have privileged access to Mexico's natural resources. Since the year 2000, the US government formulated explicitly and integrated into its National Security Strategy that Latin American Natural resources are part of US American national security and therefore of geo-strategic considerations. This is a main reason for the strong orientation towards free trade agreements. Among other things, they should guarantee unrestricted access to natural resources for foreign companies through "national treatment", market access and investment security (Ruiz Caro 2005).

### *The deregulation of the agricultural sector*

The neoliberal transformation of society which began in the 1980s comprised, among other things, the deregulation of sectors which had previously been subject to relatively strong state regulation and were at least partially protected from international competition by tariffs. For the purposes of our discussion, the Mexican agricultural sector is particularly important. Following the debt crisis of 1982, the Mexican economy was "opened up" within the framework of the so-called adjustment programmes. "In hardly any other sector did the elements of neoliberal policies have such an effect as in Mexican agriculture" (Boris/Sterr 2002: 183, our translation). Mexican agriculture with its small farmers had, first, to compete increasingly with the highly productive and highly subsidised US American agriculture.[10] Reforms in agriculture took place before the coming into force of NAFTA, especially between 1990 and 1992. For nine out of eleven basic agricultural goods (except corn and beans) import controls and guaranteed prices were resigned and the central agricultural institution, CONASUPO, weakened and then closed (Yúnez-Naude/Barceinas 2004). The previously food-exporting country of Mexico became, on balance, a food importer (Calderón Salazar 1992: 80). In addition, cultivation in Mexico itself was increasingly directed towards cash crops for export.

The reform of Article 27 of the Constitution in 1992 and the NAFTA Agreement, which came into force on 1 January 1994, are two major institutionalisations of the transformation to a competition state which had been taking place since the 1980s. It was a central achievement of the Mexican revolution that land was given to those who needed it and cultivated it. With this *Ejido* system a form of collective ownership of the land was established which was laid down in Article 27 of the Mexican Constitution of 1917 but which was not applied comprehensively until the 1930s under President Lázaro Cárdenas. Although the norm had been undermined in a masked form since the 1950s by the local *caciquismo*[11] and clientelism – that is, processes of land concentration did take place – nevertheless, the norm represented an important point of reference in the land question (Calderón 1992: 79ff.; Boris/Sterr 2002: 185). With the amendment of Article 27, the agricultural reform which had been laid down in 1917 was ended. This created a legal framework for the permeating of the Mexican agricultural sector by transnational capital. The amended article of the Constitution is not the cause of neoliberal agricultural policy, but an important part of a process which began ten years earlier, which was not clear-cut from the beginning and which was frequently controversial. Victor Toledo (2000) estimates, however, that by the end of the 1990s half of the agriculturally cultivated area in Mexico was still Ejidos or communal land (over 100 million hectares), in which three million families live in 30,000 Ejidos or communes.[12] In addition, about 80 per cent of the natural and managed forests are to be found on Ejidos, as well as most of the biosphere reserves. A large part of the *in situ* biological diversity is also to be found in arable land and forests. And finally, many reserves of drinking water exist on this land.

The restructuring of the agricultural sector must be seen in close connection with the NAFTA negotiations, for the opening of the agricultural sector was a central demand of the US representative there (Calderón 1992: 87). Due to the currency crisis in 1994 and the following economic crisis, ongoing governmental subsidies and the high percentage of small farmers in Mexico, profound took place changes in the Mexican agricultural sector. In particular, economic concentration in the agricultural sector, a collapse of corn production and more imports from the US took place (Yúncz-Naude/ Barceinas 2004: 93–95).

### *The establishment of nature protection areas*

Mexico's first nature reserve was established as a result of the Constitution of 1917 (the León desert in the north), but in the following decades one cannot speak of a clear policy of nature protection. Although many national parks were established, above all in the 1970s and 1980s, they existed largely on paper (CONANP 2007: 2). The concept was still largely propagated, however, that the protection of nature should take place without human intervention. This did not change until the 1990s.

The General Law on the Environment of 1988, in its version of 2000, lays down in Article 3 that the nation, that is the government, has sovereignty and jurisdiction over the nature reserves. By 2006 the reserves comprised more than 22 million hectares, that is more than 11 per cent of the national territory (CONANP 2007). A large number of the nature reserves have not been withdrawn from human use but is part of so-called regional sustainability development programmes; others are priority reserves and are subject to strict regulations. It is emphasised, however, that the protection of nature must be conducted together with the people who live there, which represents a change compared to the 1970s.

The main reason for the protection policies was and remains the enthusiasm of scientists, and moreover, it is an internationally attractive subject and in this field financial resources were and are available. The first National Report by Mexico to the CBD (CONABIO 1997: 14f.) illustrates that a large part of the money for the administration of the nature reserves comes from international institutions, particularly from the Global Environmental Facility (GEF), which is based at the World Bank, or from the state development agency of the USA, USAID. The GEF participates in the financing of ten nature reserves, mainly biosphere reserves (among others Montes Azules in Chiapas). The second Plan for a *Sistema Nacional de Áreas Protegidas* (SINAP II, National System of Protected Areas) got donations from the World Bank, that is the GEF, of more than 60 million USD. Conservation NGOs in particular, and some private companies, are part of it (Delgado Ramos 2004: 70–80; for a detailed study on the role of the World Bank and its "green programmes" in Mexico see Saxe-Fernández/Delgado 2005: 61–87).

Access by private actors to nature reserves is necessary because of the precarious financial situation of state authorities. According to CONANP (2001: 5) the financial backing is about 20 to 25 per cent of the international standard, a fact which illustrates the problems facing the national park administration rather than effectively helping to solve them. The second five-year plan for a *Sistema Nacional de Áreas Protegidas* therefore develops a strategy to win NGO assistance in the administration of the parks (ibid.: 16) and in view of the difficult financial situation to earn funds via the provision of services and from fees for use (ibid.: 9). This corresponds to the strategies of the financially strong nature protection NGOs, such as the *Worldwide Fund for Nature* (WWF), which favours the setting up of 233 so-called bioregions worldwide, or *Conservation International* (CI) which promotes the protection of 12 so-called megadiverse centres (something similar happened with the Mesoamerican Corridor, see below).[13]

López Ramírez (2005a: 232–234) sees the creation of parallel environmental policy institutions here which are, to a lesser extent, subject to state control but which secure access to the reserves for international actors such as the World Bank or USAID. However, the state remains responsible for the granting of prior informed consent and the negotiation of mutually agreed terms. One NGO representative said, "Mexico protects that which can be sold, which possibly has value abroad" (Mx/17, our translation).

Nature protection policy has experienced a particular change since the 1994 Zapatista uprising. Nature protection policy was enlisted here as a mechanism to pursue a state policy of combating the uprising (Mx/9; Mx/10). From this moment on, nature protection policy became a public problem. This is illustrated further below using the example of the biosphere reserve Montes Azules (*Reserva Integral de la Biósfera Montes Azules*; RIBMA) in the Lacandon Jungle in the state of Chiapas.

### *Development projects:* Puebla-Panamá Plan *and* Mexican Mesoamerican Biological Corridor

Two development projects represent an important element in terms of our ability to appraise the transformation of Mexico under competition state conditions and the importance of biodiversity policy. Both projects were already planned in the 1990s, but were only launched after Fox had become president in 2000. The *Plan Puebla Panamá* (PPP) is a central project of the new government for the south-east of Mexico. It was presented in April 2001 by Alfonso Romo Garza, a friend of the president and chief of the Mexican transnational corporation Pulsar (which does not exist any longer), and was signed by the governments of the Central American countries and Mexico on 15 June 2001. Its purpose is primarily to attract foreign investments to tax-privileged Maquiladoras, and to support the valorisation of natural resources (wood, water, oil, uranium and biodiversity) as well as telecommunications, cattle-ranching, export-oriented and technically intensive agriculture and tourism in the relatively economically poor regions which are, however, rich in natural resources and cheap labour, between the Province of Puebla near Mexico City and the country of Panama. This is to be achieved by using comparative cost advantages. The core of the plan therefore consists of "making strategic investments in infrastructure so that the region is more accessible and can use the potential benefits of Mexico's free trade agreements" (Report of the President according to López Ramírez 2005a: 216, and the outline and appraisal of the plan to be found there, our translation; cf. Barreda 2001). These public investments are to be made, in addition to private investments, in road construction, ports, railways and airports. The Inter-American Development Bank (IDB), a subsidiary of the World Bank, is financing 75 per cent of the 4 billion USD project, and the Mexican Government has contributed the rest.[14]

One special aspect of the PPP is the *centros de integración rural* (rural integration centres), by means of which rural populations are to be organised in a centralised manner in order to provide them with access to infrastructure and basic services and thus to prevent emigration. In reality – according to López Ramírez – their purpose is to more effectively control the population under the guise of social policy and also to push for the release of land for extensive projects. The Maquiladoras have a similar effect. Although the plans always mention the improvement of the living conditions of the rural population (education, housing, health), this is hardly reflected in their actual implementation. Rather, trickle-down socio-economic development effects are expected.

The second project, the *Corredor Biológico Mesoamericano* (CBM; Mesoamerican Biological Corridor) from Mexico to Panama and, as a part of that, the *Corredor Biológico Mesoamericano México* (CBM-M), was designed by the national governments, international financial and development agencies such as the World Bank and USAID and conservation NGOs. It has a time span from 2003 to 2010 (the original plan was from 2001 to 2008). The 90 million USD planned for this project are mainly funded by the GEF and the Mexican government and coordinated by the CONABIO.[15] "The global objective of the project is the conservation and sustainable use of globally significant biodiversity in five biological corridors in south-east Mexico, through mainstreaming of biodiversity criteria in public expenditure, and in selected local planning and development practices" (World Bank 2000: 2). The project consists of research into biological diversity and the causes of its erosion in five zones in the four south-eastern states (two zones are in Chiapas), in order to be able to manage the environment and resources more appropriately and as a result to be able to achieve development and the fight against poverty without environmental damage (ibid.: 3). After seven years in "focal areas", which cover 15 per cent of the corridor area, verifiable objectives should be reached (Soberón 2001: 4). The local population is to be involved in the process and will be the main beneficiary (ibid.: 13). Almost 80 per cent of the budget is earmarked for integrating the corridor region into development programmes. "This component will promote removal of institutional, technical and informational barriers that prevent the adoption, in regular rural development programs, of win–win natural resources and biodiversity management options" (World Bank 2000: 9). Another part of the programme is the development of pilot projects to support the sustainable development of biological diversity through capacity building. The concrete projects are being conducted in Chiapas with Lacandon peoples (not with Mayas), who tended to be neutral in the Zapatista uprising and – in contrast to the Mayas, who migrated there at a later point in time – have lived there since the eighteenth century and know the natural resources best (see the overview in Delgado Ramos 2004: 80–89).

López Ramírez (2005a: 230–231) highlights three major but hidden objectives of the Mesoamerican Biological Corridor: the appropriation of biodiversity, counter-insurgency, especially in Chiapas against the Zapatistas, and the interest in the enormous oil reserves in the region. The fact that Chiapas is part of the Corridor means a "preventive occupation of a strategic space for the energetic needs of the United States" because Mexico becomes more and more important for its Northern neighbour (ibid.: 242).

The difference between the PPP and the CBM-M is that the former is more of a traditional development project, which only takes up social or ecological issues in passing, while the corridor places sustainability and the fight against poverty at the centre of attention. It is explicitly directed towards biological diversity and its protection and valorisation. The corridor project certainly claims to have the potential to support regional development and therefore at first glance finds itself in tension vis-à-vis the PPP. Alfonso López Ramírez nevertheless judges the two development plans to be complementary and sees the geo-strategic interests of the US government behind them because it has declared the strategic resources oil, electricity, drinking water and biological diversity to be essential for the "national security" of the USA (López Ramírez 2005a: 223–234; see also Delgado Ramos 2002). Mexico is the second largest supplier of oil from Latin America after Venezuela and in view of its close ties to the USA, it is politically "more reliable" than the South American country. In addition to large economic projects, which do not push for domestic development but rather impel production in enclaves and an outflow of resources (particularly into the USA), the corridor project is also about an increased control of the migration of people – for this purpose the militarisation of the Mexican–Guatemalan border has been taking place for some years now – as well as being concerned with measures to combat uprisings. In view of the general conditions it is therefore likely that the CBM-M will fit into the general valorisation policies.

In recent years a strong opposition has formed in the region, particularly to the PPP (www.ciepac.org; López Ramírez 2005a: 229, 2005b: 178–179; Delgado Ramos 2004: 183–218 on the alternative proposal of an indigenous biological corridor). In July 2002 representatives of over 350 organisations met in Managua to coordinate their activities and to point out the existing and possible alternatives to the export-oriented development model based on forced valorisation in the context of the PPP, NAFTA and the planned FTAA (Declaración Foro Mesoamericano 2002). Moreover, the reorientation of US international politics and its allies such as the World Bank after the attacks of September 11, 2001 slowed the PPP project down remarkably. Some Maquiladora plants were constructed but the huge infrastructure projects such as highways, harbours and airports were not realised. This is also due to the broad criticism of and resistance against the overall and particular projects (López Ramírez 2005a: 228–229). By contrast, the establishment of a Mesoamerican Biological Corridor continues to take place. In the near future and under the recently elected president Calderón new dynamics might emerge regarding the PPP. As a "condition for success" of the projects it should not be underestimated, however, that the restructuring of the agricultural sector outlined above is accompanied by a deep crisis of the rural population. The current formulation of national development projects such as the PPP, at first sight, offer many people a sort of straw at which they can clutch in their often desperate socio-economic situation.

### *Preliminary summary*

In the outlined fields of politics it is obvious that there is a clear tendency in Mexican resource and environmental policies to increasingly valorise natural resources. This is connected with the gradual transformation to a competition state, with subordinate integration into the world market and with the increasing pressure to earn export profits for servicing foreign debt. In addition to its cheap labour, Mexico's natural resources are perceived as a decisive comparative advantage. This is also a complex process of "primitive accumulation", that is the permeation of social relationships with capitalist patterns of socialisation. Oil plays a central role today and will continue to do so in the future. Particularly due to its close ties to the USA and its formulated "national interest" in strategic resources, Mexico has become a preferred site for capital valorisation. The amendment of Article 27 of the Mexican Constitution as a decisive part of the deregulation of the agrarian sector makes possible the privatisation of land and thus access to natural resources.

NAFTA represents a central institutionalisation of free trade policies and security for foreign investors (for example via the strict securing of intellectual property rights) and has profound effects on societal relationships with nature. The same is true of the debt crisis, following which neoliberal strategies and the accelerated valorisation of nature were established. The definition of nature reserves can, in addition to the intended protection of nature, also be interpreted as a strategy to give the Mexican state responsibility and to take it away from the population living there. This could in particular gain in importance if, within the framework of bioprospecting, prior informed consent and mutually agreed terms are to be granted and benefit-sharing is to be negotiated. The national parks are also part of the privatisation strategies. At the same time there was an increased setting up of environmental policy institutions. This cannot only be understood as part of the tendency towards valorisation but is also connected with the perception on the part of the state that it must effectively deal with the erosion of biological diversity. International agreements, particularly the CBD, play an important role here.

The development plans, already discussed, are decisive because they create central preconditions for the valorisation of labour and nature. However, it is the PPP which is a conventional programme as it is intended to enable economic "development" via investments from abroad and the valorisation of labour and cheap resources. It falls back on rather traditional ideas of progress. They become particularly important in view of the fact that they are also heralded as instruments for repressing uprisings. Finally, it is obvious that all the projects (constitutional amendment, NAFTA, protection of nature policies, development projects) are developed and established largely without the participation of the population involved.

Before we discuss these projects in detail, we shall briefly show how different valorisation strategies compete with one another in the south-eastern state of Chiapas.

### *5.1.3 The federal state of Chiapas: competing valorisation strategies*

Today, Chiapas is a "strategic area" for the restructuring of Mexican society. Immense resources are to be found there, including its cultural diversity which is important for specific sectors of production such as tourism, and a large and cheap "industrial reserve army" which reproduces itself largely at subsistence level. This south-eastern state is "the clearest synthesis of that which is to be expected for the whole of Mexico both in the new international division of labour and in the use of its territory and its productive resources in the context of the worldwide restructuring of capital, the redefinition of hegemonic relationships and geo-economic integration" (Ceceña 2000: 262, our translation; Morton 2002: 30–41).

The natural riches of Chiapas had already been made into money some time ago. Directly following the Spanish Conquest, cocoa, vanilla and fine woods became important; rubber and coffee were added later, followed by extensive cattle-ranching and, since the 1970s, oil and natural gas.[16] The state was a central actor in these processes during the colonial period – in the form of the Spanish administration – and following political independence at the beginning of the nineteenth century it secured extraction by private enterprises in the "neo-colonial order" (Halperin Donghi 1991). Its role changed in Chiapas following the Mexican revolution, when the state itself became more economically active. Excellent examples are the extraction of oil and forestry. Besides state activities of resource use in the fields of agriculture, minerals and water or the legal securing of use by private enterprise, in Chiapas the declaration of nature reserves was for a long time the only environmental policy strategy of the federal and the state government. The declaration was essentially conducted by the *Instituto de Historia Natural y Ecología* (Institute of Natural History and Ecology), an institute which had existed since 1942, which has had juridical functions since the change of government in 2000 and which is a type of environmental ministry in Chiapas. The state has proportionally the largest number of reserves in the whole of Mexico, all of which have been declared in the last 15 to 20 years. There are problems with the communities living around the protected areas. Conflicts arise here partly because the population was not allowed to participate in the process (Mx/14).

Of the 7.5 million hectares of land used today, about 40 per cent is used for forestry activities, about one-third for cattle-ranching and about 20 per cent for agriculture (Pólito Barrios Morfín 2000: 62). Within Mexico, Chiapas is the top coffee producer, takes second place for beef production and third for maize. Two-thirds of the population live from a generally precarious subsistence economy and obtain wages from harvesting or transfer payments from emigrant family members. The most important strategic resources in Chiapas are oil and hydroelectric energy. There are large oil deposits in the highlands and in the Lacandon Jungle, for example in Montes Azules. Today, Chiapas produces more than 20 per cent of Mexico's oil and almost half of its natural gas. Furthermore, large, yet to be tapped deposits are also believed to exist there. Thirty per cent of the country's hydrological resources are to be found in Chiapas, and at the present time, a large proportion of Mexico's hydroelectric energy is produced there. The state's biological diversity – as part of a larger region in the south-east of Mexico and Guatemala which is called *Selva Maya* (Mayan Jungle) – is regarded as a further strategic resource, as is stated in the National Biodiversity Strategy (CONABIO 2000: 21). In particular the Lacandon Jungle is a centre of biological diversity with a high share of endemic plants and animals. On one per cent of the national territory there are 4300 plant species, which represent between 15 and 20 per cent of those known in Mexico. The indigenous peoples use 1800 plants medicinally (Ceceña/Giménez 2002; see in detail Barreda 1999). The biosphere reserve Montes Azules is today the hot-spot of biodiversity while the former hot-spot around Montes Azules – Las Cañadas, also a centre of the Zapatista uprising – was heavily damaged by logging in the 1970s and 1980s.

Since the Mexican revolution affected Chiapas only in a very partial manner and considering that the peripheral Fordist development model was not conclusively able to establish itself there, social relationships such as a large share of subsistence production and reproduction and corresponding "traditional" relationships with nature were maintained which in other regions of the country had been heavily modified in the course of the twentieth century (on the complementary relationship of the peripheral Fordist development model and subsistence production see Aboites 1989). This is connected to a large extent with the high share of indigenous people in the population, most of whom have resisted outside attempts to push them towards "modernity" with all its socio-economic trappings. Instead, they have sought to reproduce and expand upon their own multifarious value systems and lifestyles (see for example Kuppe 2002). The old state party PRI articulated itself in areas that comprised a large indigenous population with the "traditional" forms of rule dominant there, particularly the so-called *caciquismo*. The Chiapan oligarchy is strongly oriented towards the primary sector of the economy and the neofeudal relationships of rule and work are secured by a quasi institutionalised racism. Nevertheless, even in Chiapas some changes have taken place due to the complex and dynamic processes of political democratisation, the neoliberal transformation of society and the emerging post-Fordist relationships with nature. With regard to political democratisation, Chiapas can be seen as being of central importance. Democratisation in Mexico, which found its preliminary zenith in the voting of the PRI out of the federal government in 2000, would be unthinkable without the Zapatista uprising. Even in Chiapas the PRI lost the post of governor for the first time in 2000. However, the victorious candidate of the PAN (*Partido Acción Nacional*) and PRD (*Partido de la Revolución Democrática*) alliance, Pablo Salazar Mendiguchía, is hardly any different from the previous governors. Particularly in the question of indigenous rights and the position of the government towards the Zapatista uprising, nothing has changed. Against this sociocultural and natural background there exist different, partly competing and partly complementary valorisation strategies in the individual areas of Chiapas.

The Chiapan oligarchy lives from large real estate holdings which are based on both agriculture and cattle-ranching. Alejandro Toledo Ocampo (1998) names the process which has been taking place since the middle of the twentieth century – which was accompanied by the expulsion of indigenous and other peoples and the destruction of the vegetation – as the transformation of Chiapas into "a gigantic cattle pen". As we have seen, that led in recent decades to massive conflicts due to the increasing land requirements. Chiapas continues to be an important producer of coffee, cocoa, bananas and citrus fruits, all of which are largely cultivated for export (Pólito Barrios Morfín 2000).

In the future the increased extraction of the large deposits of oil and natural gas is planned, which is not being carried out by the Chiapan oligarchy but by private and state firms.[17] The state-owned company, *Petróleos Mexicanos* (Pemex), with revenues of more than 50 billion USD in 2002 and – because of the growth of the oil price – 83 billion USD in 2005, is the seventh largest oil company in the world and produced, with taxes of over 30 billion USD, more than one-third of the state's revenue. More than 9 per cent of the Mexican GDP and 40 (!) per cent of the budget of the Mexican state comes from Pemex and the exploitation and sales of oil and gas (*El Universal*, 1 September 2006). Interests in oil production are therefore directly connected with the state because Pemex is still the "government's cash cow" (*Business Week*, 13 December 2004). Although the states of Campeche, Veracruz and Tabasco have so far been more important for oil production, in Chiapas there are considerable deposits. The potential conflict especially rests on the fact that most of the oil and natural gas is to be found in the east of Chiapas, where over 90 per cent of the population is indigenous (Ceceña 2000: 269).

"External" interests also exist in land for forest plantations for the cultivation of fast-growing, commercially utilisable forests (the interest in fine woods also continues to exist). Mexican companies and international ones, above all International Papers, have direct interests here. The land question is aggravated by the fact that in the last 15 to 20 years in proportion to other states, the largest numbers of natural reserves in the whole of Mexico have been declared in Chiapas. This led, for example, to conflicts in the Lacandon Jungle because the local population was not involved. In addition Elisabeth Pólito Barrios Morfín (2000: 65) estimates that the control over and availability of drinking water will become one of the central conflicts of the twenty-first century. The availability of drinking water is incompatible in particular with oil production in the same place. The industrialisation plans of the outlined development projects do not affect the east of Chiapas but do impact upon the Chimalapas on the straits of Tehuantepec, one of the last undivided tropical natural forests in North America. The effects on Chiapas are difficult to calculate. A further economically important interest, which is however less politically articulated, is tourism, which represents an important source of income especially in the south. And finally, there is interest in the appropriation of genetic resources.

A clear line cannot be drawn between practices of the use of nature which are "protective" (indigenous life-styles, nature reserves etc.), extractive (wood, oil and natural gas) and extensive (cattle-ranching, agriculture). Nevertheless, this illustrates the potential conflicts because the different practices encroach on each other's preserve. These valorisation strategies also conflict – to differing degrees – with the fact that the state of Chiapas and particularly its resource-rich eastern part is populated mainly by indigenous people. The commodification strategies of dominant forces do not meet per se with rejection or resistance by the population, however. On the contrary, these processes and the accompanying promises of improved living conditions ("development") were seen positively by large numbers of the people living in Chiapas. At the same time it is clear particularly in Chiapas that historically and currently a large part of the confrontations between dominant forces aiming at valorisation and subordinate (i.e. indigenous and local farmers) forces is about the commodification of the living conditions of the latter: their integration as wage workers into private capitalist market processes and the changing of their specific relationships with nature.

## 5.2 The formation of a Mexican biodiversity policy

In the following section the question will be examined, using decisive points of conflict, of what form Mexican biodiversity policy has taken in recent times. This is understood as a conflictive terrain of the constitution, articulation and confrontation of interests, problem interpretations and proposals for dealing with problems, and of their political institutionalisation. In a first step different positions are identified before their relationship to one another is determined more precisely, in order then to analyse in more detail important conflicts over the strategic and institutional formation of the valorisation paradigm.

### *5.2.1 Positions in Mexican biodiversity policy: actors, problem interpretations and interests*

As a result of the studies in Mexico, five typical positions can be identified. They are intended to illustrate different interpretations with regard to the problems which must be politically regulated, different constellations of interest and alliances, and diverging concepts of what constitutes an adequate biodiversity policy. The identification of the different types helps to structure the field of Mexican biodiversity policy.

#### *The "neoliberal" position*

The neoliberal position is supported by the transnational biotechnology enterprises which have their seats in other countries. One important actor is AgroBio, an association of five transnational corporations with massive interests in promoting/gaining stronger support for the development of transgenetic plants in Mexico. AgroBio cooperates closely with the US American Biotechnological Industry Association, which is devoted in particular to the international recognition of patents. The essential power of definition of the problems lies with the actors who hold this position: they claim to represent the general interest of the country, namely the desire for progress. The new technologies are the core of development. Technology, which comes mainly from the North, benefits all. The main problem is the lack of a social consensus, which is expressed particularly in the lack of acceptance on the part of consumers. The biotechnology enterprises criticise their own failings here, as not enough information has been provided. For the new developments clearer and simpler state regulations are required, that is regulations with few transaction costs. They should correspond to "the needs of the market". The securing of intellectual property rights takes first priority here. The patenting of living material is the primary objective and imitations should be prevented. Questions of the regulation of access and above all, general stipulations with regard to benefit-sharing have a much lower priority, and should serve merely to improve the conditions of the further valorisation of biodiversity. Protection considerations are hardly to be found in their statements.

The neoliberal position and the forces supporting it are very self-confident. The generalisation of their interests is not without foundation, for in the last 20 years this position has generally been largely accepted in Mexico. It determines the basic direction, has the most direct access to the state apparatus and by far the greatest resources of power and finance. At the same time the protagonists of this position are aware of the highly conflictive nature of its subject-matter. The neoliberal position, as well as the "technocratic-nationalist" position, which will be outlined next, have been coming under pressure from the more recent public debates. Voices of dissent have complained that, "The environmental groups lie a lot, but they work well" (Mx/7).

### *The "technocratic-nationalist" position*

This position is supported by the Environmental Ministry and National Biodiversity Commission (CONABIO), consultants close to them and consultants from research institutes, and a minority in the Ministry of Agriculture. At the international level they are the promoters of the Like-minded Group of Megadiverse Countries (cf. Chapter 2). They too applaud the biotechnological valorisation of genetic resources. The main problem, in their opinion, is the efficient appropriation and further processing of the genetic resources via the development of research and industrial capacity in Mexico. Therefore, a relevant Mexican biotechnological industry must be created. An alliance with research is the necessary precondition for the appropriation of genetic resources. Interest exists particularly in international technology transfer in that the transfer of modern knowledge to Mexico is regarded as being of central importance (Mx/1; Mx/3). That is the best form of benefit-sharing. With adequate development a win–win situation can ensue. They also see themselves as being in a position to formulate Mexico's general interest.

In line with this, the most important cause of the present problems for this position is the blatantly capital-oriented state policy with regard to genetic resources (CONABIO 2000; Mx/1; Mx/3). The dominant tendencies are often criticised. The Ministry of the Economy has an "absolutely commercial attitude" and allows no criticism of the NAFTA or the WTO (Mx/1), and it is downright "religious" in its defence of the economic agreements. The business sector is regarded critically, since it has no interest in a legal framework or in national development. As far as intellectual property rights are concerned, it is stated that TRIPS and NAFTA are very effective and that powerful interests are behind them. However, the basically TRIPS compatible, rigid rights to intellectual property have led, in view of the fact that there is no effective national policy and corresponding institutions, to the situation that the genetic resources end up abroad, where they become valorised without any great benefit to Mexico. The diverse access agreements, despite or even because of their errors, are described as a necessary and important experience (Mx/3).

Conflicts over distribution and questions of participation and decision-making power are given a much lower priority. Benefit-sharing should be "pragmatically defined" (Mx/1). It is regarded as a "social construction" (Mx/3); through a dialogue between indigenous and farming communities and the state (administration and public research institutions) the advantages of the utilisation of biodiversity and procedures for benefit-sharing should be specified. The local communities must understand that in Mexico the institutional and technical capacities have to be set up for the utilisation of genetic resources (Mx/1). Of decisive importance is *capacitación* (capacity building) at the local level (communities and research institutes) and the integration of the use of genetic resources into programmes for rural development (Mx/3).

The international norms of the Food and Agricultural Organization of the United Nations (FAO) and the CBD are regarded as important, although – in the case of indigenous rights – realisable only in 10 to 15 years (Mx/1). Nuances exist, however: some consider the emerging norms to be of more "philosophical value" (Mx/3), which needs to be adjusted to the specific conditions in the individual countries, although no great hopes can be placed on this in the near future. Others consider the drawing-up and implementation of specific legislation to be unavoidable in order to guarantee prior informed consent and benefit-sharing.

As far as the existing relationships of power and rule are concerned, the "technocratic-nationalist" position can be described as pragmatic. It accepts the powerful interests of the dominant actors (businesses and Northern governments) and acts in a "tactically cooperative" manner towards the neoliberal (and the conservationist) position. By contrast, representatives of the democratic and left-wing nationalist positions are described as "extremists" and their positions as "obscurantist" and "medieval" (Mx/1). The concrete criticism of biopiracy in the framework of the bioprospecting agreements is rejected, even if it is not denied that biopiracy exists in Mexico today (Mx/1). There is no attempt on the side of the "technocratic-nationalist" position to strengthen the weaker actors, because otherwise the actors potentially cooperating with Mexican institutions would be scared off and could possibly turn to other countries.

### *The "conservationist" position*

This is a position propagated particularly by NGOs, of which the "Big Three" – the *World Wide Fund for Nature*, *The Nature Conservancy* and *Conservation International* – are the most important (on their policies in Mexico, see Delgado Ramos 2004: 33–42). Historically, they have supported the position that parts of nature should be territorially removed from human influence. This position has a long tradition in the USA where nature protection was often conceptualised on the basis of the exclusion of humans for the conservation of nature. However, this is not possible without severe conflicts in a country like Mexico where many people live directly in the designated protection areas and in a direct relationship with nature. Therefore, in recent years, the conservationist position has been guided, increasingly, by the World Bank models of *grassroots participation* and *listening to the poor*, which have been positively received by the public. They wish to have "a social component" included in legislation and its implementation as well as supporting forms of sustainable use. One example is the "bioregions" supported by the WWF and others. Conservation International Mexico also wishes to create alternative and sustainable sources of income for the local population via economic projects. This type of NGO understands itself strongly as being in the position of an interlocutor between the indigenous communities and the outside world (Mx/12; Mx/16). In addition, they are important for the generation of knowledge concerning the erosion of biological diversity and for the drawing up of inventories. Many NGOs of the conservationist position cooperate closely with enterprises. Many of them have a positive attitude to the privatisation of nature reserves, or at least do not reject it, and involve themselves in their administration ("There are NGOs which believe that privatisation is the solution"; Mx/16, our translation).

### *The "grassroots democracy" position*

Fundamentally critical positions towards current developments are only to be heard from social organisations and some academics: for example from the *Centro de Estudios para el Cambio en el campo Mexicano* (Ceccam; Study Centre for Changes in Mexico's Countryside), which was set up by independent farmers' organisations as a think-tank and lobby group, or from the internationally reputed NGO RAFI (now the ETC Group) with its office in Mexico City, or from the NGOs *Centro de Investigaciones Económicas y Politicas de Acción Communitaria* (CIEPAC; Centre for Economic and Political Investigations of Communitarian Action) and *Global Exchange* which are based in Chiapas.

According to this position, the problem lies in the dominant trend which drives forth neoliberal relationships, transforms life into a commodity and only accords worth to market values (Mx/9; Mx/10; Mx/13). The central accusation in the disputes is that of "biopiracy". The main conflict continues to be rooted in the land question (Mx/10; Mx/9). In addition, it is emphasised that the danger stems from the present trend of forcing communities to recognise property rights with regard to knowledge and nature, which impels continual community fragmentation.

The opponents of this position are, predominantly, the "power groups" (Mx/2), ruthless enterprises which want to appropriate genetic resources and the knowledge connected with them, but also natural scientists who have no sensitivity to social problems. The Ministry of the Environment is accused of using progressive language, for example by speaking of the integration of indigenous people, without putting this into practice. The law is seen as a mechanism to legitimate and indeed ensure that the interests of the dominant economic actors are upheld, irrespective of the social consequences. "The enterprises simply want the legal guarantee so that they can invest" (Mx/10). Nevertheless, there is selective reference to the room for manoeuvre which has been opened up: this is true, in particular, of the regulations on prior informed consent, benefit-sharing and Article 8(j) of the CBD, for example, in order to lend more legitimacy to the Agreement of San Andrés (also Mx/9; Mx/10; see below). The international level is also important because issues are given a political push by international partner organisations (Mx/9).

Altogether, for the "democratic" position, the most important strategy is to create awareness of current developments in the first place. Many farmers did not recognise, for example, the context of the decisive changes in the 1990s and supported the privatisation of land. The definite conviction is that enlightenment is the correct way. The most important current proposal is a moratorium on all forms of bioprospecting. However, the moratorium is not intended to be used to achieve a better position in the "sales negotiations" of Mexican biodiversity. Strategically, the point is to connect questions of human rights with those of environmental justice (*justicia ambiental*) on the basis of *poder local e indígena* (local and indigenous power) and autonomy. The grassroots democracy position has a conflictive understanding of politics. It does not trust in the good will of state actors or research institutes. Questions of self-organisation are important. Therefore, forums on specific issues are repeatedly organised at which activists exchange experience, develop common strategies and place political self-organisation on a broader base.

#### *The "left-wing nationalist" position*

The left-wing nationalist position, which is propagated in particular by a number of academics in Mexico City, is closely connected with the grassroots democracy position. The problem interpretations, lines of conflict and political concepts are similar. One major difference is that this position argues in the name of the interest "of the country". It can be interpreted as an attempt to place the particular interests of the indigenous communities and small farmers in a national context and to represent them as capable of generalisation. This position plays an important role in the public debate.

### *5.2.2 Hierarchy of the positions: dominance of the valorisation interests*

The different positions and the social forces propagating them have a hierarchical relationship to one another. The neoliberal position is dominant, the technocratic-nationalist one is highly relevant and the conservationist position is relevant to a certain degree. The public is aware of the left-wing nationalist and grassroots democracy positions, but they hardly flow into the formulation of state strategies. (There was, by the way, a surprising level of agreement in the interviews between the different protagonists in regard to their appraisal of their own position and that of others.)

One pattern of argument which is common to the technocratic-nationalist and the neoliberal positions is that the "already moving train" of the highly technical appropriation of nature must not be allowed to pass without being boarded.[18] Mechanisms for benefit-sharing should above all improve the conditions for further valorisation (particularly via the establishment of corresponding capacities in the communities and research institutes). At the same time there are considerable tensions between the two positions. In view of the increasing pressure on the neoliberal position on decisive questions, not least due to international political processes within the framework of the CBD and the FAO, the technocratic-nationalist position is more than capable of presenting alternatives to the neoliberal position. Most recently, these concern the question of necessary legislation on access. Even the non-Mexican biotechnological enterprises and research institutes, which have a considerable interest in planning and investment security, appear increasingly to speak out in favour of regulation due to the criticism of the non-regulation of access.

The technocratic-nationalist position weakens itself by its own pragmatism, because in view of the desire to have some of the technology cake it largely accepts the stipulations of the neoliberal position. Interestingly, both positions always speak of the "development of the country", and even Monsanto representatives speak of "our Mexico". Both positions wrestle to define the general interest of society in the sphere of biodiversity policy. The current conditions for a development in the technocratic-nationalist sense are realistically estimated to be unfavourable, but nevertheless, its protagonists believe that with good arguments and clever strategies they will be able to achieve at least some of their objectives. All the same, the technocratic-nationalist position is, possibly in contradiction to its intention, closer to the neoliberal position than it thinks, due to a false appraisal of the relationships of forces and the potential for change. Its strategy, which is aimed at cooperation, plays into the hands of the neoliberal position.

The conservationist perspective is propagated above all by national and international NGOs which attempt to make their protection perspective compatible with the dominant tendencies. This refers in particular to the economisation of nature protection policies. NGOs such as Conservation International or The Nature Conservancy are close, even officially, to private corporations and foundations such as the Ford Foundation (cf. 4.1). Conservation International has been accused in Chiapas of working with the military in order to expel local communities and indigenous peoples and of bioprospecting for transnational corporations (Chapin 2004: 29).[19] They also consciously accelerate the privatisation of nature reserves. Other conservation NGOs such as the WWF follow these strategies perhaps less for reasons of conducting biotechnology business than because of the necessity of funding themselves and due to their acceptance of the fundamental relationships of forces. The idea is that if financially strong actors have an interest in the concerns of nature protection, then why should they be refused? In contradiction to an often formulated claim, the conservationist position is in danger of becoming compatible with the dominant forms of the appropriation of biodiversity. Moreover, financial pressures make the needed cooperation with other NGOs and local actors difficult, that is cooperation "has lost ground over the past decade, only to be replaced by often intense competition, largely over money" (Chapin 2004: 30). The conservation NGOs therefore exaggerate their successes and hide a self-critical perspective as well as working with governments and private companies which are often responsible for the environmental problems NGOs aim to solve. In the last years, it became clear that the strict conservationist position is against the involvement and participation of local people and indigenous communities. However, the consequences of this growing awareness are yet not clear.

The left-wing nationalist and the grassroots democracy positions argue more historically and take the experiences of subordinate groups into account, particularly those of indigenous peoples, in order to criticise the win–win perspective of the other positions. They have had hardly any success, however, in the broad public sphere and have lost influence within the state apparatus. Due to the confrontations of the 1990s, above all those over the recognition of indigenous rights and culture in the course of the Zapatista uprising, shifts have taken place and the concerns of the indigenous peoples have gained in legitimacy. But the dominant state and private capitalist forces and the conservation NGOs take up these issues either highly selectively, namely through the symbolic, but not necessarily legal and material, recognition of certain demands, or they reject them. On the question of the documentary proof of property ownership there is a central difference between the first three and the last two positions. The latter reject it or at least call for precaution, while the former regard it as necessary. The dominance of the valorisation interests and the connected paradigms are illustrated most clearly here.

Yet the technocratic-nationalist, left-wing nationalist and the grassroots democracy positions coincide on four points: first, in their criticism of the appropriation of genetic resources by enterprises and research institutes, although there are considerable differences on how and by whom this should be changed. This fact has been politicised for a number of years under the accusation of biopiracy, of which the technocratic-nationalist position has a rather formal understanding and demands that prior informed consent and benefit-sharing be fulfilled. Through the perspective of the entire country and its necessary participation in biotechnological development, tensions between state authorities, research institutes and the local population are regarded as less important. The democratic and left-wing nationalist positions, by contrast, ask more precisely about the specific contents of prior informed consent and benefit-sharing for the local population. They describe biopiracy in general as the attempt to privatise nature and the knowledge related to its treatment.

Second, all three positions see themselves as confronted by powerful interests. The technocratic-nationalist position is present in the Mexican state apparatuses but in the weaker ones where the actors pursue valorisation strategies as well. This takes place in a more politicised manner than in apparatuses like the Ministry of the Economy. The other two positions face-off with representatives of the technocratic-nationalist position as well as with the antagonistic Northern actors.

Third, they all regard a broad public debate as essential for the creation of awareness and to make suitable policies possible. All three positions trust in the fact that they have the better arguments, while by contrast, the neoliberal and conservationist positions shy away from a public debate on the questions of the concrete appropriation of biological diversity.

Fourth, the problem of supply competition is recognised by all three positions. The technocratic-nationalist position attempts to deal with this by making concessionary offers of cooperation to the neoliberal position, because the latter can confer legitimacy and legal security. A second strategy is to strengthen cooperation with other *providers*, such as can be seen in the constitution of the *Like-Minded Group of Megadiverse Countries* (see Chapter 2). The grassroots democracy and the left-wing nationalist positions also regard international cooperation as important in order to obtain a picture of relationships in other countries and to be able to act together (Mx/10). At the end of 2000 the NGO CIEPAC organised a forum with representatives from six Latin American countries. "And there we discovered that the situation is similar everywhere." For Mexico, a moratorium on bioprospecting is demanded which should be as comprehensive as possible. This is diametrically opposed to the interests of the technocratic-nationalist position.

International regulations at first offer all actors the possibility of referring to them. Thus the neoliberal position emphasises the importance of TRIPS and the CBD access regulations, the technocratic-nationalist position insists on national sovereignty over biological resources and benefit-sharing, the conservationist position emphasises the stipulations concerning conservation, and the grassroots democracy and left-wing nationalist positions also emphasise benefit-sharing, as well as prior informed consent, mutually agreed terms and indigenous rights. That these regulations themselves have a hierarchical relationship to one another has already been described in detail. Important is one other fact: since the actors of the neoliberal and technocratic-nationalist positions played a major role in the formulation of the international systems of regulation, their interpretations of the problems and political strategies are most capable of relating to them.

To summarise, the neoliberal and technocratic-nationalist positions dominate. They are the protagonists of the valorisation paradigm and have by far the greatest financial and legal resources at their disposal. Furthermore, they are backed by the most powerful international agreements. The perspective that the valorisation of genetic resources is desirable and necessary in order to contribute to socio-economic progress establishes itself as by far the dominant view.

### *5.2.3 Conflicts over the strategic and institutional formation of the valorisation paradigm*

In section 5.1.2 central tendencies were shown which led to the formation of a valorisation paradigm in Mexico. The next section focuses on the appropriation of genetic resources, in particular, in reference to the question: in which state strategies is the valorisation paradigm materialised? To avoid misunderstandings it must be understood that the appropriation of biological diversity not only takes place with the aid of state policies; indeed, in many cases it occurs precisely without them. However, from the theoretical perspective already outlined, the state is the form within bourgeois society in which conflicts of interest are primarily fought out and in which the different societal groups struggle for the generalisation and institutionalisation of their interests. The valorisation paradigm does not directly, or necessarily, establish itself, but its concrete form is moulded via social confrontations which are asymmetric with regard to the available resources and access to the state apparatus. In the final section, the preliminary results of this formation are outlined and to lesser extent the confrontations themselves are described. Here we shall no longer deal separately with the policy of nature protection and the project of the Mesoamerican Corridor (cf. 5.1.2).

#### *Competencies and hierarchy of state apparatuses*

The Ministry of the Economy is the most important apparatus and acts – almost all of the interviewees and the analysed texts were in agreement on this – in accordance with the wishes of the dominant interest groups. The responsibility for intellectual property, within the framework of both NAFTA and the TRIPS Agreement, is in the hands of the Ministry of the Economy. "They defend it semi-religiously" (Mx/3, our translation). The National Institute for Intellectual Property (INPI) is responsible for implementation. Here, the Mexican biodiversity policy of the 1990s functioned with the least friction in accordance with the wishes of the dominant neoliberal position. The NAFTA treaty, with very strict intellectual property rights, was ratified in 1993 and has been in force since 1 January 1994. As a result, the Law on Intellectual Property Rights was amended in 1994. The same thing happened with the TRIPS Agreement, which came into force in 1995, to which the Mexican legislation was adjusted in 1997. At that time, there was no opposition worthy of mention to the ratification of the two international treaties.[20] The Ministry of the Economy must also examine all laws, including laws on the environment. The Environmental Ministry suggested, for example, that a condition for the granting of licenses to collect should include room for significant Mexican civilian participation. This was rejected by the Ministry of the Economy with a reference to NAFTA; everyone should be able to collect without restriction, which is the rule which won the day (Mx/1). The Biosafety Protocol, which was developed with high international priority, was also negotiated under the leadership of the Ministry of the Economy.

Responsibility for the CBD is in the hands of the Environmental Ministry (SEMARNAT). The Ministry, which was founded in 1994, has the task of introducing a sustainable development model in Mexico. In the unanimous view of those interviewed, this Ministry has had little political weight ever since it was established and under President Fox it has had even less. Issue areas such as forests and fisheries were handed over to the Ministry of Agriculture (which will have considerable consequences in view of the neoliberal position that exists there; Mx/17). There is no continuity of environmental legislation and its implementation to speak of. During the Zedillo government, SEMARNAT acted "more like an intermediary or facilitator for researchers and private corporations than as guardian of the relevant laws" (Harvey, N. 2001: 1051). The implementation of the CBD is sluggish, not only because there is too little interest on the part of articulate actors such as, for example, a Mexican biotechnology industry, but also because this apparatus has little weight. In the Environmental Ministry, only those active NGOs and experts who basically support the dominant tendencies but would like to give them a technocratic-nationalist touch are listened to (see also the national strategy, which was developed by the Environmental Ministry, CONABIO 2000). Via the National Institute of Ecology (INE), the Environmental Ministry is responsible for granting permission for the bioprospecting of "wild" flora and fauna.

Another important apparatus for biodiversity policy is the Agricultural Ministry (SAGARPA), which is responsible for agro-biological diversity. The head of the last ministry, who has been in office since 2000, is a successful agricultural exporter. One critic sees the Ministry as now being essentially closer to industry and more biotechnology-friendly (Mx/9). In January 2001, only a few weeks after the Fox Government came to power, the "First National Forum for Biotechnology and Biological Security" took place in the Mexican industrialists' club. The organiser was the inter-ministerial Commission on Biological Safety and Genetically Modified Organisms (CIBIOGEM), which had been founded a few years previously, and which has close ties to the Agricultural Ministry. The founder and chairperson of the association is Victor Villalobos, a consultant to the former transnational enterprise Pulsar, and previously a consultant to the company International Papers, who, under the newly inaugurated President Fox, was appointed to the position of Deputy Minister of Agriculture. The Forum took place in cooperation with AgroBio. The Monsanto representative spoke on the environmental risks of genetic technology, and representatives of CONABIO and CIMMYT were not even invited to contribute. A lot of rhetoric was used in reference to "the country" and about the need for "necessary social consensus" and "development", but the programme was very one-sided and provoked the protest of critical NGOs such as RAFI and Greenpeace. In the final declaration, it was stated that under the present government, it was not the Environmental Ministry but rather the Agricultural Ministry which would be responsible for ensuring sustainable development. This illustrates clearly that immediately following the change of government struggles over responsibilities broke out and the CIBIOGEM reclaimed responsibility for biotechnologies and genetically modified organisms (GMOs). The representative of the Agricultural Ministry said, in the context of the biotechnology forum, that it hardly interested anyone whether or not a forum was conducted on access or benefit-sharing (Mx/6).

On the whole, both of the houses of the parliament of the Mexican federation, the Senate and the Chamber of Deputies, play a subordinate role. They are important, however, for juridical initiatives, for example with regard to access legislation.

### *The National Biodiversity Strategy*

Mexico ratified the CBD on 11 March 1993. In Article 6 the parties to the agreement undertake to formulate a national plan, that is a type of inventory and problem synopsis, a national strategy and a plan of action. This process was very protracted in Mexico, which can be interpreted as meaning that the dominant forces are more than happy to live with the existing state of things. The national strategy which was formulated at the end of the 1990s is nevertheless an important document, because here in an extensive process a "national interest" has been outlined, which is not uniform but which nevertheless sets out a dominant perspective with regard to the problems and their treatment. Until now, a country study has been conducted under the leadership of the National Biodiversity Commission (CONABIO 1998) and a national strategy has been formulated as a compromise among different forces (CONABIO 2000). The national plan of action is still on the waiting list. The national strategy is the explicit attempt to formulate a "national interest" in the treatment of biodiversity. Experts have been involved in its formulation, but the process is also said to have been participative (CONABIO 2000: 7), a statement which, however, has been denied by critics. The state document formulates a consistent "we" throughout and thus construes a uniform interest of all Mexicans or even a "we as a country" with regard to a changed treatment of biodiversity. Following the ascertainment of the great importance of biological diversity, it is stated that, "because of its current value, its enormous economic potential, its inestimable services for the environment [*servicios ambientales*] which it provides for all Mexicans, and for ethical and moral reasons, biodiversity must be regarded as a strategic resource for Mexico" (ibid.: 21, our translation).

The main reason for the protection of biodiversity lies in its potential exchange value. The erosion is estimated to be dramatic and a great multitude of strategies are derived from that. The four central sections deal with protection and conservation, the estimation of the value of biodiversity, knowledge and the treatment of information, and the diversification of use. Interestingly, biotechnological utilisation and the regulation of access appear in the first section on "protection and conservation", in the subsection on biological safety (ibid.: 35). That subsection deals with the development of a national biotechnology programme which is to estimate economic value, support genetic technology research in connection with domestic diversity, draw up criteria for safety and "has the objective of appraising and reviving popular knowledge in order to use biodiversity selectively" (our translation). In addition, knowledge and the regulation of access to genetic resources and their utilisation are to be improved, as is the granting of patents, registers or descriptions of origin of those genetic resources which have been developed through domestication, selection or traditional manipulation by Mexican groups (indigenous groups, farmers or others) (ibid.).

Altogether, economically critical remarks, which appear in two places, are insignificant in relation to the general tenor of the strategy: primarily, a stronger awareness of problems and actions is required, particularly a greater "institutional vigilance" and environmental education, a higher evaluation of biodiversity, a legal framework and an "authentic strategy" and "coherent policies". A legal framework is regarded as necessary which defines and protects the intellectual property rights of the indigenous communities and farmers' groups. Tensions are not addressed; instead, they disappear in face of the assumed win–win strategy.

The importance of the strategy is difficult to estimate. As an important discursive element, it certainly expresses a compromise between the neoliberal and the technocratic-nationalist positions. Some references to the problems and interests of other positions are included in several passages. The long-standing director of CONABIO was surprisingly reticent with regard to these central processes and documents (Mx/1). The official strategy is, furthermore, embedded in a multitude of programmes which were formulated in the sphere of influence of the Environmental Ministry. In particular, the change of government such as that at the end of 2000 has led, in Mexico, to the formulation of a National Plan which covers various areas. Remarkable in this context is a document which is concerned with the protection and the increase of natural capital, in particular via or within the indigenous communities (SEMARNAT/INE 2002; official documents never speak of indigenous peoples). Here, it is suggested that the natural resources and in particular biodiversity should be supported by payments to the communities because this would lead to an appropriate estimation of its value, and that biological diversity should be valorised in order to encourage its sustainable use (ibid.: 13f.)

### *The failed initiative for access legislation*

The existing legal framework for the access agreements, the General Environmental Law LEGEEPA, requires, for scientific (Art. 87) and commercial (Art. 87 extended) collections of wild flora and fauna, the prior informed consent of the Environmental Ministry and of the owners of the land. The results of scientific collections must be published. In the case of commercial collections, the owners or more appropriately, the legitimate owners have the right to a fair benefit-sharing agreement. The competence for consent is clearly regulated for nature reserves: it lies with the Mexican state. Legal uncertainties exist if consent has to be given by indigenous communities, since they, as a rule, do not have clear, individual property relationships. It was clear to the experts who formulated the extension of Article 87 to prior informed consent and benefit-sharing in the mid-1990s that a specification was necessary, either in the form of a separate law or via an ordinance (Mx/8). The Environmental Law is regarded by many of those involved as very imprecise and far removed from reality, as well as being too complex. From the perspective of the local application in Chiapas, the criticism is that it has been formulated by people at the central level with little experience and without a great deal of consultation.

The Environmental Ministry, CONABIO and the Senate therefore convened a National Forum in 1997, from which the legislation process was to begin. A group of experts was formed (*Biodiversidad y Desarrollo en México*; Biodem) which was to be the driving force of the process. This group was charged by the Senate with formulating a draft for an access law. In view of the political situation there was to be no wide consultation. One of the participants of the process said that the "political situation in view of the problems in Chiapas, Guerrero and Oaxaca did not allow discussion" (Mx/8). Nevertheless, different actors were consulted. The Ministry of the Economy was not a driving force but referred to its other international obligations in the question of intellectual property, which could not be ignored. Corporations also took part in the consultations, and particularly the biotechnology firms were interested in IPR. The access legislation was only a subordinate concern for them, however, while they attempted to accelerate legislation on biological safety. At the same time, within the Agricultural Ministry – which was responsible for agro-biological diversity and which was critical of the initiative for a law – discussion began concerning a law on rural development (*Ley de Desarrollo Rural*), for which access regulations were also outlined during the discussions. These were not included in the law which was finally passed, however.

The group handed over the project to the Senate in December 1999. In the end it was laid down that the resources in the nature reserves are in the possession of the nation, independent of whether indigenous knowledge (and thus also the bearers of that knowledge) is to be found there or not. In the Senate, at the beginning of the 2000 election year, the law project was not further pursued. Officially, this was due to a lack of time, but a Mexican access law obviously was not seen as being of high priority. The protagonists of the neoliberal position had no interest in binding Mexican legislation on questions of access. From their point of view, Mexico had already fulfilled the obligation to create appropriate access legislation, which it had entered into with the ratification of the CBD.

The technocratic-nationalist position saw the necessity of filling out the Environmental Law. The state should have a more important role in benefit-sharing than is the case in the General Environmental Law. This poses the question of why these actors, who were quite capable of articulation, did not drive forward the legislation more strongly. One explanation lies in the fact that foreign companies or research institutes with which coalitions were intended should not be scared off. This position has a strong wish to participate in the current biotechnological developments. Its perception of the problem is that there is too little valorisation of resources in the country. Apparently, bilateral negotiations appear more likely to be successful than a law. Of apparent impact is the criticism by enterprises and research institutes of the legislative processes in other countries that the laws there tended to "over-regulate". The low importance of access legislation can therefore be explained by the politicisation of the subject and the sparse interest it provoked, or the other priorities of industry. The interest in planning security was apparently not accommodated by the initiative for an access law. The existing framework, in particular the reformulated environmental law, appears to be regarded as adequate. A non-decision, that is abandoning the original initiative for legislation that provided access, was the best manner of corresponding to the existing power relationship of social and political forces.

### *Failed prescription of indigenous rights and culture*

The struggle of indigenous peoples for their rights and culture is not part of biodiversity politics in a narrow sense but sheds light on the actual power constellations in Mexico which affect the conflict field of biodiversity politics. One of the most important Mexican-wide initiatives of the Zapatista rebels and the National Indigenous Congress (*Congreso Nacional Indígena*) – the autonomous association of the indigenous population which was founded throughout the country at the end of 1996 – was the attempt to prescribe indigenous rights and culture in the constitution (Hernández Navarro 1999, 2000). The "Agreement of San Andrés" was negotiated from 1994 to 1996 between the Mexican government and the Zapatistas as well as indigenous peoples in order to strengthen and constitutionally codify the rights of indigenous peoples in Mexico. A compromise was formed in March 1996 which was accepted by the indigenous population, the implementation of which was, however, refused by the government (see Higgins 2001: 886ff.). Therefore, since the end of 1996, an all-party commission (*Comisión de Concordia y Pacificación*; COCOPA) has been attempting to moderate the conflict. In order to lend emphasis to this initiative, in February/March 2001 the Commander of the Zapatistas left the area under their control for the first time and travelled through Mexico to the capital city in order to move parliament to accept the change in the Constitution. Finally, however, only a considerably weaker version of the law was passed, which in turn was rejected by the *Congreso Nacional Indígena* and the Zapatistas as being unacceptable. The changed version maintains the centralist and paternalistic role of the state towards the indigenous population, which leads Harvey to speak of a "corporatist citizenship" which has existed since the Mexican Revolution (Harvey, N. 2001: 1058), and which is actualised in the law by the greater consideration of the interests of private firms in access to resources on indigenous territory.

> In this way, the future model of citizenship proposed by the Mexican government resembles a hybrid of corporatist and market citizenship, in which the state continues to secure political order ... , while the market seeks to incorporate those who are willing and able to work, produce and consume at globally competitive rates.
>
> (ibid.)

When President Fox, in 2001, pushed for a constitutional recognition of indigenous rights, the Senate presented a very different version which was approved against the will of the representatives of indigenous peoples. The weakened version was anchored in the constitution, ratified by the necessary number of individual states and came into force in August 2001. The law was not ratified, however, by the states with the largest shares of indigenous population: Chiapas, Oaxaca and Guerrero (Harvey, N. 2001: 1049f.). Behind the rejection is a widespread unease, which does not only affect Mexico's political class.

> In the same way in which the search for national unity was tied for years to the idea of "de-Indianizing", today the fear of cultural diversity is behind the universalism of the opponents of indigenous rights. Behind the decade-long dominating idea of the inevitability of *Mestizaje* (Mestization) is both the aversion to the Other and the inability to understand the indigenous question not as a "race issue" but as a question of cultural difference ... The Agreement of San Andrés on indigenous rights and culture demonstrate that the indigenous communities exist, that they are alive and prepared to fight. Neither the old nor the new ideology of integration, whether disguised as nationalism or as universalism, could eliminate them. The agreements

> show that they are not "living relics" but political actors with a project for the future, and that in spite of their persecuted cultures they are endowed with an enormous vitality.
>
> (Hernández Navarro 2000: 179f., our translation)

After the defeat of the legal struggle of the indigenous peoples in Mexico for a constitutional recognition of their rights in spring 2001, the Zapatistas decided to stop their public activities and kept silent for more then two years.[21] In the meantime they concentrated on the reinforcement of their own political and socio-economic structures in the almost 600 Zapatista communities in Chiapas. They established *Juntas del Buen Gobierno* (Councils of Good Government) in five *caracoles* (snail-shells) as central expressions of their autonomy and self-government (Bartra 2003).

One success of the Zapatista uprising was that it reopened the discussion on the indigenous people of Mexico. It is not only a question of legal recognition but of a profound change in societal concepts of value and institutional practices. The prescription of indigenous rights and culture in the Constitution, which the indigenous population regards as central and which would have had a great symbolic meaning, could not be enforced, however. That former President Fox supported the initiative on the side of the Zapatistas can be interpreted either as a tactical ploy, or that he hoped in this way to attain more room for manoeuvre for his diverse development and valorisation projects in the South of Mexico. It is too early to say what the strategy of the current president (Calderón) is going to be.

## 5.3 Local conflicts over the valorisation paradigm in Chiapas

In this section we shall deal with the specific strategies for the appropriation of biological diversity. At the same time, other processes – such as the interests in oil or fine woods – must be understood as "the general framework", because the competing concepts and practices with regard to the formation of societal relationships with nature are important for the different strategies for the appropriation of biological diversity. To put it more pointedly: the massive interest in Chiapas's oil superimposes itself on the dispute with regard to the competing ways of dealing with biological diversity. It does not determine the latter, however; in addition, the specific interests in the appropriation of biological diversity are partly just beginning to form.

In two disputes in recent years, the conflicts and their momentum, as well as the role of the state, are concentrated like rays of light by a magnifying glass: in the dispute over the bioprospecting agreement ICBG-Maya and over the biosphere reserve Montes Azules. The ICBG-Maya project does not stand paradigmatically for the practice of bioprospecting agreements but has its own specific characteristics. It is illustrated particularly clearly here that the general strategies – ICBG projects are conducted worldwide

(Fogarty International Center 2007) – must be connected to the specific national and local conditions. The confrontations over Montes Azules can, by contrast, be taken as an example for other cases. The dominant strategies in Montes Azules show how nature protection interests and valorisation interests can be related to one another in a problematical way. Altogether – and in this the conditions in Chiapas are a special case – the conflict cannot be understood without a strong politicisation and organisation of the population. Chiapas becomes a strategic space in which very different actors operate. Central actors and their interests and strategies with regard to the appropriation of biological diversity will therefore be outlined briefly.

### *5.3.1 Central actors and their strategies*

The decisive difference between the nation-state level of politics and the local, that is that of Chiapas, is the different constellation of forces in this south-eastern Mexican state. The intensive confrontations over the establishment of the valorisation paradigm in Chiapas are due to the strong presence of actors with a grassroots democracy position. This is true in the field of biological diversity/genetic resources, but also of the appropriation of other natural resources, the valorisation of labour and the control of the population. An important point of reference is without doubt the uprising of the indigenous communities organised in the Zapatista liberation army EZLN. These were not so much guerrillas in the traditional sense as a militia which had emerged from a large number of communities and was supported by those communities (*communidades de base*). The Zapatistas not only accelerated the Mexican-wide self-organisation of the indigenous peoples through the founding of the *Congreso Nacional Indígena*, which understands itself as a counterweight to the state-corporatist indigenous association, but also gave impetus to the independent farmers' movements (Hernández Navarro 2000).

The Zapatista rebels, the communities supporting them as well as organisations in other parts of Mexico were sensitised during the conflict as to the meaning of genetic resources and traditional knowledge. Until then, the question of the natural resources and their exploitation was orientated towards the "classical" themes of water, oil and natural gas, timber and agricultural production. In February 2001 the *Congreso Nacional Indígena* at its third nation-wide meeting stated its position for the first time on the question of bioprospecting (Harvey, N. 2001: 1054). For this sensitising, the *Consejo de Organizaciones de Médicos y Parteras Indígenas Tradicionales de Chiapas* (COMPITCH, the umbrella organisation of the indigenous doctors and midwives of Chiapas, www.laneta.apc.org/compitch; Wullweber 2004: 113ff.) played an important role. The Zapatistas with their demands for self-organisation and diversity were apparently able to open a space in which new issues and lines of conflict were outlined and in which other actors could constitute themselves and gain self-confidence. COMPITCH was already in existence, but it was only through the uprising that it broke out of its previously defensive shell (Mx/21).

In local forums facts are discussed, strategies are developed and networks are created. Political demands were summarised, for example, in the declaration of San Gregorio, the final document of a forum planned by the umbrella organisation ARIC-I, Independent Rural Association of Collective Interest (*Asociación Rural de Interés Colectivo Independiente*) in the Montes Azules biosphere in May 2000 (Declaración de San Gregorio 2000). Here the accusation is repudiated that the erosion of biodiversity is due to the agricultural practices of the indigenous farmers. A moratorium on the patenting of living organisms is demanded, as is the defence of traditional knowledge, the protection of biological resources from intellectual property rights and a prohibition of the sale of GMOs by national and international enterprises. Different social groups should cooperate to formulate a law in the sense of Article 8(j) of the CBD and Article 169 of the ILO. Furthermore, land rights, health services and education, and the protection of the natural resources in the Lacandon Jungle are demanded. In November 2000, in Amador Hernández, a "forum on demilitarization, indigenous peoples, development and biological diversity", organised by the Zapatistas, took place at which similar demands were discussed, as were the planned large projects and expulsions. The "week for biological and cultural diversity" has already taken place several times, the first time in 2001 in Chiapas, with 171 participating organisations from 16 countries (Declaración de la Semana por la Diversidad 2001). Demands are raised, for example, for the prohibition of the private appropriation of biological diversity and patents on life, for the development of alternative economic projects instead of the Maquiladoras and agro-chemical agriculture, for the revision of the international instruments concerning intellectual property rights and for the recognition of the indigenous people and their autonomy.

One good example of the way in which the rebellion was connected with the confrontations over the appropriation of biological diversity is the NGO CIEPAC, mentioned above. It emerged from the National Mediation Commission CONAI (*Comisión Nacional de Intermediación*), which had been headed by Bishop Samuel Ruíz, who had been in office since 1959, in order to mediate between the government and the Zapatista rebels. In view of the intransigent stance of the government, the bishop withdrew and the commission was dissolved in June 1998. Some office workers founded CIEPAC in July 1998. Beyond questions of peace they are concerned with the complex political, economic, strategic and military processes and with the processes of expulsion and migration in Chiapas. They cooperate with COMPITCH, organise larger forums and smaller workshops in indigenous communities, and produce studies and bulletins translated into various Maya languages. CIEPAC made an essential contribution to making the issues known in the communities and to the position-finding of the communities, and is a major actor which contributed to the politicisation of the

ICBG-Maya project. This was connected not least with the great trust which the former CONAI workers enjoyed, and still enjoy today, in the communities. The international NGO Global Exchange, with its seat in San Francisco, also strongly concerned itself in the beginning with questions of human rights and has extended its work to include, among other things, the subject of bioprospecting.

In Chiapas itself, the actors with a left-wing nationalist position do not play any role. This is apparently connected with the fact that the conflict concerning the appropriation of biological diversity "at home" is much clearer and positions, which have the room for manoeuvre of weaker actors in view, cannot proceed here from within a "national interest", no matter how it may be construed. The direct confrontations of the protagonists of the grassroots democracy position take place with key state apparatuses (government, police, military) and with the local oligarchy which has close ties to them and to the paramilitary groups which are financed by them, but also with those who have a direct interest in the appropriation of genetic resources.

A central political and economic actor was for a long time the transnationally operating Mexican corporation Pulsar, which arose like many company groups in Latin American countries in the course of the privatisations of the 1980s and 1990s. Pulsar (or Savia, which was its name in Mexico) was founded in 1991 by Alfonso Romo Garza, who was one of the closest consultants of former President Fox, as well as being a consultant to the World Bank and a member of the board of Conservation International (CI). At the same time, he plays a central role in the Chiapas Fund (*Fondo Chiapas*), founded in 1994, in which Nestlé, the maize giant Maseca, the bank group Serfin and others are organised.[22] The most prominent enterprise of the Pulsar Group was the agro-biotechnology corporation Seminis, the worldwide leader in the production of fruit and vegetable seeds (22 per cent market share), with R&D establishments in 36 countries, and in Mexico in Chiapas, Las Margaritas and Taniperla. Seminis was sold in 2005 to Monsanto, which made Monsanto the largest seed company in the world and gave this US-based company access to many Mexican natural reserves.

The station in Taniperla primarily conducts genetic technological research on plants. In another field of activity, the former Pulsar and now Monsanto subsidiary, *Desarrollo Forestal* has begun to plant 300,000 hectares of eucalyptus for the production of cellulose and simple timbers in three Southern states.

The enterprise supports, together with the Ministry of the Environment and CI, three research stations in the nature reserve Montes Azules in the Lacandon Jungle. Already in 1996 Pulsar raised 10 million USD for the work of CI in the Montes Azules biosphere reserve (Harvey, N. 2001: 1054). It is suspected that bioprospecting is conducted there, which is not necessarily illegal since, jurisdictionally, Pulsar was a Mexican firm and sometimes called the "gene monopolist" (on the firms and their interests see Pólito Barrios Morfín 2000: 67–76).

Conservation International, which as mentioned above takes up a conservationist position, is a further central actor.[23] The branch of CI in Mexico was founded in 1990 and began its work in Chiapas in 1991, among other things with so-called *debt-for-nature swaps* (Harvey, N. 2001: 1054). CI-Mexico attempts in the projects to "couple" the indigenous population to the planned conservation measures. In this way, detailed studies on biological diversity and the knowledge of its treatment are conducted (Delgado Ramos 2004). Neil Harvey (2001: 1055) sees the attractiveness of the CI and Pulsar projects for the local population in the fact that the latter lacks an alternative. The interest in access to biological diversity remains dominant. However, "As in the highlands [a Zapatista stronghold, the authors] the biggest obstacle facing CIMEX and Pulsar are those communities that participate in the area's indigenous organizations." The WWF also plays a role, because it operates in Chiapas with the international project of the so-called bioregions.

The political institutions in Chiapas have little room in which to manoeuvre. Almost all of the legislative competence in the areas of the protection of natural resources and biodiversity are in the hands of the nation-state. The work of CONABIO is very centralised and there are no local consultations. The subject of intellectual property has never been discussed in Chiapas at the level of the state (Mx/14). Responsible for the implementation in Chiapas is the state government, the Institute of Natural History (*Instituto de Historia Natural*) (which has its own collections but does not conduct bioprospecting) and the Ministry for Agricultural Development (*Secretaria de Desarrollo Rural*). In view of the strong centralisation in Mexico, the other apparatuses of the state executive in Chiapas also tend to have the function of transmitting federal policies downwards.

This brief outline of the most important actors and their strategies illustrates that there are enormous differences in resources between the protagonists of the individual positions. Pulsar first, and now Monsanto and CI invest considerable sums of money in conservation activities in Chiapas, have research stations at their disposal and have access to many areas and communities. They have direct access to the government and have an excellent international network. In view of the financial problems of the Mexican government in administering the nature reserves effectively, it is dependent on personnel and financial cooperation and has formulated this explicitly as its strategy.

The most important assets of the grassroots actors are their own public education, politicisation and organisation. Few financial resources are available. Furthermore, they operate under the conditions of a strong police and military presence, which was markedly increased following the beginning of the Zapatista rebellion, and against a reactionary local oligarchy. International politics plays a role in the specific confrontations which refers to the regulations concerning the rights of indigenous peoples or prior informed consent. There is very little access to the formulation of national policies.

The bioprospecting agreement, ICBG-Maya, which will be described in the following section, is an interesting case study.

### *5.3.2 The bioprospecting agreement ICBG-Maya*

In the summer of 1992 the research programme *International Cooperative Biodiversity Groups* (ICBG) was founded by the US Health Department, the Ministry of Agriculture and the National Science Foundation, among others. This group was intended to support bioprospecting in Southern countries which was financed by the US government. A first round began in 1993 with five projects, and a second in 1998. For the latter's six projects in ten Southern countries over a period of five years, altogether 500,000–600,000 USD per annum were provided. In 2005, a third period started with an overall funding of 6 million USD. The objective is to find therapeutic substances for cancer, HIV-AIDS, tuberculosis, malaria, heart and lung disease and other illnesses. These should, according to the third call for tenders, take place via innovative approaches with regard to access and benefit-sharing. Biodiversity should be protected through the discovery of bioactive substances, that is an extremely close connection is created between the discovery of natural substances with medicinal value, protection and economic impulses (FIC 2007; NIH 2003; Delgado Ramos 2004: 75–77).[24]

Other preconditions are that prospecting in Southern countries needs to be conducted in close cooperation with research institutes based there, the project management is in the hands of a US institution and that a person named by the US government participates. Participation by a private firm is seen as desirable, in particular with regard to further utilisation and benefit-sharing (see the current call for tenders in the FIC 2002). The CBD is named as a part of a rapidly developing international framework, which is why access, prior informed consent, benefit-sharing and IPR agreements should be carefully formulated. This is connected with the uncertainties with regard to benefit-sharing and intellectual property rights in the first two rounds (see FIC 2007, Final Report). A recently founded *Centre for the Management of IP in Health R&D* is to provide guidance to the applying groups with regard to formulation (www.mihr.org). Critical voices contend that the ICBG project is less a strategy for supporting economic development in the providing countries, and more a strategy related to the US government's wishes to secure access to the genetic resources in Southern countries and to give the rights to this above all to US research institutes and firms (Ceceña/Giménez 2002). In relation to this argument, even on its own, the focus on the complex development of medication for certain diseases leaves hardly any room for manoeuvre for local developments beyond direct work for the projects. According to Delgado Ramos (2001: 491) that is precisely the function of *Bio-Maquiladoras*, that is less skilled and cheap production for export.

In Chiapas the research institute ECOSUR already participated in the tender in the first round in 1992, but was not awarded the contract. In the second tender in 1997 it was successful. The central figure in the application and in the later project was a US American anthropologist from the University of Georgia who had been operating in Chiapas since 1992. He was also a guest researcher at the US *National Institutes of Health* (NIH), which essentially coordinated and funded the ICBG projects. The implementing organisation in Chiapas was ECOSUR, an institute seated in San Cristóbal (and which was also present in three other locations). ECOSUR, like the US anthropologist, had had contact with the indigenous communities in many other projects. For the second round of applications the NIH demanded the integration of a commercial pharmaceutical enterprise. ECOSUR did not desire the participation of a large corporation and therefore chose *Molecular Nature* from Wales which had, at the time, 14 employees. One important difference to the other ICBG projects worldwide, according to a representative of ECOSUR, is that the Maya project was intended to be transparent from the beginning (Mx/11). The protocol was published, for example, and the idea that the research should be useful to the indigenous communities was also novel. The indigenous population was not included in the development of the project, however, and prior informed consent was not provided. The indigenous people's participation was to take place in the course of the project, particularly via an NGO with the name of *Promaya*, which at that time was still to be founded and which was to be involved primarily in the decision on benefit-sharing. The subsequent invitation to participate in Promaya was, however, rejected by the indigenous population.

The interest of ECOSUR was that with the ICBG-Maya project, work could be financed for five years which otherwise would not have been able to take place, particularly taxonomic work, the training of workers and the improvement of the research base, above all in terms of infrastructure, with the building of laboratories. At the beginning, it was not intended that ECOSUR should receive technical equipment for their activites, but it was successful in overcoming such restrictions in later negotiations. ECOSUR was to receive about one-third of the *premio total* (total premium) particularly for wages for over 20 *técnicos indígenas* (indigenous technicians). The equipment was bought by Georgia University and sent to Chiapas in return for a credit note. Certain research, of which ECOSUR was not capable, could have been conducted in the biochemical laboratories of this very US University. The interest of Molecular Nature was to develop potentially commercial substances and then to patent them. In a preliminary form, ECOSUR had the idea that all participants, including the indigenous organisations, should share in the intellectual property of developed products. But finally no co-patents or anything similar with ECOSUR or the indigenous communities were planned. Technology transfer was also not to take place in the form of the transfer of technology from Molecular Nature or

Georgia University to Chiapas. It would have been about "something simpler" than patents (Mx/11).

The ICBG-Maya project was granted permission in 1997, but due to the Zapatista uprising, it did not begin until May 1999 and was conducted only until October 2000. Up to that point, according to the ECOSUR representative, there was no transfer of resources to Molecular Nature. In the first year, furthermore, the concern was based on the two types of permission for the collections: "scientific" and "biotechnological". The application to SEMARNAT for bioprospecting was not approved because the prior informed consent had not been realised. Therefore, up until the end of 1999 there had only been scientific collections, and afterwards even that permit expired. Since it was not renewed, in the course of 2000 no further collections took place. Instead, more than 20 people from different communities were trained, all of whom were to serve as "contacts" between the communities and the prospectors. Among other things a play was developed in order to present the project to the communities and to inform them. This was to serve as a basis for obtaining prior informed consent. One group went into the communities and spoke to the traditional authorities. Some of these indigenous leaders then came to ECOSUR, saw the play and returned to their communities. If the community agreed, the play was presented there also. The communities – each member or the authorities – agreed in 50 cases. However, this was not yet about benefit-sharing. Although by the beginning of 2001 (between the moratorium and the end of ICBG-Maya) a proposal for benefit-sharing had been submitted, a definitive answer had yet to be given. ECOSUR's position at that point was that the indigenous communities should organise themselves, and conduct their own negotiations with the enterprises in which they would be able to say no. In September 1999, criticism of the project by NGOs, indigenous communities and critical media began. A major role was played by the association of indigenous medical practitioners, COMPITCH (Mx/14). They were considered to be the guarantee for access to indigenous communities in a legitimised manner – and turned out to be the major obstacle after they rejected it. In October 2000 the project was suspended by the announcement of a moratorium by ECOSUR. Georgia University and Molecular Nature had no other choice but to accept the moratorium because ECOSUR was the implementing organisation. The NIH paid nothing for a year. The project, such was the situation at that time, could only have been continued if an indigenous group had formed to negotiate benefit-sharing. But that did not happen. One year later, in October 2001, the project was abandoned. It is not the first ICBG project to be prematurely disbanded.

The ECOSUR representative is of the opinion that the conflict escalated because of the existence of non-indigenous advisors who had an interest in its failure and because of biased press which did not want to hear the ECOSUR arguments. Critical NGOs participated, which were sometimes right but occasionally "hit out without seeing the implications" (Mx/11).

The indigenous communities had cooperated well, which was not surprising considering the deplorable living conditions under which many of them existed. "They welcome every initiative." Therefore the criticism concerning prior informed consent was unjustified. However, ECOSUR admits that the lack of participation of the indigenous population was a mistake and names this as a central criterion for future projects. Nonetheless, this position was not heard so clearly from state actors in Mexico City. At the same time ECOSUR was criticised due to the lack of clear concepts as to how prior informed consent could in fact be achieved. The ECOSUR representative is sceptical with regard to whether, in the case of concrete prior informed consent regulations, these could do justice at all, in light of the particular cultural notion of participation held by indigenous people.

For the protagonists of the grassroots democracy position, the ending of the ICBG-Maya project is a significant milestone in the struggle against the illegitimate appropriation of biological diversity. The project was criticised because the negotiation of the agreement took place without the participation of the indigenous population, there was no prior informed consent and the collected samples were to be taken to the USA, out of the reach of indigenous people. In addition, the inherent tendency in the ICBG-Maya agreement, to turn resources and knowledge which were collective goods into private property, was also rejected. Another matter to be criticised was that it remained completely unclear as to how patents on knowledge and resources would affect the living conditions of the providers (COMPITCH 2000 according to Wullweber 2004: 117ff.). The communities would first have to understand prior informed consent. NGOs have an important educational task here. In addition to the lack of information, a methodology with which the communities could define what their benefits should be was also missing; they must define these themselves, and for the long term, independently of NGOs. The firm must listen to the community, not the community to the firm, by, for example, letting the latter define the type of *beneficios* (benefits) (Mx/16). Bioprospecting must take place in a very controlled way and be slowed down as long as these controls and clear legal relationships do not exist. Therefore, the moratorium and the abandonment of the project were regarded as positive outcomes (Mx/14).

ECOSUR's approach of placing value on transparency, at least in certain questions, displays an obvious difference in comparison to other, mostly secret, agreements. It is to be suspected that ECOSUR and the Environmental Ministry wanted to proceed more cautiously than in the case of other agreements in view of the high degree of politicisation in Chiapas. ECOSUR claims to have had good intentions and sees itself as the victim of a slander campaign. It becomes clear that bioprospecting depends decisively on the (non-)information of the population which shares knowledge and resources as well as on the legitimacy of the collectors. A lack of information in the communities was, and is, a favourable condition for the appropriation of genetic resources. If the potential providers are more informed, this increases their chances of self-determination. Although information, international and national political processes, and in particular the work of critical NGOs certainly play a role, it is also evident that they require local conditions in which this information can be spread and judged to be politically relevant. This is where the political character of the often technically understood concept of *capacity building* lies. The relationships in Chiapas, both the requirements of the access agreement and the criticisms of it, are a good example of how international processes are conveyed through local ones. The practical criticism of ICBG-Maya would have taken place anyway, according to the activists (Mx/10; Mx/13). However, the specific constellation in Chiapas, characterised by the Zapatista uprising, not only postponed the beginning of the bioprospecting project, but also "simplified" the politicisation of its subject-matter, because the social organisations were themselves politicised. NGOs which were originally active in work on human rights and peace concerned themselves with it, and the Agreement of San Andrés was a major milestone (Mx/10). In this understanding, the uprising served, in a sense, as a catalyst (Mx/13).

Bioprospecting continues to take place in Chiapas despite the abandonment of the ICBG-Maya project. Collecting very probably takes place in the research stations supported by former Pulsar and CI-Mex. So-called and self-designated "intermediaries" go into the communities and collect plants without permission. Nevertheless, the dispute over the project is very important because an organised and articulate population questioned the dominant forms of appropriation. They did not do so with the argument that knowledge should not be shared. Rather, they criticised the fact that they were not allowed to participate and that the entire context of the bioprospecting was clearly dominated by interests pushing for the privatisation of nature. Thus, future and better formulated bioprospecting agreements which take prior informed consent and benefit-sharing into account remain trapped in the dilemma that highly unequal relationships of power exist and that the dominant valorisation paradigm is rarely questioned, apart from its criticism by the grassroots democracy position. The objectively existing supply competition and the time-pressure created by dominant actors apparently leave little room for the opening-up of learning processes.

### 5.3.3 The disputes over the biosphere reserve Montes Azules

The disputes over the nature reserve Montes Azules (*Reserva Integral de la Biósfera Montes Azules*; RIBMA) in the Lacandon Jungle in Chiapas represent an example of how developments are interpreted and what interests form alliances in the process. Although RIBMA, which covers more than 330,000 ha, was declared a nature reserve in 1978, at first, none of the measures connected with the declaration were implemented. In the course of the 1980s the protection of tropical forests gained in importance and large nature protection NGOs identified the so-called "hot spots" in tropical regions which were to be protected. Since the 1990s the World Bank-based Global Environmental Facility (GEF) co-finances the administration of the RIBMA.

As is the case anywhere else, around the RIBMA, different actors follow different interests (see also Ceceña 2000: 270 and section 2.3). The area is a tropical forest in which timber companies are interested in obtaining licenses to operate. Following logging, rapidly growing eucalyptus trees are to be planted in order to produce cellulose for the production of paper. In addition there are large oil deposits in the area. Despite this fact, the partially privatised petroleum company *Petróleos Mexicanos* (PEMEX) temporarily withdrew from the area in 1994. Furthermore, the Lacandon Jungle is a centre of biological diversity, for it is believed that in 0.16 per cent of the area of the state, 20 per cent of its biological diversity is to be found. And finally, this area is located in the heartland of the Zapatista rebels. This has led, particularly since 1999, to a sharpening and politicisation of the conflict. In the region a road was to be constructed from the community Amador Hernández to one of the largest military bases. After one year of daily protests President Fox stopped the project in December 2000 and soon afterwards the military base was withdrawn.

In April 2000 there was further politicisation when the WWF alarmed the public by stating that the local indigenous farmers had caused dozens of forest fires. This was based on satellite photos which identified various fires. (It was discovered later that the fires had not been observed in the named area but rather had taken place quite some distance away.) The public dispute was about the extent to which it was necessary to resettle 12 indigenous communities in order to prevent the fire-induced destruction of the RIBMA. This was also legitimised with the argument that the reserve was not only ecologically valuable but was also the basis for the Mesoamerican Corridor (cf. 2.2) in which tourism played a significant role. Through the intervention of the Environmental Ministry it was agreed with ARIC-Independiente, an association representing the pro-Zapatista communities, to negotiate over the distribution of land and resources in the region. At first, the national government offered one hectare of land to the families who were prepared to resettle; ARIC-I responded by demanding five hectares as well as guaranteed land titles. When the government then offered the equivalent of 2000 USD per individual in the case of resettlement, ARIC-I refused the offer. They stated that they were not begging for money but wanted to see their rights accepted and suspected that the main reason for the resettlement was the attempt to weaken the Zapatistas (Harvey, N. 2001: 1055f.).

A recent development sheds light on a specific strategy: the European Union funded in 2003, with 31 million USD, a project called *Programa de*

*Desarrollo Integral de la Selva Lacandona* (Programme for Integrated Development in the Lacandon Jungle) which intended to contribute to the restoration of the destroyed parts of the Lacandon Jungle and, among other things, to educate local people to provide ecosystem services such as eco-tourism or reforestation. After the ICBG-Maya experiences the concept "bioprospecting" was taboo. For many local observers it was a hidden part of the programme, even so. For example, in order to register for an eco-tourist activity local people had to deliver information about biological resources and knowledge to state authorities. However, when the company Sanofi approached COMPITCH in 2005 in order to get consent for access to resources and knowledge about dealing with female diseases in the indigenous communities (and offered for this generics for many local diseases), this was rejected by the organised indigenous communities. Chiapas remains a highly politicised terrain for the disputes over genetic resources. But local actors are aware that the US-oriented strategy of coercive appropriation of genetic resources prevails.

Through the forced resettlements, a (supposedly) empty area would be created in which state authorities would be responsible for permitting bioprospecting and prior informed consent, thereby conveniently bypassing the need for the consent of the local population. Decisive for the issue is the fact that people can be displaced with the argument that they are destroying biological diversity. Here, Montes Azules could be seen as a further example of a specific combination of protection and utilisation strategies (for the Brazilian Amazon region see for example Acselrad 2002).

In Mexico it has now become particularly clear that protection and utilisation are conveyed in a specific way, in the context of the valorisation paradigm, at the cost of the local population. Roughly, two variants can be differentiated, both of which refer highly selectively to the CBD:

a) The "hard" variant is followed if there is interest only in resources and not in knowledge. Then protected areas can be declared, which means that the rights of disposal over the genetic resources are transferred to the state. Prior informed consent from the local population is no longer necessary and benefit-sharing is negotiated with the state. The connection between protection and the "hard" form of bioprospecting functions as follows. The Mexican state is responsible for land and resources. In a way this is a legal form of dispossession of the people who live there. This could be the reason behind the forced displacements in Montes Azules. If a dominant strategy for the protection of biological diversity is to make it economically attractive, then the declaration of nature reserves and bioprospecting are compatible. In this way the participating state actors (government bodies and research institutes) and private actors (firms, NGOs) rid themselves of the hitherto diffuse and politically vulnerable necessity of obtaining prior informed consent and of setting up benefit-sharing mechanisms. Bioprospectors such as Pulsar, or institutes financed by them, are also subject to national legislation such as the General Environmental Law.[25] The strategy also plays a role here of involving private actors to a greater extent in the administration of nature reserves. In the case of international prospectors, the Environmental Ministry would be responsible for "wild" biodiversity and the Agricultural Ministry for agricultural diversity.

b) The "soft" variant of the "reconciliation" between economy and ecology is applied when traditional knowledge is needed as a filter for the bioprospecting. In this case, prior informed consent must be obtained from the local population and a benefit-sharing mechanism must be negotiated with them. Nevertheless, the prospectors can continue to count on some systematic disadvantages for the communities. In particular, the division of the communities can be exploited, that is they are placed in a position of competition with one another. As we have shown in the example of ICBG-Maya, however, organisational processes for demanding prior informed consent are achievable. A possible bioprospecting strategy for the future could be to circumvent the legal requirements of access agreements, for example by involving "only" Mexican actors in bioprospecting. Criticism of biopiracy and bioimperialism could also become more difficult because of this. It seems that this "soft" strategy is more favoured by the European Union whereas the "hard" version, via the military and police, is the stance taken by the US government.

## 5.4 Summary

Although the use of "Fordist" resources, such as oil or timber, has not been systematically examined, the increased interest in their valorisation is also central to post-Fordism. Their valorisation is becoming more important in view of Mexico's foreign debt, economic integration with the USA and its increasing export orientation. Until now biological diversity and genetic resources have played a rather subordinate role. Nonetheless, even in view of the obvious conflicts, uncertain developments and high expectations in this field, biological diversity has been paid some attention by enterprises, state actors and social organisations. The first issue concerns the controversial possibilities and forms of valorisation and their importance for the socio-economic development of Mexico and in particular for the places in which biodiversity is to be found *in situ*. Second, specific natural geographical conditions, the erosion of biological diversity and the accompanying attempts at protection play an important role. The question is whether and if so how specific living conditions in specific places such as Chiapas become the terrain for a further round of primitive accumulation.

We can speak of a valorisation paradigm because the most important actors – those of the neoliberal and technocratic-nationalist positions together with the conservationist position – do not, in substance, question that the issue is one of valorisation and that this, furthermore, is the most effective form of protection. They are therefore able to strengthen specific perspectives and institutional practices with regard to the treatment of biological diversity in the process of the post-Fordist transformation of society.

It can be stated that at the core of the disputes is the crucial question of control over social processes. The firms attempt to control seed and pharmaceutical markets for the valorisation of capital, for which property rights are a decisive precondition; the Southern governments aim for control over their territories in order to be able to practise valorisation and protection strategies as effectively as possible; the organised local population, in turn, struggles for control over its own social relationships.

### *5.4.1 Non-simultaneous valorisation: Mexico and Chiapas*

In spite of considerable differences, a broadly shared understanding of the problems concerning the valorisation of biological diversity has developed at the *federal level*. Prominent actors accept each other as the legitimate representatives of their interests, and in spite of all differences, largely shared perspectives emerge and those actors with fundamentally different perspectives and interests are kept "outside". The national sovereignty of the state over the natural resources to be found on its territory is just as uncontroversial as the necessity of securing access to genetic resources. The protection and utilisation of biological diversity are seen from within an economic perspective. The main conflict within the valorisation paradigm is over the ranking of "Mexican" interests and the role that is to be played by general, and in particular intellectual, property rights. The possibilities of valorisation have become part of a discourse among experts, and knowledge concerning existing and possible alternatives is suppressed. The grassroots democracy and left-wing nationalist positions are only able to articulate themselves to a limited degree and they are not included in the processes through which state policies are formulated.

Although we can speak of a broadly shared understanding of the problems and mutual recognition of interests on the part of the three dominant positions, it still remains controversial and unsettled as to what forms the political institutionalisation should take. This is expressed in considerable tensions between the apparatuses. In the Ministry of the Economy a valorisation strategy dominates, or is condensed, in which alternative forms of the appropriation of nature do not play any role. Genetic resources are seen as a further resource in international competition. In the Agricultural Ministry those interests are felt more strongly which take the material problem of the erosion of plant-genetic diversity into consideration. Finally, in the Environmental Ministry, the technocratic-nationalist perspective takes priority, but weaker interests also appear to play a role. Here, if anywhere, there is an attempt to combine protection with sustainable use. It emerges, however, that here, too, the valorisation paradigm remains decisive (e.g. in the case of the Mesoamerican Corridor). In addition, the responsibilities for the implementation of international agreements are allocated correspondingly. The Ministry of the Economy is responsible for the WTO and the TRIPS Agreement, and the Environmental Ministry for the CBD. The dominance of the Ministry of the Economy is ensured, among other things, by the fact that it has a right of veto in all legislative processes. In addition, it would fall short of the mark to understand the apparatuses as the expression of competing interests. Even the Environmental Ministry contributes to the establishment of the valorisation paradigm, although in a more fractured way and taking other interests into account than the Ministry of the Economy.

The specific contents of regulations depend not only on what meaning certain actors attach to the regulations but also which actors have a privileged access to the regulating state apparatuses. This cannot simply be operationalised, however, in the sense that it can be examined whose voice is most heard. Biotechnological industry without doubt has far greater access than critical NGOs or marginalised actors. In the state apparatuses themselves, however, selectivities are embedded. A national competition state has emerged even in Mexico: state strategies are guided primarily by the production of competitiveness; attained via attracting foreign investment, pushing for privatisations and the dissolution of the material content of corporatist structures (on paper and as a clientelistic structure, they continue to exist). In contrast to the states of the capitalist metropolis, foreign debt and the accompanying "structural adjustment programmes" are a central condition of internal policies in order to establish interests. This is also true for juridical questions. In the case of regulations there are "corridors" within which certain, but not all, disputes take place. It has been shown that these regulations set a certain political emphasis and thus strengthen certain processes and interests. Furthermore, it becomes clear that the concrete shaping of policies depends on the opportunities for *power bargaining* by more powerful actors. In this context "the lack of a legal culture" was something that was continually criticised throughout the conducted interviews. The difficulties in creating political coherence are, however, not only to be understood as tensions between different state apparatuses, but these tensions are themselves expressions, or rather condensations, of contradictory societal relationships in the apparatuses. The technocratic-nationalist position recognises the neoliberal position but has a different perspective with regard to valorisation and the role of Mexican institutions. According to this position, it is the role of the state to provide a terrain on which at least the dominant interests can wrestle for, and eventually find, compromises; this does not eliminate the contradictory interests but they are made pliable. Furthermore, against completely opposing interests – such as those of the grassroots democracy position – the compromises can if necessary be enforced using violence. This is illustrated particularly clearly in the question of intellectual property: here, in future, compromises between the first three positions are conceivable. At the same time, the interest in the patentability of genetic resources continues to be diametrically opposed to collective use by small farmers and indigenous peoples. Behind the difficult to achieve compromises are also very weighty material problems, which are reflected within the different positions. A forced valorisation of nature and genetic resources is not at all compatible per se with the necessity, which has been articulated politically by various sides, of stopping the erosion of biological diversity. Mexican biodiversity policy is therefore to be interpreted as an attempt to stabilise societal relationships and the opposing interests involved in them in a specific field of conflict. This takes place in a rule-forming way such that certain interests are either not taken into account (declaration of nature reserves, formulation of the draft for an access law, ICBG-Maya formulation) or are violently suppressed (Montes Azules, the repression of the Zapatista uprising as a whole).

The social relationships of forces at the *local level*, for example in the state of Chiapas, must be differentiated from the nation-state level. The strategies for the valorisation of biological diversity compete here in concrete places *on the one hand* much more clearly with other strategies and powerful actors, who for example are more interested in the production of oil, in hydroelectric plants or in cattle-ranching and forest plantations. These follow more traditional concepts of progress. In particular, the catastrophic socioeconomic situation of a large part of the rural population strengthens the legitimacy of "traditional" concepts of progress, namely the fundamental alteration of social structures by modernisation processes induced from outside. In the discussions and processes of the formulation of policies at the national level this hardly plays any role. *On the other hand*, the grassroots democracy position is relevant here. Some of the indigenous peoples and small farmers attempt to evade the valorisation paradigm by putting the question of rights and self-determination at the centre of the disputes. One could expect an alliance between the biotechnological interests and the local population against the traditional valorisation paradigm (cattle-ranching, oil etc.). But the population, in the process of becoming politicised, attempts, due to its own historical experiences, to question the entire logic of the valorisation paradigm. This takes place through the concrete refusal of access to resources – not only genetic ones – and knowledge, but also in legal categories (indigenous and local rights). At the centre of oppositional discourses is the viewpoint that the political and social rights of the previous users of biological diversity (particularly the indigenous peoples in the biodiversity regions) are incompatible with the presently dominating perspective of valorisation. This is related primarily to local conditions, namely the high degree of politicisation and organisation of the population. Although these developments are also dependent on support from outside, for example on the provision of specific knowledge or the "translation" of demands to the national and international level, and critical NGOs play an important role here, the political "sustainability" of the grassroot's position is based on the fact that in Chiapas itself actors are organised differently.

At the end of the 1990s the question of bioprospecting gained in public importance. Against the background of the increasing criticism of the ICBG-Maya project this issue was politicised under the label of "biopiracy". This can be regarded as an important condition due to the fact that weaker actors became aware of their interests, organised themselves and developed political positions and strategies. After the threatened expulsion of the indigenous population from the nature reserve Montes Azules, in the Chiapan Lacandon Jungle in August 1999 had been broadly discussed in public, the nature protection policy came under criticism. Connections were increasingly made between bioprospecting, nature protection and indigenous rights, but also militarisation and the repression of uprisings.

The examples presented of the ICBG-Maya project and the disputes in the RIBMA are prominent, because here it was fought out in an exemplary way, what positions the indigenous communities as the bearers of knowledge and also as acting subjects have in this process. For them, in particular, the abandonment of the ICBG-Maya project is a symbolic success, because it illustrated that a "no" to unwanted bioprospecting is enforceable. For this, the organisation and politicisation, which had increased, particularly in the course of the Zapatista uprising since 1994, were important. It has already been emphasised several times that the constellation of forces in Chiapas has shifted as a result of the uprising. The representative of the Chiapan quasi-environmental ministry (Mx/14) expresses it thus: the indigenous population was quite willing to share its knowledge but with the uprising of 1994 it obtained an awareness of its marginalisation and poverty and now demands respect for its culture, its knowledge and its resources. The political and negotiating position of the indigenous population has eminently strengthened since 1994. The Zapatista uprising and – beyond the Zapatista spectrum – the negotiations of San Andrés on indigenous culture, decision-making power and autonomy worked as a sort of catalyst.

The processes outlined did not lead, however, to the grassroots democracy position being able to have its demands accepted. On the contrary: the associated movements are subject to severe repression, and bioprospecting appears to be taking place in spite of the abandonment of the ICBG-Maya project. One still unclear development in Chiapas could prove to be very important both domestically, for Mexico, and more broadly, in the international sphere: in view of the increasing criticism of bioprospecting and the possibilities of exercising criticism legitimately via the regulations of the CBD and the FAO, room could be created for the adoption of a "re-nationalisation" of collection strategies. The alliance of the neoliberal and technocratic-nationalist positions will probably end in the creation of Bio-Maquiladoras in Mexico, that is wage processing plants in which bio-workers process Mexican plant samples for a certain sum (see Delgado Ramos 2001: 491).

One further thing became clear during the study: the constitution of interests as a political process. The specific interests in the appropriation of genetic resources have developed in the light of historical possibilities as well as in disputes. In general, the interests in the valorisation of genetic resources become stronger with the biotechnological potentials for their valorisation. The historical overview of resource and environmental policies in Mexico has shown that already by the 1980s the erosion of biological diversity was the subject of state regulation. However, nature protection policy was understood as severely restricting human residence in the declared areas. However, for example in the disputes over Montes Azules, discursive use is also made of the earlier concept ("the indigenous people destroy nature").

In one further area, the thesis of the constitution of interests as a political process is substantiated. The access legislation – as a necessary part of the implementation of the CBD – apparently receives a higher ranking because in view of the increasing international, but also national and local, politicisation of the subject (in particular through the Southern governments with their developed interest in the marketing of "their" diversity), it is about legal and planning security. One difference became clear, however, between the neoliberal position, which has not been a driving force for access legislation, and the technocratic-nationalist position, which expects an explicit sharing of the "cake" of valorisation from such legislation.

### *5.4.2 The importance of international politics*

A central characteristic of the global multilevel system of biodiversity policy is the highly selective interaction of international regulation processes with those of nation-states and – fractured still further – their national and local effects. A central role is played here by the selectivities of national state apparatuses, dominant interpretations of the problems, interests and socio-economic relationships and structural conditions. The nation-state thus remains an important level of regulation in spite of all the tendencies towards internationalisation. The assimilation of intellectual property rights and the rapid process of the regulation of biological safety show that international policies have a high ranking if they are in the interest of dominant international and national actors. This is less the case with regard to the rights of indigenous peoples or benefit-sharing rules. The great importance of the valorisation paradigm has already been pointed out.

In Mexican biodiversity policy, international regulations play a role primarily in the initiation of national processes. This is also the explicit goal of framework agreements such as the CBD, which is not primarily concerned with individual projects but with the provision of an international framework for national legislation and policies. It was illustrated that the political-institutional securing of the neoliberal model – particularly through NAFTA – plays a central role. In addition there is the "interiorising" of neoliberal strategies through structural adjustment measures. Political processes remain embedded in socio-economic ones, and here the international economic interests in oil and in biological diversity in particular, play a role in Mexico. The fact that economic and political interests are closely interlinked can be seen particularly in Mexico where the US government has declared the resources there to be a part of its "national security".

These regulations are embedded in more comprehensive processes of the constitution of problems. Here, too, it is important to what extent specific perspectives can be brought to bear at different levels. In recent years a valorisation paradigm has gained priority. At the international level, perspectives of non-valorisation play a subordinate role. Other societal relationships with nature – such as those of the indigenous peoples in Mexico and Chiapas – are above all consulted in a functional manner in terms of what they can contribute to the utilisation of biological diversity. At the national level in Mexico, indigenous peoples play hardly any role in the formulation of policies. In a very specific constellation in the last ten years, the relationships of forces in Chiapas have shifted.

The "transference" of the interests in valorisation, that is the neoliberal and technocratic-nationalist positions, into legal forms is easier than for interests which are sceptical or disapproving of the dominant forms of valorisation. At the national level, the interpretations of the problems and the political strategies of those actors which were essentially involved in the formulation of the international systems of regulations, that is, again, the neoliberal and technocratic-nationalist positions, are of course the easiest to integrate. This confronts the democratic position with a dilemma: it must accept a form of dispute for the representation of its own interests which is concerned with state regulation and law (the question of property rights is absolutely central). This form of dispute is, however, itself not a neutral process but to a high degree determined by the formation of power. The indigenous people of Mexico and in particular the rebel communities in Chiapas have partially entered into these disputes (among other things with the COCOPA initiative on indigenous rights and culture), but so far they have been unsuccessful. Even if societal relationships change through daily practice in Chiapas, this does not necessarily have repercussions at the national level.

Internationally, provisions which aim to strengthen weaker actors (farmers' rights, Art. 8(j), we could also include ILO 169) have not been able to be generally accepted in recent years. Rather, a *symbolic* reference to international regulations can be observed, in order to strengthen the legitimacy of one's own position. A one-dimensional perspective would be short of the mark here, however. Very disparate processes, which have been conducted in many countries, have contributed to the development of international legal norms in accordance with the wishes of weaker actors. The international discussion on the rights of indigenous peoples was a long struggle by the peoples themselves and was not put on the international agenda by clever lawyers. In both "directions", that is the transmission from the local level to the international and vice versa, NGOs played a major role.

It was also clearly shown that the national and international levels play a very ambivalent role in Chiapas, as legally regulating but also as levels at which interests are formulated "from outside". The great importance of the local level is shown when the concrete activities concerning protection and valorisation are concerned, but precisely through this the concrete place is tied into national and international developments. In particular, supply competition plays a role here, which leads the technocratic-nationalist position to lean towards the adoption of authoritarian measures. Governments are concerned with achieving as efficient a valorisation as possible, while participative or democratic processes are regarded as a competitive disadvantage. Concrete places also remain tied to developments at a higher level because the forms in which conflicts are fought out – state and law – are in a way already given. The international level plays an important role here, because it privileges the nation-state and legal mechanisms. This is shown in that the different strategies for the legalising of appropriation and the more or less strong upgrading of weak actors can hardly avoid a fundamental tendency: the different forms of appropriation – and protection – are under pressure to identify property.

Finally, the role of international NGOs is of great importance, particularly those with a grassroots democracy position. They contribute to the clarification and politicisation of the interests of actors and their strategies with regard to local resources and to the understanding of complicated regulation questions and the meaning of international processes. They not only have an informative function in this but contribute decisively to the interpretation of specific facts through local grassroots democracy actors, which in turn influences their strategies (see in detail Görg/Brand 2001a).

### *5.4.3 Dominance and hegemony*

In spite of the dominance of a valorisation paradigm we cannot claim, either for the federal level or for Chiapas, that a clear hegemony has formed in biodiversity policy. This is connected at the national level with the continuing and considerable tensions among the dominant positions, which until now could not be brought together to form a stable compromise. At the same time, it is at this level that there exists the strongest possibility of achieving compromises based on the formulation and establishment of corresponding policies. Actors who are fundamentally critical of the developments of valorisation are more weakly represented here and can more easily be ostracised.

The valorisation paradigm is also not hegemonic to the extent that it is not accepted by some relevant local actors. This is seen particularly in Chiapas, but not in the same way as in other states such as Oaxaca or in the North of Mexico. The question of hegemony and thus of the articulation of different levels must be answered in a differentiated fashion: in certain questions – such as the valorisation of biological diversity – it can exist both internationally and nationally. But this must not be the case in all individual places and on all spatial scales. This results in a political dilemma for local actors in the form of a structural supply competition which enables international and national actors to become involved in socio-cultural and natural-geographical places in a relatively flexible – and often violent – manner. This flexibility, in turn, may allow them to appropriate genetic resources where the climate is not politicised.

# 6 Contested terrains

## Towards a neo-Poulantzian approach to International Political Economy

Scientific approaches such as regime theory and most environmental governance approaches see in the emergence of international environmental policy institutions a reaction to transboundary problems, which can only be dealt with multilaterally and in consultation with non-state actors. Due to this, the assessment of global regulation is primarily, if not exclusively, a question of whether and to what extent the institutions are appropriate in view of the given problems – such as, for example, the thinning of the ozone layer, climate change or the erosion of biological diversity. To the extent, however, that only the more or less direct effects of environmental policy institutions are studied and assessed, this understanding of the problems represents too narrow a perspective on the functioning and effects of international environmental politics. There are quite a number of reasons for this. First of all, environmental problems are not given but constituted. More recently, many approaches in political science and environmental sociology take into consideration the question of how that which is to be regarded as an "environmental problem" – or generally as a part of the "ecological crisis" – is scientifically, politically and culturally constructed. It is less often recognised, however, that in this problem constitution the interests of the participating actors play a decisive role. Many analyses only examine whether and if so how a more or less uniform understanding of the problem is reached and institutionalised in the disputes among scientific and political experts. Less attention is paid to the question of whether the results may have been achieved by neglecting or even deliberately disregarding the viewpoints of weaker societal actors. The consequence of this might be that the treatment of global problems is not linked to the (changing) capitalist mode of production and associated gender and ethnic relations, as well as societal relationships with nature and the reproduction of unequal power relations and social domination.

Therefore, a *critical* social science analysis should tread carefully when dealing with such problems and remain critical in the face of pure *problem solving theories* (Cox 1998: 39). With this study we wish to open up a perspective which avoids the usual dichotomisation between ecological and societal problems, on the one hand, and social relations of domination, on the other. Ecological problems and the institutions dealing with them affect the reproduction and re-formation of global domination, just as the treatment of them cannot be understood without taking into consideration societal interests and power relations. That both levels are closely interlinked can be illustrated by the emergence of post-Fordist "globalised" relationships with nature. In the reaction to a global "environmental problem" – the loss of biological diversity – new interests in genetic resources articulate themselves and completely new patterns of the formation of compromises and the regulation of societal contradictions emerge. In this final chapter, we summarise and discuss our arguments and emphasise theoretical consequences. Moreover, we focus on some questions which seem important to us for further research.

## 6.1 The material dimensions of globalisation and the importance of the local

The field of conflict around genetic resources shows clear evidence that even the production processes based on the latest technological innovations (as captured in the concept of a knowledge society; see Chapter 1) are dependent on a material input and in a certain manner on access to "nature". Thus, nothing is further from the truth than the belief that in the knowledge society humanity has finally emancipated itself from the material conditions of its existence. By contrast, significant changes in the relationships with nature in both their material and their symbolic dimensions are an integral part of the emerging post-Fordist mode of development. Economic interests in genetic resources, however, are not simply given, but depend on scientific-technological development strategies, which are in turn partly determined by social power relations and conflicts. At the same time, contrary views of nature are expressed in these development strategies. On the one hand there is the view that the artificial design of molecules is certainly the faster and more efficient way to the development of new products because "there is nothing special about natural products", as a representative of a biotechnology company is quoted as saying in *Nature* (Macilwain 1998). This is in contrast to the position of the representatives of other firms, who emphasise the evolutionary advantage of nature. Regarding the potentials of combinative chemistry, for example, a representative of Monsanto stated: "Nature has been trying this experiment for two billion years" (ibid.). Particularly in the context of the debates on biological diversity, "Nature's Cornucopia" (Tuxill 1999) is praised. This applauding of nature, nevertheless, often occurs not in contrast, but complementary to the use of genetic technology. "Although we have achieved unprecedented skill in moving genes around, only nature can manufacture them" (ibid.: 6). Nature as a "factory" of basic materials is the notion which most characterises the logic of the *valorisation of biodiversity*.

Nature is in fact being socialised in a new form. Whether these techno-economic strategies will be successful depends not primarily on scientific concepts and the accompanying views of nature, however, but on which strategies find the most favourable conditions on the world market. And these conditions, as shown in this study, are created and enforced politically, essentially by nation-states. Thus, whether in the future the life sciences industries will in fact be able to use the "cornucopia of nature" depends on the organisation of the "global management" of genetic resources, that is on the development of the conflicts over the internationalisation of the state.

This has important theoretical and empirical implications. Our study has demonstrated the necessity of undertaking a critical analysis of the broader process of the societalisation of nature when confronting the issue of biodiversity's preservation – that is the character of post-Fordist relationships with nature and the interests behind them, but also the accompanying conflicts and power relations. As our examination of various international treaties has shown, the process of regulating biodiversity has to be conceptualised as a struggle over the generalisation of certain interpretations of the problem (up to a broadly accepted valorisation paradigm) and their materialisation in international institutions. The result of this struggle depends largely on the power resources the different actors can mobilise, but also on the ability of powerful actors to either integrate or successfully marginalise subordinate actors. The Agreement on Trade-Related Aspects of Intellectual Property Agreement (TRIPS), especially, has to be understood as an institutionalisation of the interests of large multinationals of the life sciences industries whose economic performance is more and more dependent on an international legal framework securing them a strong protection of intellectual property rights. But also "softer" agreements such as the Convention on Biological Diversity (CBD) and the United Nations Food and Agriculture Organization (FAO) seed treaty either favour the commercialisation of genetic resources or remain rather unspecific or instrumental when they deal with the rights of small farmers and indigenous communities. Thus, in the course of and through the internalisation of the state, it is exactly those cultural interpretations, knowledge systems and agricultural practices which stand for less destructive societal relations with nature which tend to be marginalised. It is not intended here to reproduce the myth of peoples "close to nature" who live in harmony with nature. This view overlooks the fact that ecological degradation can also take place there. What is more, a further and greater danger in adopting such a perspective stems from the construction of an absolute otherness of these cultures, which often goes hand in hand with the obligation that they continue in their role as the preservers of biodiversity – which means, to functionalise them. To assume an absolute antithesis between the capitalist commercialisation of nature (valorisation) and the mythical forms of a unity with nature (represented by indigenous people), thus, reproduces the antithesis of economy and ecology. In order to reject these dichotomies, it is important to differentiate among the specific individual forms of the societalisation of nature and to analyse their ecological and socio-economic consequences. As we have seen, non-capitalist forms of treating biodiversity have been closely examined because they play an important role in the conservation of potentially valuable resources or even, as in bioprospecting, they have a particular meaning for the facilitated access to genetic resources. Rights relating to these practices are formally recognised, although within the framework of the CBD (Art. 8(j)) only in an instrumental form, in so far as they are useful for the conservation of biodiversity. Apart from that, both in the CBD and in the International Treaty of the FAO (and there in the concept of Farmers' Rights; see Chapter 3) they are subordinated to the sovereignty of the nation-states. As we could see in the example of biodiversity politics in Mexico, the decisive problem is to what extent nationally based actors have the potential at all to stand up against the interests of stronger actors on different spatial levels.

The development of post-Fordist relationships with nature, however, does not make environmental politics impossible. To some degree, the reduction of emissions, the protection of resources and the improvement of living conditions are rational even under capitalist conditions. But the way the problems are dealt with and the rationality of political regulation are shaped by the new phase of capitalist development. Therefore, environmental measures are influenced by the struggles connected with the rise of this new phase and a neoliberal concept of nature (McCarthy/Prudham 2004). In reverse, it has become clear that societal developments are imparted in a specific way through the treatment of socio-ecological problems. Thus, the individual chapters reveal that, depending on the interests and power relations involved, a highly *selective treatment* of environmental problems – such as the erosion of biological diversity – takes place.

For the *formation of post-Fordist relationships with nature* this means that it is not only a question of whether biological diversity can be conserved at all; indeed, a dramatic erosion of biological diversity has been taking place for a long time and whether this can be stopped remains extremely dubious. The really important question is *why and for whom* this diversity should be conserved. At stake is whether, in the future, only some nature reserve areas will be left, in which the remains of biological diversity are conserved, for example for aesthetic reasons or solvent ecotourism. Perhaps such nature protection areas will be inhabited by "primitive peoples" and will therefore be considered as nature-culture-reservations. They might also be used for the search for natural materials (if this increases the economic impetus for their conservation). Beside these more horrific scenarios, a more specific question is whether the *in situ* conservation of plant genetic resources for food and agriculture is indeed still necessary or whether the already existing *ex situ* collections of industry make enough material available – and under more favourable conditions – for the development of new varieties.

One central aspect in the development of post-Fordist relationships with nature refers to the possibility locally based actors have in shaping their relationships with nature. Are they able to determine their own future, or will they be functionalised by global regulation strategies, their way of life thus transformed and they themselves marginalised? This question affects not only the relationships among different levels of the formulation and implementation of politics ("scale") but also refers to the meaning of the "local" of the specific location ("place"): the extent to which the specific natural environmental conditions are taken into account in the regulation of access to genetic resources. Locally based actors are therefore not only the farming communities and the indigenous peoples but, in a strategic sense, they include all those who wish to emphasise the importance of the specific locality and the practices connected with it in the regulation of societal relationships with nature (i.e. possibly also NGOs or scientists). The decisive question here is at which levels the decision competences are institutionally anchored and whether locally based actors have competences or whether these are taken from them – perhaps with a reference to the ecological importance of the locality. The latter can be observed in the attempt to restore common goods, whereby these common goods in many cases are placed at the disposal of "global resource managers" – representatives of international organisations (such as the World Bank) and NGOs – as Michael Goldman (1998, 2005) has shown. If the question is put in this way, then it quickly becomes clear that the local place, in the sense of self-prescribed practices that help to shape societal relationships with nature, can only be strengthened against the forced tendency of valorisation at all levels jointly. Even on a local scale it is highly doubtful whether those actors who take the specific natural environmental conditions into account can in fact prevail or whether the strategies aimed at exports and the most efficient valorisation of resources will dominate (as for example in the case of the local oligarchy in Chiapas). Not all actors on the local *scale* are interested in the particular *place*. Something similar is true for the national, supranational and international scales.

## 6.2 North–South relations and the reconfiguration of scale

How scales are constituted, how they differentiate themselves from one another and in what relationship they stand to one another, is essentially determined by power relations and their institutionalisation. The process of globalisation does not at all mean the disappearance of regional, national or local scales, but their *reconfiguration*. The problem connected with this is not limited only to the more or less good cooperation between actors and institutions on different scales, which are presumed to exist. It also relates to the conflicts over their importance to each other. This has been shown in the example of the growing pressure on local actors in Chiapas. The national scale and its importance in international politics is also subjected to a permanent re-interpretation in which the strategies of the actors are very differently accentuated. The states of the South aim, particularly with regard to the disposition of genetic resources, for the strengthening of the national scale. Here, one aspect is crucial: the countries and regions which are rich in biodiversity find themselves in the situation of a *supply competition*. This means that the dominant social forces in those countries, such as the governments, have become interested in the valorisation of biodiversity and especially of genetic resources. Therefore, we consider a valorisation paradigm as hegemonic at the international level around which concrete conflicts take place, but this is also the case in other regions. The initiative of the Like-Minded Group of Megadiverse Countries is a new attempt to lend more weight to national sovereignty in the face of the supply competition between the biodiversity-rich countries. In contrast to this, the leading countries of the North would like to leave open the precise negotiation of the conditions of access (and aim here, in contrast to the countries of the South, for *voluntary* rules) but at the same time attempt to enforce internationally *binding* standards for intellectual property rights (IPRs). The reason is clear: in precisely this point the international scale is clearly moulded by the interests of the metropolitan *user countries*. Multilateralism is to remain valid here at least until more effective bilateral negotiations promise further success. Since the relationship of the scales to one another is itself characterised by power relations, the strategies of the various actors are aimed at strengthening or weakening the corresponding scales according to their own interests. There is a further aspect which complicates the problem of scale considerably. Every scale must itself be understood as a *field of forces* which is characterised by often overlapping constellations of interests and power. This is expressed, for example, in the fact that there is no simple North–South pattern. Thus, "Northern positions", for example the interest of the large or small life sciences companies, research institutes or international nature conservation organisations, can also be found in the South, and there in some cases even on a local scale. Our example showed, at least in Chiapas, that the interests of the nature conservation NGOs ally themselves with those concerned with the valorisation of genetic resources; and on a national scale the interests of the life sciences industry are also present and exercise considerable influence. And by contrast, in Northern countries certain NGOs and social movements take up the interests of the global South (if with less success than industry in the Southern countries). Each scale thus represents a field of forces in which the interests of the different actors are specifically concentrated. Certain configurations repeat themselves here – for example, the conflicts between the interests of local actors and the transnational companies (TNCs) are fought out on international, supranational, national and local scales – only with different power relations and thus with different results.

This does not make the use of the concepts North and South meaningless; rather, it makes a case for being cautious before any overly hasty generalisations are made. The case of Mexico/Chiapas cannot be generalised because of the local, national and regional peculiarities with regard to the specific processes. Certainly, the international and the regional scales play a central role concerning the opportunities of influence which can be exerted on a national and local scale. The relations between the different scales must therefore be examined in detail. But that also means that individual cases are influenced in very different ways due to their specific conditions. When the "effects" of international policies are being examined, in Mexico the natural conditions – in addition to biological diversity also the relatively easy accessibility – the relative political stability of the region, the regional differences, NAFTA and the close connection to the USA, and finally, the particular tradition of the nation-state and the national power relations, must be taken into account. In addition to the simple schematic categories of provider and user, the particular local situation must also be accorded due importance: to what extent is existing biodiversity in fact exploitable and do suitable conditions exist for this? In Mexico, for example, the long-lasting lack of recognition of indigenous cultures in the context of the state ideology of *mestizaje* (mestisation) and the recent processes of democratisation should be pointed out here. Moreover, we saw remarkable differences between Chiapas, where a strong indigenous movement and the Zapatista uprising has re-shaped power relations, and the national level. There is room here for variable and sometimes surprising coalitions and compromises – and not only between a corrupt national elite and foreign capital, as is frequently claimed, but also between locally based and global actors (social movements, NGOs, scientists).

## 6.3 International institutions as condensations of power relations

The present study focused primarily on political-institutional aspects of the internationalisation of the state. Nevertheless, the more complex creation of international hegemony as well as non-state international actors also played an important role in the reconfiguration of politics. The interests of the individual actors are not inflexibly predetermined or "derivable" from objective situations. That is revealed by the fact that in specific questions such as the regulating of access, benefit-sharing or intellectual property rights political positions emerge or change through politicisation and conflicts. Interest in secure and legalised access is growing, however, due to the politicising of the expropriation of genetic resources via international processes and via national and local criticism. Many Southern countries, to give another example, were not fully aware of the consequences of the TRIPS Agreement at the time when this was negotiated. Their interest in, for example, national control over genetic resources grew when they were confronted with the need to implement the Agreement and with the consequences the implementation would have. The formation of "Southern" interests was influenced by the demands of NGOs to prohibit the patenting of life and by the opportunities offered by the review process of Article 27.3 (b). The argument of the constellation-dependent constitution of *specific* interests should not divert our attention, however, from the fact that some *fundamental* interests do in fact remain stable, such as the interest of firms and research institutes in the appropriation of biological diversity for the purpose of its valorisation. In addition, for these actors legal and planning certainty is of central importance in an economic field which is developing in a highly dynamic fashion; although it is not always absolutely clear what this specifically means and when planning certainty is regarded as adequate. In any case, decisions continue to be made under conditions of uncertainty. For the actual degree of the valorisation of biodiversity, the political and economic framework – including the legal and cultural dimensions – is of central importance.

That brings us to the international political level and to the cooperation of different international institutions as a part of multi-level politics. One result of the study is that doubts are indeed appropriate with respect to whether the apparently attractive picture of "synergy effects" or "coherence" between the various regimes is an apposite one. In the relationship between the CBD, the WTO and the International Treaty of the FAO it is not to be expected that it will be possible to reach consensual solutions, never mind mutually supporting ones, on the really important, controversial issues. On the international scale, due to the dominance of one particular institution, the WTO, we are dealing with "*negotiations in the shadows of hierarchy*" (Scharpf 1993). By contrast to the nation-state framework, on the international scale the "hierarchy" – we prefer the concept of asymmetry in order to avoid the notion of a unilinear top-down relationship – is not anchored within the framework of the state monopoly of legitimate force at the national level. Rather, a hard power bargaining prevails. This has ambivalent effects. On the one hand power relations often have a much more direct effect on the international scale because they are less cushioned by a national legal system or by complex institutional decision-making procedures and majority decisions. On the other hand, these processes are therefore also easier to block and even dominant actors do not necessarily have the means to secure the respecting of certain regulations due to the lack of sanctioning authorities.

At all events, the asymmetry of the negotiation processes plays an important role not only for the ultimate relevance of the individual agreements but also for their formulation during the negotiation process. Above all the compatibility with the TRIPS Agreement is a central criterion for all international agreements with which certain actors – the industrial countries in general and particularly the USA – attempt to influence current negotiations. And as could be seen in the formulation of the International Treaty of the FAO, simply the threat of calling the dispute settlement body of the WTO can be decisive in disputes over certain regulations (cf. Chapter 3). However, the politicisation of certain agreements can make their usefulness for the establishment of dominant interests more difficult or even impossible. Particularly, the politicisation of the TRIPS Agreement in the revision process of Art 27.3(b) is an example of the fact that the new interests, which have now to a certain extent been accepted as legitimate, must be dealt with, for example by relegating them to other forums or, conversely, by upgrading forums such as the World Intellectual Property Organization (WIPO) for the establishment of the dominant interests. This *forum shifting* (cf. Chapter 4) is also an indication of the fact that even in view of international relations, a sanctioning authority corresponding to the nation-state is lacking and the asymmetric relationship of the agreements develops its effectiveness only indirectly and ambivalently.

The influence of international institutions and their relationships to one another cannot be examined from the perspective of a more or less successful coordination. They are themselves the condensation of antagonistic interests in global power relations. What is more, non-coordination has to be seen as the specific "rationality" of existing power relations. On the international scale struggles between highly heterogeneous actors can be observed, among them national governments as central actors. Since the strategies of these state actors are in the final analysis the result of the condensation of different power relations in the individual state apparatuses, we are dealing here with a *second order condensation of social power relations*. This is not to deny the formation of transnational alliances of social forces, from business organisations up to trans-local coalitions of indigenous people, which try to influence international institutions not only via their respective national representatives but also directly. Furthermore, as we have seen in more detail in the previous chapters, the notion of a "second order" should not be misunderstood as being a hierarchy of control. Rather, it means that the process of the condensation of particular interests to a supposedly higher general interest takes place here in an intensified form. The conflicts on the national scale precede the formulation of a national strategy and thus they also precede the opposition of interests on the international scale and accompany them in the further political process. This occurs even when international processes, as in many cases, provide the impetus for national processes. The priority of the national is thus not meant chronologically, but in the sense that national governments, in order to be able to act on the international stage, have to acquire clarity as to their national interests. In the Multilateral System of the International Treaty the impetus came, for example, from the international level, which was followed by the formulation of national interests, for example, in bilateral negotiations. Thus, the concept *second order condensation of social power relations* refers to the production of "new state spaces" (Brenner 2004) by re-scaling elements of state power and thereby creating new and more complex relationships between different scales of state politics. This is a contested process which is driven by intergovernmental negotiations as well as by the formation of transnational alliances of societal actors and by the emergence of networks between state and non-state actors on various scales (see Bieling 2006: 235f.).

The claim of a "general interest" must therefore at least be formulated in order to be useful for the analysis of the international scale. Because this general interest – which is articulated internationally as "national interest", as the condensation of particular (and antagonistic) interests – is in fact fictional, it represents an essential element of the "state fetish" (Hirsch 1994). In international processes this fiction, that is this element of the state fetish, appears to be weaker, for the contradictions between the agreements (here, conflicts over trade and IPRs; there, over environment and nutrition) are much too obvious. But this does not at all mean that the contradictions between the various agreements could in fact be used by other actors with diverging interests, for even here, in the final analysis, the fiction is upheld that a global "general interest", in the sense of a solution to global problems, must be established.

In any case, different interests can be combined to a "general interest" only to a limited extent. This is the reason why the "general interest" always remains fictional, even if the necessity of its formulation for the establishment of social compromises is real. As we have shown in this study, the issue concerns, above all, the stabilisation and thus also the reproduction of the basic structures of social relations. However, within this contentious process the way in which compromises are formed is certainly important. And it is exactly here that the asymmetry of the international systems of regulation plays a decisive role, because it is precisely the WTO-grounded TRIPS Agreement which aims at securing modern capitalist property relations. This agreement is not only dominant in the subsequent testing of existing agreements, but its dominance enters into the negotiating process (as shown in the case of the International Treaty on Plant Genetic Resources, ITPGR) and the additional shaping of international agreements and their national implementation.

A global multi-level system thus includes the condensation of power relations on different scales. These scales are not independent of one another; instead, they exert a mutual influence on each other. But they are also not directly derivable from each other; the combination of the levels is complex. Sometimes it is easier at the local scale to bring the interests of the actors based there to bear. This is not the result of a natural law, however, for local conflicts can be reshaped by the entrance of powerful external actors. Examples of the subjugation of the local scale to global or national strategies of valorisation can be found aplenty, for example in the Amazon basin, where on the Brazilian side local actors have until now hardly had a chance to exert an independent influence (see Acselrad 2002). A change presupposes the shifting of societal power relations at the national scale, which manifest themselves in state policies, among other things. This is also true of the importance of international politics for local relationships: without corresponding national laws and nationally organised capacity building, the local scale can expect to gain nothing from the Bonn Guidelines, for example (see on the problems in the case of Costa Rica: Rodríguez 2002).

A systematic implication for the process of the internationalisation of the state is that although the international scale plays an increasingly important role, the relations of the scales to one another is neither per se complementary, nor is it a type of zero-sum game in which the international scale becomes more important "at the cost" of the national scale. The very meaning of the international scale is ambivalent. Even if the CBD Article 8 (j) on the rights of local communities and indigenous peoples and the FAO concept of Farmers' Rights bring little juridical certainty, they have had great effect as an impetus for the politicisation of the problem. Also, the model law of the African Union was able to increase awareness of the problems of a national access and benefit-sharing legislation on the regional scale and beyond, and precisely the fear of this has caused the harsh reactions of the International Union for the Protection of New Varieties of Plants (UPOV) and WIPO. The international regulations thus do not make the national legislation processes less important, but confront them with a role which is partly new. And in addition, nation-states or their governments play a central role as actors, even if this role is no longer without competition on the international scale. This is meant when we speak of a *second order condensation* in the case of international institutions and the internationalised state in general. The interests and relationships of power which are already condensed in national institutions and strategies are condensed on a further scale in another form, partly with other conflicts and contents and thus with perhaps deviating results – and in this form they have an effect in turn on other scales. This inter-linkage is not really new but in the internationalisation of the state it gains particular weight.

In our research we examined the question of whether any role is played, and if so which one, by the *coherence* of national politics. In political science and in the political process itself the expectation is repeatedly expressed that a coherent state and/or inter-state politics should play a decisive role for a strategy which is appropriate to the problems and which takes into account the side-effects of political measures. In contrast to this, our example can show that because of the highly divergent and partly contradictory interests which play a role in the field of conflict, this coherence tends to be unlikely. Basically, if coherence with regard to the treatment of problems is hardly to be expected, this question nevertheless plays a decisive role in the re-formation of global domination. Furthermore, the problem of coherence shows clearly that power relations in the internationalisation of the state continue to be based at the level of the nation-state. This paradox – a *nationally shaped global domination* – can be understood using Poulantzas's concept of the state as it was introduced in the first chapter. This does not deny the actions and effects of international actors but it does argue against the assumption of a sort of "autonomising" of international apparatuses and the disputes in their "civil society environment" (Görg/Wissen 2003).

State politics can never really be "coherent" in the sense that it is no longer permeated by tensions and contradictory in itself. Understanding the state as a "strategic terrain and strategic process" (Poulantzas 1978/2002: 167) means to conceptualise state politics as a "result of a conflicting coordination of explicit and diverging micro-politics, rather than a rational formulation of a global and coherent project" (ibid.: 168, our translation). If institutionalisation takes place through the formation of compromises, this does not at all mean that a real general interest embodied by the state, in the sense of a reconciliation of opposing interests, has been established. It is precisely in the field of agro-biodiversity and the case study of Mexico that it has been shown that because of the complexity of the problems and the multitude of strategies and actors a real compensation of interests within the framework of the nation-state or even simply a consistent national strategy cannot be formulated. A study on behalf of the World Bank underlines this with surprising clarity (Petit *et al.* 2000). We must therefore differentiate between the real differences of specific interests and the nevertheless more than fictional general interest. This general interest is not simply fictional because at least the interests of the most important actors must be included in a hegemonic compromise. Dominant actors do not only comprise dominant social forces like the most powerful fractions of capital, but also refer to subaltern actors whose interests have to be taken into account, up to a certain extent, in order to prevent these actors from challenging the social power relations. Alternatively, dominant actors must be able to exclude relevant subalterns effectively. Generalised, that is hegemonic interests must also be capable of claiming that their interests correspond not only to the dominant actors but also to the dominant interpretation of the problem and that they present a central contribution to the solution of the problem (as it is understood in the hegemonic interpretation of it). If it fulfils these requirements a general interest, safeguarded by the state, can correspond to the interest of capital in the establishment and safeguarding of capitalist production and property relations.

All these points were easily recognisable in the particular case of the negotiations of the ITPGR. In spite, or even because, of opposing interests it proved possible in this field to establish a compromise in the form of a multilateral agreement which nevertheless has certain similarities to nation-state regulations and continues to require the authority of the nation-state. This was possible due to the fact that the interpretation of the problem "loss of the diversity of plant genetic resources" was strongly established, as were certain basic convictions with regard to the relevance of these resources (economic importance and nutritional security). Thus, a supposedly "global general interest" – the conservation of genetic diversity – was proposed and addressed by the establishment of an international agreement, that is by a multilateral exchange system for particularly important genetic material. In addition, precisely through this multilateral system a compromise between the conflicting interests was established which does justice both to the interest of the governments of the Southern countries in the marketing of "their" biodiversity and to the interest of the life sciences industry in the unreduced protection of its property claims and the establishment of global markets. Interests opposing these dominant interests were, however, not really respected by the compromise which was reached but (like the Farmers' Rights) were subordinated to national interests. The conflict between locally based actors and the interest in valorisation and global marketing was thus not solved in the international agreement, but was referred back to the national level. Here, the structural selectivity of the global regulation system becomes obvious.

However, multilateralism is not merely an inadequate substitute for the regulatory function of the nation-state; it does not preclude the bilateral appropriation of biodiversity either. Rather, the very construction of a public sphere and a general interest in plant genetic resources plays an important role in the commercialisation and privatisation of these resources. This can be recognised in the entire problem of intellectual property rights.[1] Both at the level of the nation-state and at the interstate level, the *structural privileging* of certain interests is obvious, namely of those which aim for the use of biodiversity within the framework of the existing social order – whether by directly propagating capitalist valorisation or alternatively, striving for the mitigating conservation of nature. It must not be forgotten, however, that this process continues to be highly controversial and a firmly established hegemony cannot be assumed. The resistance to the privatisation of plant genetic resources has not at all diminished with the passing of the International Treaty on Plant Genetic Resources, even if this agreement could be judged as being a disappointing compromise.

The lack of hegemony is even stronger in the case of the TRIPS Agreement. Here, important actors such as the governments of Brazil or India were not included in a formally established politico-legal compromise. Rather, their original resistance to the Agreement was broken by economic pressure or sanctions. Only in that way could an agreement be established which is highly structurally selective, that is strongly favouring the interests of economically powerful actors while almost completely neglecting different interests. The latent contradictions of this constellation became manifest when the crucial Article 27.3(b) was to be reviewed. In this situation those interests which were hitherto oppressed by force successfully challenged the dominant interpretation of the problem. However, they did not achieve a substantial revision of the Agreement. Rather, one can speak of a stalemate which gives rise to new strategies of the dominant actors, such as forum shifting in the direction of the WIPO or bilateral power bargaining. The latter has especially gained in importance. This is due to the fact that the strategy of establishing tighter and more substantial IPRs via the WIPO has met with the resistance of influential Southern governments such as those of Brazil and Argentina, which managed to politicise the supposedly "technical" terrain of WIPO by launching the proposal of a development agenda in 2004.

The difficulties of successfully pursuing their interests even on those international terrains which are a material condensation of Northern dominated power relations, that is the WIPO and the TRIPS Agreement, seem to have brought Northern actors to increasingly rely on bilateral bargaining. However, this does not mean that multilateralism is simply superceded by bilateralism. Just as the multilateral agreements of the 1990s cannot be understood without bilateral power asymmetries, the recent rise of bilateralism has to be seen precisely against the background of the possibilities offered by multilateral agreements. As our analysis has shown, multilateral rules like the TRIPS Agreement can provide the ground floor on which bilateral power asymmetries unfold. Far from solving "global problems" in a cooperative manner, they institutionalise international power relations and thereby enhance the opportunities to stabilise and reproduce these relations. On a symbolic level, they are the product as well as the medium of the constitution or reframing of certain issues. The TRIPS Agreement cannot be understood without taking into account the redefinition of the IPR issue against the background of technological developments in the life sciences industry, agro-business and the IT sector. In turn, it is an institutionalisation of the changed problem perception. As such it also defines the claims of what is and is not possible or even discussible in bilateral negotiations. Additionally, by establishing minimum standards of intellectual property protection without defining an upper limit, it creates the possibility for powerful Northern actors to pressure for higher standards in bilateral negotiations without violating the Agreement. Thus, rather than conceptualising bi- and multilateralism as opposed political modes, they have to be understood in their complex interrelationship.

## 6.4 The question of democracy

Nevertheless, the recent rise of bilateralism indicates a lack of hegemony on the international scale. It reflects the fact that different interests are not only articulated *within* the negotiations on multilateral terrains like the TRIPS Agreement or the WTO in general, but, more fundamentally, that the *multilateral institutions themselves* are the object of struggles. Their acceptance as legitimate terrains where conflicts are fought out and compromises are searched for is increasingly challenged by Southern governments, social movements and developmental and environmental NGOs. This constellation is largely the result of a process of interest formation on the part of the societies and governments of Southern countries, driven by the need to implement the TRIPS Agreement and to develop a position in the review of the controversial Article 27.3(b), and often supported by the expertise of NGOs. It remains an open question whether the "scale jumping" of dominant actors between multilateral and bilateral negotiations and the forum shifting between different multilateral terrains will be able to establish a precarious balance between multi- and bilateralism. Our analysis suggests a rather open situation which implies different possibilities among which such a balance is only one. Another one would be the emergence and strengthening of the scale of South–South cooperation which would limit the power of bilateral strategies of Northern governments and could contribute to further challenging multilateral terrains which are largely shaped according to Northern interests.

But even in the case of strengthening the cooperation between Southern governments, the rights of local communities to their environment would be far from guaranteed. These rights are a question of *democratic biodiversity politics*. A fundamental precondition of these politics is that the affected actors recognise the extent of the problem in order to be able to articulate their interests. In addition, *rights* to participate effectively (rights which can be established by law are further reaching than the usually vague rights of participation) and the realistic possibility of being able, de facto, to claim these rights or to reject undesirable developments must be taken seriously. The right to participate in the democratic shaping of societal developments thus refers not only to the formal decision-making mechanisms with which certain regulations are established. It also refers to the actual chances different interest groups have of their case being heard in the political process and of being able to take part to an adequate degree in deciding by which regulations their own living conditions will be structured. The fact that the process of neoliberal globalisation has decisively weakened the chances for the democratic influencing of social living conditions and of societal relationships with nature is becoming more and more clear, although there is no consensus over the reasons for, and the extent of, this. Some of the developments which are continually discussed in this context – the non-democratically controlled power of international organisations such as the World Bank and of regimes such as the WTO, and the political growth of barely controllable TNCs – are also reflected in the field of biodiversity politics.

With regard to these real possibilities of a democratic shaping of societal relationships with nature the answers vary from reservation to scepticism. While the relevance of the local is recognised on the international scale, this recognition has not been reflected in such a way that legally binding and justifiable rights have developed from it. Neither Article 8(j) of the CBD nor the principle of Farmers' Rights were shaped in such a way (the same is true for recognition within the International Labour Organization, ILO). And as far as the IGC – the Intergovernmental Committee on Intellectual Property and Genetic Resources, Traditional Knowledge and Folklore – within the WIPO is concerned, it is doubtful if a result will be produced which strengthens the position of indigenous peoples. What the rules and forums concerning indigenous and Farmers' Rights have in common is that they, mostly, do not constitute integral parts of their respective treaties and organisations. Instead, they seem to be added, and their compliance is often voluntary or subject to national regulations. Thus, they fulfil the purpose of preventing the negotiations about so-called "hard issues" such as IPR or access rules from being disturbed by subordinate interests. Of course, they are at the same time the result of social struggles and can contribute to strengthening the positions of weaker actors, at least symbolically. On the supranational scale, in our example, the picture is even bleaker, because the North American Free Trade Agreement (NAFTA) is based on neoliberal strategies and institutionalises the interests of globally operating industry. The side agreement on the environment within the framework of NAFTA tends to be of a cosmetic nature, not least because it, too, does not strengthen the interests of the locally based actors. The nature of the supranational level does not necessarily have to be so, however. The orientation of NAFTA is connected with the extreme inequalities of power between the participating states. As the example of the Organization of African Unity (OAU) model law shows, other regions are attempting to set quite different accents in their treatment of genetic resources and the local level – even if with little success so far. It was also shown that Mexico vociferously represents the interests of the indigenous peoples at the international level, but within the country itself the law for indigenous rights was finally passed only in an inappropriate form.

However, the shelving of the planned Free Trade Area for the Americas (FTAA, in Spanish ALCA) in November 2005 due to strong mobilisations and some responsive Latin American governments shows that a re-politicisation is taking place and makes further neoliberalisation at least difficult for dominant actors. A question remains how the *existing* forms of political institutionalisation can be fundamentally challenged and thereafter reformulated.

The decisive level in the process of the internationalisation of the state remains that of the nation-state. What can, in the face of the specific power relations, be implemented here in national laws or in other state measures is also decisive for developments on the local scale. This is confirmed not only by the upgrading of national sovereignty in the context of the CBD but also by similar developments in the framework of the FAO and the ITPGR. Even the advancement of bilateral instruments on the fringe of the WTO is an indication that nation-state regulations ultimately play a decisive role. This does not exclude the further development of multilateral regulating mechanisms; rather, the two instruments complement each other.

Unequal power relations not only result in the ability to establish certain rules in or through political institutions. They are also expressed in a much more comprehensive sense: in the definition of the problems, in the structuring of the terrain, in the concrete formulation of policies and in the potential of certain interests and positions to prevail. Hegemonic interests are also expressed in the fact that they bind other interest groups to a certain view of the problem and the corresponding solutions, and at the same time make concessions to them, to a certain degree. The actual inequalities of power become clear, however, when actors decide in favour of fundamentally different options (indigenous peoples, e.g., for the right to say no to the access of genetic resources) or decide principally against certain measures (e.g. in the rejection of patents on life).

The chances of locally based actors being able to shape their relationships with nature in opposition to the valorisation and the establishment of global markets for genetic resources are therefore doubtful (an analysis that Gramsci taught us) – despite the important case of Chiapas. One of the essential tendencies of post-Fordist relationships with nature is the tendency to destroy other relationships or to functionalise them. In this sense it is a continuation of primitive accumulation: the use of non-capitalist forms, which are transformed in the service of valorisation and subsumed under capitalist forms of socialisation. Our analysis in this point is in line with David Harvey's argument of an "accumulation by dispossession" (Harvey 2003; see also Swyngedouw 2005). However, it also indicates the necessity of modifying and developing further Harvey's argument as far as the role of the state is concerned. In his approach, the driving force of the continued primitive accumulation or the accumulation by dispossession is the "spatial fix", that is "capitalism's insatiable drive to resolve its inner crisis tendencies by geographical expansion and geographical restructuring" (Harvey, D. 2001: 24). Although Harvey repeatedly emphasises the role of the state in this process, his conceptualisation of the latter remains under-complex. The state in Harvey's approach appears mainly as an actor who enforces and safeguards the economic tendencies of spatial restructuring, which result from the inner contradictions of capitalism. Thus, Harvey's approach lacks a concept of the state as a terrain of conflict where economic tendencies are not simply enforced, but where disparate and particular strategies for the resolution of the crisis of over-accumulation struggle for their generalisation, some strategies being structurally and strategically privileged over others. As our analysis has shown, taking these conflicts into account, on various terrains and geographical scales of the state, is crucial in order to understand the establishment of the valorisation paradigm, its character and the contradictions it implies as well as the chances to challenge it in favour of democratising the control over natural resources.

The valorisation paradigm constitutes the overall framework for specific institutional developments in order to deal with the environmental crisis in general and the erosion of biodiversity in particular. This tendency is crucially supported by legal rights, primarily by rights to "intellectual property". In the establishment of these rights nation-states, as the proprietors of the "monopoly of legitimate physical force" and with their material as well as knowledge resources and their experiences, play a central role. The process of primitive accumulation presents itself at this level as the conflict between different forms of rights: "modern" IPR such as patents versus Farmers' Rights or collective rights. But these rights do not stand alone. They must be seen in connection with certain societal practices and they cannot simply be located in the assumed contradiction of economy versus ecology. The protection of nature is also rational within the framework of the life sciences industry as long as it does not cause disproportional costs and above all as long as access (and thus transaction costs) are not bureaucratically over-regulated. The conservation of biological diversity to a certain degree would therefore be possible if it were not made difficult by other interests which are also oriented towards capitalist utilisation: the extractive industry in Chiapas and elsewhere, but above all the increasing use of land both in the South – meadows, the felling of trees and so on – and in the North – settlements, roads and so on.

Opportunities for democratic influence at the international level – as a precondition for the future shaping of democratic societal relations – arise above all where the contradictions between institutions such as the CBD and the TRIPS Agreement of the WTO have become obvious and can be exploited. That is why the resistance to the patenting of genetic resources correctly stands at the centre of many campaigns. But beyond that a more comprehensive understanding of the contradictions and conflicts is necessary. It is fatal if problems of regulation are misunderstood as disputes purely between experts and the problems referred to are technocratically foreshortened. Behind the apparently neutral questions of fact are far-reaching conflicts of interest and complex power relations, which must be taken into account if the treatment of the problem is to be appropriately judged. A critical theory of the internationalisation of the state must examine the emerging forms of global domination and power relations without appearing to be disinterested in the multitude of details of socio-ecological conflicts and alternative strategies. These are often of decisive importance precisely with regard to the shaping of post-Fordist relationships with nature – for the chances of emancipative action as well as for the material dimensions of societal reproduction. In this sense there is still a great deal of material for further empirical research in the restructuring of societal relationships with nature and the internationalisation of the state.

# Notes

## Introduction: genetic resources and the internationalisation of the state

1 The aim of our study is not to contribute to the debate concerning whether or not post-Fordism as a new stable mode of development has emerged and how regulation theory can analyse that with its theoretical concepts (see Albritton *et al.* 2001; Brand/Raza 2003; Candeias 2004; Jessop 2002). We use the concept of post-Fordism in a heuristic manner, that is to shed theoretical and empirical light on the societal background of the conflicts about genetic resources.

## 1 The regulation of nature in post-Fordism

1 This aspect of the specific materiality of nature is neglected by Smith (1984), when he criticises the notion of the "domination of nature" as being based on a dualistic conception of nature and society and instead introduces the notion of the "production of nature", rendering the difference between first and second nature obsolete.
2 Nevertheless, even in Germany there exist different approaches which make use of the concept; some of them start with a constructivist perspective (see Weingarten 1998; Demirovic 1997), while others do not (see Becker/Jahn 2006).
3 We use the concepts in a heuristic way and on an abstract level. We do not intend to capture all the developments in real life but want to figure out general trends.
4 RAFI stands for Rural Advancement Foundation International, ETC for Erosion, Technology and Concentration, naming the three most important problems in the field of agriculture and nutrition (cf. ETC Group 2005a).
5 The consequences of the patents were that Indian famers could no longer export the Neem seed or any associated Neem products and that those who traditionally bought the seeds could not afford to do so any longer because the market price rose tenfold.
6 For a survey of the latter aspect cf. Kütting 2000; Paterson 2000; Lipschutz 2004; Axelrod *et al.* 2005; Betsill *et al.* 2006; on the role of the state in environmental politics Barry/Eckersley 2005.
7 Cf. Smith 1995; Delaney/Leitner 1997; Swyngedouw 1997; Marston 2000; Brenner 2001, 2004; Brenner *et al.* 2003; Novy 2003; Wissen 2007; on the inadequate conceptualising of the term "global" in many contributions cf. also Shaw 2000, Chapter 1.

## 2 On the value of nature: the Convention on Biological Diversity and the commercialisation of genetic resources

1 Cf. for example the enormous list of participating non-state actors at the eighth COP in March 2006 in Brazil (Secretariat of the CBD 2006b: 5–10); but this was also due to an intense mobilisation of Brazilian farmers and organisations against the introduction of genetically modified seeds.

2 Methodologically, it is a reconstruction of positions which we have extracted from numerous interviews and documents.
3 On ambiguities in the concept of megadiversity cf. Caillaux/Ruíz 2002, footnote 1.
4 Interestingly enough, the Nigerian government, which has introduced a benefit-sharing model in the country, in a top-down process without participation, hardly makes an appearance at the negotiations.
5 Another major issue for the biotechnological industry is, aside from access and intellectual property rights, the so-called genetic use restriction technologies (GURTs) or, as a politicised concept, Terminator Technology. Officially, there is an accepted moratorium which was questioned by the governments of New Zealand, Australia and Switzerland in the beginning of 2006, following an initiative by Canada, allowing a "case-by-case assessment". But this proposal was rejected and the moratorium renewed at COP 8 (ENB 2006a: 15–16).
6 Cf. Görg/Brand 2003; Brand 2000: 203ff. with a classification of six types of NGO. On empirical studies, especially of environmental NGOs, cf. Brunnengräber *et al.* 2001. On a critical socio-theoretical and socio-political placement of NGOs cf. Demirovic 2003; Hirsch 2003.
7 Cf. for example the homepages www.iucn.org, www.panda.org, www.wri.org, www.etcgroup.org, www.grain.org, www.twnside.org.sg.
8 The IIFB is supported by European NGOs which have an important share, both financially and administratively, in the organisation of the forum. On the one hand, this undermines the political self-understanding of the IIFB, but on the other hand, the selection of the participants also does not take place from within a self-organised structure of the IIFB. For this situation, which is regarded by the indigenous peoples as problematic, there exists no solution at present despite an intensive search which is seen as promising future success. By itself, the arduous task of translating the most important documents into the four "major languages" is one which the IIFB can hardly master. In the periods outside the official meetings of the IIFB, the forum, which has no secretariat and has to exist without a permanent representative, is virtually non-existent. From the point of view of the participants, personal contacts are lost between the meetings; new methods of communication are often unavailable due to financial conditions or bad infrastructure in their home regions. Different offices and tasks in the forum are not assigned permanently to particular persons. This creates certain flexibility in the administration – especially if a participant is hindered – but it is accompanied by the additional trouble of a renewed training programme and a possible loss of know-how.
9 Since the 1990s representatives of the San people from Angola, Namibia, Botswana and South Africa have met in order to discuss common problems and promote solutions. In 1996, WIMSA was established to give these efforts continuity, concerning issues such as human rights violations, ownership rights or the documentation of their oral traditions. Moreover, the collaboration was important in raising self-consciousness of the San people (Hoering 2004: 11).
10 They are called Bonn Guidelines because in October 2001 the Open-Ended Ad Hoc Working Group on Access and Benefit-sharing gathered in Bonn and adopted the Guidelines.
11 Quantitative aspects could play a role if, for example, Southern countries agreed to cultivate certain genetically modified products within the framework of the international division of labour. This aspect does not play any role in the negotiations and was mentioned in the interviews only by one NGO activist (CBD/17).

### 3 Limits to commercialisation? Genetic resources in agriculture and the conflict over a multilateral exchange system

1 In the following section, we concentrate only on plants; but there is a rising awareness that genetic diversity of animals is just as important.

2 Crosby (1986) even sees a causal correlation here: diseases and the European relationships with nature were partly responsible for the conquest of America. For a critical comment see Arnold 1996: 74ff.

3 Although it seems that due to the increasing privatisation of research and development on plant genetic resources some International Research Centres today have changed their strategies and present themselves as being part of the public sector and involved in the production of public goods.

4 The general irritation must have been great, for even the normally rather reticent Earth Negotiations Bulletin speaks not only of an icy atmosphere but even of the preparedness of some delegates to jump into the Lake of Neuchatel (cf. ENB 2000b, 2000d).

5 Binding payments are only required in the case of the commercialisation of the material from the Multilateral Exchange System. IPRs are only mentioned in a general way, with reference to the limitation of access; details must, however, still be regulated by the Governing Body of the International Treaty; cf. ITPGR 2001: Art. 13.2(d).

6 Information given privately to the authors at the sixth extraordinary meeting of the Commission on Genetic Resources for Food and Agriculture, in June 2001, in Rome.

7 Although it must be remembered that these regulations are being violated more and more in practice and that patents on genetic resources are repeatedly registered and granted which are neither new nor an invention. This points to a fundamental problem concerning genetic resources: to what extent can these in fact be invented? More details on this later.

8 In the sense of international law, in the International Treaty the same formulations have been used as were already used in the Biosafety Protocol to regulate its relationship to other agreements. While they were still extremely controversial in the Biosafety Protocol and almost led to the failure of the entire process (Gupta 1999; cf. Görg 2003a: 272ff.), within the framework of the International Treaty they were relatively unproblematic and were taken over with reference to the apparently good solution of the Biosafety Protocol. But the problem with the phraseology taken from the Biosafety Protocol is that these formulations are not really unequivocal, but that the solution consists of listing the different objectives one after the other.

## 4 Politicising intellectual property rights: the conflicts around the TRIPS Agreement and the World Intellectual Property Organization

1

> The previous governance structure for intellectual property had included 135 states as members of the WIPO, although of the eighteen conventions administered, even the Paris convention, with the most signatories, had been ratified by only 108 countries. With the exception of the Bern convention (95 signatories), other agreements overseen by WIPO only had between 20 and 50 signatories.
>
> (May/Sell 2006: 162f.)

2 Cf. the debate on the internationalisation of the protection of intellectual property in the second half of the nineteenth century (May/Sell 2006: 115ff.).

3 See CIPR (2002a: 2), ETC Group (2001: 3f.), Byström/Einarsson (2001), Correa (2000: chapters II and VI), Federal Ministry of Education and Research/OECD (2002), Perelman (2003), Stilwell (2001).

4 Cf. Chapter 2 on this.

5 The following passages are largely based on an analysis of communications from national governments to the WTO concerning the TRIPS Agreement and biodiversity, on the summaries of the positions articulated in these communications which are periodically issued by the WTO and on the newsletter "Bridges Trade BioRes – Trade and Biological Resources News Digest", which is produced by the International Centre for Trade and

Sustainable Development (ICTSD) in collaboration with IUCN – The World Conservation Union. The latter is available at www.ictsd.org/biores/index.htm, the WTO documents under http://docsonline.wto.org. Individual documents are only indicated in the case of citations. For a summary of the positions and the state of negotiations as of spring 2006 see the notes of the Secretariat of the TRIPS Council in WTO (2006b) (concerning the review of Article 27.3(b) of the TRIPS Agreement) and WTO (2006a) (concerning the relationship between the TRIPS Agreement and the Convention on Biological Diversity).

6 The UPOV lays down the criteria for the protection of plant varieties: a plant variety must be new, differentiable, homogeneous and enduring in order to be able to be protected. The UPOV was set up in 1961. Revisions were made in 1972, 1978 and 1991. Until April 1999 non-UPOV members had the opportunity to join the Convention in the version of 1978. Since then it is only possible to join the version of 1991, which, compared to the version of 1978 strengthens the property rights of plant breeders:

> The 1991 revision has brought the Convention more into line with patent law. The 1991 Act requires the authorisation of the right-holder for production or reproduction, conditioning for the purpose of propagation, offering for sale, selling or other marketing, exporting or importing, and stocking for any of these purposes.
> (Leskien/Flitner 1997: 61)

7 See a communication from Kenya on behalf of the African Group from 1999 (WTO 1999a: paragraph 21):

> The review process should clarify that plants and animals as well as microorganisms and all other living organisms and their parts cannot be patented, and that natural processes that produce plants, animals and other living organisms should also not be patentable.

8 The prescribed period for industrial countries ended one year after the coming into force of the TRIPS Agreement, that is on 31 December 1995. The least developed countries originally should have implemented the Agreement by 31 December 2005. At the Ministerial Conference of the WTO in December 2005 in Hong Kong, this period was prolonged until the summer of 2012.

9 However, its effects are more symbolic than material. Even if it remains in the framework of the TRIPS Agreement, the international as well as the domestic relationships of power in many African states are likely to prevent the model law from being implemented. This is shown not least in the reactions of WIPO and UPOV. Both organisations saw considerable deficits in the draft with regard to patent protection and the protection of varieties. They rejected the attempt to establish inalienable community rights and to strengthen farmers' rights vis-à-vis the rights of breeders.

10 See for example a communication from Brazil from 2000 (WTO 2000c), in which it is stated, "Traditional knowledge may bring significant benefits for several industrial sectors, in particular the pharmaceutical and agricultural sectors." See also Dutfield (2002: 17), Ribeiro (2002b: 127f.) and Bridges Trade BioRes 2002: "For indigenous groups, TK protection is a rights-based issue, while many governmental representatives regard it as a matter of international equity or national development."

11 For recent initiatives, see the submission of a group of developing countries (WTO 2006c) and of Norway (WTO 2006d) from June 2006, both available under http://docsonline.wto.org. For a summary of both initiatives and the corresponding reactions, see Bridges Trade BioRes 2006b. For a discussion of disclosure requirements as legal instruments to prevent biopiracy and to ensure benefit-sharing, see Chouchena-Rojas *et al.* (2005).

12 Communication from the USA from 2006 (WTO 2006e: 2).

13 See WTO 2001 and WTO 2002b. On this condition, the European Union has declared its willingness to negotiate in the TRIPS Council on a multilateral system for the disclosure

of the origin of genetic resources in patent applications. The disclosure requirement is to be limited to the origin of the resources and should not include a proof of compliance with the regulations on access and benefit-sharing.

14 For a detailed discussion of the SPLT see Tvedt (2005).

15 In the case of a clarification under the WTO's dispute settlement body, the economically stronger countries have an advantage, simply because they can more readily afford the necessary juridical competence than can their weaker country counterparts and because they possess more effective opportunities for retribution. Due to the power asymmetries which can themselves develop to the full in bilateral negotiations, a legalistic argumentation which emphasises the flexibility of the TRIPS Agreement, such as that to be found in Correa (2000) and in Plahe/Nyland (2003), falls short of the mark.

16 Another factor, which however lies beyond the scope of this analysis, is the possibility of the convergence of social movements, that is the chance to establish links between the struggles against the strengthening of intellectual property protection in the area of biodiversity and similar struggles which take place for example in the area of software. See GRAIN (2005b) and Gerstetter/Kaiser (2006) on this issue.

### 5 The relevance of the national and the local: the disputes over a valorisation paradigm in Mexico and Chiapas

1 Scientific contacts with Mexico existed since the mid-1990s and were intensified between 1998 and 2000 under the EU-financed exchange programme ALFA (*América Latina de Formación Académica*).

2 The PRD is less a centralised, Western European-type party, being more a collection at the level of the provinces of very differing, mostly left-wing groups, which were able to participate in the elections through the PRD.

3 In 2005 imports of goods were worth 232 billion USD and exports 214 billion USD which is far more than the imports and exports of Brazil, for example (WTO 2006f: 11).

4 The Maquiladora industry carries out one particular step in the production process. The inputs can be imported tax-free to Mexico from other countries, for example the USA, and are then exported again tax-free after processing in the Maquiladoras. Important for the producers is cheap labour (and therefore weak trade unions). Laws on this have existed since the 1950s in Mexico; since 1972 the Maquiladora industry has been growing, above all at the border between the USA and Mexico.

> The exploitation of people and nature is the secret of the dynamic Maquiladora industry on the northern border of the country above all since the currency devaluation of 1995, while at the same time in the rest of Mexico crisis and cut-backs in employment – at least since 1997 – were clearly dominant.
>
> (Boris/Sterr 2002: 200, our translation)

The Maquiladora production fell in 2002 for the first time since the 1990s (EIA 2003: 1).

5 We did considerable research on the NAFTA Side Agreement on Environment but it plays hardly any role in the present issue area beside symbolic politics. Therefore, we leave it aside.

6 A representative of the Ministry of Agriculture (Mx/6) points out that this development was not at all uncontroversial, but that, within the Ministry ever since the 1950s, violent disputes over the concentration on high yields and the Green Revolution, as well as the reclaiming of intellectual property, took place.

7 One result is the massive change in the structure of the local economy, particularly because it becomes more attractive for the local population to work in the oil sector instead of in the (to a large extent subsistence-oriented) agricultural sector.

8 Due to the rapid growth of the population from 3 million inhabitants in 1960 to 7 million in 1970 and the extreme growth in the number of cars in the Mexico of the

1970s, which was transforming itself into an oil-producing country, the air pollution in the 2000 metre high capital city surrounded by mountains was the most discussed problem, above all because a large part of manufacturing industry was also based there (Russell 1994: 235ff.) Between 1979 and 1987 emissions of pollutants multiplied by a factor of eight (ibid.: 232).

9 Social groups gained in importance in the 1980s, particularly following the earthquake in Mexico City and the disastrous crisis management by the PRI in 1985. Apart from the main theme of the democratisation of the political system, environmental themes were taken up. The intellectual association "Group of one hundred", centred around Octavio Paz, Carlos Fuentes and Elena Pontiatowska, issued statements on various environmental issues. In the Mexican environmental movement (*Movimento Mexicano para el Medio Ambiente*) 63 national and regional groups were united in the Alliance of Mexican Ecologists (*Alianza de Ecologistas Mexicanos*), primarily those from the capital city (Russell 1994: 233). The Mexican Green Ecological Party (*Partido Verde Ecológico Mexicano*; PVEM) is unimportant, although it participated in the presidential elections in 2000 together with the PAN (*Partido de Acción Nacional*) and was among the victors (but remained without a ministerial post).

10 John Warnock (1995: 191) quotes a study which illustrates the difference: in the USA 7 tonnes of maize are harvested from one hectare of land, and in Mexico 1.7 tonnes; for the production of a tonne of beans in the USA 4.8 hours of labour are required, and in Mexico 50.6 hours. Oxfam (2003: 9) points to the important fact, however, that the main problem of Mexico's small producers is the immense power of the purchasing concerns and not so much the high productivity of the US farmers.

11 Already in the pre-Columbian world there were intermediaries between the indigenous communities and the centralised political power, whereby the relationships were based on mutual acceptance. During the colonial period this mutual acceptance was lost. In pre-Columbian times the cacique was to all intents and purposes the tribal chief who mediated between his tribe and the imperial power (Incas or Mayas). During and after colonialism, however, the cacique became the *political* chief who mediated between the tribe (village, town) and the nation-state and who aquired power as the political voice of his/her community before the state, thereafter using this power to advance his/her familial and/or groups interests, and thus becoming central for the reproduction of those relationships of power which suppressed the communes. It is today a member of the commune, but does not represent the external power of the state there.

12 In comparison to the beginning of the 1990s, this is not a dramatic shift. Rather, it is an indication not only that the changed legal conditions lead to sales but that the privatisation of the Ejido land is a complex process. The Ejiditarios do not sell immediately and there is generally no direct or huge interest from buyers.

13 Conservation International has a quite different view to the WWF:

> Initiatives led by CI and partners brought peace to the Montes Azules Biosphere Reserve in Chiapas state, site of a long-simmering conflict between government and local communities over indigenous rights. Through sustainable land-use planning, the promotion of alternative livelihoods, and the voluntary relocation of more than 10 squatter groups, previously violent disputes were resolved without a single incident of bloodshed.
>
> (CI 2006: 10)

14 A detailed analysis of the PPP would have to examine to what extent there are competing strategies. In addition to the outlined dominant strategy – which in the final analysis is defined according to the provision of resources – there is, for example, a paper by the Inter-American Development Bank and the CEPAL which proposes several initiatives within the objective of the creation of competitiveness, which emphasise domestic economic connections and sustainable development (see Grupo Técnico Interinstitucional 2001).

15 When the project was announced as comprising over 90 million USD, the GEF share of almost 15 million USD was only small, however, at the beginning of the project the GEF figure remained the same while the rest – particularly the share of the Mexican Government – had shrunk from over 70 million to 2.6 million USD (*La Jornada*, 3 March 2003).
16 Connected with this were specific labour relationships and a constitutive racism (Ceceña 2000).
17 Ana Esther Ceceña (2000: 268) asks the interesting question whether the exploitation of the oil reserves in Chiapas will wait until the privatisation of the Mexican oil industry.
18 Again, at the CBD Conference of the Parties in March 2006, in Curitiba, the positions of the Mexican delegation were very close to that of the US Government, for example concerning weak indications of transgenic food. And as the only government of all megadiverse countries it tried to promote Genetic Use Restriction Technologies (GURTS, *La Jornada*, 7 April 2006).
19 On important differences between conservationist and indigenous perspectives, see Nigh/Rodríguez 1995: 177–202.
20 It is paradoxical that from 1991 to 1994 the possibility of being granted patents on living organisms had already existed. This regulation was revoked in the course of the NAFTA implementation under pressure from the Ministry of Agriculture, and today UPOV is in force (Mx/6).
21 On the strategy and effects of the "silencios zapatistas" (Zapatista silences), that is their rejection of any public discourse, see the excellent essay of Rajchenberg/Héau-Lambert (2004).
22 The main objective of the members of the *Fondo* is the economic development of Chiapas through a wide range of investments in labour-intensive agricultural, industrial and tourist sectors, all of which hinge upon, among other things, the creation of industrial parks for Maquiladora industries and support for industries close to agriculture.
23 It may be of interest to note that CI, in the year 2005, raised funds of more than 92 million USD and invested more than 114 million USD (CI 2006: 18–19). The list of the supporters and financiers of CI reads like a "Who's Who" of the international economy. Alfonso Romo of Pulsar, who we have already mentioned, belongs to the Board; among the sponsors are Monsanto, Coca-Cola, McDonald's, Intel Corporation, Ford Motor Company and Starbucks (ibid.: 25–26). Unfortunately, in spite of numerous attempts, it was not possible to interview representatives of Pulsar or Conservation International during our research.
24 The seven programmes which began in 2005 take place in Costa Rica, Panama, Papua New Guinea, Fiji, Madagascar, Laos/Vietnam, Uzbekistan/Kirghizia.
25 It remains unclear whether the fact that the Mexican company Pulsar-Seminis was sold in 2005 to the US-based Monsanto has influenced the possibilities open to action in Mexico.

### 6 Contested terrains: towards a neo-Poulantzian approach to International Political Economy

1 Of particular relevance is the fate of the existing collections; see Chapter 3.

# Bibliography

Aboites, J. (1989) *Industrialización y desarrollo agrícola en México, un análisis del régimen de acumulación en el largo plazo: 1939–1987*, Mexico City: Universidad Autónoma Metropolitana-Xochimilco.

Acselrad, H. (2002) "Die ökologische Herausforderung zwischen Markt, Sicherheit und Gerechtigkeit", in Görg, C. and Brand, U. (eds) *Mythen globalen Umweltmanagements, Rio +10 und die Sackgassen "nachhaltiger Entwicklung"*, Münster: Westfälisches Dampfboot.

Aglietta, M. (1979) *A Theory of Capitalist Regulation, The US-Experience*, London: New Left Books.

Agrawal, A. (1995) "Dismantling the divide between indigenous and scientific knowledge", in *Development and Change*, vol. 26: 413–439, The Hague: Blackwell Publishing.

Aguilar, A.G. and Graizbord, B. (2006) "Latin America: a region of shared loyalties and persistent dependencies", in Geyer, H. S. (ed.) *Global Regionalization, Core Peripheral Trends*, Cheltenham: Edward Elgar Publishing.

Akhter, F. (2001) "Die Nayakrishi-Kampagne: Saatgut in die Hände der Frauen!", in Klafenböck, G., Lachkovics, E. and Südwind Agnetur (eds) *Biologische Vielfalt. Wer kontrolliert die genetischen Ressourcen*, Frankfurt/M.: Brandes & Apsel.

Albritton, R., Itoh, M., Westra, R. and Zuege, A. (eds) (2001) *Phases of Capitalist Development. Booms, Crises and Globalization*, London: Macmillan.

Alnasseri, S. (2003) "Ursprüngliche Akkumulation, Artikulation und Regulation, Aspekte einer globalen Theorie der Regulation", in Brand, U. and Raza, W. (eds) *Fit für den Postfordismus?*, Münster: Westfälisches Dampfboot.

—— (2004) *Periphere Regulation, Regulationstheoretische Konzepte zur Analyse von Entwicklungsstrategien im arabischen Raum*, Münster: Westfälisches Dampfboot.

Alnasseri, S., Brand, U., Sablowski, T., Winter, J. (2001) "Raum, Regulation und Periodisierung des Kapitalismus", in Albritton, Robert *et al.* (eds) *Phases of Capitalist Development. Booms, Crises and Globalization*, London: Macmillan.

Altvater, E. (1993) *The Future of the Market, An Essay of the Regulation of Money and Nature after the Collapse of "Actually Existing Socialism"*, London: Verso.

Altvater, E. and Mahnkopf, B. (1999) *Grenzen der Globalisierung, Ökonomie, Ökologie und Politik in der Weltgesellschaft*, 4th edn, Münster: Westfälisches Dampfboot.

Anderson, J. and Goodman, J. (1995) "Regions, States and the European Union: Modernist Reaction or Postmodernist Adaptation", in *Review of International Political Economy*, vol. 2, 4: 600–631.

Arnold, D. (1996) *The Problem of Nature*, Oxford: Blackwell Publishing.

Arts, B. (1998) *The Political Influence of Global NGOs, Case Studies on Climate and Biodiversity Conventions*, Utrecht: International Books.

Axelrod, R. S., Downie, D. L. and Vig, N. (eds) (2005) *The Global Environment: Institutions, Law, and Policy*, Washington, DC: C. Q. Press.

Ayales, I., Solís Rivera, V. and Madrigal, P. (2002) "Las mujeres en la conservación de la biodiversidad", in Heineke, C. (ed.) *La Vida en Venta: Transgénicos, Patentes y Biodiversidad*, San Salvador: Ediciones Heinrich Böll.

Bache, I. and Flinders, M. (eds) (2004) *Multi-Level Governance*, Oxford: Oxford University Press.

Baguio Declaration (2004) *Baguio Declaration of the 2nd Asian Indigenous Women's Conference*, 8 March 2004, Baguio City, Philippines, www.tebtebba.org/tebtebba_files/gender/aiwcdec.html

Barreda, A. (1999) *Atlas Geoeconómico y Geopolítico del Estado de Chiapas*, Dissertation: UNAM.

—— (2001) "Los peligros del Plan Puebla Panamá", Manuscript.

Barry, J. and Eckersley, R. (eds) (2005) *The State and Global Environmental Crisis*, Cambridge, MA: MIT Press.

Bartra, A. (2003) "Descifrando la treseava estela", in *Observatorio Social de América Latina*, 12: 279–292, www.observatoriosocial.com.ar/revi12.html

Beck, U. (1992) "Risk Society: towards a New Modernity", London: Sage Publications.

Beck, U., Giddens, A. and Lash, S. (1994) "Reflexive modernization: politics, tradition and aesthetics in the modern social order", Stanford, CT: Stanford University Press.

Becker, E. (2001) "Die postindustrielle Wissensgesellschaft – ein moderner Mythos?", in *Zeitschrift für Kritische Theorie*, vol. 7, 12: 85–106.

Becker, E. and Jahn, T. (1987) "Soziale Ökologie als Krisenwissenschaft", Frankfurt/M.: ISOE.

—— (eds) (2006) "Soziale Ökologie: Grundzüge einer Wissenschaft von den gesellschaftlichen Naturverhältnissen", Frankfurt/M. [u.a.]: Campus-Verlag.

Behrens, M. (ed.) (2005) *Globalisierung als politische Herausforderung: Global Governance zwischen Utopie und Realität*, Wiesbaden: VS Verlag für Sozialwissenschaften.

Bell, D. (1967) "Die nachindustrielle Gesellschaft", Frankfurt/M.: Campus.

Benton, T. (1989) "Marxism and natural limits", in *New Left Review* 178: 51–86.

Berne Declaration (2001) "Keep Crop Seeds Patent-Free in Order to Maintain World Food Security", open letter signed by 255 NGOs, Media Release, 20 January 2001, www.evb.ch/bd/press/20_04_01.htm

Betsill, M. M., Hochstetler, K. and Stevis, D. (eds) (2006) *International Environmental Politics*, Basingstoke: Palgrave.

Bieler, A. and Morton, A. (2004) "A critical theory route to hegemony, world order and historical change: neo-Gramscian perspectives in International Relations", in *Capital & Class*, 82 (Spring): 85–113.

Bieling, H. (2003) "Internationale Politische Ökonomie", in Schieder, S. and Spindler, M. (eds) *Theorien der Internationalen Beziehungen*, Opladen: Leske + Budrich.

—— (2006) "Europäische Staatlichkeit", in Bretthauer, L., Gallas, A., Kannankulam, J. and Stützle, I. (eds) *Poulantzas lesen, Zur Aktualität marxistischer Staatstheorie*, Hamburg: VSA.

BIO – Biotechnology Industry Organization, www.bio.org/ip/

Blank, K. (2002) "Das Internationale Indigene Forum zu Biodiversität im aktuellen Politikprozess der Konvention über biologische Vielfalt", *Forschungsbericht zum VI. Internationalen Indigenen Forum zu Biodiversität*, Manuscript.

BMU – Bundesministerium für Umwelt, Naturschutz und Reaktorsicherheit (ed.) (1992) *CBD, Convention on Biological Diversity, Übereinkommen über die biologische Vielfalt*, Vertragstext, Bonn: BMU.

Bödeker, S., Moldenhauer, O. and Rubbel, B. (2005) "Wissensallmende. Gegen die Privatisierung des Wissens der Welt", *AttacBasisTexte 15*, Hamburg: VSA-Verlag.

Borg, E. (2001) *Projekt Globalisierung, Soziale Kräfte im Konflikt um Hegemonie*, Hannover: Offizin.

Boris, D. (1996) *Mexiko im Umbruch, Modellfall einer gescheiterten Entwicklungsstrategie*, Darmstadt: Wissenschaftliche Buchgesellschaft.

Boris, D. and Sterr, A. (2002) *Foxtrott in Mexiko, Demokratisierung oder Neopopulismus?*, Köln: Neuer ISP-Verlag.

Boyer, R. (1986) *La Theorie de la Régulation: Une Analyse Critique*, Paris: Editions La Découverte.

Braithwaite, J. and Drahos, P. (2000) *Global Business Regulation*, Cambridge: Cambridge University Press.

Brand, U. (2000) *Nichtregierungsorganisationen, Staat und ökologische Krise, Konturen kritischer NRO-Forschung, Das Beispiel der biologischen Vielfalt*, Münster: Westfälisches Dampfboot.

—— (2004) "Hegemonía fragmentada", in Sader, E. and Brie, M. (eds) *Para além do capitalismo neoliberal: concepcões, atores e estratégia*. Sao Paolo: Expressão Popular.

—— (2005) "Order and regulation: Global Governance as a hegemonic discourse of international politics?", in *Review of International Political Economy*, vol. 12, 1: 155–176.

—— (2006) "The Internationalised State and its Functions and Modes in the Global Governance of Biodiversity", paper presented at the conference *Science, Knowledge Communities and Environmental Governance: Global–Local Linkages* at Rutgers University, May 2006.

Brand, U. and Ceceña, A. E. (eds) (2000) *Reflexionen einer Rebellion, "Chiapas" und ein anderes Politikverständnis*, Münster: Westfälisches Dampfboot.

Brand, U. and Görg, C. (2007) "Sustainability and Globalisation: A Theoretical Perspective", in Conca, K., Finger, M.and Park, J. (eds) *Sustainability, Globalization and Governance*, London: Routledge (forthcoming).

Brand, U. and Kalcsics, M. (eds) (2002) *Wem gehört die Natur? Konflikte um genetische Ressourcen in Lateinamerika*. Frankfurt/M.: Brandes & Apsel.

Brand, U. and Raza, W. (eds) (2003) *Fit für den Postfordismus?*, Münster: Westfälisches Dampfboot.

Brand, U., Brunnengräber, A., Schrader, L., Stock, C. and Wahl, P. (2000) *Global Governance, Alternative zur neoliberalen Globalisierung?*, Münster: Westfälisches Dampfboot.

Brenner, N. (2001) "The limits to scale? Methodological reflections on scalar structuration", in *Progress in Human Geography*, vol. 25, 4: 591–614.

—— (2004) "New State Spaces, Urban Governance and the Rescaling of Statehood", Oxford: Oxford University Press.

Brenner, N., Jessop, B., Jones, M. and Macleod, G. (eds) (2003) *State / Space: A reader*, Malden, MA: Blackwell Publishers.

Bridges Trade BioRes (2002) "Misappropriation of Traditional Knowledge Discussed at CBD Working Group", vol. 2, 3, www.ictsd.org/biores/02-02-21/story2.htm

—— (2004) "WIPO Committee Resumes Work on Genetic Resources and TK", vol. 4, 6, www.ictsd.org/biores/04-04-02/story1.htm

—— (2006a) "WTO Disclosure Talks Try to Clarify CBD–Trips Relationship", vol. 6, 5, www.ictsd.org/biores/06-03-17/story4.htm

—— (2006b) "Discussions on CBD–Trips Gain Momentum with New Proposals", vol. 6, 11, www.ictsd.org/biores/06-06-16/story3.htm

—— (2006c) "WIPO Committee Considers Mechanism to Protect TK", vol. 6, 9, www.ictsd.org/biores/06-05-19/story4.htm

Brockway, L. H. (1988) "Plants, Sciences and Colonial Expansion: The Botanical Chess Game", in Kloppenburg, J. R. (ed.) *Seeds and Sovereignity, The Use and Control of Plant Genetic Ressources*, Durham, NC: Duke University Press.

Brühl, T. (2001) "Mehr Raum für unbequeme Mitspieler? Warum Nichtregierungsorganisationen verstärkt in internationalen (Umwelt-)Verhandlungen mitwirken dürfen", in Walk, H., Klein, A. and Brunnengräber, A. (eds) *NGOs als Legitimationsressource*, Opladen: Westdeutscher Verlag.

Brunnengräber, A. and Walk, H.(eds) (2007) *Multi-Level-Governance. Klima-, Umwelt- und Sozialpolitik in einer interdependenten Welt*, Baden Baden: Nomos.

Brunnengräber, A., Klein, A. and Walk, H. (eds) (2001) *NGOs als Legitimationsressource. Zivilgesellschaftliche Partizipationsformen im Globalisierungsprozess*, Opladen: Leske & Buderich.

Brush, S. B. (1993) "Indigenous Knowledge of Biological Resources and Intellectual Property Rights: The Role of Anthropology", in *American Anthropologist*, vol. 95, 3: 653–669.

BSR – Biodiversity Synthesis Report (2005) *Millennium Ecosystem Assessment: Ecosystems and Human Well-being*, Washington, DC.: World Resources Institute.

Buckel, S. (2003) "Global 'Non-State', Überlegungen für eine materialistische Theorie des transnationalen Rechts", in Buckel, S., Dackweiler, R. and Noppe, R. (eds) *Formen und Felder politischer Intervention, Zur Relevanz von Staat und Steuerung,* Festschrift für Josef Esser, Münster: Westfälisches Dampfboot.

BUKO – Kampagne gegen Biopiraterie (2005) *Grüne Beute, Biopiraterie und Widerstand – Argumente, Hintergründe, Aktionen*, Grafenau: Trotzdem-Verlag.

Buntzel-Cano, R. (2000) "Aufruf zu Aktionen", in *Forum Umwelt & Entwicklung*, Rundbrief 3, 24.

Burchardt, H. (2004) *Zeitenwende. Politik nach dem Neoliberalismus*, Stuttgart: Schmetterling.

Burrows, B. (ed.) (2005) *The Catch, Perspectives in Benefit Sharing*, Edmonds: The Edmonds Institute, www.edmonds-institute.org/thecatch.pdf

Buttel, F. H. (1992) "The 'Environmentalization' of Plant Genetic Resources: Possible Benefits, Possible Risks", in *Diversity*, vol. 8, 1: 36–39.

Buttel, F. H., Dickens, P., Dunlap, R. E. and Gijswijt, A. (2002) "Sociological Theory and the environment: An overview and introduction", in Dunlap R. E. *et al.* (eds) *Sociological theory and the environment: classical foundations, contemporary insights*, Lanham, MD: Rowman & Littlefield Publishers.

Byström, M. and Einarsson, P. (2001) *TRIPs, Consequences for developing countries, Implications for Swedish development cooperation, Consultancy Report to the Swedish International Development Cooperation Agency (Sida)*, www.grain.org/docs/sida-trips-2001-en.PDF

Caillaux, J. and Ruíz, M. (2002) "Legislative Experiences on Access to Genetic Resources and Options for Megadiverse Countries", paper presented at the *First Meeting of Like-Minded Megadiverse Countries*, February 2000, Mexico, www.megadiverse.org/armado_ingles/PDF/five/five5.pdf

Calderón Salazar, J. A. (1992) *El TLC y el desarrollo rural*, Mexico City: Ed. CEMO.

Camp, R. A. (2003) *Politics in Mexico, The Democratic Transformation*, 4th edn, New York: Oxford University Press.

Candeias, M. (2004) *Neoliberalismus, Hochtechnologie, Hegemonie, Grundrisse einer transnationalen kapitalistischen Produktions- und Lebensweise*, Hamburg: Argument-Verlag.

Candeias, M. and Deppe, F. (eds) (2001) *Ein neuer Kapitalismus?*, Hamburg: VSA-Verlag.

Casares, E. S. and Sobarzo, H. (eds) (2004) *Diez años del TLCAN en México, Una perspectiva analítica*, Mexico City: Fondo de Cultura Económica.

Castree N. and Braun, B. (1998) "The construction of nature and the nature of construction", in Braun, B. and Castree, N. (eds) *Remaking Reality: nature at the Millennium*, London: Routledge.

Castro Escudero, T. and Oliver Costilla, L. (eds) (2005) *Poder y política en América Latina*, El debate latinoamericano, vol. 3, Mexico City: Siglo XXI/UNAM.

Castro Escudero, T., Mussali Galante, R. and Oliver Costilla, L. (2005) "Revisitando al Estado, Los Estados populistas y desarrollistas: Poner las cosas en su lugar", in Castro Escudero, T. and Oliver Costilla, L. (eds) *Poder y política en América Latina,* El debate latinoamericano, vol. 3, Mexico City: Siglo XXI/UNAM.

Catton, W. R.and Dunlap, R. E. (1978) "Environmental Sociology: A New Paradigm", in *The American Sociologist*, vol. 13: 41–49.

CBD – Convention on Biological Diversity (1992) www.biodiv.org/convention/convention.shtml

—— (2001) *Report of the Ad Hoc Open-Ended Working Group on Access and Benefit-Sharing*, (UNEP/CBD/COP/6/6), www.biodiv.org/doc/meetings/cop/cop-06/official/cop-06-06-en.pdf

—— (2002) "Access and benefit-sharing as related to genetic resources", *Decision VI/6 of the CBD*, www.biodiv.org/decisions/

—— (2006) "International regime on access and benefit-sharing", (UNEP/CBD/WG-ABS/4/CRP.1/Rev.2), 3 February 2006,

www.ip-watch.org/weblog/index.php?p = 213&res = 1280_ff&print = 0

CBD-Handbook (2001) *Handbook of the Convention on Biological Diversity*, London: Earthscan.

CEAS (2000) "Study on the relationship between the agreement on TRIPS and biodiversity related issues", *Center for European Agricultural Studies, Final Report for DG TRADE European Commission*, europa.eu.int/comm/trade/miti/intell/ceas.htm

Ceceña, A. E. (2000) "Die Grenzen der Modernität. Kämpfe um strategische Ressourcen", in Brand, U. and Ceceña, A. E. (eds) *Reflexionen einer Rebellion, "Chiapas" und ein anderes Politikverständnis*, Münster: Westfälisches Dampfboot.

— (ed.) (2006) *Los desafíos de las emancipaciones en un contexto militarizado*, Buenos Aires: CLACSO.

Ceceña, A. E. and Barreda, A. (eds) (1995) *Producción Estratégica y Hegemonía Mundial*, Mexico City: Siglo XXI.

Ceceña, A. E. and Giménez, J. (2002) "Hegemonía y bioprospección, El caso del International Cooperative Biodiversity Group", in Brand, U. and Kalcsics, M. (eds) (2002) *Wem gehört die Natur? Konflikte um genetische Ressourcen in Lateinamerika*, Frankfurt/M.: Brandes & Apsel.

Chambers, W. B. (2004) "Use of Genetic Resources: New Global Rules or a Rocky Path to Nowhere?", in *Environmental Policy and Law*, vol. 34, 2: 61–64.

Chapin, M. (2004) "A Challenge to Conservationists", in *World Watch Magazine*, vol. 17, 6: 17–31.

Chouchena-Rojas, M., Ruiz-Muller, M., Vivas, D. and Winkler, S. (eds) (2005) *Disclosure Requirements: Ensuring mutual supportiveness between the WTO TRIPS Agreement and the CBD*, Gland: IUCN/ICTSD, www.iucn.org/en/news/archive/2005/12/disclosure_requirements_publication.pdf

CI – Conservation International (2006) *Annual Report 2005*, Washington, DC: CI www.conservation.org

CIEL – Center for International Environmental Law and WWF – World Wide Fund for Nature (2001) *Biodiversity and Intellectual Property Rights: Reviewing Intellectual Property Rights in Light of the Objectives of the Convention on Biological Diversity*, Joint Discussion Paper, Gland, Geneva.

CIEPAC – Centro de Investigaciones Económicas y Políticas de Acción Comunitaria, www.ciepac.org

CIPR – Commission on Intellectual Property Rights (2002a) "Integration von geistigen Eigentumsrechten und Entwicklungspolitik", *Bericht der Kommission für geistige Eigentumsrechte (Executive Summary)*, London, www.wissensgesellschaft.org/themen/wemgehoert/PropertyRightsCommissiondt.pdf

—— (2002b) "Integrating Intellectual Property Rights and Development Policy", Report of the Commission on Intellectual Property Rights, London, www.iprcommission.org

CLACSO – Consejo Latinoamericano de Ciencias Sociales (ed.) (2004) *Observatorio Social de América Latina*, issue on "movimientos sociales y desafíos políticos" and "resistencias continentales frente al 'libre comercio'", Buenos Aires: CLACSO.

Comor, E. A. (1999) "Governance and the Nation State in a Knowledge-Based Political Economy", in Hewson, M. and Sinclair, T. J. (eds) *Approaches to Global Governance Theory*, Albany: State University of New York Press.

COMPITCH – Consejo de Organizaciones de Médicos y Parteras Indígenas Tradicionales de Chiapas (2000) "Buletín informativo", San Cristóbal de las Casas.

CONABIO – Comisión nacional para el conocimiento y uso de la biodiversidad (1997) "Situación Actual sobre la Gestión, Manejo y Conservación de la Diversidad Bioogica de México", First National Report of the Conference of the Parties of the CBD, Mexico City: Conabio, www.biodiv.org/doc/world/mx/mx-nr-01-es.pdf

—— (1998) "Diversidad Biológica de México: Estudio de País", Mexico City: Conabio.

—— (2000) "Estrategia Nacional sobre Biodiversidad de México", Mexico City: Conabio.

CONANP – Comisión Nacional de Áreas Naturales Protegidas (2007), www.conanp.gob.mx.

Conca, K., Finger, M. and Park, J. (eds) (2007) *Sustainability, Globalization and Governance*, London: Routledge (forthcoming).

Cooper, D., Engels, J. and Frison, E. (1994) "A multilateral system for plant genetic resources: imperatives, achievements and challenges", Issues in Genetic Resources, 2, Rome: IPGRI.

Correa, C. (2000) "Intellectual Property Rights, the WTO and Developing Countries, The TRIPs Agreement and Policy Options", London: Zed Books; Penang (Malaysia): Third World Network.

—— (2001) "Pro-Competitive Measures under the TRIPs Agreement to Promote Technology Diffusion in Developing Countries", paper presented at the *Oxfam International Seminar on Intellectual Property and Development "What Future for the WTO TRIPs Agreement?"*, Brussels, 20 March 2001.

—— (2003) Interview, *Seedling*, January, www.grain.org/seedling/seed-interview-c.cfm

—— (2004) "Bilateral Investment Agreements: Agents of new global standards for the protection of intellectual property rights?", www.grain.org/briefings_files/correa-bits-august-2004.pdf

Correa, C. and Musungu, S. F. (2002) *The WIPO Patent Agenda: The Risks for Developing Countries*, South Centre, www.southcentre.org/publications/wipopatent/toc.htm

Cox, R. W. (1987) *Production, Power and World Order, Social Forces in the Making of History*, New York: Columbia University Press.

—— (1993) "Gramsci, hegemony and international relations: an essay in method", in Gill, S. (ed.) *Gramsci, Historical materialism and international relations*, Cambridge: Cambridge University Press.

—— (1998) *Weltordnung und Hegemonie – Grundlagen der "Internationalen Politischen Ökonomie"*, Marburg: Universität Marburg.

Crosby, A. W. (1972) *The Columbian Exchange*, Westport, CT: Greenwood Press.

—— (1986) *Ecological Imperialism: The Biological Expansion of Europe 900–1900*, Cambridge: Cambridge University Press.

Crucible Group (1994) *People, Plants and Patents*, Ottawa u.a.

Crucible II Group (2000) "Seeding Solutions, Policy Options for genetic resources", vol.1, co-published by IPGRI, Dag Hammarskjöld Foundation and IDRC, Rome: Litopixel.

—— (2001) "Seeding Solutions, Options for national laws governing control over genetic resources", vol. 2, co-published by IPGRI, Dag Hammarskjöld Foundation and IDRC, Rome: Litopixel.

de Boef, W., Amanor, K. S., Wellard, K. and Bebbington, A. (eds) (1993) *Cultivating Knowledge*, London: Intermediate Technology Publications.

Declaración de la Semana por la Diversidad (2001) *Declaración de la Primera Semana por la Diversidad Biológica y Cultural*, www.laneta.apc.org/biodiversidad/semana/declaracion.htm

Declaración de San Gregorio (2000) "Foro por la Defensa de la Vida, la Tierra y los Recursos Naturales", in *Chiapas*, 10: 195–197, www.ezln.org/revistachiapas/No10/ch10gregorio.html

Declaración Foro Mesoamericano (2002) "Frente al Plan Puebla Panamá el Movimiento Mesoamericano pol la Integración Popular", www.foromanagua.com.ni/documentos/declaracion-III-foro-mesoamericano.PDF

Delaney, D. and Leitner, H. (1997) "The Political Construction of Scale", in *Political Geography*, vol. 16: 93–97.

Delgadillo Macías, J. (1994) "Recursos naturales y ecología: bases para un desarrollo sustentable", in *Problemas del Desarrollo*, vol. 25, 96: 125–161.

Delgado Ramos, G. (2001) "Biopir@terie und geistiges Eigentum als Eckpfeiler technologischer Herrschaft: Das Beispiel Mexiko", in *Das Argument*, vol. 242: 481–494.

—— (2002) *La amenaza biológica, Mitos y falsas promesas de la biotecnología*, Barcelona: Plaza y Janés.

—— (2004) *Biodiversidad, desarrollo sustentable y militarización, Esquemas de saqueo en Mesoamérica*, Mexico City: Plaza y Valdéz/UNAM.

Demirovic, A. (1997) *Demokratie und Herrschaft, Aspekte kritischer Gesellschaftstheorie*, Münster: Westfälisches Dampfboot.

—— (2003) "NGOs, the State, and Civil Society: The Transformation of Hegemony", in *Rethinking Marxism*, vol.15, 2: 213–236.

de Paranaguá Moniz, P., Doctorow, C. and Rezende, P. A. D. (2005) "Manifesto for Transparency, Participation, Balance and Access", an open letter to the WIPO, www.cic.unb.br/docentes/pedro/trabs/wipo-en.html

Dieter, H. (2003) "Die Demontage der multilateralen Wirtschaftsordnung durch die Dritte Welle des Regionalismus", in *PROKLA*, vol. 33, 4: 599–624.

Dolata, U. (1999) "Die Bio-Industrie, Märkte, Unternehmen, politische Alternativen", in Emmrich, M. (ed.) *Im Zeitalter der Bio-Macht*, Frankfurt/M.: Mabuse-Verlag.

Drahos, P. (2001) "Negotiating Intellectual Property Rights: Between Coercion and Dialogue", paper presented at the *Oxfam International Seminar on Intellectual Property and Development "What Future for the WTO TRIPs Agreement?"*, Brussels, 20 March 2001.

Drahos, Peter and Braithwaite, John (2003) "Hegemony Based on Knowledge: The Role of Intellectual Property", *Law in Context*, vol. 21: 204–223.

Dutfield, G. (2001) "Biotechnology and Patents: What Can Developing Countries Do About Article 27.3 (b)?", in *Bridges Monthly Review*, vol. 5, 9: 17–18, www.ictsd.org/monthly/bridges/BRIDGES5-9.pdf

—— (2002) "Protecting Traditional Knowledge and Folklore: A review of progress in diplomacy and policy formulation", *UNCTAD/ICTSD Capacity Building Project on Intellectual Property Rights and Sustainable Development*, www.ictsd.org/iprsonline/unctadictsd/docs/Dutfield2002.pdf

ECLA/CEPAL – Comisión Económica para América Latina y el Caribe (2004) *Desarrollo productivo en economías abiertas*, Santiago de Chile: CEPAL.

Egziabher, T. B. G. (2002) "Bedrohte Ernährungssouveränität, internationales Recht und *Farmers' Rights* in Afrika", in Görg, C. and Brand, U. (eds) *Mythen globalen Umweltmanagements. Rio+10 und die Sackgassen "nachhaltiger Entwicklung"*, Münster: Westfälisches Dampfboot.

—— (2005) "Benefit Sharing", in Burrows, B. (ed.) (2005) *The Catch, Perspectives in Benefit Sharing*, Edmonds: The Edmonds Institute.

EIA – Energy Information Administration (2003) "Mexico", *Country Analysis Briefs*, Washington, DC, www.eia.doe.gov/emeu/cabs/mexico.html

Ekpere, J. A. (2000) *The OAU's Model Law, an Explanatory Booklet*, Lagos, www.grain.org/publications/oau-en.cfm

El-Said, H. and El-Said, M. (2005) "TRIPS, Bilateralism, Multilateralism & Implications for Developing Countries: Jordan's Drug Sector", in *Manchester Journal of International Economic Law*, vol. 2, 1: 59–79, www.electronicpublications.org/stuff.php?id = 106

ENB – Earth Negotiation Bulletin (1997) vol. 9, 68, www.iisd.ca/download/pdf/enb0968e.pdf

—— (2000a) vol. 9, 161, www.iisd.ca/download/pdf/enb09161e.pdf

—— (2000b) vol. 9, 164, www.iisd.ca/download/pdf/enb09164e.pdf

—— (2000c) vol. 9, 165, www.iisd.ca/download/pdf/enb09165e.pdf

—— (2000d) vol. 9, 166, www.iisd.ca/download/pdf/enb09166e.pdf

—— (2000e) vol. 9, 167, www.iisd.ca/download/pdf/enb09167e.pdf

—— (2001) vol. 9, 213, www.iisd.ca/download/pdf/enb09213e.pdf

—— (2002) vol. 9, 235, www.iisd.ca/download/pdf/enb09235e.pdf

—— (2006a) vol. 9, 363, www.iisd.ca/download/pdf/enb09363e.pdf

—— (2006b) vol. 9, 364, www.iisd.ca/download/pdf/enb09364e.pdf

Enquete-Kommission Globalisierung der Weltwirtschaft (2002) "Schlussbericht. Berlin", BT-Drucksache 14/9200, www.bundestag.de/gremien/welt/glob_end/glob.pdf

Escobar, A. (1996) "Constructing Nature. Elements for a poststructural political ecology", in Peet, R. and Watts, M. (eds) *Liberation ecologies. Environment,development, social movements*, London: Routledge.

Esser, J., Görg, C. and Hirsch, J. (eds) (1994) *Politik, Institutionen und Staat, Zur Kritik der Regulationstheorie*, Hamburg: VSA-Verlag.

ETC Group (2001) "New Enclosures: Alternative Mechanisms to Enhance Corporate Monopoly and BioSerfdom in the 21st Century", Communiqué, November/December 2001, www.etcgroup.org

—— (2005a) "Global Seed Industry Concentration – 2005", Communiqué, www.etcgroup.org

—— (2005b) "Oligopoly, Inc. – Concentration in Corporate Power", www.etcgroup.org

Executive Secretary of the CBD (2005) "Annotations to the provisional agenda", UNEP/CBD/COP/8/1/Add.1, www.biodiv.org/doc/meeting.asp?mtg = cop-08

FAO – Food and Agriculture Organization (1989) *International Undertaking on Plant Genetic Resources*, Rome: FAO.

FAO Focus (2006) "Women: users, preservers and managers of agro-biodiversity", www.fao.org/FOCUS/E/Women/Biodiv-e.htm

Federal Ministry of Education and Research/OECD (2002) *Short Summary Report of the Workshop on Genetic Inventions, Intellectual Property Rights and Licensing Practices*, held in Berlin, Germany, 24 and 25 January 2002, www.oecd.org/EN/home/0,EN-home-27-nodirectorate-no-no–27,00.html

Fogarty International Center (FIC) (2002): Announcement for Third Round of Research Grants. http://grants2.nih.gov/grants/guide/rfa-files/RFA-TW-03-004.html

—— (2007) *International Cooperative Biodiversity Groups*, www.fic.nih.gov/programs/research_grants/icbg/index.htm

Finston, S. K. (2005) "Commentary: An American BioIndustry Alliance Perspective on CBD/TRIPS Issues in the Doha Round", in *Global Economy Journal*, vol. 5, 4: 1–8.

Fischer, F. (2000) *Citizens, Experts, and the Environment, The Politics of Local Knowledge*, Durham, NC: Duke University Press.

Flitner, M. (1995a) *Räuber, Sammler und Gelehrte. Die politischen Interessen an pflanzengenetischen Ressourcen*, Frankfurt/M.: Campus.

—— (1995b) "Genetische Ressourcen: Von der Entstehung und den Konjunkturen eines Begriffs", in Mayer, J. (ed.) *Eine Welt – Eine Natur?*, Loccumer Protokolle, Loccum: Evangelische Akademie Loccum.

—— (1999) "Biodiversität oder: Das Öl, das Meer und die 'Tragödie der Gemeingüter'", in Görg, C. *et al.* (eds) *Zugänge zur Biodiversität*, Marburg: Metropolis.

Flitner, M., Görg, C. and Heins, V. (eds) (1998) *Konfliktfeld Natur, Biologische Ressourcen und globale Politik*, Opladen: Leske + Buderich.

Fowler, C. (1995) *Unnatural Selection: Technology, Politics, and Plant Evolution*, Yverdon: Gordon and Breach.

Frankel, O. H. and Bennet, E. (eds) (1970) *Genetic Resources in Plants – their Exploration and Conservation*, International Biological Program Handbook, 11, Oxford: Science Publications.

Frein, M. and Meyer, H. (2005) "The concept of benefit-sharing: A step forward or backward?", in Burrows, B. (ed.) *The Catch, Perspectives in Benefit Sharing*, Edmonds: The Edmonds Institute.

GAIA/GRAIN (2000) *Biodiversity for Sale, Dismantling the Hype about Benefit Sharing*, April 2000, www.grain.org

García, A. (2006) "Es gibt Alternativen zur ALCA, ALBA und warum der ALCA-Prozess gestoppt wurde", in *PROKLA*, vol. 142: 81–94.

García de León, A. (1985) *Resistencia y utopía, Memorial de agravios y crónica de revueltas y profecías acaecidas en la provincia de Chiapas durante los últimos quinientos años de su historia*, 2 volumes, Mexico City: ERA.

Gaskell, G., Allum, N., Stares, S. *et al.* (2003) "Europeans and Biotechnology in 2002", Eurobarometer 58.0, a report to the *EC Directorate General for Research from the project "Life Sciences in European Society"* (QLG7-CT-1999-00286 – 2nd edn: 21 March 2003), http://europa.eu.int/comm/public_opinion/archives/eb/ebs_177_en.pdf

GBF – Global Biodiversity Forum (2002) *Statement of the 16th Session of the Global Biodiversity Forum to the 6th meeting of the Conference of the Parties to the Convention on Biological Diversity*, The Hague, 5–7 April 2002.

Germain, R. D. and Kenny, M. (1998) "Engaging Gramsci: international relations and the new Gramscians", in *Review of International Studies*, vol. 24, 1: 3–21.

Gerstetter, C. and Kaiser, G. (2006) "Gemeinsam die Allmende verteidigen?! Ansätze und Formen des Widerstands gegen die Ausdehnung geistiger Eigentumsrechte in den Bereichen pflanzengenetische Ressourcen und Software", in *Peripherie*, vol. 26, 101/102: 69–98.

Gibbs, D. (2000) "Globalization, the Bioscience Industry and Local Environmental Responses", in *Global Environmental Change*, vol. 10: 245–257.

Giddens, A. (1988) *Die Konstitution der Gesellschaft, Grundzüge einer Theorie der Strukturierung*, Frankfurt/M.: Campus.

Gill, S. (2000) "Theoretische Grundlagen einer Neo-Gramscianischen Analyse der europäischen Integration", in Bieling, H.-J. and Steinhilber, J. (eds) *Die Konfiguration Europas. Dimensionen einer kritischen Integrationstheorie*, Münster: Westfälisches Dampfboot.

—— (2003) "Power and Resistance in the New World Order", New York: Palgrave.

Goldman, M. (ed.) (1998) *Privatizing Nature. Political struggles for the global commons*, London: Routledge.

—— (2001) "Constructing an Environmental State: Eco-governmentality and other Transnational Practices of a 'Green' World Bank", in *Social Problems*, vol. 48, 4: 499–523.

—— (2005) *Imperial Nature, The World Bank and struggles for social justice in the age of globalization*, New Haven, CT: Yale University Press.

Goodwin, M. and Painter, J. (1997) "Concrete Research, Urban Regimes, and Regulation Theory", in Lauria, M. (ed.) *Reconstructing Urban Regime Theory, Regulating Urban Politics in a Global Economy*, Thousand Oaks, CA: Sage Publications.

Gordon, D. M. (1994) "The Global Economy: New Edifice or Crumbling Foundations?", in Kotz, D. M., McDonough, T. and Reich, M. (eds) *Social Structures of Accumulation, The political economy of growth and crisis*, Cambridge: Cambridge University Press.

Görg, C. (1994) "Der Institutionenbegriff in der Theorie der Strukturierung", in Esser, J., Görg, C. and Hirsch, J. (eds) (1994) *Politik, Institutionen und Staat, Zur Kritik der Regulationstheorie*, Hamburg: VSA-Verlag.

—— (1998) "Die Regulation der biologischen Vielfalt", in Flitner, M., Görg, C. and Heins, V. (eds) (1998) *Konfliktfeld Natur, Biologische Ressourcen und globale Politik*, Opladen: Leske + Buderich.

—— (1999a) *Gesellschaftliche Naturverhältnisse*, Münster: Westfälisches Dampfboot.

—— (1999b) "Erhalt der biologischen Vielfalt: zwischen Umweltproblem und Ressourcenkonflikt", in Görg, C., Hertler, C., Schramm, E. and Weingarten, M. (eds) *Zugänge zur Biodiversität*, Marburg: Metropolis.

—— (2003a) *Regulation der Naturverhältnisse, Zu einer kritischen Theorie der ökologischen Krise*, Münster: Westfälisches Dampfboot.

—— (2003b) "Nichtidentität und Kritik, Zum Problem der Gestaltung der Naturverhältnisse", in Böhme, G. and Manzei, A. (eds) *Kritische Theorie der Natur und der Technik*, München: Fink-Verlag.

—— (2004) "Stichwort 'Inwertsetzung'", in *Historisch-kritisches Wörterbuch des Marxismus*, vol. 6, II, Hamburg: Argument-Verlag.

Görg, C and Brand, U. (2001) "Patentierter Kapitalismus, Zur politischen Ökonomie genetischer Ressourcen", in *Das Argument*, vol. 242: 466–480.

—— (2002) "Konflikte um das 'grüne Gold der Gene', Access, geistiges Eigentum und Fragen der Demokratie", in *PROKLA*, vol. 129: 631–652.

—— (eds) (2002) *Mythen globalen Umweltmanagements, Rio+10 und die Sackgassen "nachhaltiger Entwicklung"*, Münster: Westfälisches Dampfboot.

—— (2003) "Post-Fordist Societal Relationships with Nature: The Role of NGOs and the State in Biodiversity Politics", in *Rethinking Marxism*, vol. 15, 2: 263–288.

—— (2006) "Global Regulation of Genetic Resources and the Internationalization of the State", in *Global Environmental Politics*, vol. 6, 4, Cambridge: MIT Press.

Görg, C. and Wissen, M. (2003) "National dominierte globale Herrschaft, Zum Verhältnis von Uni- und Multilateralismus in der 'Neuen Weltordnung'", in *PROKLA*, vol. 133: 625–644.

Görg, C., Hertler, C., Schramm, E. and Weingarten, M. (eds) (1999) *Zugänge zur Biodiversität*, Marburg: Metropolis.

Grahl, J. and Teague, P. (2000) "The Regulation School, the employment relation and the financialization", in *Economy and Society*, vol. 29, 1: 160–178.

GRAIN – Genetic Resources Action International (2000) For a Full Review of TRIPS 27.3 (b) and an update on where developing countries stand with the push to patent life at WTO, www.grain.org/publications/tripsfeb00-en-p.htm

—— (2001a) *IPR agents try to derail OAU process*, www.grain.org/publications

—— (2001b) "A Disappointing Compromise", in *Seedling* December 2001, www.grain.org/publications/seed-01-12-1-en.cfm

—— (2001c) "'TRIPs-plus' through the back door, How bilateral treaties impose much stronger rules for IPRs on life than WTO", GRAIN publications July 2001, www.grain.org/publications/trips-plus-en.cfm

—— (2002a) "Biopiracy by another name? A critique of the FAO-CGIAR trusteeship system", in *Seedling* October 2002, www.grain.org/seedling /seed-02-10-2-en.cfm

—— (2002b) "WIPO moves toward 'world' patent system", GRAIN publications July 2002, www.grain.org/docs/wipo-patent-2002-en.doc

—— (2005) "Convergence Zone?", in *Seedling* October 2005, www.grain.org/seedling_files/seed-05-10.pdf

Gramsci, A. (1991) "Gefängnishefte", edited by Bochmann, K. and Haug, W., Hamburg: Argument-Verlag.

Graz, J. (2006) "Hybrids and Regulation in the Global Political Economy", in *Competition & Change*, vol. 10, 2: 230–245.

Grupo Tecnico Interinstitucional para el Plan Puebla–Panamá, BCIE-BID-CEPAL (2001) "Plan Puebla-Panamá, Initiativas mesoamericanas y proyectos", *San Salvador*, 15 June 2001, www.iadb.org/ppp/files/projects/OTRO/OT_PPP_PPP1_ES_PP.doc

GTZ – Deutsche Gesellschaft für Technische Zusammenarbeit (2002) *The Convention on Biological Diversity, Ensuring Gender-Sensitive Implementation*, Eschborn: GTZ.

Guillén Romo, H. (1997) *La contrarevolución neoliberal en México*, Mexico City: Era.

Günther, K. and Randeria, S. (2001) *Recht, Kultur und Gesellschaft im Prozess der Globalisierung*, Bad Homburg: Werner-Reimers-Stiftung.

Gupta, A. (1999) "Framing 'Biosafety' in an International Context: The Biosafety Protocol Negotiations", *ENRP Discussion Paper E-99-10*, Kennedy School of Government, Harvard University.

Gupta, A. and Falkner, R. (2006) "The Influence of the Cartagena Protocol on Biosafety: Comparing Mexico, China and South Africa", in *Global Environmental Politics*, vol. 6, 4: 23–55.

Halperin Donghi, T. (1991) *Geschichte Lateinamerikas von der Unabhängigkeit bis zur Gegenwart*, Frankfurt/M.: Suhrkamp.

Haraway, D. (1991) *Simians, Cyborgs and Women: The Reinvention of Nature*, New York: Routledge.

Harry, D. and Kanehe, L. (2005) "The BS in Access and Benefit Sharing (ABS), Critical questions for Indigenous Peoples", in Burrows, B. (ed.) *The Catch, Perspectives in Benefit Sharing,* Edmonds: The Edmonds Institute.

Harvey, D. (1996) *Justice, Nature & the Geography of Difference*, Malden, MA: Blackwell Publishing.

—— (2001) "Globalization and the 'Spatial Fix'", in *Geographische Revue*, vol. 3, 2: 23–30.

—— (2003) *The New Imperialism*, Oxford: Oxford University Press.

Harvey, N. (2001) "Globalisation and resistance in post-cold war Mexico: difference, citizenship and biodiversity conflicts in Chiapas", in *Third World Quarterly*, vol. 22, 6: 1045–1061.

Haug, W. F. (2001) "Fragen einer Kritik des Biokapitalismus", in *Das Argument*, vol. 242: 449–465.

Hecht, S. B. (1998) "Tropische Biopolitik – Wälder, Mythen, Paradigmen", in Flitner, M., Görg, C. and Heins, V. (eds) (1998) *Konfliktfeld Natur, Biologische Ressourcen und globale Politik*, Opladen: Leske + Buderich.

Heigl, M. (2006) "Private is beautiful? Zu den Auseinandersetzungen um die Eigentumsrechte im mexikanischen Energiesektor", in *Peripherie*, vol. 101/102: 147–165.

—— (2007) "Anfang vom Ende? Zum Zustand des neoliberalen Projekts in Lateinamerika", in *Journal für Entwicklungspolitik*, vol. 22, 1: 125–146.

Heineke, C. and Wolff, F. (2004) "Access to Genetic Resources and the Sharing of Benefits: Private Rights or Shared Use for Biodiversity Conservation?", in *ELNI Review – Environmental Law Network International*, vol. 2: 26–33.

Heins, V. (2001) *Der Neue Transnationalismus*, Frankfurt/M.: Campus.

Heins, V. and Flitner, M. (1998) "Biologische Ressourcen und Life Politics", in Flitner, M., Görg, C. and Heins, V. (eds) (1998) *Konfliktfeld Natur, Biologische Ressourcen und globale Politik*, Opladen: Leske + Buderich.

Held, D. and McGrew, A. (2002) *Governing Globalization, Power, Authority and Global Governance*, Cambridge: Polity.

Hepburn, J. (2002) *Negotiating intellectual property: Mandates and options in the Doha Work Programme*, Quaker United Nations Office, Occasional Paper 10, Geneva.

Hernández Navarro, L. (1999) "El laberinto de los equívicos: San Andrés y la lucha indígena", in *Chiapas*, vol. 7: 71–93.

—— (2000) "Der Kampf um indigene Rechte im Labyrinth der Doppeldeutigkeiten, Die Dialoge und Abkommen von San Andrés", in Brand, U. and Ceceña, A. E. (eds) *Reflexionen einer Rebellion, "Chiapas" und ein anderes Politikverständnis*, Münster: Westfälisches Dampfboot.

Hertler, C. (1999) "Aspekte der historischen Entstehung von Biodiversitätskonzepten in den Biowissenschaften", in Görg, C., Hertler, C., Schramm, E. and Weingarten, M. (eds) *Zugänge zur Biodiversität*, Marburg: Metropolis.

Hewson, M. and Sinclair, T. (eds) (1999) *Approaches to Global Governance Theory*, Albany, NY: SUNY Press.

Higgins, N. P. (2001) "Mexico's stalled peace process: prospects and challenges", in *International Affairs*, vol. 77, 4: 885–903.

Hirsch, J. (1990) *Kapitalismus ohne Alternative?*, Hamburg: VSA-Verlag.

—— (1993) "Internationale Regulation, Bedingungen von Dominanz, Abhängigkeit und Entwicklung im globalen Kapitalismus", in *Das Argument*, vol. 198: 195ff.

—— (1994) "Politische Form, politische Institutionen und Staat", in Esser, J., Görg, C. and Hirsch, J. (eds) *Politik, Institutionen und Staat, Zur Kritik der Regulationstheorie*, Hamburg: VSA-Verlag.

—— (1995) "Der nationale Wettbewerbsstaat, Staat, Demokratie und Politik im globalen Kapitalismus", Berlin: Edition ID-Archiv.

—— (1997) "Globalization of capital, nation-states and democracy", in *Studies in Political Economy*, vol. 54: 39–58.

—— (2001) "Postfordismus: Dimensionen einer neuen kapitalistischen Formation", in Hirsch, J., Jessop, B. and Poulantzas, N. (2001) *Die Zukunft des Staates*, Hamburg: VSA-Verlag.

—— (2002a) *Herrschaft, Hegemonie und politische Alternativen*, Hamburg: VSA-Verlag.

—— (2002b) "Wissen und Nichtwissen, Anmerkungen zur 'Wissensgesellschaft'", in Brüchert, O. and Resch, C. (eds) *Zwischen Herrschaft und Befreiung*, Münster: Westfälisches Dampfboot.

—— (2003) "The State's New Clothes: NGOs and the Internationalisation of States", in *Rethinking Marxism*, vol. 15, 2: 237–262.

—— (2005) "Materialistische Staatstheorie: Transformationsprozesse des kapitalistischen Staatensystems", Hamburg: VSA-Verlag.

Hirsch, J. and Roth, R. (1986) "Das neue Gesicht des Kapitalismus, Vom Fordismus zum Postfordismus", Hamburg: VSA-Verlag.

Hirsch, J., Jessop, B. and Poulantzas, N. (2001) *Die Zukunft des Staates*, Hamburg: VSA-Verlag.

Hoering, U. (2002) "Früchte der Vielfalt, Globale Gerechtigkeit und der Schutz traditionellen Wissens", Bonn: Evangelischen Entwicklungsdienst.

—— (2004) "Biopirates in the Kalahari? How indigenous people are standing up for their rights – the experience of the San in Southern Africa", Bonn: EED/WIMSA.

Hooghe, L. and Marks, G. (1999) "The Making of a Polity: The Struggle over European Integration", in Kitschelt, H. (ed.) *Continuity and Change in Contemporary Capitalism*, Cambridge: Cambridge University Press.

Horkheimer, M. and Adorno, T. W. (1987) "Dialektik der Aufklärung, Philosophische Fragmente", in Horkheimer, M. *Gesammelte Schriften*, vol. 5, Frankfurt/M.: Fischer.

Howard, P. L. (2003) "Preface and Women and the Plant World: An exploration", in Howard, P. L. (ed.) *Women & Plants, gender relations in biodiversity management and conservation*, London: Zed Books.

Hübner, K. (1989) *Theorie der Regulation, Eine kritische Rekonstruktion eines neuen Ansatzes der Politischen Ökonomie*, Berlin: Sigma.

ICBG – International Cooperative Biodiversity Group (2002) "Call for Papers for the 3rd Round", http://grants2.nih.gov/grants/guide/rfa-files/RFA-TW-03-004.html

ICC – International Chamber of Commerce (2002) *Roadmap 2002*, www.iccwbo.org/home/intellectual_property/IP%20roadmap/Roadmap-2002/roadmap2002-REV.doc

ICTSD – International Centre for Trade and Sustainable Development (2005a) "WIPO members create new forum to discuss development agenda", www.grain.org/bio-ipr/index.cfm?id = 451&prints = yes

—— (2005b) "Governments meet in Geneva to discuss WIPO development agenda", in *Bridges Weekly Trade News Digest*, vol. 9, 12, www.ictsd.org/weekly/05-04-13/story1.htm

IFC – International Finance Corporation (2006) *A Guide to Biodiversity for the Private Sector: Why Biodiversity Matters and How It Creates Business Value*, www.ifc.org/ifcext/enviro.nsf/Content/BiodiversityGuide.

IIED – International Institute for Environment and Development (2001) "The Future is Now, For the UN World Summit on Sustainable Development", vol. 1, London: International Institute for Environment and Development, www.iied.org/pubs/pdf/full/9000IIED.pdf

Imbusch, P. (1988) "Mexiko – entwicklungsstrategische Alternativen", Marburg: Verlag Arbeiterbewegung und Gesellschaftswissenschaft.

IPGRI – International Plant Genetic Resources Institute (1996) "Access to plant genetic resources and the equitable sharing of benefits: a contribution to the debate on systems for the exchange of germplasm", in *Issues in Genetic Ressources*, vol. 4, www.bioversityinternational.org/Publications/Pdf/467.pdf

IP-Watch (2005) "Nations clash on future of WIPO Development Agenda", 11 April 2005, www.ip-watch.org/weblog/index.php?p = 42

ITPGR – International Treaty on Plant Genetic Resources (2001), ftp://ftp.fao.org/ag/cgrfa/it/ITPGRe.pdf

IUCN – The World Conservation Union (1989) "Draft articles prepared by IUCN for inclusion in a proposed convention on the conservation of biological diversity and for the establishment of a fund for that purpose", Gland/CH, June 1989.

IWBN – Indigenous Women's Biodiversity Network (2006) *Introduction to the IWBN and History*, www.nciv.net/spaans/iwbn/IWBN.htm

Jacob, M. (2005) "Boundary Work in Contemporary Science Policy: a Review", in *Prometheus*, vol. 23, 2: 195–207.

Jänicke, M. and Jacob, K. (2004) "Lead Markets for Environmental Innovations: A New Role for the Nation State", in *Global Environmental Politics*, vol. 4, 1: 29–46.

Janssen, J. (1999) "Property Rights on Genetic Ressources: Economic Issues", in *Global Environmental Change*, vol. 9, 4: 313–321.

Jasanoff, S. (ed) (2004a) *States of Knowledge: The Co-Production of Science and the Social Order*, London: Routledge.

—— (2004b) "Ordering knowledge, ordering society", in Jasanoff, S. (ed.) *States of Knowledge: The Co-Production of Science and the Social Order*, London: Routledge.

Jasanoff, S., Markle, G. E., Petersen, J. C. and Pinch, T. (eds) (1995) *Handbook of Science and Technology Studies*, Cambridge, MA: Harvard University Press.

Jenson, J. (1989) "'Different' but not 'exceptional': Canada's permeable fordism", in *Canadian Review of Sociology and Anthropology*, vol. 26, 1: 69–94.

Jessop, B. (1990) "Regulation Theories in Restrospect and Prospect", in *Economy and Society*, vol. 19, 2: 153–216.

—— (1997) "Die Zukunft des Nationalstaates – Erosion oder Reorganisation? Grundsätzliche Überlegungen zu Westeuropa", in Becker, S., Sablowski, T. and Schumm, W. (eds) *Jenseits der Nationalökonomie? Weltwirtschaft und Nationalstaat zwischen Globalisierung und Regionalisierung*, Hamburg: Argument-Verlag.

—— (2000) "The State and the Contradictions of the Knowledge-Driven Economy", in Bryson, J. R., Daniels, P. W., Henry, N. and Pollard, J. (eds) *Knowledge, Space, Economy*, London: Routledge.

—— (2002) *The Future of the Capitalist State*, Cambridge: Polity.

—— (2003) "Postfordismus und wissensbasierte Ökonomie, Eine Reinterpretation des Regulationsansatzes", in Brand, U. and Raza, W. (eds) *Fit für den Postfordismus?*, Münster: Westfälisches Dampfboot.

Kaiser, G. (2003) "Wenn Leben zur Ressource wird", in *ila – Zeitschrift der Informationsstelle Lateinamerika, 263*, Bonn: Informationsstelle Lateinamerika e.V..

—— (2006) "Schieflage mit System – die WTO in Hongkong", in links-netz, www.links-netz.de/K_texte/K_kaiser_wto.html

Kay, L. E. (2000) *Who wrote the book of life? A History of the Genetic Code*, Stanford, CT: Stanford University Press.

Keil, R. and Mahon, R. (eds) (2008) *The New Political Economy of Scale*, Vancouver: University of British Columbia.

Keystone Dialogue (1991) *Oslo Plenary Session: Consensus Report*, Keystone: The Keystone Center.

Khor, M. (2001) *Competing Views on Competition Policy in WTO*, www.twnside.org.sg/ title/ views-cn.htm 07.04.01

Klaffenböck, G., Lachkovics, E. and Südwind Agentur (eds) (2001) *Biologische Vielfalt, Wer kontrolliert die genetischen Ressourcen*, Frankfurt/M.: Brandes & Apsel.

Kloppenburg, J. R. (ed.) (1988) *Seeds and Sovereignty, The Use and Control of Plant Genetic Ressources*, Durham, NC: Duke University Press.

Kloppenburg, J. R. and Balick, M. J. (1996) "Property Rights and Genetic Resources: A Framework for Analysis", in Balick, M. J., Elisabetsky, E. and Laird, S. A. (eds) *Medicinal Resources of the Tropical Forest*, New York: Columbia University Press.

Kloppenburg, J. R. and Kleinman, J. L. (1988) "Plant Genetic Resources: The Common Bowl", in Kloppenburg, J. R. (ed.) *Seeds and Sovereignity, The Use and Control of Plant Genetic Ressources*, Durham, NC: Duke University Press.

Kongolo, T. (2001) "New Options for African Countries regarding Protection for New Varieties of Plants", in *The Journal of World Intellectual Property*, vol. 4, 3: 349–371.

Krasner, S. (ed.) (1983) *International Regimes*, Ithaca, NY: Cornell University Press.

Krebs, M., Herkenrath, P. and Meyer, H. (2002) *Biologische Vielfalt zwischen Schutz und Nutzung, 10 Jahre Konvention über Biologische Vielfalt*, Bonn: EED und Forum Umwelt und Entwicklung.

Krohn, W. (2001) "Knowledge Societies", in Smelser, N. J. and Baltes, P. B. (eds) *International Encyclopedia of the Social and Behavioral Sciences*, Amsterdam: Elsevier.

Kuppe, R. (2001) "Der Schutz des traditionellen umweltbezogenen Wissens indigener Völker", in Klaffenböck, G., Lachkovics, E. and Südwind Agentur (eds) (2001) *Biologische Vielfalt, Wer kontrolliert die genetischen Ressourcen*, Frankfurt/M.: Brandes & Apsel.

—— (2002) "Indigene Völker, Ressourcen und traditionelles Wissen", in Brand, U. and Kalcsics, M. (eds) *Wem gehört die Natur? Konflikte um genetische Ressourcen in Lateinamerika*. Frankfurt/M.: Brandes & Apsel.

Kütting, G. (2000) *Environment, Society and International Relations, Towards More Effective International Environmental Agreements*, London: Routledge.

Lander, E. (2006) "La Ciencia Neoliberal", in Ceceña, A. E. (ed.) *Los desafíos de las emancipaciones en un contexto militarizado*, Buenos Aires: CLACSO.

Lasén Díaz, C. (2005) "Intellectual Property Rights and Biological Resources, An Overview of Key Issues and Current Debates", Wuppertal Papers, 151, Wuppertal: Wuppertal Institute for Climate, Environment and Energy.

Latour, B. (1993) *We Have Never Been Modern*, Cambridge, MA: Harvard University Press.

—— (1999) "Is Re-modernization Occurring – And If So, How to Prove It?", in *Theory, Culture & Society*, vol. 20, 2: 35–48.

Laurell, A. C. (1992) "Democracy in Mexico: Will the First be the Last?", in *New Left Review*, vol. 194: 33–54.

Le Buanec, B. (2005) "Plant genetic resources and freedom to operate", in *Euphytica*, vol. 146: 1–8.

Le Prestre, P. G. (2002) "The Operation of the CBD Convention Governance System", in Le Prestre, P. G. (ed.) *Governing Global Biodiversity, The evolution and implementation of the Convention on Biological Diversity*, Aldershot: Ashgate.

Le Prestre, P. G. (ed.) (2002) *Governing Global Biodiversity, The evolution and implementation of the Convention on Biological Diversity*, Aldershot: Ashgate.

Leskien, D. and Flitner, M. (1997) "Intellectual Property Rights and Plant Genetic Resources: Options for a Sui Generis System", in Issues in Genetic Resources, 6, Rome: International Plant Genetic Resource Institute, www.bioversityinternational.org/Publications/Pdf/497.pdf

Liebig, K, Alker, D., Chih, K., Horn, D., Illi, H., Wolf, J. *et al.* (2002) "Governing Biodiversity, Access to Genetic Resources and Approaches to Obtaining Benefits from their Use: the Case of the Philippines", *Reports and Working Papers 5/2002*, Bonn: German Development Institute.

Ling, C. Y. and Khor, M. (2001) "International Environmental Governance, Some Issues from a Developing Country Perspective", *Working Paper by the Third World Network*, www.twnside.org.sg/title/ieg.htm

Lipietz, A. (1985) "Akkumulation, Krisen und Auswege aus der Krise, Einige methodologische Bemerkungen zum Begriff "Regulation"", in *PROKLA*, vol. 58: 109–138.

—— (1987) *Mirages and Miracles, The Crisis of Global Fordism*, London: Verso.

Lipschutz, R. (2004) *Global Environmental Politics, Power, Perspectives, and Practice*, Washington, DC: CQ Press.

López Ramírez, A. (2005a) "Los intereses geoestratégicos de Estados Unidos en el Plan Puebla Panamá y el Corredor Biológico Mesoamericano", in Castro Escudero, T. and Oliver Costilla, L. (eds) *Poder y política en América Latina,* El debate latinoamericano, vol. 3, Mexico City: Siglo XXI/UNAM.

—— (2005b) "La globalización en el sur, La apropiación de los recursos de la biodiversidad en el Corredor Biológico Mesoamericano", in *Estudios Latinoamericanos*, vol. 45: 165–182.

Luna Martínez, S. and González Nolasco, E. (2004) "Libre comercio y convergencia, La macroeconomía del TLCAN", in Casares, E. S. and Sobarzo, H. (eds) *Diez años del TLCAN en México, Una perspectiva analítica*, Mexico City: Fondo de Cultura Económica.

McCarthy, J. and Prudham, S. (2004) "Neoliberal nature and the nature of neoliberalism", in *Geoforum*, vol. 35: 275–283.

McCauley, D. (2006) "Selling out on nature", in *Nature*, vol. 443, 7 September.

McCormick, J. (2005) "The Role of Environmental NGOs in International Regimes", in Axelrod, R. S., Downie, D. L. and Vig, N. (eds) *The Global Environment: Institutions, Law, and Policy*, Washington, DC: C. Q. Press.

McGraw, D. M. (2002) "The Story of the Biodiversity Convention: From Negotiation to Implementation", in Le Prestre, P. G. (ed.) *Governing Global Biodiversity, The evolution and implementation of the Convention on Biological Diversity*, Aldershot: Ashgate.

Macilwain, C. (1998) "When rhetoric hits reality in the debate on bioprospecting", in *Nature*, vol. 392, 6676: 535–540.

MacLeod, D. (2004) *Downsizing the State, Privatization and the Limits of Neoliberal Reform in Mexico*, Pennsylvania: Pennsylvania State University Press.

Malcher, I. (2006) "Private Regulierung der Weltwirtschaft", in *PROKLA*, vol. 142: 127–143.

Marglin, S. A. and Schor, J. B. (eds) (1990) *The Golden Age of Capitalism, Reinterpreting the Postwar Experience*, Oxford: Claredon.

Marston, S. A. (2000) "The Social Construction of Scale", in *Progress in Human Geography*, vol. 24, 2: 219–242.

Marx, K. and Engels, F. (1978) "Die deutsche Ideologie", MEW, 3, Berlin: Dietz.

MASR (2005) *Millennium Ecosystem Assessment Synthesis Report*, Washington, DC: Island Press, www.millenniumassessment.org/

Maus, I. (1991) "Sinn und Bedeutung der Volkssouveränität in der modernen Gesellschaft", in *Kritische Justiz*, vol. 24: 137–150.

May, C. (2000) *A Global Political Economy of Intellectual Property Rights, The New Enclosures?*, London: Routledge.

May, C. and Sell, S. K. (2006) *Intellectual Property Rights, A Critical History*, Boulder, CO: Lynne Rienner Publishers.

Mayntz, R. (2005) "Governance Theory als fortentwickelte Steuerungstheorie?", in Schuppert, G. F. (ed.) *Governance-Forschung, Vergewisserung über Stand und Entwicklungslinien*, Baden-Baden, Nomos.

Mekouar, M. A. (2001) "Treaty Agreed on Agrobiodiversity: The International Treaty on Plant Genetic Resources for Food and Agriculture", in *Environmental Policy and Law*, vol. 32, 1: 20–25.

MIHR – Centre for the Management of Intellectual Property in Health Research and Development, www.mihr.org

Mitsch, F. J. and Mitchell, J. S. (1999) *AG Biotech: Thanks, But No Thanks? Analyse der Deutsche Bank*, 12 July 1999, www.biotech-info.net/Deutsche.html

Mittermeier, R. and Goettsch Mittermeier, C. (1992) "La importancia de le diversidad biológica de México", in Sarukhán, J. and Dirzo, R. (eds) *México ante los retos de le biodiversidad*, Mexico City: Conabio.

Mooney, P. (1981) "Saat-Multis und Welthunger, Wie die Konzerne die Nahrungsschätze der Welt plündern", Reinbek: rororo aktuell.

—— (1998) "The Parts of Life, Agricultural Biodiversity, Indigenous Knowledge, and the Role of the Third System", Special Issue of *Development Dialogue*, Uppsala: Dag Hammerskjöld Foundation.

Morton, A. D. (2002) "'La Resurrección del Maíz': Globalisation, Resistance and the Zapatistas", in *Millennium*, vol. 31, 1: 27–54.

—— (2003) "Structural change and neoliberalism in Mexico: 'passive revolution' in the global political economy", in *Third World Quarterly*, vol. 24, 4: 631–653.

Mulvany, P. and Meienberg, F. (2000) *Seeds for All, Call to Action in the International Undertaking in Plant Genetic Resources*, www.ukabc.org

Mulvany, P. and Redding, I. (2000) "Ours to have and to hold, How world food resources are threatened by seed patents", *Society Guardian*, 6 December.

Murphy, C. (2000) "Global governance: poorly done and poorly understood", in *International Affairs*, vol. 76, 4: 789–803.

New Delhi Ministerial Declaration of Like-Minded Megadiverse Countries on Access and Benefit Sharing (2005) Expert Working Group Session (17–19 January 2005) and Ministerial Session (20–21 January 2005), Delhi, India, www.undp.org/biodiversity/docs/Summary_Report_Dehli_Megadiverse_17_21Jan05.doc

Nigh, R. and Rodríguez, N. J. (1995) "Territorios violados, Indios, medio ambiente y desarrollo an América Latina", Mexico City: Instituto Nacional Indigenista.

NIH – National Institutes of Health (2003) *Third Round Awards are Announced Under Interagency Biodiversity Program*, Press Release, 16 December, www.nih.gov/news/pr/dec2003/fic-16.htm

Novy, A. (2003) "Politik, Raum und Wissen, Zentrale Kategorien eines erneuerten radikalen Reformismus am Beispiel Brasilien", in Brand, U. and Raza, W. (eds) *Fit für den Postfordismus?,* Münster: Westfälisches Dampfboot.

Nuss, S. (2002) "Download ist Diebstahl? Eigentum in einer digitalen Welt", in *PROKLA*, vol. 32, 126: 11–35.

—— (2006) *Copyrights & Copyriot, Aneigungskonflikte im geistiges Eigentum im informationellen Kapitalismus*, Münster: Westfälisches Dampfboot.

O'Connor, J. (1988) "Capitalism, Nature, Socialism, A Theoretical Introduction", in *Capitalism, Nature, Socialism*, vol.1, 1: 11–45.

Oberthür, S. and Gehring, T. (2006) "International Interaction in Global Environmantal Governance, The Case of the Cartagena Protocol and the World Trade Organization", in *Global Environmental Politics*, vol. 6, 2: 1–31.

Oetmann-Mennen, A. (1999) "Biologische Vielfalt in der Landwirtschaft – Luxus oder Notwendigkeit", in Görg, C., Hertler, C., Schramm, E. and Weingarten, M. (eds) *Zugänge zur Biodiversität*, Marburg: Metropolis.

Oh, C. (2000) "The OAU draft law and convention: A model for protecting community rights and access", www.twnside.org.sg/title/co1-cn.htm

Oldham, P. (2004) *Global Status and Trends in Intellectual Property Claims: Genomics, Proteomics and Biotechnology*, www.cesagen.lancs.ac.uk/resources/docs/genomics-final.doc

Olivier Costilla, L. (2005) "Revisitando al Estado, Las especifidades del Estado en América Latina", in Castro Escudero, T. and Oliver Costilla, L. (eds) *Poder y política en América Latina,* El debate latinoamericano, vol. 3, Mexico City: Siglo XXI/UNAM.

Ornelas Bernal, R. (1994) "Nueva Ley de inversiones extranjeras, Cambios y repercusiones de cara al Tratado de Libre Comercio", in *Problemas del Desarrollo*, vol. XXV, 98: 259–266.

Oxfam (2003) "Comercio con Justicia para las Américas, Agricultura, Inversiones y Propiedad Intelectual, tres razones para decir no als ALCA", *Oxfam Briefing Paper*, January, www.oxfam.org/es/files/pp210103_37_FTAA.pdf/download

Palan, R. (2003) *The Offshore World, Sovereign Markets, Virtual Places, and Nomad Millionaires*, Ithaca, NY: Cornell University Press.

Panitch, L. (1994) *Globalisation and the State*, in *Socialist Register*, vol. 30: 60–93, http://socialistregister.com/socialistregister.com/files/SR_1994_Panitch.pdf

Parsons, T. (1975) *Gesellschaften. Evolutionäre und komparative Perspektiven*, Frankfurt/M.: Suhrkamp.

Paterson, M. (2000) *Understanding Global Environmental Politics*, New York: Palgrave.

Pelegrina, W. (2001) "Die Grüne Revolution und ihre Hinterlassenschaft", in Klaffenböck, G., Lachkovics, E. and Südwind Agentur (eds) *Biologische Vielfalt, Wer kontrolliert die genetischen Ressourcen*, Frankfurt/M.: Brandes & Apsel.

Perelman, M. (2003) "Intellectual Property Rights and the Commodity Form: New Dimensions in the Legislated Transfer of Surplus Value", in *Review of Radical Political Economics*, vol. 35, 3: 304–311.

Petit, M., Fowler, C., Collins, W., Correa, C. and Thornström, C. (2000) *Why Governments Can't Make Policy, The Case of Plant Genetic Resources in the International Arena*, www.fao.org/docs/eims/upload/207144/GFAR2003.PDF

Plahe, J. K. and Nyland, C. (2003) "The WTO and patenting of life forms, Options for developing countries", in *Third World Quarterly*, vol. 24, 1: 29–45.

Plehwe, D., Walpen, B. and Neunhöffer, G. (eds) (2006) *Neoliberal Hegemony, A global critique*, London: Routledge.

Pólito Barrios Morfín, E. (2000) "El capital nacional y extranjero en Chiapas", in *Chiapas*, 9: 61–81, www.ezln.org/revistachiapas/No9/ch9polito.html

Posey, D. (ed.) (1999) *Cultural and Spiritual Values of Biodiversity*, London: UNEP and Intermediate Technology Publications.

Potvin, C., Revéret, J., Patenaude, G. and Hutton, J. (2002) "The Role of Indigenous Peoples in Conservation Actions", in Le Prestre, P. G. (ed.) *Governing Global Biodiversity, The evolution and implementation of the Convention on Biological Diversity*, Aldershot: Ashgate.

Poulantzas, N. (1978/2002) *Staatstheorie, Politischer Überbau, Ideologie, Autoritärer Etatismus*, Hamburg: VSA-Verlag; English version (1980) *State, power, socialism*, London: Verso.

—— (1980) *State, power, socialism*, London: Verso.

—— (2001) "Die Internationalisierung der kapitalistischen Verhältnisse und der Nationalstaat", in Hirsch, J., Jessop, B. and Poulantzas, N. (2001) *Die Zukunft des Staates*, Hamburg: VSA-Verlag.

Purcell, M. (2002) "The state, regulation, and global restructuring: reasserting the political in political economy", in *Review of International Political Economy*, vol. 9, 2: 298–332.

RAFI (2000a) *DeCoding the Clinton/Blair Announcement*, www.etcgroup.org

—— (2000b) "Seedy Squabble in Switzerland", *RAFI News Release*, 20 November, www.etcgroup.org

—— (2001a) "Frequently unasked Questions about the International Undertaking on Plant Genetic Resources", RAFI-Communiqué, 18 April 2001, www.etcgroup.org

—— (2001b) "Supporting the Treaty to Share the genetic Commons", *RAFI News*, 19 April 2001, www.etcgroup.org

Rajchenberg, E. and Héau-Lambert, C. (2004) "Los silencios zapatistas", in *Chiapas*, vol. 16: 51–63, www.ezln.org/revistachiapas/No16/ch16rajchenberg.html

Raustiala, K. and Victor, D. G. (2004) "The Regime Complex for Plant Genetic Resources", in *International Organization*, vol. 58, Spring: 277–309.

Reinicke, W. H. (1998) *Global Public Policy, Governing without Government?*, Washington, DC: Brookings Institution.

Ribeiro, S. (2002a) "A Different Perspective", in Heinrich Böll-Stiftung (ed.) *Comments on the Jo'burg Memo*, World Summit Papers of the Heinrich Böll Foundation, 18: 36–41, www.boell.de/downloads/rio+10/wsp18.pdf

—— (2002b) "Biopiraterie und geistiges Eigentum – Zur Privatisierung von gemeinschaftlichen Bereichen", in Görg, C. and Brand, U. (eds) *Mythen globalen Umweltmanagements, Rio+10 und die Sackgassen "nachhaltiger Entwicklung"*, Münster: Westfälisches Dampfboot.

—— (2005) "The traps of 'benefit sharing'", in Burrows, B. (ed.) *The Catch, Perspectives in Benefit Sharing*, Edmonds: The Edmonds Institute.

Ribot, J. C. and Peluso, N. L. (2003) "A Theory of Access", in *Rural Sociology*, vol. 68, 1: 153–181.

Rifkin, J. (2000) *Access, Das Verschwinden des Eigentums*, Frankfurt/M.: Campus.

Risse, T. (2003) "Konstruktivismus, Rationalismus und Theorien Internationaler Beziehungen – warum empirisch nichts so heiß gegessen wird, wie es theoretisch gekocht wurde", in Hellmann, G., Wolf, K. D. and Zürn, M. (eds) *Die neuen Internationalen Beziehungen, Forschungsstand und Perspektiven in Deutschland*, Baden-Baden: Nomos.

Rittberger, V. (ed.) (1993) *Regime Theory and International Relations*, Oxford: Clarendon Press.

Ritter, E. H. (1979) "Der kooperative Staat", in *Archiv des öffentlichen Rechts*, vol. 104, 3: 389–413.

Robinson, W. I. (2001) "Social theory and globalization: The rise of a transnational state", in *Theory and Society*, vol. 30, 2: 157–200.

Robles, A. C. (1994) *French Theories of Regulation and Conceptions of the International Division of Labour*, New York: St Martin's Press.

Rodríguez, S. (2002) "Biodiversitätspolitik und lokale Gegenmacht – Das Beispiel Costa Rica", in Görg, C. and Brand, U. (eds) *Mythen globalen Umweltmanagements, Rio+10 und die Sackgassen "nachhaltiger Entwicklung"*, Münster: Westfälisches Dampfboot.

Rosenau, J. N. (1995) "Governance in the Twenty-First Century", in *Global Governance*, vol. 1, 1: 13–43.

Rosendal, G. Kristin (2001) "Impacts of Overlapping International Regimes: The Case of Biodiversity", in *Global Governance*, vol. 7, 1: 95–117.

Ruggie, J. G. (1982) "International regimes, transactions, and change: embedded liberalism in the postwar economic order", in *International Organization*, vol. 36, 2: 379–415.

Ruiz Caro, A. (2005) "Recursos naturales y aspectos ambientales en los Tratados de Libre Comercio con Estados Unidos", in *Observatorie Social de América Latina*, vol. 17.

Russell, P. L. (1994) *Mexico under Salinas*, Austin, TX: Mexico Resource Center.

Sachs, J. (2000) "A New Map of the World", *Economist*, vol. 24: 81–83.

Sánchez, V. and Juma, C. (eds) (1994) *Biodiplomacy, Genetic Resources and International Relations*, Nairobi: ACT Press.

Saxe-Fernández, E. E. (2005) "La 'gobernabilidad-gobernanza' como ideologema neoliberal globalista", in Castro Escudero, T. and Oliver Costilla, L. (eds) *Poder y política en América Latina,* El debate latinoamericano, vol. 3, Mexico City: Siglo XXI/UNAM.

Saxe-Fernández, J. and Delgado, G. C. (2005) *Imperialismo económico en México, Las operaciones del Banco Mundial en nuestro país*, Mexico City: Random House Mondadori.

Scharpf, F. W. (1993) "Positive und negative Koordination in Verhandlungssystemen", in Héritier, A. (ed.) *Policy-Analyse: Kritik und Neuorientierung*, Opladen: Westdeutscher Verlag.

—— (1999) *Regieren in Europa*, Frankfurt/M.: Campus.

Schedler, A., Diamond, L. and Plattner, M. (eds) (1999) The Self-Restraining State: Power and Accountability in New Democracies, Boulder, CO: Lynne Rienner Publishers.

Scherrer, C. (1998) "Neo-gramscianische Interpretation internationaler Beziehungen. Eine Kritik", in Hirschfeld, U. (ed.) *Gramsci-Perspektiven*, Berlin: Argument-Verlag.

—— (2003) "Internationale Politische Ökonomie als Systemkritik", in Hellmann, G., Wolf, K. D. and Zürn, M. (eds) *Die neuen Internationalen Beziehungen. Forschungsstand und Perspektiven in Deutschland,* Baden-Baden: Nomos-Verlag.

Secretariat of the CBD – Convention on Biological Diversity (2006a) "Global Biodiversity Outlook 2", Montreal, www.biodiv.org/GBO2

—— (2006b) *Report of the 8th Meeting of the Parties to the Convention on Biological Diversity*, Curitiba, Brazil, 20– 31 March 2006, UNEP/CBD/COP/8/31, www.biodiv.org/doc/meetings/cop/cop-08/official/cop-08-31-en.pdf

Secretariat of the CBD – Convention on Biological Diversity / United Nations Environmental Programme (2001) "Global Biodiversity Outlook", Montreal.

Seibert, T. (2003) "Die neue Kriegsordnung, Der globale Kapitalismus und seine barbarisierte Rückseite", in Azzellini, D. and Kanzleiter, B. (eds) *Das Unternehmen Krieg, Paramilitärs, Warlords und Privatarmeen als Akteure der Neuen Kriegsordnung*, Berlin: Assoziation A.

Seiler, A. (1999) "Das TRIPS-Abkommen und die für 1999 vorgesehene Überprüfung von Art. 27.3.(b)", in *Nord-Süd-aktuell*, vol. 2: 312–314.

—— (2000) "Die Bestimmungen des WTO-TRIPs-Abkommens und die Optionen zur Umsetzung des Art. 27.3(b): Patente, Sortenschutz, Sui Generis", *Studie im Auftrag der GTZ*, www2.wto-runde.de/uploads/GTZ-TRIPS-Studie.pdf

—— (2003) "Der Internationale Saatgutvertrag der FAO – ein Ansatz zur Sicherung des nachhaltigen Umgangs mit pflanzengenetischen Ressourcen?", in *BUKO-Agrarinfo*, vol. 120.

Seiler, A. and Dutfield, G. (2001) "Regulating Access and Benefit Sharing, Basic issues, legal instruments, policy proposals", BfN-Skripten, 46, Bonn: Bundesamt für Naturschutz, www.bfn.de/fileadmin/MDB/documents/access.pdf

SEMARNAT (2002a) "Access to Genetic Resources and Fair and Equitable Sharing of Benefits: Building a Common Agenda", paper presented at the *First Meeting of Like-Minded Megadiverse Countries*, February, Mexico, www.megadiverse.org/armado_ingles/PDF/five/five1.pdf

—— (2002b) "Programa estratégico para conservar los ecosistemas y su biodiversidad", Mexico City: Semarnat, www.semarnat.gob.mx/programas/documentos/index.shtml

SEMARNAT/INE (2002) "Programa estratégico para detener y revertir la pérdida del capital natural", Mexico City: Semarnat www.semarnat.gob.mx/programas/documentos/index.shtml

Shaiken, H. (1994) "Advanced Manufacturing and Mexico: A New International Division of Labor?", in *Latin America Research Review*, vol. 29, 2: 39–71.

Sharma, D. (2005) "Selling Biodiversity, Benefit sharing is a dead concept", in Burrows, B. (ed.) *The Catch, Perspectives in Benefit Sharing,* Edmonds: The Edmonds Institute.

Shaw, M. (2000) *Theory of the Global State, Globality as an Unfinished Revolution*, Cambridge: Cambridge University Press.

Shiva, V. (2005) "Bioprospecting as sophisticated biopiracy", in Burrows, B. (ed.) *The Catch, Perspectives in Benefit Sharing,* Edmonds: The Edmonds Institute.

Siebenhüner, B. and Suplie, J. (2005) "Implementing the Access and Benefit Sharing Provisions of the CBD: A Case for Institutional Learning", in *Ecological Economics*, vol. 53: 507–522.

Siebenhüner, B., Dedeurwaerdere, T. and Brousseau, E. (2005) "Introduction and overview to the special issue on biodiversity conservation, access and benefit-sharing and traditional knowledge", in *Ecological Economics*, vol. 53, 4: 439–444.

Smith, N. (1984) *Uneven Development: Nature, Capital and the production of Space*, Oxford: Blackwell Publishing.

—— (1995) "Remaking Scale: Competition and Cooperation in Prenational and Postnational Europe", in Eskelinen, H. and Snickars, F. (eds) *Competetive European Peripheries*, Berlin: Springer.

Soberón, J. (2001) *Corredor Biológico Mesoamericano-México*, Mexico City: Semarnat, www.semarnat.gob.mx/programas/documentos/index.shtml

South Bulletin (2002) *Agendas on Patents*, 48, 15 December, www.southcentre.org/info/southbulletin/bulletin48/bulletin48.pdf

Stehr, N. (1994) *Arbeit, Eigentum und Wissen, Zur Theorie von Wissensgesellschaften*, Frankfurt/M.: Suhrkamp.

Stilwell, M. (2001) "Review of Article 27.3(b)", paper prepared under the *CIEL/South Centre joint project*, Geneva, www.ciel.org/Publications/Article273b_Review.pdf

Stoll, P. (1999) "Werte der Vielfalt aus rechtlicher Sicht", in Görg, C., Hertler, C., Schramm, E. and Weingarten, M. (eds) *Zugänge zur Biodiversität*, Marburg: Metropolis.

Svarstad, H. (1994) "National sovereignty and genetic resources", in Sánchez, V. and Juma, C. (eds) *Biodiplomacy, Genetic Resources and International Relations*, Nairobi: ACT Press.

Swanson, T. (1997) *Global Action for Biodiversity: an international framework for implementing the convention on biological diversity*, London: Earthscan.

Swyngedouw, E. (1997) "Neither Global Nor Local, 'Glocalization' and the Politics of Scale", in Cox, K. (ed.) *Spaces of Globalization, Reasserting the Power of the Local*, New York: Guilford Press.

—— (2004) "Social Power and the Urbanization of Water, Flows of Power", Oxford: Oxford University Press.

—— (2005) "Dispossessing $H_2O$: The Contested Terrain of Water Privatization", in *Capitalism Nature Socialism*, vol. 16, 1: 1–18.

Taba, S. (ed.) (1995) *Current Activities of CIMMYT Maize Germplasm Bank*, Mexico City: CIMMYT.

Takacs, D. (1996) "The idea of biodiversity. Philosophies of paradise", Baltimore, MD: John Hopkins University Press.

Teichman, J. A. (1995) "Privatization and Political Change in Mexico", Pittsburgh, PA: University of Pittsburgh Press.

—— (2002) "Private sector power and market reform: exploring the domestic origins of Argentina's meltdown and Mexico's policy failures", in *Third World Quarterly*, vol. 23, 3: 491–512.

ten Kate, K. and Laird, S. A. (1999) "The Commercial Use of Biodiversity", London: Earthscan.

Thomas, U. P. (2002) "The CBD, the WTO and the FAO: The Emergence of Phytogenetic Governance", in Le Prestre, P. G. (ed.) *Governing Global Biodiversity, The evolution and implementation of the Convention on Biological Diversity*, Aldershot: Ashgate.

Toke, D. (2001) "Ecological Modernisation: A Reformist Review", in *New Political Economy*, vol. 6, 2: 279–281.

Toledo, V. M. (2000) "La Paz en Chiapas, Ecología, luchas indígenas y modernidad alternativa", Mexico City: Quinto Sol and UNAM.

Toledo Ocampo, A. (1998) "Hacia una economía política de la biodiversidad", in *Chiapas*, vol. 6: 7–40, www.ezln.org/revistachiapas/No6/ch6toledo.html

Touraine, A. (1972) *Die postindustrielle Gesellschaft*, Frankfurt/M.: Suhrkamp.

Trommetter, M. (2005) "Biodiversity and international stakes: A question of access", in *Ecological Economics*, vol. 53, 4: 573–583.

Tuxill, J. (1999) "Nature's Cornucopia: Our Stake in Plant Diversity", *Worldwatch Paper*, 148, Washington DC: Worldwatch Institute.

Tvedt, M. W. (2005) "How Will a Substantive Patent Law Treaty Affect the Public Domain for Genetic Resources and Biological Material?", in *The Journal of World Intellectual Property*, vol. 8, 3: 311–344, www.fni.no/doc&pdf/mwt-JWIP-2005.pdf

TWN – Third World Network (2004) *Launch of Geneva declaration on future of WIPO and joint NGO statement supporting a development agenda in WIPO*, www.twnside.org.sg/title2/twninfo163.htm

Underhill, G. R. D. (2000) "State, market, and global political economy: genealogy of an (inter?)discipline", in *International Affairs*, vol. 76, 4: 805–824.

Van der Pijl, K. (1999) *Transnational Classes and International Relations*, London: Routledge.

Velinga, M. (ed.) (1998) *The Changing Role of the State in Latin America*, Boulder, CO: Westview.

Wade, R. H. (2001) "The Rising Inequality of World Income Distribution", in *Finance and Development*, vol. 38, 4: 37–39.

Warnock, J. (1995) *The Other Mexico, The North American Triangle Completed*, Montréal: Black Rose Books.

WBGU (1999) "Welt im Wandel: Erhaltung und nachhaltige Nutzung der Biosphäre", *Wissenschaftlicher Beirat Globale Umweltveränderungen*, Jahresgutachten 1999, Berlin: Springer.

Weingarten, M. (1998) *Wissenschaftstheorie als Wissenschaftskritik: Beiträge zur kulturalistischen Wende in der Philosophie*, Bonn: Pahl-Rugenstein.

Wendt, A. (1999) *Social Theory of International Politics*, Cambridge: Cambridge University Press.

Wichterich, C. (2000) *The Globalized Woman: Reports from a Future of Inequality*, London: Zed Books.

Wiener, A. (2003) "Konstruktivistische Brückenstationen und ihre Zukunft", in Hellmann, G., Wolf, K. D. and Zürn, M. (eds) *Die neuen Internationalen Beziehungen, Forschungsstand und Perspektiven in Deutschland*, Baden-Baden: Nomos.

Wilson, E. O. (ed.) (1988) *Biodiversity*, Washington, DC: National Academy Press.

—— (1992) *The Diversity of Life*, New York: W.W. Norton.

Wingens, M. (1998) *Wissensgesellschaft und Industrialisierung der Wissenschaft*, Wiesbaden: Deutscher Universitäts-Verlag.

WIPO – World Intellectual Property Organization (2001) *Intergovernmental Committee on Intellectual Property and Genetic Resources, Traditional Knowledge and Folklore*, first session, Geneva, 30 April to 3 May, document prepared by the Secretariat, WIPO/GRTKF/IC/1/3, www.wipo.int/documents/en/meetings/2001/igc/doc/grtkfic1_3.doc

—— (2002) *WIPO Patent Agenda, Options for the Development of the International Patent System*, Memorandum of the Director General, A/37/6, www.wipo.int/eng/document/govbody/wo_gb_ab/doc/a37_6.doc

—— (2004) "Proposal by Argentina and Brazil for the establishment of a development agenda for WIPO", *WIPO general assembly*, 31st (15th Extraordinary) Session, Geneva, 27 September to 5 October, document prepared by the Secretariat, www.wipo.int/documents/en/document/govbody/wo_gb_ga/doc/wo_ga_31_11.doc

—— (2005) "Proposal by the United States of America for the Establishment of a Partnership Program in WIPO", *Inter-sessional Intergovernmental Meeting on a Development Agenda for WIPO*, 1st session, 11–13 April, document prepared by the Secretariat, www.wipo.int/edocs/mdocs/mdocs/en/iim_1/iim_1_2.doc

Wisner, R. N. (2005) "The Economics of Pharmaceutical Crops, Potential Benefits and Risks for Farmers and Rural Communities", *Union of Concerned Scientists*, Cambridge: UCS Publications, www.agmrc.org/NR/rdonlyres/AFC20F70-ADEA-48D8-A367-CCFD72ACBD41/0/economicspharmacrops.pdf

Wissen, M. (2000) "Die Transformation des regionalen Staates. Diskurse und Institutionen regionaler Entwicklung im Wandel", in *Das Argument*, vol. 236: 374–386.

—— (2007) "Politics of Scale, Multi-Level Governance aus der Perspektive kritischer (Raum-)Theorien", in Brunnengräber, A. and Walk, H.(eds) (2007) *Multi-Level-Governance. Klima-, Umwelt- und Sozialpolitik in einer interdependenten Welt*, Baden Baden: Nomos.

World Bank (2000) "Project Appraisal Document on a Proposed Grant from the Global Environmental Facility Trust Fund ( . . . ) for a Mexico Mesoamerican Biological Corridor Project", *Project-ID MX-GE-60908*, Washington, DC, www.conabio.gob.mx/institucion/corredor/doctos/PAD-ingles.PDF

WTO (1999a) *Preparations for the 1999 Ministerial Conference, The TRIPS Agreement. Communication from Kenya on behalf of the African Group*, WT/GC/W/302 http://docsonline.wto.org

—— (1999b) "Review of the Provisions of Article 27.3(b), Communication from the United States", IP/C/W/162, http://docsonline.wto.org

—— (2000a) "Review of Article 27.3(b), Communication from Brazil", IP/C/W/228, http://docsonline.wto.org

—— (2000b) "Review of the Provisions of Article 27.3(b), Further Views from the United States", IP/C/W/209, http://docsonline.wto.org

—— (2000c) "Review of Article 27.3(b), Communication from Brazil", IP/C/W/228, http://docsonline.wto.org

—— (2001) "Review of the Provisions of Article 27.3(b) of the TRIPS Agreement, Communication from the European Communities and their member States", IP/C/W/254, http://docsonline.wto.org

—— (2002a) "Review of the Provisions of Article 27.3(b), Summary of Issues and Points Made", Note by the Secretariat, WTO IP/C/W/369, http://docsonline.wto.org

—— (2002b) "Communication from the European Communities and their Member States, The Review of Article 27.3(b) of the TRIPS Agreement, and the Relationship between the TRIPS Agreement and the Convention on Biological Diversity (CBD) and the Protection of Traditional Knowledge and Folklore", A Concept Paper, IP/C/W/383, http://docsonline.wto.org

—— (2003) *Annual Report*, Geneva: WTO.

—— (2006a) *Council for Trade-Related Aspects of Intellectual Property Rights, the Relationship between the TRIPS Agreement and the Convention on Biological Diversity, Summary of Issues Raised and Points Made*, Note by the Secretariat, IP/C/W/368/Rev.1, http://docsonline.wto.org

—— (2006b) *Council for Trade-Related Aspects of Intellectual Property Rights, Review of the Provisions of Article 27.3 (b), Summary of Issues Raised and Points Made*, Note by the Secretariat, IP/C/W/369/Rev.1, http://docsonline.wto.org

—— (2006c) *Council for Trade-Related Aspects of Intellectual Property Rights, Article 27.3 (b), Relationship between the TRIPS Agreement and the CBD, and the Protection of Traditional Knowledge and Folklore*, IP/C/W/469, http://docsonline.wto.org

—— (2006d) *General Council, Trade Negotiations Committee, Doha Work Programme – The Outstanding Implementation issue on the Relationship between the TRIPS Agreemnet and the Convention on Biological Diversity*, WT/GC/W/564/Rev.1, http://docsonline.wto.org

—— (2006e) *General Council, Trade Negotiations Committee, Council for Trade-Related Aspects of Intellectual Property Rights, The Relationship between the TRIPS Agreement, the Convention on Biological Diversity and the Protection of Traditional Knowledge Amending the TRIPS Agreement to Introduce an Obligation to Disclose the Origin of Genetic Resources and Traditional Knowledge in Patent Applications*, WT/GC/W/566, http://docsonline.wto.org

—— (2006f) *World Trade Report 2006*, Geneva: WTO, www.wto.org

Wullweber, J. (2004) *Das grüne Gold der Gene. Globale Konflikte und Biopiraterie*, Münster: Westfälisches Dampfboot.

Young, O. R. (2002) *The Institutional Dimensions of Environmental Change: Fit, Interplay, Scale*, Cambridge, MA: MIT Press.

Yúnez-Naude, A. and Barceinas, F. (2004) "El TLCAN a la agricultura mexicana", in Casares, E. S. and Sobarzo, H. (eds) *Diez años del TLCAN en México, Una perspectiva analítica*, Mexico City: Fondo de Cultura Económica.

Zedan, H. (2005) *Statement by the Executive Secretary of the CBD to the Meeting of the Ad Hoc Open-ended Working Group on the Review of Implementation of the Convention*, Montreal, 5–9 September.

# Interviews

## Interviews at COP 5 in Nairobi in May 2000

CBD/1: IPGRI
CBD/2: SEARICE, Philippine NGO
CBD/3: IUCN Law Centre
CBD/4: Swiss Government
CBD/5: OECD
CBD/6: PRONATURA, Mexican NGO
CBD/7: Third World Network, international NGO
CBD/8: Indigenous Australian Peoples
CBD/9: Japanese Government
CBD/10: Japanese Biotechnology-association
CBD/11: Kalpavriksh, Indian NGO
CBD/12: member of the Peruvian delegation
CBD/13: private company Novartis
CBD/14: FAO CGIAR
CBD/15: World Resources Institute (WRI)
CBD/16: member of the Malaysian delegation
CBD/17: ECOROPA, European NGO
CBD/18: CBD Secretariat
CBD/19: EU Commission
CBD/20: member of the Ethiopian delegation
CBD/21: RAFI / ETC Group, international NGO
CBD/22: member of the German delegation
CBD/23: member of the Brazil delegation
CBD/24: Indigenous Biodiversity Information Network (IBIN)

## Working Group on Access and Benefit Sharing, October 2001, Bonn

CBD/25: Representative of indigenous peoples from Costa Rica
CBD/26: Representative of indigenous peoples from Argentina

## Extra

CBD/27 Expert interview German Environment Ministry, Bonn, Mai 2001

## Interviews at the 6th Extraordinary Session der CGRFA in June 2001 in Rome

FAO/1 GRAIN, international NGO
FAO/2 ASSINSEL, international plant breeder association (2 persons)
FAO/3 WIPO, World Property Organisation
FAO/4 BIO, international lobby alliance of the biotech-industry
FAO/5 IPGRI/FAO
FAO/6 Greenpeace, international NGO
FAO/7 ITDG, English NGO

## Interviews in Mexico

### Mexico-City, January 2001:

Mx/1: Mexican Biodiversity Commission (CONABIO)
Mx/2: Greenpeace Mexico
Mx/3: Councelor of CONABIO
Mx/4: Scientist at the National Ecological Institute
Mx/5: Councelor of the Commission on Environment and Resources of Mexican House of Representatives
Mx/6: Ministry of Ariculture
Mx/7: AgroBio, Association of Mexican Biotechnological Industry
Mx/8: Lawyer close to Mexican Ministry of Environment
Mx/9: Think Tank of independent Farmers Organisation CECCAM

## San Cristóbal de las Casas, Chiapas, January 2001

Mx/10: independent NGO CIEPAC, close to Zapatistas
Mx/11: Research Cenre ECOSUR
Mx/12: Conservationist NGO PRONATURA
Mx/13: international NGO Global Exchange
Mx/14: Instituto de Historia Natural, in fact Ministry of Environment in Chiapas

## Different Places in Mexico, February 2001

Mx/15: International Agricultural Research Centre CIMMYT (near Mexico City)
Mx/16: WWF Mexico
Mx/17: NGO DASSUR, Xalapa, Bundesstaat Veracruz
Mx/18: SEMARNAT, Dept. Of International Relations, Mexico City
Mx/19: Member of a NAFTA Commission on Environment
Mx/20: Member of the Group of Biodiversity Experts in Mexico BIODEM, Expert in IPR
Mx/21 RAFI, Office in Mexico City
Mx/22 Member of the Group of Biodiversity Experts in Mexico BIODEM, Expert in access to genetic resources

# Index

For Product Safety Concerns and Information please contact our EU representative GPSR@taylorandfrancis.com
Taylor & Francis Verlag GmbH, Kaufingerstraße 24, 80331 München, Germany

www.ingramcontent.com/pod-product-compliance
Ingram Content Group UK Ltd.
Pitfield, Milton Keynes, MK11 3LW, UK
UKHW022015180926
475701UK00003B/374
*9780415586795*

# Governance of Security and Ignored Insecurities in Contemporary Europe

Edited by
SALVATORE PALIDDA
*University of Genoa, Italy*

Routledge
Taylor & Francis Group
LONDON AND NEW YORK

First published 2016 by Routledge

2 Park Square, Milton Park, Abingdon, Oxfordshire OX14 4RN
52 Vanderbilt Avenue, New York, NY 10017

*Routledge is an imprint of the Taylor & Francis Group, an informa business*

First issued in paperback 2020



*British Library Cataloguing in Publication Data*
A catalogue record for this book is available from the British Library

*Library of Congress Cataloging-in-Publication Data*
Governance of security and ignored insecurities in contemporary Europe / by Salvatore Palidda.
pages cm. — (Advances in criminology)
Includes bibliographical references and index.
ISBN 978-1-4724-7262-5 (hardback) — ISBN 978-1-4724-7263-2 (ebook) — ISBN 978-1-4724-7264-9 (epub) 1. Crime prevention—Europe. 2. Internal security—Europe. 3. Law enforcement—Europe. 4. Europe—Social conditions. I. Palidda, Salvatore, editor.
HV6938.5.G68 2016
364.4094—dc23
2015025170

ISBN: 978-1-4724-7262-5 (hbk)
ISBN: 978-0-367-59683-5 (pbk)

Typeset in Times New Roman
by Apex CoVantage, LLC

# Contents

# List of Figures

# List of Tables

# Acknowledgements

This book contains contributions that formed part of a research project for the GERN programme on 'Ignored Insecurities' and also for an Italian research programme (PRIN) on the professions of security governance and control agencies, including the police forces. I would like to thank both the officers from the different control agencies, and the authors of the chapters in this book, as well as the colleagues who participated in workshops and debates organized by Fabienne Brion at the Catholic University in Louvain-la-Neuve (UCL-Belgium), by Rita Haverkamp at the Max Planck Institute for Foreign and International Criminal Law in Freiburg im Breisgau (Germany) and by myself at the Department of Educational Sciences in Genoa (Italy). A particular thank to Hans-Jörg Albrecht.

# Notes on Contributors

**Harald Arnold** is psychologist and senior researcher at the Max Planck Institute for Foreign and International Criminal Law in Freiburg (Germany). He is a member of the Security Research Dialogue initiated by the German Federal Ministry of Education and Research.

**Barbara Giovanna Bello** is a researcher in criminology at Milan University, Italy; her main topic is the persecution of Roma people and the aporias of the European governance on this phenomenon.

**Cecilia Blengino** is a researcher in criminology at Turin University, Italy; her work concerns in particular the qualitative research on some aspects of judicial trial.

**Monica Colombo**, PhD, is a researcher in sociology at the Department of Psychology of the University of Milano-Bicocca, Italy. Her research interests include racist and anti-immigration discourses, ethnic discrimination, urban security, critical discourse studies and political discourse. On these topics she has published several journal articles and chapters in edited volumes. She has been guest editor of a special issue of the *Journal of Language and Politics* on discourse and politics of migration in Italy.

**Luigi Ferrari** is full professor of Economic and Work Psychology at the Department of Psychology of the University of Milano-Bicocca (Italy) where he teaches Economic and Work Psychology and Financial Psychology. His research interests include: individualism/collectivism, behavioural finance, stock market psychology, money psychology, tax psychology, business cycle and unemployment, history of work and economic behaviour. He is a clinical psychologist and a psychotherapist.

**Rita Haverkamp** is professor of crime prevention and risk management at the University of Tübingen. From 2008 to 2013 she was a senior researcher at the Max Planck Institute for Foreign and International Criminal Law. In this capacity she mainly worked on terrorism and security research. Her recent publications include: 'New responses to sexual offenders: Recent developments in legislation and treatment in Germany' in *Monatsschrift für Kriminologie und Strafrechtsreform* (2013, with Gunda Wößner); 'The prognosis of terrorist attacks – limits of scientific findings' in *Crime, Law and Social Change* (Vol. 62, No. 3, 2013) 'Electronic monitoring' in *Encyclopedia of Criminology and Criminal Justice* (2013, edited by Gerben Bruinsma and David Weisburd, New York, Springer).

**Mikael Kazgandjian** is a PhD student at Dipartimento di Scienze della Formazione of Genoa University, Italy and at Ecole des Hautes Etudes en Sciences Sociales of Paris, (France). The principal topic of his research regards the judicial system and the ethnography of prisons, according to a Foucaultian and Goffmanian approach.

**Susanne Knickmeier** is a German lawyer who also holds a master‘s degree in criminology. She is a researcher at the Max Planck Institute for Foreign and International Criminal Law working on the FIDIUCIA project. Her research interests include the trafficking of goods and trust in the justice system, as well as European criminal and security policy with a particular focus on migration, border security, cross-border police cooperation and the policing of mega events.

**Tim Lukas** is a sociologist and criminologist. He is head of the Department for Object Security at the Institute for Security Systems and lecturer at the Faculty for Safety Engineering at the University of Wuppertal. His research focuses on the interface of urban resilience and crime prevention.

**Salvatore Palidda** is professor at the Department of Education, University of Genoa, Italy. He holds a PhD in Sociology and European Studies from the Ecole des Hautes Etudes en Sciences Sociales, Paris. He has been co-director and director of several European research projects. His main books are: *Racial Criminalization of Migrants in the 21st Century*; *Migrations Critiques* (2011, Farnham: Ashgate); *Migrations critiques* (ed. with Dal Lago, 2010, Paris: Kartala); *Conflict, Security and the Reshaping of Society: The Civilisation of War* (2008, London: Routlege); *Mobilità umane* (2000, Milan: Cortina); *Polizia postmoderna* (2000, Milan: Feltrinelli).

**Antonello Petrillo** is professor (sociologist) at Suor Orsola Benincasa University, Naples (Italy). His recent publications include: *Il silenzio della polvere*; *'Razze informali' al lavoro. Naturalizzazione della 'plebe' e postfordismo nella trasformazione del territorio napoletano* (2011, Milan: Mimesis); Napoli globale: discorsi, territorio e potere nella 'città plebea', in *Città Mediterranee e deriva liberista*, (edited by S. Palidda, 2009, Messina: Mesogea); *Biopolitica di un rifiuto. Le rivolte anti-discarica a Napoli e in Campania* (2009, Verona: Ombrecorte).

**Fabio Quassoli**, PhD, is associate professor of sociology at the Department of Sociology and Social Research, University of Milano-Bicocca, where he teaches qualitative research methodology and intercultural relations. His research interests focus on immigration policies, immigrant criminalization, ethnic discrimination, multiculturalism and intercultural communication. On these topics, he published a book and several papers in international journals and edited volumes. He is currently writing a book on immigrant control and criminalization in Italy. He has also been involved in a research project on urban governance and safety policies in Italy over the last two decades.

**Pietro Saitta** is senior lecturer in Sociology at the University of Messina (Italy). His research interests concern city, environment and critical criminology. He has worked with several universities and research institutions in Italy and elsewhere. He is the author of several books and articles published with international publishers. Among his works in English is *Getting by or Getting Rich* (2013, co-edited with J. Shapland and A. Verhage, The Hague: Eleven Publisher).

**Giovanni Torrente** is researcher (criminologist) at Turin University, Italy. The principal topic of his research regards the judicial system and the ethnography of prisons.

# Preface

After completing this book, my mind went to the words of famous authors about the wars and many disasters following the Second World War, and how countries had declared 'never again'. In a poem, alluding to inurement to evil (as Hanna Arendt wrote after), Bertolt Brecht writes: 'The first time it was reported that our friends were being butchered there was a cry of horror. Then a hundred were butchered. But when a thousand were butchered and there was no end to the butchery, a blanket of silence spread. When evil-doing comes like falling rain, nobody calls out "stop!" When crimes begin to pile up they become invisible. When sufferings become unendurable the cries are no longer heard. The cries, too, fall like rain in summer.' And, always with reference to Brecht (and also to the parrhesia theorized by Foucault), this book also brought to mind the phrase in *Life of Galileo*: 'Nowadays, anyone who wishes to combat lies and ignorance and to write the truth must overcome at least five difficulties. He must have the courage to write the truth when truth is everywhere opposed; the keenness to recognize it, although it is everywhere concealed; the skill to manipulate it as a weapon; the judgment to select those in whose hands it will be effective; and the running to spread the truth among such persons.'

# Chapter 1
# Introduction[1]

Salvatore Palidda

> Demographic bomb, ecological bomb, migration bomb, health bomb, bloody terrorists and so on. For over 20 years now we've been bombarded by continual fears sent to terrify us ... and then, every day when we go to work in the factory, it's like we are going to the front, going to war! Yes! Because work has become a war! We never stop risking our lives because of all the various accidents or toxic contamination. Not to mention the fact that the foreman never stops pressuring us to speed up the pace of work so we're just increasing production and the profits for those bastards, at any cost! What has become of this world? I don't think we're facing any more risks than before. I think the real issue is that they don't care! It's like they say 'Samson dies with all the Philistines'. They don't die; they have enough money to pay for any type of care, even for human organs! So what should we be thinking? Is this a Nazi dominant class? And why do people seem to be anaesthetized? How is it that the dominant ones do what they want and we are no longer able to respond? That is the real, the biggest, the most important disaster! And the end of the world as we know it will follow! The rich don't care because they'll be going to live on Mars or I don't know where. Unless we get to catch up with them before they leave in their shuttles for outer space ... I see myself not giving up on the idea of reacting ... but all the others just seem like the workers in Fritz Lang's film Metropolis ...
>
> (From a conversation with a 62-year-old worker interviewed during research into the Governance of security and ignored insecurities.)

Over the last 20 years a great deal of literature and many scholars (including ourselves) have set out to describe and analyse the different aspects of security governance, in particular after 1990, and more specifically after the September 2001 attack.[2] However, none of this research has focused on some of the most important consequences of security governance: the proliferation of ignored insecurities that affect most of the population, the state's tax income and the general damage to the *res publica.*

These insecurities include: risks to health and environmental disasters (frequently exacerbated by the effects of climate change) in particular the increase

---

1 English revised by Deborah Kay.

2 See the bibliography. Among our main publications see Dal Lago and Palidda (eds. 2010); Palidda (ed. 2011) (also published in French, Spanish and Italian).

in cancer rates and other diseases;[3] the consequences of the growth of shadow economies (principally neo-slavery, the reproduction of irregular immigration for these activities and the inferiorization/overexploitation of all workers – including nationals- tax and social contribution evasion or fraud); the hybridization between legality and illegality and also the increase in corruption within control agencies – labour and health inspectorates, civil protection services, national and local police forces and so on, that is, all the institutions responsible for preventing these types of risk, the pursuit of crimes in this field and also of reclamation/regularization.[4]

The common objective of all the contributions in this book is to examine how the securitization of society has exacerbated these serious phenomena, and to try to answer the following questions: what is stopping recovery and consequently the implementation of devices and the establishing of prevention practices? Why are the majority of victims and also the control agency professionals seemingly resigned to these ignored insecurities?

Almost all the contributing authors have spent at least 20 years working on research projects, at a national and international level, into the changes in the posture, devices and practices of police forces, control agencies and security governance. The analysis of the results of these projects was the catalyst for a new research perspective from 2010, into the growth of ignored insecurities as

---

3 Currently about 8.2 million people die from cancer worldwide every year, among them, about a million people are aged 30 to 69 years. According to IARC (2015), based on current knowledge, nearly half of all cancer cases worldwide could be prevented. There were approximately 14 million new cases and 8.2 million cancer-related deaths in 2012. The number of new cases is expected to rise by about 70 per cent over the next two decades (WHO, 2015; AIRC, 2014). By 2030, 27 million new cases of cancer will have been recorded, along with 17 million deaths – again, almost 65 per cent. During the same period, 80 per cent of all deaths (52 million) will be caused by non-communicable diseases (NCDs) (cancer, diabetes, mental illness, and heart and pulmonary diseases). In 1990, the developed and developing nations shared the global burden of cancer equally, but by 2020, developing countries are expected to be experiencing almost a doubling of their share of cancer incidence, with a proportion of deaths ranging from 150 per cent to 180 per cent higher than that seen in the developed world. The World Economic Forum estimates that the five major NCDs will cost US$47 trillion over the next 20 years (Sutcliffe, 2012). There is another even bigger problem: the protection of pharmaceutical lobbies has permitted the abuse of antibiotics: the European Society of Clinical Microbiology and Infectious Diseases (ESCMID) and a group of leading scientists have warned that Europe could face more than a million deaths in an impending ‘Antibiotic Armageddon’ unless more is spent on developing new cures, rapid diagnostics and preventative measures to combat the spread of drug resistant diseases. In 2050 the number of dead could reach 10 million a year, at a cost to the European Union (EU) of around 1.5 billion euro: ‘It’s the result of 40 years of the inappropriate use of antibiotics’ (ESCMID, 2015).

4 In Europe these institutions are part of the state or of local administrations such as regions, municipalities and departments.

a consequence of a security governance that does not respect the fundamental rules of the democratic state of law (that is, the protection of health and other fundamental rights).

As illustrated in the various chapters, we believe that the primary cause of the increase in ignored insecurities is the consequence of the neoliberal turn in security governance. In fact, we think that the neoliberal posture is the antithesis to the liberal–democratic theories and policies that some of the most important authors, such as Durkheim, Weber, Simmel, Schumpeter, Keynes, Polany, Arendt, Galbraith and Mills and so on hoped for. And this is because its main aim is the prosperity *hic et nunc* of the small minority of dominants, who become increasingly rich, and no posterity, that is, no future, for the greater majority of humanity. This is one of the main causes behind the continual negligence towards the risks and insecurities described in this book.

Thus, the consequences of the neoliberal turn, and in particular the governance of security which then produces more and more ignored insecurities, can be considered a 'total political fact', because they concern all humanity, almost all aspects of the political organization of all societies and their futures. For this reason, we have approached our research with due caution[5] in order to achieve a rigorous multidisciplinary approach, not an easy challenge as far as the philosophical, political, cultural, economic and social aspects are concerned. This approach implies an articulation between the diachronic and synchronic, macro- and micro- political and social sciences and also the comparison between different case studies. The objective of our work, therefore, is to gain a better understanding of the future of the political organization of contemporary society after more than 25 years of neoliberal development.

After analysing some of the aspects dealt with in this book we have been able to show that since 1990 the governance of security has actually increased ignored insecurities. This situation has worsened during the current financial crisis, which has itself become a justification for the lack of interest in recovery, in the implementation of prevention measures, in helping victims, and involving them in their own recovery and regularization. This discourse has acquired an extraordinary force, that is, a biopolitical effect, which is consequently interiorized by the majority of people.[6] But the apparent paradox is that ignored insecurities are also the result of the behaviour of the victims. Sometimes the

5 This concepts refers to 'total social fact' coined by Marcel Mauss (1950/1999), one of the fathers of anthropology; he suggested adopting this definition of the 'social fact' theorized by Durkheim (his uncle) because it included economic, social, cultural and political aspects. Here I propose calling Mauss's concept a 'total political fact' because, effectively, 'political' includes all of the aspects that concern both society and individual behaviours.

6 Obviously, I am referring here to the theory of 'biopolitic' as proposed by Foucault (2004a) who, along with Goffman, is the main reference for many of us.

securitization provided for health or environmental disasters is conceived and organized as a technical operation and also as a business opportunity for private actors, but rarely as a recovery programme in the effective interests of the victims and of the *res publica*. We often criticize the abuse of the term 'resilience' when it appears ambiguous, and especially when it leads us to transform a 'total political fact', such as health and environmental disasters and shadow economies, into an individual problem which is approached by a 'psychologism' embedded in a neoliberal orientation.

The biopolitic of neoliberal security governance has produced a consensus towards penal–repressive practices (in particular the criminalization of the poor and immigrants), by distracting public opinion, and also by forming a discourse regarding the individual causes of insecurity and vulnerability, for example climate change, and which all lean towards a more technical approach and private business opportunities. In addition, some of the victims of neo-slavery or occupational diseases find it impossible to imagine their own safety or emancipation, because they are unable to see any way of achieving it; more often after they have no access to any form of guidance or protection and after years of this 'way of life' they lose any idea of, or capacity for, collective action. This shows how successful neoliberal development is in achieving one of its most important goals: the erosion of public action, that is, collective commitment for the *res publica,* or the vital interest of each human being, because people do not always understand exactly what the *res publica* is, or it is distorted or totally overturned, notoriously in favour of a private *res*.[7] This erosion also corresponds to the frequent corruption found within public administrations and in particular among control agencies and police forces, because neoliberal actors would prefer fewer controls and rules of law, and more freedom to do what they want. Obviously, the corruption within the public administration, and in general within state bureaucracy, favours a widespread loss of respect for the rules of law, a loss of the idea of *res publica* and the idea that it is impossible for victims to receive protection. This helps to explain the fact that we are not facing a simple problem of ethics or morality, or of Weberian professionalism, but an economic, social and cultural, that is, political situation, that risks destroying the real democratic rules of law (frequently through the heterogenesis of democratic references).

This book proposes an innovative approach to security governance, not only through a serious analysis of the balance of the costs and benefits, but also by highlighting what we have termed the 'ignored insecurities', which are often caused by distraction tactics that have led the security agencies to exclusively prioritize targets and issues that are not actually a priority. In our analysis we also take into account the measures that purport to combat the current financial

---

7 In this regard the abuse of oxymorons and euphemisms such as: humanitarian war, exportation of democracy, collateral effects (to cover the massacres of civilians by bombing raids aimed at striking the enemies of the democratic world), the sacrifice of liberty to have security and so on, is emblematic.

crisis, adopted at international, European and national level. These measures do not substantially limit financial speculation and do not help the people who suffer the consequences of the crisis most acutely, and who are the most exposed to the various ignored insecurities. Thus, by analysing the European research programme 'Horizon 2020' (ch.1) it becomes clear that no serious programme has been envisaged to tackle the various ignored insecurities mentioned above.

Beyond the research into the political and social sciences embedded in neoliberal governance, we feel that the existing circumstances produce a form of scepticism in those carrying out rigorous research with intellectual honesty. But we argue that it is possible to develop a new research perspective for rethinking a truly democratic security governance which fully respects the *res publica,* while still working with all those people who want to resist, and if necessary build a way to reorganize society *ex-novo*, and in full compliance with all the fundamental human rights. At the same time it is a form of 'parrhesia', a rethinking of ways to create a political organization of society that go beyond the outlines or paradigms linked to modernity and the industrial period, and beyond the trivial postmodern mainstream. It is true that liberal–democratic knowledge has shown itself to be weak when facing the neoliberal think tanks; in particular, criticism of neoliberal security governance has been inadequate because it is frequently limited to strong criticism – often correct[8] – of the most evident consequences (the metamorphosis of social questions, world poverty, the turn from welfare to workfare, punishment of the poor, governing through crime, the culture of fear, mass imprisonment, racist criminalization, police violence, the exacerbation of permanent war, security as a means for reshaping society, mass distraction and so on). But, we think that this criticism remains unheard not only because of the strength of the neoliberal powers in all fields, but also because ignored insecurities have been seriously neglected, and more generally, the neoliberal powers have tried to destroy the *res publica.* It seems particularly relevant here to note that criticism of power or dominion brought important thinkers, such as Hanna Arendt and Foucault, to almost total scepticism. This is because dominion is primarily imposed through knowledge. The success of dominion is the result of the discourses of the human, social political sciences and also of the devices and forces of the reproduction of these. It is most successful when this discourse is interiorized by the dominated people, that is, becomes socially shared or the 'banality of evil'. Therefore, this criticism of dominion becomes a radical and systematic deconstruction of science and discourse of reproduction of power,

8 There is a rich literature of criticism of neoliberalism that provides valuable tools and also a literature on risks. In Germany the father of studies on risk is Luhmann (1993, 1996) followed by Beck (1992, 1998/2009); on the economic, social and political aspects, in France we should first mention Bourdieu (ed. 1993) and Castel (1991, 2003, 2009, 1995); in the Anglo Saxon world, Bauman (1995, 1997, 1998, 1999, 2000, 2004) and also Rose and Miller (2006) and some other authors quoted in the different chapters of this book.

a parrhesiastic counter-knowledge, a task that excludes to give an alternative, does not promise a possible reclamation of the evils of the world and even less a future just society. 'Beware of dealers of hopes,' wrote Pasolini after many years of comments to the conspiracies of power, to corruption, to speculations, scams and violence always to the detriment of the *res publica*. However, it seems useful to note that a rigorous criticism against the sciences and the discourse of domain first is based on what the domain denies: the fundamental rights of all humans: first and foremost the protection of the physical and moral safety of all persons and therefore of the environment and also the correct satisfaction of their needs; in short, all that is generally understood as the indispensable foundations of an equitable, sustainable and inclusive society, all included in the concept of *res publica*.

## About the 'State of Art'

The history of the human race has been punctuated by all kinds of disasters.[9] From ancient Greece to the present day, political philosophy and later the political and social sciences have considered these as events that governments and populations are occasionally forced to deal with, and the way they are handled (either well or badly) depends on the quality and ability of the ruling class. It is interesting to note that the development of the political and social sciences is fairly recent; they seem to propose an interpretation and analysis perspective or approach that can be considered as a special branch of sociology, but also of political science, which can't explain the political nature that is at the heart of the governance of disasters, and frequently at the origin of these, even more so with the development of modernity and now of postmodernity. In fact, many of the studies in the fields of sociology and the political science of disaster focus on the link between 'social solidarity' and the 'vulnerabilities' exposed by disasters. The majority of scholars in this field have observed that such events produce social solidarity (Barton, 1969; Drabek, 1986; Dynes, 1970; Taylor, 1977); subsequently, other scholars also observed the social conflict generated by disasters (Bolton, 1997; Fradkin, 2005; Henderson, 2005) and, more importantly, some authors have recently pointed out that disasters demonstrate and exacerbate the inequalities inherent in the social order by exponentially multiplying the effects (Tierney, 2007). Unfortunately, despite these perfectly correct observations and the often excellent work of researchers in this field, disaster research tends to focus more on the solidarity in the aftermath of disasters, and on seeing disasters as the consequence of our inability to adapt to hazardous environments. Therefore, we have a posture that tends to place responsibility for good or bad prevention

9 See Barkun (1977). On the history of all 'natural disasters', see: http://en.wikipedia.org/wiki/List_of_natural_disasters_by_death_toll; and for industrial disasters: http://en.wikipedia.org/wiki/List_of_industrial_disasters, accessed on 4 September, 2015.

and disaster governance with the population in general. It is true that frequently the population is co-responsible because it shares the attitudes and behaviours that result from their interiorization of the dominant discourse and choices. But, as we have tried to explain in this book, it is notably the neoliberal turn that has created an exacerbation of risks from disasters, and their consequences and victims.

Our interpretation and analysis focuses on the aporias, the darker sides, the ominous consequences of neoliberal security governance, which are responsible for the continued growth of disasters in today's world, and in particular in Europe. Although Webb (2002) does not actually mention neoliberalism as being one of the main reasons for this deterioration, in a chapter in a sociology manual on disasters, he also writes that 'the vulnerability to extreme events is increasing'. This is also tragic because knowledge about the origins of disasters has been developing over many decades, and geologists, biologists and other scientific researchers have been underlining the deterioration of the ecological system since the end of the nineteenth century. It is evident that natural disasters are worsened (for example, number of victims and damage) by the bad governance of the territory and more particularly by negligent or non-existent prevention plans.

The sociology and political science of disasters was developed in the United States, in Germany and Italy,[10] in particular towards the end of the sixties when major disasters occurred (the disaster in Seveso – near Milan – led to the Seveso Directive, which was issued by the European Community and imposed much harsher industrial regulations).

The more prominent researchers in the sociology of disasters (linked to environmental sociology and political sciences in these fields) are Stallings (2006) and Quarantelli (ed. 1998), an American sociologist pioneering in this field. In his presidential address on 'Sociological Theory and the Sociology of Disasters: Moving from Periphery to Center', in 2006, Stallings (president of the International Research Committee on Disasters) he said:

> I believe that the sociology of disasters has had too little impact on the discipline of sociology. I further believe that findings from disaster research are invaluable for developing general sociological theory.

It is paradoxical that this idea has had so little impact, especially when there has been a continued rise in risks from disasters; but it also reveals the dominant aim

10 The political and social sciences on disasters are linked to risk and ecological studies. In the United States, Stallings, Quarantelli, Dynes (see bibliography); in Italy the researchers of ISIG – Istituto di Sociologia Internazionale di Gorizia – (among them De Marchi, 1991 and Nocenzi, 2002, 2005). The most important journals are Massemergencies, stopped in 1979, *International Journal of Mass Emergencies and Disasters* (www.ijmed.org/) and Disasters (Wiley).

of the private and public powers to conceal their responsibility for the origins and the management of disasters, and in general, for the prevention of risks regarding health and the environment. For example, there is not only negligence of some of the more relevant risks, but an evident complicity with the European authorities in favour of lobbies, despite the dangers to the health of the population. That is the case of the Transatlantic Trade and Investment Partnership (TTIP), the trade agreement between the US and Europe.[11] According to many organizations for consumer protection, the TTIP would facilitate the entry of genetically modified organisms (GMOs) that are difficult to trace, the increased use of pesticides and the main concentration of production in the hands of the few biggest companies. At the same time, the 'European Directive' on the 'trade secrets' remains an unclear or non-transparent procedure[12]. According to some members of committees charged to examine the texts known by members of the European Parliament (MEPs) and non-government organizations (NGOs), these texts are chilling from the point of view of real public interests and transparency. That is also the case for so-called 'shadow banking'. Information laws, labour laws, health laws, consumer laws and environmental laws are whole areas of law and regulations that are under threat. In a historic reversal, Europe, which previously was strong in promoting transparency and information for consumers and citizens, now sets up laws in an atmosphere of shadowy secrecy and obscurity.[13]

Another emblematic fact is that in Italy, as in some of the other European countries, it has been discovered that a lot of meat on the market is treated with illegal substances. This practice has been uncovered by a number of judicial inquiries, which found that young bovines had been treated with illegal substances with the complicity of corrupt doctors from the various health agencies. A serious and far-reaching inquiry into the wide diffusion of illegal substance usage on farms has also revealed that European policies in this field have permitted the addition of anabolic substances to artificial milk, including carcinogenic substances. Controls are still ineffective because, so far, they have not included protocols and proper techniques for detecting these substances. And yet all the European States claim that, for some time now, there has been perfect security in place for all food systems (Rai Report, 2015).

11 Josling and Tangermann (2014); CorriereTv (2015).

12 According to Corporate Europe Observatory (2015) http://corporateeurope.org, in Brussels there are about 1,500 companies and 30,000 professional lobbyists who work to influence all European legislative process. Among the most powerful lobbies is that of the banks. Lobbying is not a crime if it respects the rules of transparency and does not practice the corruption. But so far, European governments have sabotaged transparency. This is not what has been called 'Americanization of Europe', but the triumph of neoliberalism which gives more freedom to the stronger actors and more proliferation of conflicts of interest in the United States, as in all European countries.

13 See Orange (2015) and Chapter 6 by Mikael Kazgadjian.

## The Book

In Chapter 2 we aim to analyse the frame of the rise of 'postmodern' security governance and the proliferation of ignored insecurities and their victims. After considering some of the theoretical and methodological aspects, we also give a description of the different aporias of this governance for the period 1990–2015, with particular reference to some of the most important ignored insecurities and some of the most eloquent cases, despite almost 30 years of disquieting reports by experts to both world and European authorities. In this chapter we point out the particular significance – decisive and not only determinant – of the neoliberal turn in the rise in disasters and all ignored insecurities.

In Chapter 3, Harald Arnold and Rita Haverkamp propose the adoption of the 'security quadrate' as a means of categorizing ignored securities. Security incorporates several apparent paradoxes and conflicts, including the cases when security may arouse greater feelings of insecurity, or when an individual or group security might collide with the security interests of others, or again, when there is conflict between the security issues of wealthy consumers and non-consumers (the poor/impoverished/indigent). The ambivalence that exists between objective and subjective security will be explained by the so-called 'security quadrate' which helps to illustrate the differences and similarities between different security issues. It also helps to ascertain 'real risks and hazards', and 'fears and worries', as well as their mutual relationship, and to develop measures to classify levels of security. This chapter presents the results of a representative study about security and quality of life conducted at the Max Planck Institute for Foreign and International Criminal Law. The analysis will be embedded in a framework for the future research of ignored securities.

In Chapter 4 Tim Lukas analyses the governance of crime in the risk society as a way of moving from 'designing out crime' to 'built-in resilience'. Resilience has become a fashionable buzzword in recent security discourse, and the concept of urban resilience promises answers to a range of security problems in cities. The chapter discusses to what extent crime prevention through urban planning can benefit from the discourse about urban resilience. The question is whether the urban resilience perspective can provide the basis for a new paradigm of crime prevention through urban planning. Since the concept of resilience is often used in a narrow technical understanding, the chapter concludes that the concept of urban resilience can primarily be considered promising in the context of crime prevention through urban planning, if questions of security are addressed not only in terms of construction and technology, but also within their wider social context.

In Chapter 5, Monica Colombo and Luigi Ferrari discuss resilience discourses, the ingenuity or ambiguity or abuse that exists between the various interpretations, and the neoliberal myth of self-contained individualism. Through a critical review of some of the selected literature on this field, this chapter analyses how the concept of resilience has, in the recent past, rapidly 'infiltrated', 'proliferated' and 'infused' vast areas of the social sciences, as demonstrated by the impressive increase in publications across a variety of fields – from psychology, sociology, geography, anthropology to public health, ecology, technology and communications. Several criticisms of resilience discourses have become increasingly ubiquitous not only across a variety of disciplines, but also in politics and policy documents. As some colleagues suggest, resilience can be so far regarded as 'the organising principle in contemporary political life'. In this chapter, the authors attempt a critique of resilience, in particular on the proximity between resilience discourses, self-contained and economic individualism, and contemporary neoliberalism. Taking critical discourse analysis as a broad theoretical framework they review and discuss the most convincing works by several scholars in the ongoing sociological and psychological debate to unveil their ideological implications. Starting the discussion with the 'highly political nature of resilience', particular attention is paid to individualization, which has undergone a dramatic acceleration in the time of economic crisis and a shift from social (economic) insecurity to individual/personal insecurity. Then the authors analyse discourses on psychological resilience as directly stemming from the neoliberal myth of self-sufficiency and from economic individualism. Some empirical cases are also discussed.

In Chapter 6, Mikael Kazgandjian describes the case of the decriminalization of business law in France. It is a revealing case because of its similarity to cases in many other countries, that is the new impunity of white-collar crime. The case refers to the period 1990–2014, which was characterized by the considerable growth in the criminalization of low deviancy assigned to immigrants and marginal people. Also relevant here is the fact that during this period there has been extensive growth in the distance between poverty and wealth. In the framework of this book, this is an important chapter since it reveals the real nature of the neoliberal turn, that is, the interests of the dominant classes to the detriment not only of the poor, but also of the future.

In Chapter 7, Fabio Quassoli and Monica Colombo analyse the case of the continuity and change in the governance of urban security in Milan. After a critical review of the studies in this field, the authors show how the technical governance that emerged in the eighteenth century is intrinsic to the neoliberal turn. This produces an expansion of municipal by-laws targeting the urban poor and ethnic minorities, and the aggressive enforcement of these by-laws and other typical aspects that derive from it. Thus, the 'new culture of control' relies upon the conception of a community whose identity is defined in purely negative terms. Also in Milan, fear-centred media discourse has become a powerful device for the manufacture of consensus and the legitimation of bio-power control practices. Like many other big cities, Milan has been governed by a right-wing administration for 20 years, and one that has made profitable discourse and carried out practices based on racist criminalization. But, at the same time it is interesting to observe that this same administration has increased the exploitation of regular and irregular immigrants, whose numbers have risen. In other words, Milan can be seen as an example of zero tolerance and as a strong system for the inferiorization/exploitation of immigrants, and one that is very profitable for a large part of the population as well as companies, banks, real estate companies, business leaders and also some members of the police force (because regular and irregular immigrants are easily blackmailed).

In Chapter 8, Cecilia Blengino and Giovanni Torrente describe and analyse how courts choose which crimes to pursue, a case study which also reveals similar judicial practices in many other countries. Their case study shows that, over the last two decades, the discernment and discretion of the courts have been biased by 'civil society's' demand for justice, which is frequently confused with public opinion (understood as the media) and also by the neoliberal turn, justified as a new efficiency and productivity. Thus, we can observe a general enlargement of the criminal area, and secondly, new habits which have been formed according to 'economic thinking' applied to criminal justice. The 'fair trial' is connected to its speed in responding to the claim for justice: a claim legitimated by public opinion, but not necessarily a just/legal demand. The rise, also in Italy, of the new 'managerial justice' follows the crisis of the paradigm of judicial 'ritual' and the definition of new criteria for the accountability of justice and judicial decisions. Prosecution represents a very interesting issue in helping to underline the relationship between the application of public management principles to the judicial system and the direction of criminalization processes. The prosecutor has the power to decide the starting point of the criminal trial, and the pre-trial prosecutor's inquiries represent a focal point in secondary criminalization. Prosecutor's offices are situated at a vital point between instances of 'civil society's' repression of crime and the start of the trial. Therefore pre-trial inquiries are a key phase for the reception and interpretation of people's fears and security concerns.

In Chapter 9, Barbara Giovanna Bello discusses the governance of Roma people in Italy with comparison to other cases which have raised a great deal of controversy in other European countries. In this contribution, she suggests testing Roma governance at a local level on the basis of Roma human security, and that we should enquire into its implications from the perspective of Newman's Critical Human Security Studies (Newman, 2010). Starting with a description of the main developments in the neoliberal governance of the Roma people over the last few years in Italy, the chapter reverses the question of security, from public and state security to human security, in order to delve into the rarely explored topic of the Roma people's human (in)security in the Europe of today. Through some examples at the local level, the chapter concludes that Newman's work – still embryonic – provides insights for embarking on transformative research and action beyond the 'policy-centric and institution-centric approaches' related to Roma activation and participation at the local level.

In Chapter 10, Susanne Knickmeier discusses crime prevention and security in the matter of refugees. In particular since 1990 – the implementation of the Schengen treaty – the mainstream discourse on migrations has mostly concerned security, but not the security of the refugees, and is increasingly connected to crime and is seen as a risk factor and threat to Europe's internal security and welfare. This chapter proposes a review of several texts adopted by the European Commission to analyse the evolution of the European Union's (EU) policy towards refugees. The challenging topics of migration and security have become an indivisible unit high on the political agenda of the EU. The European Commission has emphasized the importance of appropriate measures to prevent and combat cross-border crime (for example, illegal entry) at external European borders. The European internal security framework identifies (illegal) migration as a security risk that has to be controlled, ignoring the fact that most migrants aim to attain protection from insecurities in their home countries. This chapter, based on theoretical insights and legal analysis, discusses: (1) the dichotomy of national and human security, (2) European security policy, (3) border security measures and (4) the consequences of these policies. It is, for example, questionable, whether migration constitutes a real or perceived risk/threat for Europe's security, or whether the chosen measures are an effective prevention strategy to reduce (irregular) migration or just an act of securitization and criminalization. A massive security operation is supposed to keep migrants out of the EU; yet who protects their security?

In addendum to the chapter on EU crime prevention and security in the matter of refugees, the author proposes a synthesis of today's refugee drama. The text highlights the main causes and responsibilities for this drama, that is, the reproduction of permanent wars provoked by the production and trading of arms, and the prohibitionism of migrations. The risk of a twenty-first century quasi-genocide is very real.

In Chapter 11, Pietro Saitta describes and discusses some of the industrial disasters that have taken place in Italy with a comparative perspective. For decades, this country has witnessed a systematic erosion of its territory and a pollution process stemming from the mainly unregulated diffusion of industrial sites and from a process of housing-related speculation. The chapter addresses three issues: epidemiological rates, the social effects of the environmental issue, and the resistance processes to speculative land-uses. The author suggests that the results of epidemiological investigations are clear: environment-related diseases are one of the most important causes of mortality in the country. Even so, no serious policy has been undertaken, and the Italian agenda does not seem to be greatly concerned with the environment. As a consequence, Italian courts have tried to compensate for the lack of political leadership in these matters. However, these attempts have failed to produce the desired effects, and it is unlikely that these types of issue will be solved by means of judicial decisions. The environment is too strictly intertwined with development and the economy, and the solution to problems needs to be, first and foremost, political. Perhaps the Italian case may appear as the worst, but there are many other similar cases in the rest of Europe, all concealed by the attitude of EU, which chooses to minimize or ignore the most dangerous risks.

In Chapter 11, Antonello Petrillo describes and analyses the case study of a natural disaster, which is subsequently used to turn the rule of law into corrupt governance, thus favouring private companies. In fact, this anamorphosis of the rule of law becomes very profitable both for the political powers and for the development of business. This generates a new local ruling class that manages to conquer a significant position, even on a national level, through friendship circles that include politicians, businessmen and even police officers, members of the armed forces and the intelligence services as well as the national bureaucracy.

# PART I
# The Change of Security Governance Since 1990

Chapter 2

# The Rise of Postmodern Security Governance and the Proliferation of Ignored Insecurities and Their Victims[1,2]

Salvatore Palidda

## Introduction

Since 1990, but particularly after 11 September 2001, the concept and practices of security governance in all countries have made the fight against insecurity- understood as street or urban delinquency, the presence of the Roma (or 'gypsies'), irregular immigration, terrorism, and, only sometimes, the mafias- a priority.[3] The popular legitimacy of this approach has been significant thanks to a form of 'mass distraction' instigated by 'governance through fear' and the continual alarms it has raised (crime, terrorism, pandemics and so on). During this period, much of the media, as well as TV and film productions, have made substantial contributions to the success of this approach (Furedi, 1997, 1999, 2002, 2003, 2005, 2013; Altheide, 2003; Maneri, 2010, 2011).

The public and private resources allocated to this securitarian governance (but not in the case of more serious prevention) have increased considerably. As

1 English revised by Deborah Kay.

2 This text is also the product of my first contribution to the PRIN (DM 1152/ric's 27/12/2011), Italian National Project 2013–2016: 'The professions of public space beyond the crisis'.

3 The 'securitarian posture' prevailed firstly in the United States in the 1980s, then in the 12 EU countries but also in Japan, and later among the middle and upper classes of all cities worldwide (gated communities, video surveillance, the growth of police forces and private security guards, the universal use of metal detectors and so on). Our first European research on racial criminalization started in 1994 (our first publication on the subject by Palidda (ed. 1996) with many contributors, among them Sayad and Brion). The last publication by Palidda (ed. 2011) with contributors from the main European countries and also United States, published also in Italian, Spanish and French. The literature in this field is abundant; among others, see Furedi, quoted; Zedner (2003, 2007, 2010, 2012); Simon et al. (2007); Cartuyvels and Mary (eds. 1999); Christie (1993); Palidda (2000): *Conflitti globali* (2007); Collective (2006); Fassin and Fassin (eds. 2006); Bigo, Bonelli and Deltombe (eds. 2008); Dal Lago and Palidda (eds. 2010); Mucchielli (ed. 2008) and others quoted in the various chapters of this book.

some authors have understood so well, security has become the business of the twenty-first century (Glassner, 1999). The increasingly rapid development and application of new technologies, of knowledge and expertise, of law enforcement, of devices and equipment for use by the police and the military forces (both public and private) have reached unimaginable levels. This was especially true between 1960 and 1980, decades of relative progress, and social peace and welfare in the G7 countries, after the collapse of the Soviet bloc.[4] The pervasiveness of security research has been extended to all fields, first and foremost in terms of devices, techniques, technologies and 'postmodern' professionalism. Therefore, in the view of mainstream opinion, today's world is much safer than it was decades ago (with the exception of those moments of terrorism or criminal acts that are so useful for relaunching the security ideology, permanent wars and related business[5]).

Unfortunately, as we shall try to show in the following pages, the large-scale development of public and private security governance, at local, national and international levels, has not reduced our 'ignored insecurities'. These are: the increase in the health and environmental risks created by disasters, the shadow economies linked to undeclared work, neo-slavery, tax evasion and the interrelationship between licit, semi-legal and also criminal practices, accidents at work and occupational diseases. Therefore the victims of these insecurities are likely to be ignored even though they make up the majority of the population in almost all the EU countries, as well as the rest of the world.

The EU and its founder states have repeatedly maintained that they give their utmost attention to all insecurities. However, these good intentions have been translated into research programmes and into a distribution of financial resources that have primarily promoted insecurities attributed to crime, to illegal immigration and terrorism; furthermore, some lavishly funded projects are almost entirely biased towards a more technical point of view, so that they benefit lobbies for new technologies and for those private enterprises specializing in various fields. In fact, the prevailing orientation always confirms a protectionist and prohibitionist posture, that is, a perspective of protection of a European citizenship and *res publica* quite opposed to the universal aims and rights, and a prohibition

4 The rise of the 'securitarian posture' coincides with the increase in contemporary wars, for example in the Balkans, and then in the permanent wars and the many atrocities all over the world.

5 According to some authors, economic development after the First World War seems more biased by the military-police affairs lobby and thus inevitably by its interest in reproducing wars. The dominant powers (United States and NATO versus USSR) externalized the wars in Korea, Indonesia, Central and South America, Africa, the Middle East and provoked the permanent war between Israel and the Palestinian people. Now there is a new relaunching of this reproduction by the stronger lobbies in the US and NATO (North Atlantic Treaty Organisation), by Putin and by all criminal arms dealers. The internal continuum of these wars in the wealthier countries is the racist criminalization of immigrants, marginal people and protesters (among others see Hobsbawm (2007, 2008); Flores (2005); Joxe (2002, 2012).

of migrations that feeds the reproduction of the irregularity and also continuous hundreds of deaths (see below).

What sort of assessment and analysis can be made of security governance, also in terms of costs and benefits, from the point of view of the *res publica* (that is, the majority of the population)? Since 1990, has security regarding public health, the environment and fundamental rights improved? What has been done to prevent disasters and their economic and social consequences, and the growth of poverty and marginalization? What remediation for risky situations and victimization has been achieved?

In order to try and answer these questions, I begin with the concept of security that has been adopted by the EU governments, and go on to describe the practices involved in its application, and the results.[6] When looking at the results it is important to analyse the effects of the imbalance, that is, the non-articulation between: the protection of victims/social, health and environmental prevention and protection (by the various prevention and control agencies and also by the police forces)/ sanctions and penalties/recovery and rehabilitation.[7] I go on to analyse the ignored insecurities and their victims, that is: a) the shadow economies (moonlighting, illegal work, neo-slavery, tax fraud and so on); b) the reproduction of irregular migration; c) accidents at work and occupational diseases; d) serious damage to public health and increased cancer mortality rates; f) environmental disasters, often connected to the lack of preventive measures in industrial activities, and to waste treatment, caused by the failure to reclaim sites after devastation caused by speculation, by public works and by contamination from military sites. I will focus on the fact that the vast majority of victims of such insecurities do not receive any

6 My approach is first to deconstruct the dominant *discourse* on security by referring in particular to Foucault. See also Boltanski (2008, 2012); Dillon and Neal (eds. 2009); Dal Lago and Palidda (eds. 2010).

7 This articulation should be considered the main foundation of a government's security in a democratic rule of law. It is interesting to note that at the beginning of the 1970s, regarding changes in the police forces of several democratic countries, it was observed that the activities of these institutions no longer involved only repression but also prevention; of course during those years, prevention was accompanied by the cooperation of the social services. Since 1990 police prevention has become 'proactive' and has also led to the social services taking on a more auxiliary role. Postmodern proactivity has also led to the myth of a 'Minority Report' style of policing and also to the programmes financed by the Seventh Framework Programme FP7 – including INDECT; see Knaebel (2012). About 'postmodern controls' see Ball, Haggerty and Lyon (eds. 2012); Deleuze (1992); Dubber and Valverde (eds. 2006); Lyon (1994, 2001); Marx (2002); Norris (2009); *Conflitti globali* (2007). The obsessive introduction of new technologies is also frequent in prevention devices in all sectors where the protection of health and the environment is necessary. But frequently this innovation is a scam: the postmodern devices are not put to good use or do not function, in particular in the industries because the employer doesn't want stop production (Palidda, ed. 2009).

protection, due also to corruption,[8] abuse and violence on the part of the stronger actors and some of the staff within control agencies, including the police forces,[9] which is a much neglected phenomenon.[10] I would like to thank the reader for having the patience to read the numerous references to other researches and also to video-reports concerning the aspects analysed here.

## Security and Safety in Accordance with the European Union

Horizon 2020, the biggest EU Research and Innovation Programme, has been provided with nearly €80 billion of funding, made available over seven years (2014 to 2020), in addition to the private investments that it will attract. It promises more breakthroughs, discoveries and world-firsts by taking great ideas from the research laboratories to the market (European Commission, 2014). Among the 'societal challenges' of this programme, are:

Secure societies – protecting the freedom and security of Europe and its citizens. This challenge is about undertaking the research and innovation activities needed to protect our citizens, society and economy as well as our infrastructures and services, our prosperity, political stability and wellbeing. The primary aims of the Secure Societies Challenge are:

- to enhance the resilience of our society against natural and man-made disasters, ranging from the development of new crisis management tools to communication interoperability, and to develop novel solutions for the protection of critical infrastructure;
- to fight crime and terrorism, ranging from new forensic tools to protection against explosives;
- to improve border security, ranging from improved maritime border protection to supply chain security and to support the Union's external security policies including conflict prevention and peace building;

8 About corruption and his effects in the different countries, see the important Report of Institute for Economics & Peace (2015).

9 It is emblematic that no data exists in any country on deviancy and crimes among police forces employees. Thanks to confidential data received by a high-ranking officer, I calculate that in Italy the rate of this deviancy is nine time more than that among the male people aged 18–65 years and the rate of deviancy among the politicians is even more (30 times); see Palidda (2012).

10 Here I am referring especially to the victimization surveys that only use samples of people with a landline phone; this excludes not only the nomads, the majority of regular and irregular immigrants, but also a large number of nationals that no longer use a landline phone; this section of the population is the most at risk of being a victim of crime, abuse, violence and so on. In addition, these surveys never administer questions concerning offences committed by public officials at local or national level, including members of the police.

- to provide enhanced cyber-security, ranging from secure information sharing to new assurance models.

To explain these 'primary aims' the document adds some self-evident phrases (such as 'securing society against disasters is one of the central elements in the functioning of any society') but it also introduces the term 'resilience' that seems to have become the latest word in the new think tanks' vocabulary. This 'concept' appears somewhat ambiguous because, in fact, for some authors it reduces the approach to psychological practices, both individual and group: it is the usual trick that is used to remove the political value of this huge problem (see Chapter 5). It goes without saying that in no European document transpires the doubt that the safety and the future of humanity may be a field antithetical to the choices and priorities of the dominant lobbies within the EU, which include military, securitarian controls, oil, nuclear, speculative use of new technologies and so on. Indeed, immediately after, the document states:

> Fighting crime and terrorism requires new technologies and capabilities for fighting and preventing crime (including cyber-crime), illegal trafficking and terrorism and cyber-terrorism, including understanding and tackling terrorist ideas and beliefs to also avoid aviation-related threats. The protection of the European borders requires the development of systems, equipment, tools, processes, and methods for rapid identification. This includes supply chain security in the context of the EU's customs policy. ... On Digital Security, this Challenge focuses on increasing the security of current applications, services and infrastructures by integrating state-of-the-art security solutions or processes, supporting the creation of lead markets & market incentives in Europe, following an end-user driven approach, including, for instance, law enforcement agencies, first responders, operators of critical infrastructures, ICT service providers, ICT manufacturers, market operators and citizens. This Challenge should bring together all security stakeholders: industry – including SMEs, research organisations, universities, as well as public authorities, non-governmental organisations and public and private organisations in the security domain. The active involvement of end-users is of high importance (European Commission 2010a).

So, the 'Societal Challenges – Secure Societies – Protecting Freedom and Security of Europe and Its Citizens' declared by Horizon 2020, confirm the section on security already proposed in the Seventh Framework Programme for Research (FP7 of the period 2007–2013).[11] On closer inspection, the intrusiveness of discourse concerning military, policing and postmodern controls in society is much more pervasive, thorough and 'microphysical' (as Foucault might have said). FP7 had the same strategic objectives as Horizon 2020, and the same suggestions regarding

11 See European Commission FP7 (2009a and 2009b).

applicants; in particular, suitable partnership participants include: universities or private research institutes, companies, small or medium-sized enterprises (SMEs), SME associations or groupings, public or governmental administration (local, regional or national), institutions running research infrastructures of transnational interest, international organizations and civil society organizations. So, in many of the already funded projects, we can find ministries of defence, of the interior, national or local police forces, public and private military industries (among these, the producers of an increasing number of devices for postmodern control and also many gadgets useful for the security business, but not for the prevention and contrast of the ignored insecurities and the protection of their victims).[12] Among these gadgets, the last is Women's Security App for smartphones,[13] a new civil application of what has been experimented with in the military-police field in the name of the defence of women, insinuating, obviously, against the 'bloody barbarism of terrorisms generated among peoples uncivilized', a product among many others proposed in the securitization market of the high and middle classes of all countries.[14]

Apart from the – *de facto* – abolition of any support for basic research, and in general to an effectively independent research, inevitably, as was the case with the FP7, it seems most likely that Horizon 2020 will also fund only projects that affect the practices of a security governance in continuity with what we have known since 1990 up to the present. We are sorry to say that this is a programme that will produce the recrudescence of ignored insecurities, and so create a real risk for the *res publica* of a very democratic rule of law.

## Regarding Postmodern Global Security and its Hypocrisy or Complicity with Those Responsible for the Greatest Menaces to Humanity

According to some authors on 'postmodern' security thinking, that is, 'global security and the new menaces', today, the concept of security must take many new aspects into account, some already known and others that are new.[15] Above all, states and their military forces are no longer the only actors; these authors go on to talk about the various menaces, proposing a sort of hierarchy; war in space, cyber-warfare, arms of mass destruction, the bankruptcy of some states, terrorism, insurrections, organized crime, the privatization of policing and the rise of

12 A critical analysis of the part of EC FP7 concerning security is proposed in p.7–11 of Dal Lago and Palidda (eds. 2010).

13 See https://play.google.com/store/apps/details?id=com.zayaninfotech.security, accessed on 4 September, 2015.

14 See https://www.google.it/?gfe_rd=cr&ei=iEVLVY_1LKKO8Qeir4CwDg#q=women+security, accessed on 4 September, 2015.

15 Florini and Simmons (1998); Baylis and Smith (eds. 1997); Buzan and Hansen (2009).

contractor companies, and civil wars. With great nonchalance, these authors state that while in the past it was the strongest states that posed a threat to international security, now the principal dangers come from the smaller states and the violent organizations centred there (Andreatta, 2014). In other words, the biggest powers of the world are not responsible for the proliferation of war thanks to the arms they produce and sell, also illegally. The proof of this is blatantly exposed by these states' absurd and criminal squandering of their budgets, first and foremost by the United States – during the period of economic crisis – when Congress declared that it could not pay for the health cover for 50 million citizens without income:

> '$40-Billion Missile Defence System Proves Unreliable and The Pentagon's $ 10-billion bet gone bad' (Willman, 2015); another 'scandal' and 'fiasco' concerns the F35 Joint Strike Fighter Program, evaluated by many qualified experts of the same US Defence as unreliable weapon system and a senseless financial waste.[16] According to Quigley (2010), in this year the US spends 708 billion dollars on war and another $125 billion for Veterans Affairs – over $830 billion (see also SIPRI, 2015).

However, frequently, in international, national and local institutions, the various departments and the employees responsible for health and environmental problems do meticulous and relevant work. But they are also relegated to marginal positions, too less important roles, with little or no power regarding decisional spheres or towards public opinion (except the episodic echoes thanks to some ecologist journalists). It is emblematic that to this great expertise in monitoring health and environmental disasters and risks (Schipper and Pelling, 2006), there is no equivalent programme for the reclamation and creation of an efficient framework for prevention. As different research has reported, there is a certain amount of perplexity over some NGOs, but, even more concerning several of the larger institutions such as FAO and UNICEF, UNEP, UN-Habitat and so on (whose officers receive high wages and plenty of benefits).[17] This is proof that there is little or no experience in the fields of reclamation and prevention, due to the (criminal) negligence of all governments, and especially those who have consistently refused to sign the Kyoto Protocol agreements and other urgent measures to protect the world. In December 2015, the world summit for the fight against climate deregulation (Cop 21) took place in Paris. The target of reducing emissions so as to slow the greenhouse effect, which has been proposed by the states involved, aims to halt the rise in global temperature by 2°C, beyond which threshold our climate would be permanently damaged by chain reactions. The disastrous consequences of the climate threat are obviously much more serious

16 Among many articles and also the Rand Corporation Reports, see: Bender, Rosen and Gould (2015); Engineer Sprey (2012); McGarry (2014).

17 For a theoretical and methodological approach of the criticism of the institutions see Boltanski (2008); Lecompere, 2014; Combes, 2014; Pech and Padis (2004).

than the worst-case terrorist or mafia threats (there are many more deaths due to cancer and other diseases caused by pollution and environmental disasters). But even the highest authorities concerned with environmental problems have a rather subordinate attitude: they know from the outset that the threats to health and the environment will never be taken seriously … it is a well-known fact that the major powers responsible for these threats have refused to reduce their CO2 emissions. In reality, it is the discourse about the dangers of climate change that is biased if not actually wrong. It is also clear that greenhouse gas emissions are directly associated with many industrial productions and activities that are dangerous to human life: arms trade and production, nuclear power, oil production and the abuse of its derivatives, agricultural production based on the use of poisonous chemical substances and fertilizers and so on.

EU policies and research into security are not very different from those adopted by the United States and, in general, other NATO countries, in particular after the Revolution in Military Affairs (RMA), and the rise of 'governing through fear' and the 'crime deal', the business of 'postmodern controls' and so on. In the name of 'protecting Europe's citizens and society from harm, while enabling its economy to recover from man-made or natural disasters' (European Commission, 2010a, 2010b and 2012a), research and financial resources are increasingly allocated under the general field of security programmes; there has been a progressive process of militarization and policing of everything that concerns protection, safety and the 'securization' of the EU countries (always in the name of the citizens' 'happiness').[18]

The most obvious examples of this process can be observed during health or environmental disaster management. In these cases, civil protection is more often than not under the direction of military or police chiefs and becomes a military-policing operation (similar to keeping or enforcing 'peace'). As we have seen in the wake of Hurricane Katrina (Johnson, ed. 2011; Whitehead, 2005), or after earthquakes in Italy and in many other cases, the military-policing style of management authorized is at the sole discretion of this type of governance, and is the ideal type of management from a neoliberal point of view (see Chapter 12). This sort of management is always given the special status of emergency and is therefore beyond any control. This means that in a democratic rule of law it is possible to pass to a state of emergency or exception whenever and however the powers that be wish to, and without a change in the rule of law.[19] So, for example, if the local or national government wants to carry out major public works

18 Among other facts about the processes of militarization see the creation of EUROGENDFOR now enshrined with the Velsen treaty: Statewatch (2007).

19 Therefore we can say that there is always coexistence between exceptionalism and democratic rules of law, as authoritarian and peaceful practices. For a critique of the lecture proposed by Agamben on state of exception see Guareschi and Dal Lago in Dal Lago and Palidda (eds. 2010).

without encountering any obstacles (such as protests or fiscal controls by the court, and to the benefit of certain lobbies) they declare a 'state of emergency' for this work.[20] The same kind of rationale has imposed an economic and fiscal policy in the EU that has produced a serious increase in unemployment and poverty, in the name of fighting the crisis (Gallino, 2011). Thus, not only have those responsible for the financial collapse paid nothing, but enormous amounts of money have been distributed to the banks, and the rich have grown wealthier.[21] Can this be seen as good practice for governance that aims 'to protect Europe's citizens and society from harm, while enabling its economy to recover from man-made or natural disasters'?

According to Stenson and Sullivan (2001):

> For more than three centuries, or since John Locke, liberal states have practiced a schizophrenic form of criminal justice. One form of such justice is for the rich, enlightened, and occasionally blessed. Quite another is for the 'noxious' and 'degenerate' underclass that have declared themselves quit of the principles of human nature.

But, until the beginning of the last neoliberal revolution (in 1970, and especially after 1990) there has been a penalization linked to the growth of the role of the state, both with the least authoritarian regimes and the absolutist or dictatorial ones.

As Kazgandjian illustrates in Chapter 6 about 'Decriminalization in Business Law in Neoliberal France', a case that is common to all the most important democratic countries, there is a legalization of 'tolerable illegality' versus stiffer penalties for lesser crimes, in particular after 1990.

The neoliberal frame is legitimized by the discourse summed up in the famous phrase 'less state, more market', and therefore more freedom of action for the economic actors (and of course for the strongest). The evolution of business laws is more or less the same – sometimes worse – not only in Italy under the Berlusconi Governments but also under those of the 'centre-left' or 'national unity'. The same could be said for all major democratic countries such as the United States since Reagan and Bush Jr., and the United Kingdom since Thatcher (Teachout, 2014; Purdy, 2015).

The current economic crisis is just one detail of the most serious consequences of this process. The Obama administration gives the impression of wanting to stop or limit the damage, especially after the Madoff scandal, subprime and so

---

20 In Italy the Berlusconi Government adopted this stratagem for the preparation of the G8 in 2009 (see also Chapter 12); there are also some other similar cases in the past and in recent periods in many other countries; for example, the exception adopted for Katrina disaster management (see United States. Dept. of Homeland Security. Office of Inspector General, 2006).

21 Among others, see Di Palma (2013).

on. But we can note that the so-called output measures of the crisis have almost always tried to save the banks and the financial companies responsible for the crisis, while increasing the impoverishment of the planet's population. The proof can be seen in the constant rise in the distance between rich and poor and the growing freedom of action for financial speculators, a freedom that is possible in all countries thanks to transnational games (with tax heavens even within the European space). Over the last few years, concern about the economic and social malaise of a large part of the population, and the consequent risk of riots, has made all governments promise to pay more attention to social problems. But, as the official statistics show, expenditure for military and police affairs continues to remain high (see SIPRI, 2015) at a moment when so-called problems of criminality are dwindling. All democratic countries have seen a decrease in the number of serious crimes such as homicides and robberies. However, in some courts, serious punishments have been handed out for lesser crimes, such the case of a homeless man, with no police record, being punished for the attempted theft of four euro. It is important to note that this ruling was adopted by the appeal court, despite the fact that the prosecutor requested an acquittal. The defence lawyers of the supermarket insisted that a punishment was administered for the petty theft and were pleased when the court administered an exemplary condemnation of the accused, even if he had no criminal past. It is implied that this will prevent the increase in thefts by the homeless, a habitual and classic penal solution to a social problem (Persano, 2015). So, police and courts spend their time on such 'serious crimes'. At the same time, crimes of corruption are on the increase but punishment for these types of offence is rare and generally light.

The real legitimacy for the expenditure on military and police affairs lies in the permanent war against pseudo-Islamic terrorism. As some authors suggest (see Joxe, 2015, Mini, 2015) after al-Qaeda, IS or Daesh, the so-called islamic state is the ideal enemy for the revival of the permanent postmodern war; it is well-known that this terrorism is bankrolled by the Emirates,[22] who are flattered allies of the NATO countries. In the face of these and many other facts, it is hard not to think that the revival of the war against terrorism is a new and formidable action of mass distraction that is very useful for diverting our attention away from the ignored insecurities aggravated by the financial crisis (that has not been caused by its victims).

---

22 Chivers (2014). At first, ISIS armed itself with weapons seized from the battlefield in Iraq and Syria. The group also reports that it received funding from wealthy individuals in Saudi Arabia, Kuwait, Turkey and Qatar and then used the money to buy arms on the black market. According to a December 2013 study by the Brookings Institution, donations are funnelled through Kuwait and make their way to IS groups. See also Harling (2014). Former Iraqi prime minister Nouri al-Maliki has publicly accused Qatar and Saudi Arabia of supporting IS (ISIS Explained, 2014). Some other sources show that the weapons and funds of ISIS come from the Emirates, allies of the United States and NATO countries.

In an article entitled 'The curious case of the fall in crime' (20 July 2013) *The Economist* (accessed 05/12/2015), and after that other journals in 2014, have discovered that since 1995 there has been an evident decrease in the trends in criminality. Murders, aggravated assault and damage to property (car theft, burglary, robbery) are clearly falling: all *per capita* crime rates have undergone a very significant drop in all the so-called developed Western countries: in North America and throughout Northern, Western and Southern Europe (Italy has one of the lowest homicide rates in the world). Instead, since 1990, and even today, the media have consistently raised alarms over rising crime rates by giving voice to certain criminologists and other 'scientists' who are advocates of zero tolerance. In fact, as some cases show, the statistics are often distorted; but these trends cannot be explained as mechanically determined by the growth of poverty or, conversely, to the increase in wealth, much less to increased immigration (a trivial arithmetic calculation shows that this corresponds to a reduction in crimes) (see infra).

## What is the Meaning of Security Governance in a Democratic Rule of Law?

Security has always been a total political fact (see Introduction, footnote 5), crucial for ensuring the reproduction of society, but sometimes also responsible for its destruction. Security concerns devices, tools and human resources devoted to the 'defence-defensive'[23] towards an external enemy, and the sole and unquestionable protection of collective goods, the safety of human beings, and therefore the protection of the *res publica* as well as the integrity of the political entity (state). However, security can lead to war and consequently to the destruction of much of the same population and property when it becomes a strategy and a tactic subordinate only to the interests of those in authority and the dominant warmongers. In other words, the concept of security and its application are a crucial aspect for any government of society, and critical and decisive for its future: the first concern of the ruling class.

The changes in the concept and in the practice of security governance are always dictated by the balance of power between the main actors. As noted by some authors (Delumeau, 1978/1991 and 1989; Robin, 2004), from Hobbes to the police theorists and even the democratic thinkers of the twentieth century, security concerns the protection of the integrity of the state and the protection of the lives and property of citizens. In fact, security almost always becomes the protection practice of the more powerful and dominant actors. At times, this goes

23 The idea of 'defence-defensive' is proposed by Alain Joxe; it corresponds also to the JFC Fuller (1945) reflexion: when the arms system produces the superiority of offensive capacities over defensive ones, it favours the constitution of great states; on the contrary, the superiority of defence over offensive capacities leads to political parcellization, that is, to the formation of small states. Then, the 'defensive-defence' excludes arms for attack and does not lead to war, except for pure defence purposes. Joxe (1991; 2012).

against the majority of the population, especially when the 'laborious classes' become the 'dangerous classes' and are seen as subversives, as occurred in 1848 in many European countries (Chevalier, 1958/1984; Senellart, 2007). Well beyond the auspices of the early police and 'pastoral governmentality' theorists (Foucault, 2004b), security governance practices are not concerned with what, late on in the Second World War, became known as 'social security', 'health care' or environmental protection or safety. Thanks to the democratic achievements of the 1960s and 1970s, in the richer countries, attention to safety was extended to all sectors of society. In Europe, until the end of post-Second World War reconstruction and even during the economic boom of the 1960s, prevention and safety in the workplace, in schools, in health, in housing and in urban settings, in transport, in the various infrastructures, in the environment and also in other areas, was neglected. The student protest of 1966–1968, and later the workers' protest, had no ecological sensibility. Even now the unions accept compensation in exchange for risks from disease or accidents. The same can be said for 'social prevention', in particular the work of street educators and social workers, who work with marginal people and more especially with juvenile deviancy or delinquency. As some of the good professionals in this field say: 'The culture of prevention has been understood by few people and rather rarely practised.'[24]

Awareness to these issues, which are of vital importance to the future of humanity itself, has been very rare or only episodic. The evolution of the various laws and regulations in these areas shows that it was not until the 1970s that a kind of 'pastoral governance' began to develop, one that took care of all the aspects mentioned above, in particular social policies and the welfare state (in the US in the 1960s with Johnson, and after in Europe). But it should also be remembered that while a part of the labour movement and public opinion claim greater protection, another part is favourable towards the monetization of risks concerning health, economic and urban development or major public works, even when they are increasingly the cause of damage that will continue to get worse because there is no prevention or recovery.

The pervasiveness of promises of wellbeing and happiness, of progress at all costs, has generated a huge consensus. This is particularly so among the population that does not realize that it is most likely to be the first victim. The prevalence of consumerism leads many people to an abusive use of gadgets and products that are often disposable and even harmful. But, frequently, the waste treatment industry is part of the same financial group involved in the commercial distribution and packaging sector, and who are often linked to city mayors.

24 Thanks to Calabresi for explaining to me the importance of this question on the basis of his long experience, and thanks also to other friends who work in the difficult field of social prevention, of police prevention and also in the increasingly rare field of social reclamation, especially after the progressive erosion of the welfare state over the last 20 years.

As Calvino had imagined in his novel *Invisible Cities*, published in 1972, the city and practically the entire planet are in danger of being overwhelmed by toxic and non-toxic waste. The American satirical science fiction comedy film *Idiocracy* (2006) may be considered a caricature of the dystopian society towards the risks of today: the triumph of idiocy nourished by all the bad and ignoble aspects partially described, unfortunately, in this book. As some reviewers write, the movie is a 'pitch-black, bleakly hilarious vision of an American future' … (not only American!) 'Like so much superior science fiction, *Idiocracy* uses a fantastical future to comment on a present.' Well beyond the comical imaginary, the concrete reality has already landed tragically devastating aspects producing more and more victims.

Expenditure for waste disposal has grown exponentially, and illegal disposal is often required, becoming one of the mafia's most profitable activities. According to some estimates (such as by the United Nations Office on Drugs and Crime – UNODC), most of the profits made by contemporary mafias are linked to the trafficking of money (recycling), drugs, weapons, human beings and toxic waste (see below).

## From Illusions of Democratic Progress to False Security after 1990

Some authors, including Egon Bittner (1990), who is considered one of the fathers of the sociology of policing, thought that social peace had been achieved in the 1970s; this led them to predict that an armed branch of the state with a 'monopoly of legitimate violence' would not be necessary in the future. This belief stemmed from the evolutionist view of 'democratic civilization', which was progressively victorious (and was inspired, sometimes arbitrarily, by Norbert Elias). In fact, during the 1970s and 1980s, in the more important, wealthier Western countries, there was a significant development in social work, welfare and also a weakening of repression and penalties (for example, the development of social policies, until 1990, and therefore the welfare system and other achievements concerning civil and labour rights, the reduction in the number of prisoners, the relative democratization of the police, the judiciary and so on).[25] The Western European nations, North America and Japan, have managed to maintain peace within their national territories despite their direct or indirect involvement in the wars that they have always reproduced or fostered elsewhere.[26]

25 See Quassoli (1999); Palidda (2000; ed. 2011).

26 Before and after the end of colonialism, one of the main aspects of the delocalization of the affairs of dominant countries was the export of arms, which is more and more linked to all other sectors, and then the direct or indirect participation in all the wars around the world; see Joxe (2002; 2012); Palidda (2007).

However, after a period of development between the 1960s and 1970s, the last 'great transformation' has caused the progressive demise of industrial society as it was known in the nineteenth century up to the 1970s. The consequences of this were understood only partially or very late on by the political and social sciences as well as by the political powers.[27] Since then, the economic, social, cultural and political disruption has been just as shocking as the industrialization of the nineteenth and twenty centuries. And, as we try to show in this book, we are still unable to see, at the present time, how it might be possible to organize society so that it becomes a new, 'stable, peaceful and prosperous' structure, as the liberal–democratic thinkers such as Schumpeter, Keynes, Polanyi, Mills, Galbraith, Elias, Robert Castel, Bourdieu, Bauman, Beck, Wallerstein, Sassen and others hoped it would become.

After the triumph of the neoliberal turn (in 1979 and 1981 with Thatcher and Reagan) and the conversion of the left in the same direction, a short time after Mitterrand goes to power and even more so under Blair, almost all countries became dominated by the neoliberal orientation.[28] In this way neoliberalism wins, weakening the rule of law in favour of the private interests of the dominant classes, and leading to many cases of corruption and vulgar illegalities in all countries.[29] The idea of the *res publica* seems to have become almost unheard of, the conservative think tanks are more efficient than the intellectuals of the left who appear obsolete or fascinated by neoliberal thinking, and the pervasiveness of liberal discourse has been highly successful; it appears to be the most effective solution for governing the new great transformation. However, beyond the liberal facade of Thatcher and

27 Among others Castel, Bourdieu and Bauman. Their analysis proposes a kind of reconstruction of the welfare state, of sociability (or community according to Bauman, who tends toward a sort of theory of illusions and some 'hydraulic' theorems), that is, the salient features of an economic, social and political structure that may no longer exist, because postmodern society has undergone a continuous and heterogeneous segmentation, a permanent destructuration; moreover there is the production of an almost perpetual anomie. Here, then, is the first cause of so-called insecurity (in the polysemic sense): in every segment or location of society the actors with the abilities to manage conflicts or simple quarrels have been weakened or have disappeared and this relationship with the social control agencies is missing: and after consequently we seldom find actors able to understand and to govern the new situation; in the past there were unions, popular associations, parishes, sports clubs and so on (see Palidda, 2000).

28 The criticism to neoliberalism started with Foucault in his 1978–1979 lectures on 'biopolitics'. Bourdieu talked of 'neoliberal doxa' and adopted an almost 'militant' and rigourous approach (but he never quoted Foucault). After this criticism is developed by many famous authors of today (partially quoted in our bibliography). Bourdieu and Wacquant (1998).

29 See the cases of corruption and different crimes committed by various important leaders, such as Craxi, Mitterrand, Gonzalez, Blair, Schroeder and others (in the US, Clinton), not unlike those more numerous cases concerning conservative leaders, in particular relating to almost all the members of the Bush administration, the Christian Democrats and after, Berlusconi in Italy, Pompidou, Giscard d'Estaing, Chirac and so on.

then Blair, some aspects of welfare persist in the UK, and the same can be said of France, Germany and even more so of Scandinavian countries;[30] instead, in the euro-Mediterranean countries, there is only a trace of the welfare system; they went through a long period of Fascism (even if – in Italy – followed by a strong labour movement) and later the rise of mass corruption through clientelism.

In almost all countries, the new security think tanks, supported by a powerful media, have successfully legitimized the drift towards 'more repression, more penalties' (Davis, 1990, 1998; Palidda, ed. 1996; Altheide, 2003; Simon, 2007; Maneri, 2010, 2011). This approach has triggered a strong and sustained reduction in social prevention (in the general erosion of the welfare state) while less and less attention is paid towards the insecurities that affect the majority of the population. Social workers and officers from the various agencies for prevention and control have been isolated, rendered helpless and forced to undergo cuts in personnel (for example, labour inspections) and in some sectors, such as health and social services, (see, for example, juvenile delinquency or drug addiction) privatization has been favoured through outsourcing, which is often disguised as voluntary work or interventions by NGOs.[31]

Among the ranks of the police forces, the normal practices of prevention and 'peaceful disorder management' have often been denigrated as being outdated and inappropriate, especially in producing results (that is, growth in the numbers of arrests and so on).[32] The cult of the 'postmodern' and the obsession with productivity, measurement and evaluation invades all sectors. Inevitably, this sort of adaptation or modernization often produces misleading failures and even disasters. As we will see later, in the field of control and the prevention of health risks and environmental disasters, the neoliberal turn of control agencies and policing has produced serious material consequences and many victims.

Over the last few decades there have been continuous repetitions of the riots in the French banlieues, riots in the UK, in the United States and elsewhere, the reproduction of so-called street crime or so-called 'urban incivility', but also the weakening of investigative activities, an inability to manage the 'rules of disorder' – passing from operations that seem more military than police in style, until there is chaos in this field and petty violence against nomads, migrants and protesters (see also against farmers, environmentalists, workers, students, in addition to the 'massacre' at the 2001 G8 summit in Genoa, and in other countries, and also the return of police murders of black people in the United States in 2014–2015). The frequent slip to violent management of social disorder and more into racial

30 I allude to the minimum wage, the various forms of direct and indirect aid to young people and families, direct and indirect aid to drug addicts, the disabled, the mentally ill, and so on.

31 Among others, see: Carbonnier (2015); Pech and Padis (2004); Lashaw (2013); Fassin (2009, 2011a).

32 Among others, see the works of J. Simon (quoted in bibliography) and the contributions into Mucchielli (ed. 2010) and into Palidda (ed. 1996, ed. 2011).

criminalization of the marginal and immigrants seems the evident consequence of the neoliberal turn in police forces; a turn powered by a militarization of these forces and a strong diffusion of a mainstream discourse that mixes racism, zero tolerance and neo-fascism. A seemingly innocuous example is in the meaning of the message that passes through apparently banal advertising: a man goes in a SUV and says: 'I can finally look everything and everyone from top to bottom.' Clearly, this advertising doesn't serve to sell the SUV but to send the message that only those who have the possibility and capacity to own a SUV may have superiority over the people who do not.

Also, ordinary police officers know that crime statistics and the way they are used are often a sham; the political entrepreneurs and other actors involved falsify or interpret the statistics arbitrarily, depending on whether they are in opposition to or in support of the government (local or national). When they need more resources for the police and when they want to gain easy consensus in the name of the 'war for security' they increase the crime data; alternatively, in order to show that their activity has created a fall in crime, they reduce the data. In some cities, police leaders have encouraged local citizens to cooperate by making complaints and reporting suspicious events and people. This has created an abnormal increase in the number of calls and letters to the police and to local authorities, often concerning petty squabbles between neighbours or false accusations and so on.[33] There are also 'political forgeries'; among others, we should remember the false statistics published by the New York police at the beginning of Giuliani and Bloomberg's 'zero tolerance' period, a deception that was discovered thanks to the investigation of two criminologists into 20 years of police activity (Rashbaum, 2010). The same sort of deception was discovered in France.[34]

As suggested by Kitsuse and Cicourel (1963), the statistics on various crimes are nothing more than a measure of the productivity of the police forces, that is, what they want or are able to detect and pursue. It is a well-known fact that variations of these statistics can be manipulated.

If you assign police to a neighbourhood, you certainly obtain an increase in the number of complaints and crimes reported; on the contrary, if there are no police, fewer crimes are reported and then calculated. In other words, the police only see crime when they want to and when they can. Firstly, they measure their production, to justify its existence, for the 'needs' of the authorities and to satisfy the demand of the inhabitants, who, from a police point of view, deserve to be

33 It is important to observe that after 1990 there was an increase in the research into 'calling the police'. When this research is not embedded within the police authorities, it reveals that, in democratic countries, police forces often show greater care towards citizens for a better legitimization of their securitarian orientations; see in particular the large bibliography proposed by Brodeur (2003).

34 One report reveals an organized system existed under Sarkozy for rigging crime statistics: Tourancheau (2014) and Bordenet (2014).

satisfied. As one policeman says: 'It is easy to build evidence of a crime, especially when we have arrested a thug who has certainly a record.'

It is important to note that the escalation in the numbers of calls made to the police has not fallen with the proliferation of toll-free numbers to call for help or assistance, or to receive advice or help over the phone; in addition to toll-free numbers for health emergencies, the non-emergency numbers have multiplied thanks to the work of hundreds of specific associations providing assistance for minors, women, the disabled and even mistreated animals, along with to the numerous offers by psychologists or religious associations (there are also many videos on YouTube for the promotion of resilience). The years 1990–2010 were marked by a huge increase in the numbers of actors and devices promising an adequate response to insecurities of all kinds: public and private police, psychologists, social workers and even philosophical and sociological counsellors. To paraphrase U. Beck, we can say that we have seen the triumph of a society terrified by insecurity risks and deluded by the idea of total protection (also by the ideology of insurance for any type of risk). This is contrary to the auspices of some classic authors, against a utopia of generalized safety, of a 'universal asepsis or the immunization of the body and the spirit against all uncertainties and all dangers', an 'ideal' strongly criticized by Raymond Aron (Delumeau, 1990).

Wealthy societies have given the illusion of being able to guarantee 'total security' and the insurance companies who exacerbate this are certainly interested in exploiting fears and insecurities; but what insecurities? As noted by Bauman, insecurity has assumed the polysemous meaning of *Unsicherheit* (insecurity, uncertainty and unsafety) (Bauman, 1999). Unfortunately, as we shall see below, this idea concerns only the insecurities useful for the entrepreneurs of a securitarian 'way of life', that is, those intended to emphasize the threat of an alleged crime and therefore its presumed authors. The answer to the demand for security has 'naturally' led to zero tolerance towards the enemy (that is, the convenient or useful enemy); and all discourse on insecurities leads us to ignore the prevention and the many insecurities that are not attributable to crime or the imaginary nightmares fed to us by the media; during this period many movies and television series have provided strong support for this escalation of 'governance through fear'. Within this frame are the periodic alarms aroused by 'epochal epidemics' (avian flu, Ebola, and so on), which assert that they will be the same as the 1918–1920 Spanish flu epidemic; it is also relevant that many in the media have stated that the current economic crisis is worse than that of 1929. As a result, and especially after the 9/11 attack, the exacerbation of fear has become the foremost political game of the postmodern era on a local, national and global level; it is an efficient promotion of the involvement of zealous citizens in postmodern controls[35] and represents a

35 See the journal *Conflitti globali* (2007) and in particular the articles by Marx, Heilmann and Sabesta (some papers by G.T. Marx are available at: http://web.mit.edu/gtmarx/www/garyhome.html, accessed on 6 September, 2015).

boom for those companies offering security measures and devices of all kinds, and above all it is huge business for the insurance companies.

Obviously, the powers always try to show their ability to master disorder and the various problems and emergencies; but, these problems are reproduced as a result of bad governance, or collusion with the interests of the ruling or dominant classes, that is, the first consequences of the neoliberal turn. For this reason, they have the necessity to renew an extraordinary form of mass distraction, thanks to their control of almost all the communication mediums, the opinion makers and also because of the weakness of democratic intellectuals. The alibi created by an alleged disproportionate increase in criminality is always helpful in concealing the responsibility of power for economic and social problems. As history has shown, a society plagued by insecurity (real or fabricated by power and exacerbated by the media) is easily led to granting more power to the police and therefore to repressive and punitive answers (Delumeau, 1978; Robin, 2004).

Even during periods when the official statistics have shown much higher crime rates than today (from 1945 to 1990, armed robberies and murders were much more numerous) there was no exacerbation of alarm. Even during periods of internal terrorism (in the UK, Italy, Spain and other countries) the response by the powers that be did not bring about the general policies and practices that reshape society, that we have known since 1990, and more especially since the 9/11 attack. The explanation that seems most plausible is that this last period can be seen as the frame marked by the rise of domination strategies and practices that aim to 'maximize' the economic and political benefits of repressive and criminal policies. There has also been a formidable and progressive subjugation of the majority of the population, a biopolitical success: the interiorization of the dominant discourse and the consequent erosion of any possible political action. In other words, the population is becoming increasingly unable to understand and participate in the management of disorder and in general of *res publica*. As some workers say:

> It is as if we are drugged by everything, by concerns about consumerism, by fear of losing our jobs, by fear of some misfortune, by not being able to understand our children or even our wives, by being unable to understand anything about the union or about politics, in other words, by our impotence in the face of this world that is crushing us and to which we can only surrender.[36]

Since 1990, there have also been many innovations: postmodern securitarian governance has been able to adopt a softer character; in this way it creates a higher level of popular consensus along with the relevant business that is nourished by it. The population can interiorize the discourse of those in power because it is simple and there are no credible alternatives; as a consequence there has been real participation in 'postmodern' controls, through various devices or gadgets (such

36 From an interview with a worker during research on ignored insecurities in Italy (carried out by the author in 2014).

as the iphone or ipad, meaning that anyone can film a scene or record speech) (Marx, 2006).

The crime that took place between the years 1945–1990 was tacitly tolerated as a kind of collateral cost and was not considered a serious problem, thanks to the strong economic and democratic progress of the time; now, people suffer the many serious ignored insecurities as an obligatory cost imposed by the powers that be for the safety of 'democratic western civilization' (some philosophers justify the sacrifice of liberties for this 'supreme goal', legitimizing the oxymoron of the 'just war', or of the 'authoritarian democracy', both in the past and in the present). Under previous forms of social organization, the management of any disorder or social problems was carried out as 'the least bad' by different social actors: the family, neighbourhood relationships, schools, associations, parishes, unions, political parties, social workers, and even the paternalist policeman. Under the newest forms of social structure (which are unstable, discontinuous, marked by a weak or non-existent social life) the social or institutional figures who are able to mediate have almost disappeared (especially the informal arrangements); therefore there is a clear absence of help to prevent or limit marginality, impoverishment, conflicts, violence and delinquency. The pervasiveness of securitarian mainstream discourse has delegitimized the welfare state and cancelled out almost all hope of paternalism or 'pastoralism' that 'thinks about everything, from the cradle to the grave'. Even the democratic guarantees that were achieved at the end of the 1960s have been eroded, which explains the revival of NGOs that have largely become 'multinationals of the heart' (Pech and Padis, 2004). Aside from the tens of thousands of sincere volunteers who do the work of social workers, the majority of NGO members work as social policy subcontractors following the neoliberal logic, that is, with the priority of making profits for their organizations. It is therefore not a coincidence that criminals have infiltrated these NGOs, and make profits at the expense of the homeless and refugees. A case was discovered in Rome in January 2015, but there have been many other similar cases in other countries, particularly where the NGOs work in partnership with the military and police forces on war missions, often called peace-keeping or peace-enforcing.

As one NGO employee stated:

> For me there is no difference between the work of an NGO officer and an armed contractor working alongside the military or police forces on peacekeeping missions. The risk of being kidnapped, injured or killed is the same. And here, in our rich countries, the refugees we manage are the same people we're dealing with in Afghanistan, Iraq, Kosovo, Libya, Syria etc., i.e. uncivilized people who are unable to live in a modern and democratic country.[37]

37 From an interview with an NGO officer during research on police and ignored insecurities in Rome, Italy (carried out by the author in 2014).

From the US to Europe, security discourse is then connected to an idea of prevention, combined with repression and sanctions;[38] in this way the practice of surveillance and subsequently of punishment prevails (also in the field of occupational safety) to the detriment of prevention and recovery. The productivity of repressive action has therefore been enhanced, and earned glory for Giuliani, the sheriff–mayor of New York, who fostered the business of his zero tolerance and the falsification of statistics on crime realized during his mandates, and 'has made school' among the mayors of many cities around the world.

In France, Sarkozy achieved success by railing against the *racaille* in the banlieues (referring to the youths as 'scum'); in Italy certain judges have protested about the high numbers of illegal arrests carried out by the police, while all governments have paved the way for a prohibitionism towards immigration that feeds racist criminalization.[39] In the UK, Blair did not hesitate to espouse the cause of the 'war on juvenile delinquency' and then argued in favour of the reintroduction of corporal punishment for 'deviant' children (Maccanico, 2011), along with a growing anger against marginal people and immigrants (Bosworth and Guild, 2008), as well as an exponential growth in video-surveillance, despite the fact that numerous experts, even moderate ones, have revealed its very limited effectiveness and that it is an enormous waste of resources (see European Parliament, 2009; Heilmann, 2011). Fassin (2011b) and Rigouste (2012) have shown that, both in the France of Sarkozy and now of Hollande-Valls, a special police unit has been created simply to carry out systematic persecution against 'illegitimate children' in the country. In many cities in Italy, a not-official special squad from the ranks of the municipal police has been created to 'chase the nomads' or 'blacks' (Palidda, 2000, 2015b); some of those

38 This shift has been studied by several authors, among others Garland (2001), but, it seems to me that little attention has been paid to the most important consequences of the destructuration of the various social institutions exercising social control, and therefore the articulation between prevention, repression and recovery/reintegration. In particular, the worst effect of this destructuraction is the erosion of the possibilities and the capacities of public action. This is not an effect of the violent triumph of the neoliberal 'way of life', but a softer success, that is, the penetration of the discourse of the dominant classes into the minds and bodies of the dominated, according to Foucault, biopolitics (unknown or less well known by almost all authors including Garland, Wacquant, and also Fassin and many others). See Palidda (2000 and ed. 2011, 2015b).

39 See Palidda (2000). In Italy the ex-communist leader, D'Alema, became head of government in 1999; he made an important shift towards the securitarian choice, thanks also to the suggestions of Violante (1995); this ex-magistrate and later leader of the ex-communist party, suggested that the fight against street crimes should be given just as much importance as those against terrorism and the mafias. For their part, many left-wing mayors adopted practices similar to those of right-wing mayors. One of the leaders of the ex-communist party is the founder and president of the ICSA Foundation (Intelligence Culture and Strategic Analysis), which includes high-ranking officers from the police forces, secret services, army, and three academicians coherent with the securitarian doxa.

stopped or arrested are tortured and killed by policemen in prison or in the street.[40] Even simple gatherings of young people who wish to spend the evening in a public square, or a party of Erasmus students, or *botellón* are assumed to be dangerous, wild and unrestrained, creating 'urban incivility', and therefore need to be dispersed by force without concession (a little like the mods and rockers in Brighton in 1964 (Cohen, 1972); but then governments, in their own way, were still trying to build a future for society and also for young people).

Since the 1980s, in the whole of democratic Europe, there has also been a growing consensus towards leaders more inclined to racism and authoritarian regime (Le Pen and Sarkozy in France, Haider in Austria, the Northern League and Grillo in Italy and so on); for their part, almost all the leaders of the left (Blair in UK, Valls in France, Amato in Italy) have thought about winning this competition by adopting the practices of zero tolerance and 'more police, more penalties'; this type of competition has been frequent also amongst mayors, members of the left-wing or people whose political success was previously unimaginable. As a result, the prison population is continuously on the increase (Garland, ed. 2001; Palidda, ed. 2011). The discourse on insecurity has come about due to a short circuit between US and European conservative think tanks, moral security entrepreneurs, the police, almost all the media (Maneri, 1996, 2010, 2011), and also some judges; all these have benefited as far as their careers are concerned, and some have profited thanks to their ties with security businesses; less visible are the insurance companies who have also made huge gains.[41]

Therefore, security can be defined as a fundamental 'value' and also as good for the consumer-citizen. As noted by Baratta (2001), this security does not refer to the security and safety of all human beings. Those responsible for insecurity are almost always singled out as the children of immigrants, or worse,

40 See also images and descriptions in Sironi (2013); Crispino (2014); Guadagnucci (2015) and the numerous videos on Genoa G8 2001 (http://www.processig8.org/, accessed on 6 September, 2015).

41 Since the late 1990s, even in Europe, the economics media has become increasingly interested in the field of security (video surveillance, remote control devices, sophisticated devices, private police, sometimes also for the surveillance of disabled or older people through handhelds and so on). For their part, the interest of the insurance companies is straightforward: people require more insurance (those who can) when they hear more and more about the risks of theft, muggings, robberies, assaults, and so on. It is amazing to see regular surveys sponsored by insurance companies about the perception of insecurity. These surveys are published with emphasis by the most important media that frequently distort the results to show that crimes such as theft in apartments or in the street are becoming more and more frequent and violent. The sociologist responsible (a known democratic intellectual) for these surveys wrote, in a comment on his work: 'Of course, surveys collect only signs. Solicited, sometimes even extorted' (Diamanti, 2008). The company of survey for which Diamanti and his team work is Demos, always financed by a most important assurances company linked to the Cooperative League and the Left.

the Roma, irregular immigrants, marginal people or the homeless, who are all criminal suspects.

The rise of the permanent war against pseudo-Islamic terrorism has favoured a sort of continuum between the global war and the local war in the richer countries. According to some of the authors quoted above, the media has frequently adopted the same sort of warlike language to describe the fight against internal or local enemies, and against the 'Islamic terrorists', legitimating suspicion about any 'Arab' or 'Muslim', that is, the enemy of moment.[42]

The origin of the constant rioting in the French banlieues, the riots in England and elsewhere, lies precisely in the early 1980s with the increasing economic, social, cultural and political disruption of the previous society: the consequence of the neoliberal revolution. This is also a good starting point for beginning to understand the (rare) cases of young people who have become pseudo-Islamic terrorists. This is a period at the end of the reproduction of industrial society, in which immigrants and their children are being integrated into the lower and middle classes. Neoliberalism has begun to dismantle welfare, and along with it the possibility of peaceful social and cultural integration, assimilation, triggering instead racist criminalization. The riots and revolts are clearly a reaction against the neoliberalism that makes the young people 'ill-timed' or 'inopportune posterity',[43] not useful for the neoliberal, except for bad and low-paid jobs or for some illegal activities. The answer to these riots is devastating and criminogenic and little tempered by the 'candy' (incentives) distributed to some of the more 'docile' young people (a small minority of young marginalized people). The subcontractors working on this distribution consist of educators, social workers, psychologists, Islam specialists, anthropologists, sociologists, political scientists and various NGOs; they are used by neoliberal governance, by privatizing welfare rather than helping with any actual recovery of the causes of unemployment, of the shadow economies, and economic and social disintegration. Thirty years of this 'carrot' approach for a minority of 'not opportune posterity' and of systematic and often very violent racist criminalization for the majority of young people, has ended up by producing monsters, such as the perpetrators of the Charlie Hebdo massacre. The stigmatization of these young people, the fact that they feel rejected, insulted and without any kind of future, pushes some to seek recognition, gratification or even glory in their own self-destruction, so they are sacralized in the media. Jihadism, like any other form of terrorism, gives the illusion of

---

42 The Socialist Minister Valls has seemed an extreme case in the securitarian way (he seems to want to 'do better' than Sarkozy), but all countries have some left-wing personalities, at a national and local level, who have begun to adopt tones and behaviours that could compete with those of Le Pen or the Northern League or fascists, all stigmatized many times by the European Commission on Human Right for their attitude towards Roma, immigrants and marginal (see: http://www.echr.coe.int/Pages/home.aspx?p=home&c=, accessed on 6 September, 2015).

43 This expression paraphrases Sayad (Palidda, 2015a).

worldwide recognition, as if it were an alternative to marginality. There is a short distance to becoming the sort of 'loose cannon' who finds refuge in jihadism or other radical proposals, taking the nihilistic role of the 'suicide bomber', not much different of the auto-destruction, for example suicide by overdose.[44] This is not a new form of behaviour in the history of humanity, and is typical of those who do not see any possibility of peaceful action and negotiations that might help meet some of the demands for improving their lives. The asymmetry of power that has developed along with neoliberalism has eroded the possibility of peaceful collective action. The phenomenon of pseudo-Islamist radicalism, like other forms of radicalism or even self-destruction and 'postmodern' suicide, are indicators of a political organization of society that is dominated by the ruthless logic of profit maximization for the few, at the expense of the majority.

## Democratic Security and the *Res Publica*

In a democratic rule of law there should be no distinction between the security of the state and that of its inhabitants (citizens, foreigners or city users); in other words, the *res publica* encompasses the two.

In several articles contained in the Constitutions of democratic countries, the definition of security and safety, even if not always entirely explicit, corresponds to a concept of *res publica* including all the rights and duties of individuals and institutions and the equality between them. According to Ferrajoli (2001 and 2008), there are four classes of fundamental rights: (1) human rights (the right to life and of the integrity of the person); (2) public rights (freedom of movement, rights of workers, rights and security for the disabled); (3) civil rights (powers of negotiation, bargaining and freedom of enterprise, privacy); and (4) political rights. The application of these rights and duties is partly the responsibility of the state, partly of local administrations and also of all people. The different aspects of security in the broadest sense are usually the subject of 'concurrent legislation' between the state and local powers. Among these aspects we can find: health, protection and occupational safety, food security, civil protection, the government of the territory, transport and communication. It is therefore possible to state that partial security and safety show a lack of respect for the Constitution, a serious violation of the rules of law, that is, a crime. But, almost none of those responsible for various disasters have ever been seriously condemned to pay the damage. There have been some cases of trials for health and environmental disasters with condemnations, but you can observe that there has never been a proportion of human and material damages and penalties/sanctions and there are many cases

44 Similar cases may be observed in almost all countries in particular after 1990 precisely because the erosion of welfare has reduced the social-health assistance. I myself have been a witness on these cases during my research in the banlieues of Paris in the 1980s and then in Milan in the years 1990–2005.

without condemnations because it is difficult to find proof that will be accepted in court, sometimes because of corruption, deviation and so on. For example, who paid for the thousands of deaths and sickness cases from asbestos and for the reclamation often unrealized of the polluted areas? (See in the next pages the data on deaths and diseases by asbestos and also by other toxic substances.)

Examples of flagrant ambiguities, distortion or gross negligence in all the different aspects of public security are frequent, even from the point of view of some constitutionalists who are considered masters of democracy.[45] Throughout his entire discourse about security in the city, one of these distinguished constitutionalists[46] refers only to crime, and mentions nothing about health, the environment, protection from over-exploitation in the shadow economies and so on. It is no wonder then, that both conservative and left-wing governments have adopted policies and practices of security governance that generate a serious increase in ignored insecurities, neither do they hesitate to support the Giuliani experience.[47]

### *Attempt at a Democratic Orientation on Security*

The European Forum for Urban Security (EFUS) (http://efus.eu/en/) is the most important attempt at a democratic orientation towards security governance at a local level. It has autonomous branches in all Western European countries and has more than 300 adhesions from local administrations. The main objective proclaimed by EFUS is: 'to support and encourage crime prevention policies to reinforce and promote the role of Local Authorities in the development of national and European policies'. Therefore it is an approach that stems primarily from the desire to avoid or counteract the repressive-penal orientation by proposing the enhancement of social and preventive activities and for a 'softening' of police enforcement. EFUS cooperates with the International Centre for the Prevention of Crime (ICPC) and is a member of the International Scientific and Professional Advisory Council (ISPAC) of the United Nations Crime Prevention and Criminal Justice Programme and other international democratic institutions.

After almost 30 years, what assessment can be made of this attempt? Unfortunately, it is very disappointing. Since 1990, local governments (often considered as being democratic or left-wing, though not always) have not hesitated

45 For example: see G. Amato in his foreword to the book edited by Pajno (ed. 2010).

46 In addition to eminent constitutional expert, Mr Amato was also the vice Craxi; several times minister and also head of government, he is a friend of Blair and Clinton and a member of the most important clubs which bring together the most influential people in the world.

47 When Amato was in charge of the ministry of the Interior (Security and Police) in an interview with some of the most important media (17 August 2007), he said: 'We like Giuliani zero tolerance in our cities' (see Corriere.it, 2007). For Amato on the field of prohibitionism, and also military support and policing of immigration, see Vassallo Paleologo (2008, 2013, 2015).

to compete with the conservatives in the practice of zero tolerance, and encourage business connected to these practices. Furthermore, the EFUS has not requested from its members the rigorous respect for a security concept consistent with the *res publica* and then the true protection of the rights of all city inhabitants, and protection from all insecurities. EFUS neglects the fact that social control always oscillates between 'carrot and stick' and then depends on the exercise of discretion. The hope for social prevention thus becomes an illusion: the conservatives and also many mayors of the left are engaged in false security governance because this produces easy consent and is unlikely to upset the upper and middle classes. Almost all the mayors think of security only as a way of contrasting so-called 'urban incivilities', marginality and deviancy, and they ignore what should be a coherent articulation between prevention, discipline, punishment and recovery. The main aim of the majority of mayors (or local administrators) is to be popular; today, consensus depends, above all, on portraying an image of a strong mayor who is more like a sheriff. If a mayor is unable to mobilize the population so that they participate in the government of problems concerning urban life, giving the correct importance to ignored insecurities, the only problems on the everyday agenda will be those imposed by an easy securitarian and racist discourse. For a more democratic approach, above all, it is necessary to monitor the daily practices of the municipal administration personnel, and in particular of the police. It is extremely difficult, but the only way to achieve a more democratic approach is through the everyday fight to overthrow mainstream discourse, always starting with those who feel less protected and at a local level. This idea of democratic governance may seem impracticable, or one that harks back to past dreams of militancy; however, it could well be the most rational solution, also because the social ills and problems present in local neighbourhoods and those faced by the more vulnerable members of the population, inevitably affect the city. In fact, in many cases we can observe that in order to satisfy upper and middle class expectations, city mayors often divert their resources, equipment and forces towards false or less important insecurities, and favour devices that will be profitable only to the securitarian business. It is interesting to observe that the fall of Bloomberg is a failure of the conservative securitarian model favoured by Giuliani, as well as a problem of real estate speculation; this model neglected the serious problems of the majority of the population, who then elected De Blasio with a historically large majority.

The EFUS has failed to promote any changes in the training and practices of local and national police forces, and their effective coordination with the various prevention and control agencies, not only as far as social prevention is concerned, but also the prevention of health and environmental disasters, the new slavery in the shadow economies, and the abuse and violence of the stronger actors to the detriment of many.

At a theoretical level, the most obvious problem with the EFUS, and also with other attempts at a democratic government of security, lies in reducing the criticism of conservative governance in the matter of 'harm reduction', that is, the connection between prevention and the social treatment of those who are responsible for deviant behaviour. That is correct, but insufficient, and also inevitably intended for failure and to a risk of diversion or corruption of prevention services (because it forces a government to be enslaved to a repressive and penal orientation). Regarding the posture of EFUS, it can be observed that: (1) there is no criticism of mass distraction and of skills of which the conservative escalation nourished itself; (2) there is no criticism of the fact that security governance has spawned more and more ignored insecurities and their victims; (3) there is no criticism of the fact that the governance of security has become a cynical business; (4) there is no understanding of the fact that today's conservative securitarian governance is not the same as that of the past, but it is forged in the neoliberal frame, a frame that leads to the violent and cynical management of order and disorder, maximizing the profits of the dominants and increasing the racial criminalization. The problem, then, is to build, *ex-novo*, a governance of disorder from a micro-sociological level, through practices of equality, solidarity and respect for the fundamental rights of all people, that is, respect for the *res publica*. The devastating effect of the neoliberal trend is the ignorance of what the *res publica* means, and then, what it means to organize society with this in mind.

Below is an emblematic example, just one of many other cases.

In 2006, in the historic centre of Genoa, a woman was killed with a broken bottle. The murder took place in one of the nightlife areas on a busy evening, in a small street next to a square equipped with good video-surveillance. Witnesses reported that the woman was very upset and had been going around for at least an hour stopping people and asking them insistently and quite aggressively for a dose of drugs or money. The most appropriate action in this case would have been for some social-health worker to take her to a centre where they could provide adequate treatment. But there is no provision for these types of educators and social workers, even in situations and locations frequented by drug addicts. This simple consideration has not occurred to the local authorities: the prefect, superintendent, commander of the police and the mayor (left-wing) had no hesitation in signing a statement in which they called for more resources to increase video-surveillance and police presence. Nothing however has been done to increase the health and social services. Since 1990, thanks to liberal measures, these social services have been progressively closed down or reduced to the point that they still open only to distribute doses of methadone, but they do not take care of the addicts who return to shoot up and stay on the street (this is also the case in many other cities). Social workers are still in service but work for only a few hours a week with very low-paid and precarious contracts; therefore they are forced to look for other jobs and cannot care for the patients, as they should. There are also psychologists with 24-hour contracts, but they are expected to take on up to 80 patients (18 minutes per case); of course they can only take care of a few, but the municipality states that the greater part of the population is in the care of the social services.

Another example is a somewhat grotesque and tragic situation, which again took place in Genoa under the same left-wing administration; in some cases, in right-wing municipalities, it is possible to observe mayors who participate or promote in the pursuit of marginal people, nomads and immigrants.[1]

In their enthusiasm for security, the mayor and her councillors launched in 2011 a 'bracelet' for the safety of tourists, women and the elderly: their idea was that this bracelet would make these people feel completely 'secure'; by simply pressing a red button they would alert the police and receive their immediate intervention (Dentini, 2009). It is important to note that even tour operators have rejected the project because it is obviously counterproductive: 'If a tourist feels that you have to give them a bracelet to feel safe, then they think that the city is riddled with crime, and consequently it would be advisable not to go there.' After a year of experimentation, in an interview regarding the results of this programme, the councillor responded, with great aplomb, that they are very satisfied. But he also declared that for the whole year they had received only one call … about a car accident. However, he also stated that the success of this programme lay in the fact that many major cities around the world have asked for information on this prestigious bracelet with a view towards making the same choice for their municipal councils. Some people say that the mayor can boast that he helped promote a new product produced by a local industry specializing in this sector, in joint venture with American partners and the local university.[2]

Unfortunately, we must tragically remember that this mayor and her councillor's zeal for crime prevention have meant that they have not given the necessary attention to the risks from environmental disasters that are so frequent in Genoa and Liguria,[3] to the point that shortly after the 'bracelet performance' on 4 November 2011, there were six deaths in the city (including three children) when people were swept away by the sudden flash-flooding of a stream which had been restricted by the concrete foundations on both banks created to accommodate housing . In 2013, the court indicted the mayor and councillor for multiple murder and manslaughter, and for allegedly falsifying the minutes of the intervention meeting during this disaster.

---

1 In Italy racist attitudes and actions are no less frequent than in other Western European countries. Despite the power of the Catholic Church and despite the fact that many artists, intellectuals and the media proclaim they are not racist, it is not rare to see people slip into a racist frame of mind or make a racist comment as a form of occasional behaviour. After, people say: 'But I am not racist.' This attitude is also blatant within the media that repeats the same racist description of facts whenever an immigrant is suspected or effectively author of a crime. (see Maneri (2010 and 2011); for a wider documentation see websites of Cronache di ordinario razzismo, Meltingpot; Naga).

2 In a report, Cavallaro (for *l'Infedele*, 15 September 2008, PauraS.p.a. Chefatturati!), explicitly alludes to bribes used to persuade many mayors, including those from smaller municipalities, to adopt programmes including video-surveillance and 'reassurance' measures.

3 As shown by the research of some historians, since the nineteenth century, urban development in Genoa, as in many other cities, has caused damage and will inevitably lead to periodic environmental disasters. A glimpse of these studies is proposed by Torre (2004). See also Palidda (ed. 2011), Salzano (eddyburg.org) and archipelagomilano.org.

Subsequently nothing has been done to prevent similar disasters. In fact, in October and November of 2014, Genoa was hit by a new flood with one death and a lot of serious damage. According to the National Research Council, from 1970 to the present there have been 101 deaths due to flooding in Genoa, while during the previous 300 years there were only 180 in the whole region. This utterly outrageous fact is the result of 70 years of local and national governments, right or left-wing, who have authorized the worst real estate speculation, overbuilding and destruction of the ecological balance. And even today, after this latest disaster, there is no regeneration project or indeed any organization aimed at serious prevention. The courts are busy investigating in what way these facts are the responsibilities of today's administrators, but meanwhile, they continue unhindered in a city where all the most important personalities have been indicted for extremely serious financial crimes and corruption.

The Genoa case is by no means unique (see similar cases in Naples, Messina, Palermo, but also some in France, in the US and elsewhere). It is a city sandwiched between the mountains and the sea, and the rivers that descend from the hills are covered by illegal speculative building that was mostly erected in the 1960s and 1970s. All the local governments that have taken office since then have done nothing for the safety of these urban structures; according to Istituto Superiore per la Protezione e la Ricerca Ambientale (ISPRA) (2013), set up by the Ministry of the Environment, it is the area most at risk in Italy. After the last flood in Sardinia (November 2013), Italian geologists have pointed out that there are more than six million people at high risk from flash flooding in Italy.

At present, the economic crisis does not allow sufficient resources for recovery, so the likelihood is that there will be further disasters. In order to avert responsibility for any disasters, the mayors send out increasing numbers of emergency alerts, closing schools and ordering blackouts, sometimes when there is no real risk. Careless and wanton real estate speculation have been added to the devastating effects of industrialization, in total disregard for public health and the environment, and thanks to a public opinion that has been completely insensitive for many decades. Aspiration towards prosperity has pervaded the discourse on economic progress and as a result there is no attention to the protection of human life and posterity. Italy is the country where a very great percentage of housing has been built without respecting even the most elementary safety norms, thanks to the complicity of local administrators and the police. It is a very popular lawlessness, that is, socially legitimated.

Unfortunately, plans to carry out major public works are still being proposed, although it seems unlikely that they would create more jobs than restoring the territory's health and environment would create, or indeed any alternative projects, which would, however, produce more satisfactory results, geared towards real needs.

## Shadow Economies and Neoliberal Economic Development

After 1970 and even more so during the 1980s and 1990s, neoliberal development converted the so-called informal economies into shadow economies, often characterized by the occasional combination of legal and illegal and sometimes also criminal relationships in almost all sectors (Palidda, 2013). In fact, the neoliberal turn has imposed a weakening of the democratic rule of law in favour of the freedom of the stronger economic actors; this has favoured the collaboration between legal and criminal actors, the increase of neo-slavery, a crisis in state revenue and in particular of funding for social policies.

However, at EU level, undeclared work is defined simply as 'any paid activities that are lawful as regards their nature but not declared to public authorities, taking into account differences in the regulatory system of Member States' (European Commission, 2014b). A joint management project started with the International Labour Organization (ILO, 2013) on 'Labour Inspection strategies for combatting undeclared work in Europe'. This Report stated:

> The European Commission has highlighted that undeclared work, if not properly confronted, threatens to undermine the EU's ability to meet its employment targets for more and better jobs and stronger growth. Undeclared work is a form of social dumping that introduces unfair competition between firms on the basis of low wages and the nonpayment of social security benefits. Above all, it leads to working situations that violate the rights and dignity of workers. In this regard, the ILO, like the EC, emphasizes the need to encourage transitions from informal to formal work as a prerequisite for achieving decent work.

According to ILO's approach to undeclared work:

> is situated in the context of the broader notion of the informal economy, which it defines as 'all economic activities by workers and economic units that are – in law or in practice – not covered, or insufficiently covered, by formal arrangements'. This definition includes the notion of undeclared work as understood by the EC, while also covering workers who sometimes fall outside the coverage of labour legislation (e.g. domestic or agricultural workers).

As we try to show below, mainstream approaches to shadow economy produce a poor knowledge of the phenomenon and above all a lack of analysis of the multiplicity of interactions and consequences. These gaps (sometimes not casual omissions) then lead to hypotheses of measures to fight shadow economies which are often ineffective.

'Social dumping' has acquired great significance in several European countries. Thanks to 'cascade' subcontracting, many small and middle-sized companies also offer their services in other countries, frequently bringing their workers with them, with false documents, or with contacts from their country of origin, at much lower wages and guarantees.

This phenomenon of undeclared work has been known about since 1980 (Wallraff, 1986; Giudice et al., 1989).

At the end of the 1970s, on Paris building sites, the majority of the irregular workers were Portuguese (arriving in France just after the 'carnation' revolution). In Germany, during the seventies, eighties and nineties, there were still many Italians who preferred to be employed as black labour because they were under the illusion that they could earn more and therefore return more quickly to their country of origin.[48] After the collapse of the USSR (and still today all over Germany, as in the rest of the Western European Union) there were many irregular workers from different Eastern European countries (Broeders and Engbersen, 2007; Engbersen and Broeders, 2009).

What all these cases have in common is the importance of the role of the foreman, the recruiter and team leader who also commonly benefits from the complicity of corrupt police officers. Sometimes when payday arrives the foreman will raise the alarm: 'Quick – the police are coming – run!' After everybody has run off, the foreman disappears, but not before he has given part of the wages to his police accomplices. This sort of activity has been observed on some construction sites in Germany and elsewhere, starting from 1989. In Europe, since the 1990s, there has been an ethnicization of the labour market. This has become important in helping to create a system of subcontracting by which it is possible to make higher profits, especially when the management of the labour force is guaranteed by a boss or leader (also the day recruiter) who is able to gain almost total control over his workers.[49] For many entrepreneurs it can be very convenient to hand over the control for different parts of the construction to these boss–subcontractors, because they take charge of all the costs. This is made possible by the fact that the workers are trapped by their ethnicity, from which it is almost impossible to escape; they have no papers and no housing, except what is offered by the boss, frequently at an exorbitant price. The boss, who is generally of the same origin as the workers, is often protected by some form of religious authority (Catholic, Jewish, Islamic, and so on). In this way, the company discharges all the 'dirty work' and all the risks to this boss, who in turn exploits his fellow workers. The boss also plays the role of a kind of 'pastoral guide', and shows his capacity to discipline and punish the sheep in his flock. He may also be a sort of 'power-broker' (Blok, 1973), such as the mafia bosses known to the Italian immigrants in the US or in the North of Italy, or in Germany and elsewhere (Palidda, 2008).

The social dumping phenomenon is often discussed by the European authorities,[50] but it appears evident that they have difficulty in concealing their embarrassment over the hidden refusal of all countries to fight against social dumping, as well as against the shadow economies and even against 'EU fraud' (by brands who delocalize their entire production activity and distribution of products). It is evident that this attitude also corresponds to the defence of

---

48 In the early 1970s I was a militant for an association of Italian workers in Germany and later I also worked as a labourer in France, between 1978–1981. Later, I carried out research on the immigrants working, in particular, in the construction industry.

49 Among others, see the work of Perrotta and Sacchetto (2013); Sacchetto and Andrijasevic (2013); Sciurba and Carnemolla (2013); Foschini and Pleuteri (2006); inImmigrazione Onlus (2015); Asud'Europa (2015); Fo and Cazacu (2015); Legrande (2008).

50 See the website of the European Federation of Building and Woodworkers; Bernaciak (2014).

prohibitionism of migration and of protectionism, simply because they constitute an advantage for the financial and economic big actors.

On a global scale, delocalizations are all too often social dumping as well as the illegal export of capital and other crimes. According to research conducted by the Richard Murphy group, EU-wide tax avoidance amounts to 850 billion and 150 million euros.[51]

> In March 2013, three Belgian ministers presented a formal complaint against Germany, who they accused of tolerating the fact that 7.5 million people (mostly Romanians and Bulgarians) were working for a monthly salary of only 450 euro, hired by fictitious companies for 3–4 euro per hour, for 10 hours per day) and without any social security contributions (European Parliament 2013).

In his analysis of the 'transfer' (*de facto*) of accidents at work and delocalizations, in particular within the textile industry, to the so-called 'third world countries', Calabresi observes 'the last major disaster was on April 24th, 2013 in Bangladesh, in which 1,200 workers died: more than those who die in one year in almost all Italian production sectors'.[52] The Rana Plaza building which collapsed, employed about 5,000 people as subcontract workers for garment factories. Among the brands for which these labourers worked were Benetton, Bonmarché, the Children's Place, El Corte Inglés, Joe Fresh, Monsoon Accessorize, Mango, Matalan, Primark and Walmart. Just after the collapse, all these brands declared they were not involved … but the friends of the victims have documented proof. This practice of delocalization is usual for many other well-known brands in all production sectors, despite promises to rigorously respect the deontological code signed by all. Among others cases, see that of the Apple factory in China, called the 'suicides factory' (Bilton, 2014), and the emblematic case presented in the film documentary *China Blue* (by Peled).[53]

Almost all the big brands violate the European Community rules: they subcontract their production of finished goods into third world countries outside the customs areas (that is, offshore) when Community rules only permit the production of unfinished goods.[54] But they are very seldom worried by controls, recrimination or

---

51 Murphy (2012). See also OECD (2011) and Palidda (2013).

52 Calabresi (2013); about the delocalization in Bangladesh see Presa Diretta RAI 2014.

53 Chine Blue (2005); ABC News (2012); Inside China's Slave Labour Toy Factories (Various, 2011); Glawogger et al. (2007); http://it.wikipedia.org/wiki/China_Blue_(film 2005); Apple Chinese Factory Foxconn Nightline: http://www.youtube.com/watch?v=h1_XAuJ8qCc, accessed 19 September, 2015; Santa's Workshop - Inside China's Slave Labour Toy Factories: http://www.youtube.com/watch?v=yF8jUDzz5bE, accessed 19 September, 2015; Glawogger, M. (2005).

54 We observed these practices already in 2006 during the ECOBAZ (a FP5/FP4 European Research project) in particular the subcontractors of Benetton in Tunisia. The

punishment by the police, the judicial system or the political authorities in Europe, and much less so in the areas to which they are delocalized, which constitute neo-colonial areas with the complicity of local regimes and new power-brokers. In fact, these violations of the law, such as tax evasion and the illicit transfer of capital to tax havens, are the forms of illegality that are typically tolerated by the powers that be, who are always respectful of the dominant classes. Part of the production by the richest brands in European countries takes place within the shadow economies dispersed throughout their territories. The consequences of this are very serious for European workers, who are either resigned to unemployment or to pre-retirement (paid for by public funds and not by the company who delocalizes). But it seems astonishing that neither the trade unions nor the Parliament have ever asked for a law to control delocalizations.

We should remember that almost every country has a Constitution that provides for the defence of national property, which also includes private production structures. It is therefore surprising that some entrepreneurs remove all their machinery, sometimes overnight, others all their financial resources. In many cases these brands have received a good deal of financial aid from the state over the years. But what good is this protectionism?

In Italy, since 1980, almost all the most known authors in social sciences have exalted the development of districts, of 'made in Italy' of the 'third Italy', as a formidable and extraordinary new age of development, without mentioning the fact that there has also been a growth in the shadow economy, of neo-slavery, of the reproduction of irregular immigration, of evasion of social contributions and tax fraud.[1] The phenomenon had been known also in France and in other developed countries;[2] but its consequences and its multiples interactions are neglected, in particular: the signification of neoliberal conversion of informal economies, the articulation legal–illegal–criminal and with externalization at short, medium and long distance, the neo-slavery, the start of growth of irregular immigration, the development of corruption of some agencies of control and police forces, of executives of banks, local authorities and public administration.

1 In particular see Bagnasco (1981); Beccattini (2002).

2 If you research 'shadow economy' via in the internet you obtain 7,700,000 results, proof of the importance acquired by this topic. The literature on this field is vast, but there are rarely strictly committed authors to describe and analyse all consequences from the point of view of the victims also, and to try real hypothesis of protection and regulation of these.

To understand why, for over 40 years, Europe has suffered or accepted delocalizations it is necessary to understand the multiple social and political complicity. The unions have accepted the situation not only because they have

same type of delocalizations are observed by some researchers in different countries because, frequently, the brands 'play' or 'continually move' their delocalizations around the world due to blackmail or so as to pay less.

become partially corrupt and the employers' accomplices, especially during the 1970s and 1980s, but also because some workers believed they could benefit from this great turning point. These, in fact, thought it was convenient to be laid off and also do an irregular job, which amounted to almost two salaries. Others have accepted retirement and created a small irregular company or a regular one in the name of a family member. Almost no politicians and judicial authorities pay attention to the dismantling of the national and European industrial heritage.

From 1990 until 2015, irregular immigrations have been an extraordinary resource for the economies of all countries of immigration, especially for the US, the Arab Emirates and Western Europe. The real interest of the prohibitionist policy is to gain a substantial work force that is easily enslaved and easily got rid of once it has been used. Prohibitionism excludes any protection: illegal immigrants are seen as disposable objects. Moreover, the countries of immigration also have the possibility to show the native people that they are not treated as neo-slaves and the exploitation of immigrants is beneficial to all. So, prohibitionism is socially shared; many nationals can profit from it, and you can also see that the police rarely prosecute those who (illegally) exploit illegal immigrants but are ready to stop these if a citizen asks for it (Palidda 2015b). After almost 900 migrants drowned on 19 April 2015 while attempting to reach the Italian island of Lampedusa, the European countries faced the choice of using military operations to bomb traffickers and their barges with drones or aircraft. Those actions would inevitably end up hitting the refugees because they would probably be used as 'human shields' by the traffickers, and because it is impossible to distinguish between the traffickers, and especially their affiliated members and the refugees. So-called 'intelligent' wars have produced more 'collateral damage', that is, many massacres of civilians and also of soldiers.[55] In 2003, the EU's European Security Strategy, adopted by the European Council in Brussels in December 2003, stated as a fact that 'since 1990, almost 4 million people have died in wars, 90% of them civilians'. Many other individuals and institutions have made similar statements (European Union Institute for Security Studies, 2003).

Here, as elsewhere, it is clear that if you want to reform underground activities a crackdown would not suffice: the activities would reappear elsewhere because they are necessary to the neoliberal economy, and if there were no Neapolitans or Chinese or others to contribute to this, there would be other natives, immigrants, semi-regular and irregular workers or ethnic brokers. It is obvious that in order to recover the whole economic fabric, it is indispensable to begin providing the victims with effective protection and regularization, rather than expulsion (which is a double penalty). It is also true that in many cases the semi-irregular or totally 'black' workers (in Italy the majority are nationals) agree to be subjected to mobbing and harassment, and even rape in the case of women, because they do not perceive any alternative, nor any effective protection, and they are afraid of losing their income even from the most miserable, poorly paid, heavy and dangerous jobs.

---

55 Larson (2007); Human Rights Watch (2002); CIVIC (2009); Roberts (2010).

**A Case Study (Among Others) about the Consequences of Shadow Economies**

The tragic fire in a building near the textile area of Prato-Florence (where seven undeclared workers died on 1 December 2013) is an example of a 'neoliberal delocalization'. It is an area where there were once only autochthonous artisans, but since 1980 the majority are now Chinese workers. They sometimes produce artefacts (somewhere between legal and illegal) for the biggest brands and also for popular Italian and European markets. For quite some time now (rare) joint operations (by police, work inspectorates, health controls and so on) have tried to counter these submerged activities, but as some of them say, 'it is like trying to empty the ocean with a bucket'.

In fact, they are only episodic operations, and the national and local authorities have never had any intention of planning an effective reform of the shadow economies in this and other areas of Italy (and Europe). For his part, the left-wing president of the region has said that the Chinese mafia controls these illegal activities. This statement appears to be very singular: it does not explain why, for at least two decades, the forces involved in the fight against the mafia have not managed to do much, or indeed why the same regional authority (the same political party) has never said anything about it. Indeed, a few months before the disaster, the research centre (IRPET) organized a conference on the economic crisis in the area and also noted the extraordinary resilience of the district of Prato thanks to Chinese entrepreneurship and thanks to its 'underground' activities, without any allusion whatsoever to the mafia (IRPET, 2013; 2015). Obviously, when a tragedy like this happens it is convenient to attribute blame to the 'damned foreigners'. In fact, it is well-known in Italy, the Chinese mafia has tried to develop its activities primarily at the expense of Chinese workers and small business owners (as the 'ethnic' mafias have always done in all the countries to which they have emigrated, starting with the United States – Palidda and Sanfilippo, eds. 2013). But, criminal organizations are able to penetrate the fabric of economic activities only if they have the possibility of agreements or joint ventures with local economic actors and even the politicians in the areas where these activities take place. It is easy, therefore, to understand why the joint operations of all the control agencies and the fight against the mafia are unable to go beyond episodic actions without ever seriously contrasting the irregular conduct of bank managers and other major economic and political figures in the area.[1] For his part, the councillor for security for the Municipality of Prato (an ex-police officer and also a former member of the secret services, elected on the right) accuses the unions and the left because they have opposed his fervent fight against undeclared activities; in four years he claims to have shut down 1,200 illegal businesses.[2] He states that the number of actions to contrast these phenomena were 233 in 2009; 287 in 2010, 365 in 2011, 333 in 2012 and 345 in 2013. But, the participation of the state police forces (working at a local level) has decreased (from 44 per cent to 17 per cent). This testifies that insufficient attention is paid to the shadow economies by security governance, despite its relevance and importance in this area. In fact, the actions carried out by all the

1 Quarrata news (2012); Corriere della Sera (20 October 2013).

2 Panorama (2014) and also *La Nazione* (2014).

control agencies are episodic and depend only on the choices made by the local police agencies; there is no established programme for these actions at a national level for all the control agencies, and there has never been a programme for the implementation of stable prevention and continued recovery.[3] Finally, the regional government has signed an agreement with the court and also set up a programme for the formation of 74 prevention technicians (but only for three years). Ten million euros have been allocated for this programme, and the goal is to carry out 3,000 systematic controls each year (by work and health inspectors and the new technicians). This programme has created great perplexity because its funding is expected to come from sanctions; in fact, this aspect corresponds to a concept of prevention as repression and not as regularization and recovery of irregular activities for its workers and entrepreneurs. Furthermore, common to all training courses for prevention technicians, is training as health security experts in labour laws and production safety. No training is provided for a professional with security governance abilities concerning production sites or indeed the environment, or (most importantly) governance for the regularization of shadow economies. However, even in this case, two years later, nothing has changed. In March 2015, research carried out in the region and about its businessmen communicates with enthusiasm that, thanks to the Chinese, the economy of the area has not suffered during the crisis, on the contrary. The authorities have simply made it clear to all, including the Chinese businessmen, that it is necessary to respect the law. However, the inflated statistics concerning growth refer to the period of the fire that left seven dead, a period in which there were many accidents, illnesses and deaths, perhaps even some of which were hidden, as some observers suspect.

3 I interviewed the local police chief on 3 March 2014; after, he also sent me the statistics for the 'interforze' actions, that is, the actions carried out with the participation of almost all the control agencies. The local police is part of the municipality administration.

Take for example the case of black or semi-black labour in a small factory or building site or in agriculture.[56] The workers involved are all victims except for a minority who want to earn higher pay than if they were regularized; but for these the illusion that they are taking advantage of irregular activities continues until there is an accident or illness; then, all of a sudden there is just poverty. The majority of semi-undeclared and undeclared workers are forced to accept miserably low wages, dangerous, heavy and dirty jobs; and for the women also harassment or sexual violence; some think that a viable alternative is prostitution and go on to that, but many don't dare.

56 There are some very eloquent documentaries on youtube.com on the neo-slavery in small factories, in agriculture and in the building: special reports by some journalists of public TV (in English, French and Italian) (see references on Shadow Economies/ Delocalizations and Neo-Slavery).

> A 'colonial set' in the twenty-first century: some small business owners recruit clandestine Romanian women for work in greenhouses. These owners get the labour they need at bargain prices and 'disposable workers', but also the bodies of these women, on which they can vent with impunity the dual purpose of male domination. As in a yet new form of the banality of evil, the conditions for neo-slavery have been created here for the benefit of some autochthons, tacitly shared by a large part of the population, thanks to the negligence of the authorities and of police. Separations and divorces are on the increase, but the numbers of Romanian women seeking abortions has increased hugely. This situation is known to all and tolerated by the majority of local society. If the Romanian women want to work, they must agree not only to a very low salary, squalid accommodation, long and stressful hours and no safety at work (in particular because of the poisonous chemicals used in many greenhouses), but they are also at the mercy of the sexual needs of the master of the moment and maybe even his friends. In addition, the victims are forced to keep absolutely silent, which is easy enough, since there is a total absence of effective protection offered by the police and the justice system (Sciurba and Carnemolla, 2013).

The resignation of the workers clearly corresponds to the fact that there are no alternatives and also because they are in a condition of terrible impotence: they are alone and at the mercy of their employer, recruiter or foreman. Also, the workers with a fixed term contract are forced to live with a high risk of injury and illness, in perpetual fear of being fired (Palidda, ed. 2009). Their families and almost all the people in the same situation are resigned to this state. The control agencies, and in particular the police, even those with a minimal knowledge of the area, are more than aware of the phenomenon. But very rarely do they intervene because they risk taking action against the people who make their living only from these kinds of resources. They also know that the young people who refuse to go into undeclared work and neo-slavery often become drug dealers or thieves, sometimes affiliated to gangs or mafias, and some of them may even be tempted into terrorism. Moreover, there are always corrupt policemen and control agency operators who may be involved in the management of undeclared activities; at the beginning, they start out with the moral justification that they need to 'supplement their income' or to buffer unexpected family expenses, for 'well meant' reasons and so on. Sometimes this corruption also affects the various local authorities, work inspectors and so on. This phenomenon has always been widespread in all democratic countries and its growth is due to neoliberal development and, currently, to the economic crisis. But often, the police, the control agencies and the local authorities do not bring these issues into the open; they are aware that the necessary recovery for these aspects of the shadow economies is the responsibility of the government and its policy choices, and those of the European powers and of world financial dominion; they realize that their activities cannot offer solutions. Therefore, the police in particular just carry on doing their routine work, reassuring people by their presence, primarily the elderly, the local administrators, the 'good people'; they choose the exemplary option, especially when it comes to protecting the people who 'matter most' from the ones that arouse most fear. It is a well-known fact that the police's 'habitual prey' are rarely those actually responsible for creating anxiety and problems in local society; but, the police have learned that an 'enemy' is necessary both to them and to 'politics' (a 'double' legitimacy: both social and relative to the authorities).

Neither labour inspectors nor local health authorities, among others, receive many reports of illegal economic activities in their areas of jurisdiction. After about 20 years of neoliberal governments in favour of this sort of development, it is possible to observe the success of 'mass distraction', which is so useful for hiding the insecurities that have been created; it is quite evident, therefore, that this new focus on combating tax evasion does not concern the shadow economies and is inevitably doomed to further failure. Someone from the authorities has anonymously stated:

> Now is not the moment for chasing utopias ... Legalizing the shadow economy would take a real revolution! Today many people take advantage of undeclared money, starting with the banks ... also a chief of government has said that the shadow economy is an evidence of vitality, the vitality of an important part of the population and of the national and international economy. There is a sort of chain of people at the different local, national and international level, each one works for himself but he knows his accomplices ... and among these there are all sort of persons of the public institutions, enough to exclude a recovery of shadow economies, of corruption etc. Now, do you understand why I no longer believe in anything?[57]

## Health and Environmental Disasters

Despite the apparent increase in awareness about the need for environmental protection, actually, in the whole of Europe there is a prevailing negligence, especially regarding the number of sites that are deemed high risk because of pollution or environmental disasters, and because of a decades-long disinterest towards the reclamation of contaminated land. Yet as early as 1998, a report on the sites at risk in Europe underlined:

> Past and present human activities have caused the contamination of soil and groundwater at some 750,000 sites in Europe. Some of these sites may endanger water resources, the ecosystem and/or human health.

57 This interview is with a senior official who has worked in various sectors of the state, and whom I have known since 1998. His remarks reminded me of the scepticism of Foucault also referring to Nietzsche (Veyne, 2008).

The health statistics on the population living in these areas do not always show very high rates of mortality and cancer-related diseases because, up to now, the monitoring of these aspects has been uncertain, or of doubtful reliability, or non-existent.[58] Health disasters have been reproduced for decades, in particular with the development of the chemical and steel industries after the Second World War. The case of Taranto (in Puglia, Italy) appears today as the most dramatic,[59] also because it created great conflict between the local population and the workers; they were faced with the dilemma to 'save health and lives, or save jobs?' a dilemma with no way-out, because both the economic crisis and the neoliberal logic exclude any large-scale investments for reclamation (see Chapter 11 and Chapter 12). The historic irony is that the first article of the Italian Constitution establishes that 'Italy is a democratic Republic based on work'; article 2 states that: 'The Republic recognises and guarantees the inviolable rights of the human being' and art. 32 that: 'The Republic guarantees health as the fundamental right of each individual and in the interests of the collective.'

Of all the Constitutions belonging to the democratic countries, the Italian Constitution can be considered the most democratic, even though it was written in 1947 – as anywhere else, in this context, after war, the first goal is reconstruction and economic renewal – so, it is easy to understand why the right to protect health is not placed in the first article along with the right to work, and why the duty to protect the environment is not mentioned at all. But, it is emblematic that in some countries, and also in Italy, governments have eroded the effective guarantee of health and the environment, beyond the official rhetoric discourse on the protection of the 'quality of life'. According to some experts, Italy is the European country in which there have been the most disasters: some schools and other public buildings have collapsed, along with bridges, viaducts and tunnels. These disasters have increased and become more frequent, not only due to negligence in maintenance and checks, but also because these works – mostly recent – were carried out without meeting the required criteria as regards safety and quality of materials and techniques, in order to maximize the profits. This is a type of criminal practice that has produced many victims and a great deal of serious economic damage. Obviously, these disasters are aggravated when they coincide

58 Recent statistics on deaths caused by occupational accidents and occupational diseases show that in Europe there is a decline; but, according to some experts this is partly false because: (1) these statistics don't always consider injuries and illnesses directly or indirectly connected to various activities (for example, during travel) and don't consider accidents among the self-employed; (2) they don't calculate the injuries and illnesses of irregular workers; (3) they don't calculate the number of deaths after retirement from an active life as a consequence of work-related diseases; (4) the decrease in deaths and injuries corresponds to huge decreases in workers in large, medium and small enterprises (with the 'transfer' of accidents and occupational diseases in enterprises delocalized to third world countries – see Calabresi, 2013).

59 See also Report RAI 18 November 2013 and RAI Blu note (2012).

with so-called 'natural' or industrial catastrophes (recently experienced during earthquakes, floods, explosions in factories and so on – Fois and Forino, 2014).

> According to a WHO report (2015), about 125 million people in the world are exposed to asbestos at the workplace; more than 107,000 deaths each year are attributable to occupational exposure to asbestos; almost 15,000 lives are lost each year in Europe, especially those of workers, from diseases caused by exposure to asbestos (at a cost of almost 1.7 billion euro a year). Italy and Britain the most affected countries with almost 3,000 deaths each year (UN, 2013b; WHO, 2013; WHO, 2015).

The case of Marseilles is emblematic. This city has long been classified as a place with a high rate of criminal violence, a city in which the criminalization of French 'southerners' reminds us of a similar situation in the South of Italy. Mr Valls, previously as Minister of the Interior and now as Chief of Government, sends special forces for the 'war on crime' into the Marseilles 'banlieues' without considering the need for the rehabilitation of a police force contaminated by corruption. And what is worse, the government and also the local authorities neglect the fact that this city and its outskirts are seriously affected by a health and environmental disaster, with a particularly high rate of cancer victims in the area of Berre-Fos. As for many of the other areas of France that are affected by cancer, the state also imposes secrecy on the epidemiological statistics (Médiapart, 2014).

It is true that almost the entire population is at least partially responsible for the ruin of the world, but we also know that those most responsible are the big companies and the governments of the largest world powers, the same ones that decide whether or not they will sign the Kyoto Agreement and so on. So, there is an evidently gigantic asymmetry between the people that demand a just and sustainable world and those who are only interested in profits at all costs and are against humanity. This is the irony of history: modernity, the age of enlightenment, the sciences and continual progress, seem to have brought us to the brink of ruin[60] (and where the origins of the pessimistic views of some intellectuals such as Benjamin, Hanna Arendt, Pasolini, Foucault, after the two world wars, are found).

Investigations and also some video-reports on the Eco mafia exist,[61] but it seems that there is never any serious opposition to the network either on land or sea; there are suspicions that many 'toxic ships', laden with toxic waste, have been sunk in the Mediterranean area, in the Baltic and in the oceans. According to some special reports, it is a well-known fact that, with the involvement of certain secret service agents, toxic waste has been dumped in several African countries

60 On all these cases there are some books and reports and also a video on youtube.com (see RE Inchieste).

61 *Toxic Europe* (2011) is a documentary on the different ways organizations and countries are involved in the trafficking of toxic waste.

with the complicity of local armed gangs, also involved in the weapons trafficking (the journalist Ilaria Alpi and the cameraman Miran Hrovatin were assassinated because they uncovered this kind of trafficking and complicity).[62]

> The biggest landfill for (toxic) electronic waste is Agbogbloshie, a suburb of Accra (Ghana). This so-called 'e-waste ecosystem' legally (but also frequently illegally) receive from western countries televisions, monitors, computers, keyboards, and other computer parts etc. which are exported to Africa at a lower cost than legal disposal in designated landfills in the rich/democratic and ecologically minded countries (*The Guardian*, 2014).

In another report from 2011 by the European Environment Agency (EEA, 2014 and 2015), based on 2009 data, there are 622 industrial sites considered as the most 'toxic' on the continent. The list, based on the costs of damage accumulated by country, shows that at the top are Germany, Poland, the UK, France, Italy, Romania, Spain, Czech Republic, Bulgaria, the Netherlands, Greece and Belgium. According to different databases on the causes of mortality, it is possible to estimate that in the EU countries, before 2020, cancer is likely to be the first cause of mortality. Apart from a lack of interest in the remediation of industries that produce carcinogenic pollution, one of the biggest problems in Europe, and in the entire world, is nuclear. But, despite the promises made after the Fukuyama disaster, no countries have wound down or are contemplating winding down their nuclear programmes (this particularly applies to France, Germany and the UK).

It should also be noted that one of the most dangerous and most ignored sources of health and environmental risk are the military sites dispersed throughout Western Europe, as well as all over Eastern Europe, which are polluted by the enormous and obsolete nuclear arsenal that belonged to the USSR. The most evident proof of this danger is the increasing number of deaths from cancer, and patients with the disease, among them both soldiers and the local population. A great many of these people have accepted and still accept this situation because they receive benefits (jobs, services, compensation). Over the last few years there has been a relevant increase in cancer among soldiers who have participated in exercises on military sites or in overseas missions. The cause is the depleted uranium that has been increasingly used by NATO since 1990.[63]

According to the Association of Victims of Uranium (see website), among Italian soldiers there have already been more than 318 (perhaps more than 350) deaths and more than 3,700 sufferers as a result of contamination by this substance; these should be added to the cases of people from NGOs who have been

62 See Ilaria Alpi website: http://www.ilariaalpi.it/, accessed on 17 September, 2015

63 See website: http://nomuosniscemi.it, accessed on 17 September, 2015; video on Quirra: https://www.youtube.com/watch?v=uq1Xk0EOrEc, accessed on 17 September, 2015.

into the different theatres of war, and the people who live in the vicinity of military sites where depleted uranium was used.[64] Evidently, such a phenomenon concerns soldiers from all the NATO countries as well as the population in the areas where the most recent wars have taken place in the world. According to some testimonies by US veterans, and also English soldiers, almost 36 per cent of the military who participated in different missions are affected by cancer caused by depleted uranium, and the number of dead amounts to thousands. And yet NATO has not produced any information about this phenomenon. This attitude is similar to that shown by the authorities of all the NATO countries, who refuse to recognize the phenomenon and only partially recognize it even when confronted by the clamorous evidence.

> According to the data published also by Wikipedia, OECD and others, the stock of depleted uranium in the world is currently (in tonnes) 480,000 (USA), 460,000 (Russia), 190,000 (France), 30,000 (UK), 16,000 (UK, Germany and The Netherlands), 10,000 (Japan), 2,000 (China), 200 (South Korea), 73 (South Africa), 60 (Singapore); Total: 1,188,273. It is evident that the powers with more nuclear sites produce more depleted uranium. *Ergo*, there is a strong link between nuclear and military production.
>
> We note that no international organization provides current and complete information on all the states probably holding depleted uranium because of their nuclear power plants or weapons.

## The Diversion of Public Opinion and the Distraction of Skills

As we have seen, the ignored insecurities should concern the skills of the various control agencies, but it is evident that these agencies are unable to cope with these problems with their limited resources; in any case they need the cooperation of the police forces, as in many cases there are also criminal offences involved. There are numerous cases in which inter-cooperation, with the participation of almost all the control agencies and all police forces, have managed to carry out some effective interventions (activities against sources of environmental damage, moonlighting and other offences). But, these cases are almost always episodic punitive actions and, above all, unfortunately they don't lead to reclamation, or to the implementation of a stable prevention programme; they don't provide measures to regularize the workers involved (who remain without a job and income). The negative effect of all this is that the workers are then contrary to intervention. In some European countries, unfortunately, the ignored insecurities and their victims are frequently the product of what we might call the real 'distraction' or deviation of police skills, as these are directed instead towards other goals (that is, towards the repression of 'easy preys', such as those who are socially condemned as being responsible for all the ills and problems in society). This diversion of skills is fuelled by the

---

64 see http://espresso.repubblica.it/inchieste/2014/07/21/news/inchiesta-uranio-impoverito-1.173721 and http://espresso.repubblica.it/inchieste/2015/12/31/news/uranio-impoverito-la-strage-silenziosa-continua-1.244981?ref=HRBZ-1

distraction of public opinion in favour of more 'comfortable insecurities' (useful for governance through fear and the business of the securitarian posture). In effect, it can be said that there has been a sort of capitulation to the democratic rule of law, and sometimes a sabotaging of the duties of the public institutions as far as the protection of the fundamental rights of the majority of the population is concerned, and therefore the *res publica*.

The damages connected to the costs for the victims of ignored insecurities, and for much of the population, and for public resources, have been and will be relevant and unbearable and perpetual. The weakening of prevention, of control and the distraction of the control agencies has been imposed by mainstream securitarian discourse, but is also due to the fact that the majority of the population and the victims themselves are unable to act. Now, these costs are expected to rise because the economic–financial crisis means that governments are not investing in reclamation (although this would be the wisest choice to boost sustainable development and create jobs ... but that is a Keynesian (and Rooseveltian) point of view, antithetical to the neoliberal way).

But there is worse: in all democratic countries there have been many cases of violent police repression against those who dare to revolt or protest over the growing incidence of cancer-related diseases, or the criminal disposal of toxic waste, super-exploitation and major public works that provoke new health disasters, and the squandering of public funds.[65] The persecution of people fighting against these serious problems, as if they were terrorists or a threat to the rule of law, appears particularly absurd. It is confusing to the tens of thousands of ordinary people who claim the protection of public health, or are hostile to the use of GMOs or are opposed to nuclear power, to see them and treat them as eco-terrorists (Maccanico, 2012; Statewatch, 2012–2015), which is shameful, cowardly and even dangerous. This approach will favour the unrealistic and completely insane inclinations of a very few youngsters whose behaviour is an obvious reaction to the brutality practised by the authorities.[66]

65 See NOTAV history (see website); NO MUOS (see website); http://antoniomazzeoblog.blogspot.it/; Petrillo (ed. 2009) and the video *Una montagna di balle* (2009); Bonifacio et al. (2013).

66 It is likely that some young people (as it always happens during every historical shift) are radicalized, because they consider the high risks threatening the population living in some sites like a near-genocide, for example, the disasters of Seveso or Casale Monferrato and, worse still, at many other sites such as Chernobyl, Bhopal, Fukushima. But, some judges, politicians and secret services think these radicalized young people are terrorists. This seems horribly ridiculous; for example, it takes a great deal of imagination to liken the so-called NOTAV people to criminals or terrorist organizations such as al-Qaeda. The major works such as the tunnel through the Alps and others like it are, first and foremost, financial speculations, a waste of public resources and also a serious risk to public health and the environment while there are viable alternatives that are much less expensive, non-hazardous and can provide good results compared to real needs (see proposals by the WWF, Legambiente and others environmental organizations (see RAI PresaDiretta, 2013).

## The Habitual Practices of the Police Forces

In all countries the police responsible for the control of the territory rarely reserve any attention for ignored insecurities. For example, police officers working on the street are not likely to be on the lookout for illegal waste disposal, or for black work on sites with no safety regulations, or for other health-related or environmental risks. Only when judges are carrying out an important investigation, they can request checks on the variations in the amounts of water, electricity or gas being consumed, or checks on the sewage system or the transport of goods and movement of people. All this is very useful in helping to uncover the shadow economies at a particular site when they are not immediately visible. But unless someone reports these aspects, generally there is no investigation. Actually it is a lack of control agency forces and also a inadequate training of personnel, and even bad work organization. In all countries the police force generally has a sufficient number of officers, devices and resources, but the different control agencies never seem to be enough. The question that can then be asked by the experts is: why, if the police and military forces are abundant in number, are they not regularly used for the prevention and reclamation of sites at risk from health and environmental disasters? Evidently this is a naïve question, but it exposes the dominant reasoning of today's powers, a form of reasoning that, in fact, neglects the most serious insecurities, and the ones that are a danger to the majority of the population.

Government security that is able to guarantee the prevention, control, recovery and regularization necessary to combat all insecurities, and protection for all, does not need more policemen; it needs more operators in a variety of fields, and above all the collaboration of all the people. Without this collaboration it is impossible to ensure a secure society. But this is possible only when the government implements the effective participation of the population. Today, the government resembles the allegory of 'bad government' painted by Lorenzetti in 1338, in Siena, or the image of Jurisprudence painted by Klimt (Bottero, 2016). The collaboration of the people has become impossible, because nothing is done to create the population's trust in the effective protection of the state.

## The Real Political Organization of Society

Beyond abstract visions of the state and its local and national apparatuses, the real political organization of society works thanks to the roles played by the various social institutions (the family, associations, churches, unions, political parties, schools, transport, local authorities, social workers, the police, the courts, but also labour and health inspectors, local authorities, the media, and so on). In order to function well (without too many failures) there needs to be good cooperation between all these institutions; this is quite rare, because there is often competition or, even serious conflict. When the relationships between these institutions are not good then we are likely to find a disorganized society (that is, the anomie), because no institution can meet the expectations of the population by itself, providing good or less bad solutions to the various everyday problems reproduced by society. This is even more obvious when we consider one segment of society; then it is possible to understand how, in certain cases, criminal organizations – such as mafias – can provide (apparently) satisfactory answers to at least a portion of the local population (especially as regards the offer of fast and 'easy money' by supplying services that appear to be risk free). In some cases, there is consensus regarding various illegalities, which therefore become socially shared; this frame is at high risk of corruption, abuse and violence among the police forces. A degree of illegality is common to all society, but it can be higher among the people in power because they are endowed with discretion that can easily slip into arbitrariness.[67] The police share the informal social rules or, in any case, they respect them because they belong to different social circles. Their discretion is forged in those circles through their primary and secondary socialization, their interiorization of positive and negative categories. Then, the boundary between tolerated or non-tolerated illegalities is established in coherence with what they consider legitimate, while these can be illegal but shared by the social circles that ensure the social and moral legitimacy of each public and private act. The legitimacy is not always legal; in different situations, contexts or conjunctures, there are traditions and customs, habitual behaviours that are socially shared rules even if they are illegal, especially in certain segments or circles of a social class (and even more so among the stronger social actors).

## Police and Other Control Agency Practices

The repeated attempts by control agencies to govern the everyday disorders reproduced in society, in 'the best' or the 'least bad' way, are dictated by their adaptation to a reality which is a continuous experimentation of the social life of human beings. In particular, the police are continually testing out (in a good or bad way) new ways to intervene, but these are destined to fail, especially when the tests are not carried out with the real intention of adapting to any type of situation in an impartial way, but in a much more pragmatic spirit (much as a craftsman would). The art of governing disorder, like any other craft, is rarely the practical translation of a theory, but it is the implementation of practical knowledge that requires constant renewal through adaptation, and then the necessary mediation between informal rules and rules of law. Continuous experimentation is the only way to avoid failure, precisely because the political organization of society needs never to cease to update (and, as in the Goffmanian theatral metaphor, each actor has to play his role).[68]

67 In a investigation into the Italian Guardia di Finanza approximately 5,000 corrupt civil servants were identified with an estimated three billion euros for the corruption of government officials (see Corriere, 2013).

68 Here I refer to Goffman (1959) and his description of everyday life through the metaphor of the theatre where each social actor plays different roles according to the different frames.

The real ability of a social institution (or a social actor) to participate in the real government of society (and even more to appear to be the main author) does not depend so much only on investiture from above, nor only on physical (military) force or wealth, but first and foremost on their ability to be recognized by a large part of the population as able to offer relatively satisfactory answers to the problems, needs and expectations of society (or possess the ability to sway the population into believing that good solutions are being created).[69] Power is not exercised simply 'to discipline and punish', it can also be paternalistic or 'pastoral' (Foucault, 2004b). Police forces need to be recognized by at least a part of the population that counts as the institution that makes an important contribution to the management of the permanent disorder that society reproduces. But, above all, it is important to understand that this disorder is caused by the imbalance between wealth and poverty, dominant and dominated, and frequently the action of the police force worsens the situation because it is unable to offer good remediation and, *de facto*, ensure its reproduction.

'Within society, the police are nothing more than a mechanism for the distribution of force justified by a situation' (Bittner, 1990); therefore force distributed through the adaptation of either violent or exemplary or *laissez-faire* (non-intervention) action for social regulation.

The 'practical knowledge' of the police roughly follows some basic rules:[70] (1) to gain respect from those who matter and thus act more or less in the way these people want (at the same time gaining the approval of some of the people at the top and of the local and national political authorities); (2) to be feared by the entire population, showing that when they want to they can be very 'bad' but also 'good'; and (3) to try to get by between socially shared legitimacy and legality (avoiding the excesses of arbitrary acts that are inevitable but always minimized for the public).

The relationship of the police with the different actors of local society is decisive (and not only determinant) because it is the source of the 'daily plebiscite' (Palidda, 2015b).

In democratic countries the media have become powerful thanks to the oversaturation of surveys on the 'perception of insecurity', victimization and so on. The police, especially the chiefs at local level, are greatly influenced by this 'measure of popularity' and try to make use of the media and marketing to help

69 The mafia-type organizations have always been able to gain the consensus of certain social segments primarily because they give proof of 'eating and feeding'; in other words, the population knows they are criminals, but have no alternative because the legal institutions are unable to give a satisfactory response to the people; sometime the practices of official institutions are similar to those of the mafia; the mafia give concrete answers (employment, income, protection) (Santino, 2015).

70 In any case, police knowledge is learned primarily by placing younger officers with older or more experienced colleagues. The police schools offer mainly generic training, almost always only superficial legal training and even the simple use of weapons. The real learning for police officers is done on the street or even in different structures and offices, but never through theoretical learning or in a more 'pedagogical way'. Most of the chiefs themselves know nothing of police theory.

their careers.[71] Of course, the police staff forms a part of society, that is, of the different social circles; in the police force we find almost the same percentage of democrats and conservatives, racists and 'humanitarians', as there are in the rest of society (except where conservatives or authoritarians prevail and push for the selection and recruitment of reactionaries and 'Rambos').[72]

For the police, the 'daily plebiscite' is the ability to stay 'on the street', within the multiple social relations that exist there. This means that the police are necessarily encouraged to tolerate or not illegalities, even to violate the rules of law. The regulation of current illegalities within the various segments of society, as well as within police ranks, is reproduced not according to the rules of law, or the political programmes, or by the design of a 'big brother', but through the relationship between the police and the various social actors and the authorities. As Foucault says, 'the law is made … to understand how to get around it' (Droit, 2004). In other words, we can say that, despite the jurist *credo*, the rule of law allows for anamorphosis, that is, the possibility of a continual passage from legality to illegality and back again,[73] but this possibility is accorded only to those who have a good understanding with the police, the judges, the authorities – of course it is not an opportunity for the 'miserable' or 'useful enemies'.[74]

In this sense, Foucault's words 'the police are a permanent coup'[75] should be understood as a kind of eternal redefinition of what the random application of the rule of law is, that is, the concrete political organization of society. Here is the 'anamorphosis of the rule of law', that is to say, the use of discretionary power to move from legality to illegality and vice versa, which is currently practised in the switch from legality to tolerated illegalities, from democracy to authoritarianism, from violence to the respect for democratic rules (Palidda, 1992).

## Conclusion

The governance adopted, in particular after 1990, seems to have become an important part of the democratic rule of law in a state of the market. According

71 The bargain between journalists and police officials has always existed, but today it seems exacerbated by the frantic rush to appear as if this has become the sure means to making a career and also to have the possibility of selling this popularity and getting various benefits (from every sort of social actor). The same sort of bargain also concerns police chiefs and bank directors, that is, the (illegal) contamination in the collection of information between these two institutions, despite the Privacy Authority, who actually control nothing.

72 Unfortunately, really independent and rigorous research into racism in the police forces is rare; those carried out are always timid or even embedded.

73 As with distorting mirrors, anamorphosis allows you to switch from legality to illegality and vice versa; on the use of this metaphor see Palidda (1992).

74 In *The Betrothed* by Manzoni (grandson of C. Beccaria) the lawyer, dubbed 'shyster', says to 'make the law find the loophole'.

75 Foucault (2004a, 347); see also Valverde (2008a, 2008b).

to neoliberal reasoning, all the political organizations in society must produce profits at all costs, and in all fields, including health, public education, prevention, social, police and justice. To legitimize this orientation the dominant think tanks have forged a popular discourse, and the majority of the media and opinion leaders have contributed to a formidable mass distraction and provided new oxymorons. Like Orwell's 'war is peace' it does not matter if the war is not real, or when it is, that victory is not possible. The war is not meant to be won. It is meant to be continuous. The essential act of modern warfare is the destruction of the produce of human labour. Over the last decade the dominant discourse has been about 'humanitarian war', peacekeeping missions instead of actual war operations, the export of democracy, the defence of democracy with the sacrifice of freedom and so on. In the field of prevention and security governance, the definition of insecurities and their management have also helped to conceal the true causes and also the victims.

The damages connected to the costs for the victims of ignored insecurities and for much of the population and public resources have been and will continue to be relevant, and at times unbearable and perpetual.[76] The weakening of prevention, of control and of the distraction of control agencies has been imposed by mainstream securitarian discourse, but also because of the erosion of the ability of much of the population and of the victims to act. Now these costs are expected to rise because of the economic–financial crisis experienced by all governments, who do not invest in reclamation (although this would be the wisest choice in order to boost sustainable development and create jobs … but that is a Keynesian perspective antithetical to the neoliberal way).

The commitment in this research is also very painful because many people die every day and also because it shows that we live in a universe of risks threatening all people everywhere in the world.

Through this research you can better understand the pessimism of some of the famous authors (Benjamin, Hanna Arendt, Foucault) who became pessimists because of the terrible things humanity has had to endure because of the greater powers (as well as the two world wars and the reproduction of all the wars until today). But, as Foucault shows through the Socrates example, the exercise of parrhesia is the only possibility for those who want to work, even modestly, for the interests of humanity. Perhaps this opportunity for public action would provide the chance to re-evaluate the responsiveness of those sincere democrats who have built, and who still want to build from scratch, and from a micro-social level, actions for prevention and reclamation, as well as mobilization against the disasters that afflict all corners of the earth.

76 See in particular the photos of Salgado in Sahel and in other areas of the world where populations are exterminated by wars and environmental disasters, almost all of which are caused by the biggest economic and political actors.

Chapter 3

# Ignored Securities? On the Diversity of Securities and a Proposal for Their Integration[1]

Harald Arnold and Rita Haverkamp

The twenty-first century has thus far been characterized by terrorist attacks like 9/11 as well as technical and natural disasters such as earthquakes in Europe, hurricanes in the US, the Fukushima nuclear incident in Japan and the Deepwater Horizon oil spill in the Gulf of Mexico. Due to these rare and extreme incidents,[2] security research has gained special attention, especially scenario-orientated research. In addition, practical efforts aim to turn hazards into risks, in order to gain control over fate and nature, and to allocate responsibility when damaging incidents occur.[3] However, this approach may collide with academic predictions: one striking example is the 2012 conviction of six geologists in Italy who were found guilty of negligent homicide because they trivialized the hazard[4] of an earthquake.[5] Moreover, when combined with technological innovations, scenario-orientated research seems to offer practical solutions to guarantee security; however, it also tends to neglect fundamental questions concerning security in democratic societies.

Yet before discussing security any further, the future-orientated term needs to be explored as it is multifaceted and dynamic. Over the last decades, securitization (Buzan, Wæver and de Wilde, 1998) has become an important term to grasp the advent of different security topics (for example, the assessment of offenders as dangerous). In this context, the concept of 'subjective security' has emerged

1 The authors would like to thank Chris Murphy for his editorial assistance.

2 Incidents of low probability but high impact, that is, dread risks; cf. Gigerenzer (2004).

3 According to the terminology of Niklas Luhmann (1993): While the individual cannot do anything against hazards, a risk can be influenced by a human being; see also Ulrich Beck (1992).

4 As to Luhmann's definition (see footnote 3 above), the judges assessed the earthquake as a calculable risk rather than as a hazard.

5 According to scientists, an earthquake is not predictable, http://www.spiegel.de/wissenschaft/mensch/prozess-von-l-aquila-protest-von-erdbeben-forschern-gegen-urteil-a-862670.html, accessed on 10 June, 2015.

and gained significant traction in the social sciences. Attention has also turned to differences between the objective situation and the subjective feelings of individuals, because 'being secure' and 'feeling safe' do not necessarily correspond with each other. So-called 'ignored securities' might be observed on the objective as well as the subjective side, though before these can be further analysed, the notion of what an ignored security might encompass needs to be clarified. Moreover, security incorporates several apparent paradoxes[6] and conflicts: on the one hand, a higher level of security may arouse greater feelings of insecurity;[7] on the other hand, the need for the security of an individual or group might collide with the security interests of another person or group.[8] Against this backdrop, the so-called 'commodification' of security (Goold, Loader and Thumala, 2010, 3 ff.; Loader, 1999, 373 ff.) supports the security issues of wealthy consumers in order to ensure their interests and protect them from non-consumers (the poor/impoverished/indigent) (Shearing and Stenning 1981, 193 ff.).

In this chapter, the ambivalence that exists between objective and subjective security will be explained by the so-called 'security quadrate'.[9] The security quadrate will offer a means to illustrate differences and commonalities concerning diverse security issues. It should also help to ascertain 'real risks and hazards' and 'fears and worries', as well as their mutual relationship. Measures will be developed to classify levels of security, using the example of natural disasters and terrorism as modern hazards. In addition, subjective security will be analysed in more detail, based on the security quadrate differentiating between personal and societal respects. In order to depict the methods used, findings of a representative study about security and quality of life[10] conducted at the Max Planck Institute for Foreign and International Criminal Law will be presented. Finally, the analysis will be embedded in a framework for the future research of ignored securities.

---

6 See Zedner (2003, 155 ff.) about the six paradoxes of security.

7 Because one fears the future loss of achieved security.

8 One example is the police law in the case of averting a danger: When the police ban beggars from a given premises, ordinary people may feel safer. In contrast, the beggars could feel less safe and unprotected by the law. If the priority in security matters is led by the majority, the security interests of the minority will be disregarded.

9 The security quadrate was developed in the project 'Barometer Sicherheit in Deutschland – BaSiD' (Barometer of Security in Germany) which was funded by the German Federal Ministry of Education and Research; for more information about the project see http://www.mpicc.de/ww/en/pub/forschung/forschungsarbeit/kriminologie/basid.htm, accessed on 10 June 2015.

10 Dr Dina Hummelsheim designed and carried out the study at the Max Planck Institute for Foreign and International Criminal Law; she is currently analysing the collected data.

### On the Term 'Security'

*General Annotations*

Security is an elementary basic need of human beings[11] that encompasses their thinking and emotions as well as security provisions (conative measures).[12] Over the centuries, conceptions of security have changed among individuals, societies and nations. Security is a relative concept that touches upon every part of our lives and, indeed, goes beyond. The term alludes to the absence of personal and societal threats. Parallel to this negative definition, security also consists of positive elements such as quality of life and contentment, freedom from fear and worry, coziness (literally: Geborgenheit[13]), and trust in other persons and societal institutions (Ruhne, 2003, 61). Security is based on the trust that risks will not materialize at a micro or macro level, and that if they do, society and the individual will be able to cope with the harm they may cause (resilience) (Albrecht, 2009). Trust, in turn, cannot be equated with security because trust implies the possibility of disappointment and therewith a moment of insecurity.

The future-orientated dimension of security means uncertainty and, thus, an unattainable state devoid of risks and dangers.[14] The resultant instability attests to a societal problem situation which raises comprehensive and wide-reaching questions. The imponderability of the future and its inherent dynamic can only result in structural stability for a specific reference object and a limited time frame (Giebel, 2012, 59). The related contingency of security partially explains the security paradox that can generate both more security and more insecurity.[15] As a result, security represents a complex, emotional and normatively charged construct.[16]

---

11 Cf. Maslow's (1954) hierarchy of needs.

12 That means the cognitive, affective and conative (behavioural) constituent.

13 Cf. Hutta (2009, 252): '... the German concept of Geborgenheit [ ], which is commonly translated as 'security' but actually evokes an immediately positive sense of shelteredness, nested-ness, and well-being.'

14 Fritzsche (1986, 9 f.) about definition problems and four constituents of the term security based on J. Conradi: 'Existence of a danger against which security can be offered, the efficacy of the protection, and finally the trust of the person concerned in this efficacy' [Note: translation of the authors].

15 Kaufmann (2012, 10 ff., 157 ff.); in this context, the relationship between security and freedom is mapped out: a proportion of exclusion is often assumed (more security would lead to less freedom); compare Würtenberger and Tanneberger (2009, 225 f.). This hypothesis may be the case in particular, but it is not a question of generally valid conclusion: A stable democracy constitutes security which promotes the claim of constitutional protected rights of freedom.

16 For an early systematic approach dating from 1970, see Kaufmann (2012, 340 f.), [reprint].

## 'Negative' Definition of Security – A Social Constructivist Perspective

In accordance with numerous theoretical works[17] on security, security is understood here as a social construction.[18] The objective situation of security and insecurity consists of a bundle of diverse societal risks and hazards like catastrophes and terrorism as low-probability, high-impact incidents, as well as crime as a ubiquitous phenomenon. The quantifiability of these phenomena suggests objectivity. However, this objectivity is relative (constructivist perspective). For example, crime statistics cover only those offences known to the police but not those offences that have taken place and remain unreported. On their own, the other mentioned threats show selectivity and only touch a section of objective security. Although a variety of elements constitute security, health, accidents and stable economic conditions are typically mentioned.

Negative connoted incidents of damage and emotions are the initial point of several studies on security.[19] These emotions are commonly named as hazards, risks and threats in an objective regard and as anxieties, fears and worries in a subjective regard. The negative pre-emphasis traces back to a corresponding orientation in criminology as well as sociology on the one hand and, on the other hand, to the operationalization of security. The operationalization comes from the absence of insecurities because – as already explained – positive components of security are independently complex. Due to this, the subjective elements of trust, wellbeing and life contentment serve as explanatory factors for feelings of security in an already referred to representative study of the Max Planck Institute 'Security and Life Quality' (see footnote 10) as part of the mentioned joint project (BaSiD). This is the reason for the chosen negative definition: security means the absence of insecurity.[20]

---

17 Cf. for example, Giebel (2012, 25 ff.).

18 In German, we use the term 'objectified' to emphasize the constructivist character; 'objectified' means the preferably figurative gathering of actual securities by being aware that absolute objectivity cannot be achieved, see Berger and Luckmann (1966).

19 The already mentioned project Barometer of Security in Germany (BaSiD). In order to operationalize the term in a qualitative and a quantitative survey, security was explored based on negative expressions; the researchers of BaSiD monitor and analyse perceptions, reports, conditions and expectations that relate to the notion of security and rely on a combination of basic data collection, methodology development and application-orientated group analysis. http://www.mpicc.de/ww/en/pub/forschung/forschungsarbeit/kriminologie/basid.htm, accessed 10 June 2015; see also Haverkamp (2012).

20 Or as Zedner (2009), 14 formulates: 'security is the condition of "being protected from threats" – whether through their neutralization, through avoidance, or through their non-exposure to risk.'

### *Central Terms to Categorize Security in a 'Negative' Sense*

#### *Hazard and risk*

The negative direction resorts to essential expressions in order to describe and capture security. From a sociological perspective, the central words 'danger' and 'risk' are joined because they also refer to the future as security. In contrast to risk, hazard[21] always adheres to a negative connotation. Hazard is understood as the possibility of the occurrence of a negatively weighted incident (Kaufmann, 2012, 264). Though this negative and unwanted incident may appear as an apparition, the time of occurrence is uncertain.[22] The missing controllability is independent of decisions and complicates plans to avert the hazard by means of counterstrategies and protection measures.

In contrast to hazard, risk is associated with negative as well as positive effects.[23] In general, risk is considered as the product of the probability of an occurrence of an incident and the amount of damage that will ensue in case of its occurrence.[24] The calculability of risk enables the imputability of risk causer and the assignment of responsibilities (Bonß, 2011, 55). The outcome of this is an action and decision-orientated controllability which opens up a rationale intercourse with insecurity including strategies of prevention.[25]

21 The German police laws apply different terms of dangers: danger, concrete or abstract danger, (subjective) apparent danger, (subjective) imaginary danger, (only indications) suspected danger, exigent circumstances and present danger; the common term danger is understood as a condition which gives reason to expect harm with sufficient probability according to general life experience; such a danger is concrete when indications for the existence of a danger are given in a concrete individual case; the danger is abstract when the harm generally occurs with sufficient probability in a certain group of circumstances; the situation of danger is sharpened by exigent circumstances and a present danger. In contrast to the sociological conceptualization, the police laws assume that the start of the danger can be avoided or the harm can be attenuated by police intervention in time. The differentiation between dangers serves to choose and apply the allowed measures for averting dangers according to the quality of intervention.

22 For example, waiting for the next big earthquake in California ('The Big One') caused by the San Andreas Fault.

23 For example, risk appetite, progress, innovation and entrepreneurship are connected with taking risks; the term risk implies the chance for success as well as the possibility of failure.

24 Giebel (2012, 33); Renn (2010, 153) addressing technical prevention: 'Technical security philosophy is principally interested in decreasing the probability of a given failure occurring, so that the likelihood and damage caused is as small as conceivably possible' [Note: authors' translation].

25 The calculability of a risk is complex and the result of cognitive construction efforts, see Giebel (2012, 37).

In this context, Luhmann's differentiation between danger and risk is of interest. While danger refers to an external attribution act, risk belongs to an internal attribution act:

> In the case of self-attribution it is a question of risks, in the case of external attribution of hazards. Cancer is only a risk for smokers; it is still a danger for other persons [Note: author's translation].[26]

Risk signalizes subjective controllability and personal decision-making ability; the occurrence of harm seems to be a consequence of independent actions. In contrast, the emerging damages are dispossessed of one's own influence and control in the case of the realization of a hazard (for example, earthquake, decisions of other persons).

*Threats and worries*

The aforementioned representative study on security and life quality[27] applies worries as initial points to comprise securities. Additionally, and in contrast, the expression threat was chosen to tackle insecurities in another study using a qualitative approach.[28] In short, the differences between threat and worry will be pointed out.

Within the sociological meaning, a threat is seen as a substantiated danger or an impending risk.[29] This culmination characterizes a temporal and apparently conceivable closeness to the occurrence of a damaging incident which displays a damaging exposure for the health or life of a human being (Giebel, 2012, 32).

In comparison to a threat, everyday worries have a rather psychologically-orientated and alleviated meaning. Worries are understood as cognitive phenomena that trigger unpleasant feelings due to the reference to actual and future negative incidents (Siegl, 2006, 13). Uncertainty about their occurrence combined with the not necessarily successful search for solutions accounts for feelings of loss of control that are accompanied by negative affective conditions like fear.[30] In a

---

26 (Luhmann (1990, 22 f.).

27 Dina Hummelsheim is in charge of the study and carried out a representative survey (n=2,525) in autumn 2012 (see footnote 10); the first findings are presented in Hummelsheim and Oberwittler (2014).

28 This study 'Subjective perceptions of security/insecurity' was also carried out in the framework of the joint project BaSiD by Baldo Blinkert and Hans Hoch from the Albert-Ludwigs University Freiburg; face-to-face interviews (n=405) were conducted in the cities Mannheim and Leipzig as well as in two neighbouring rural areas.

29 In this context, the differentiation between self and external attribution according risk and danger is relevant.

30 According to Siegl (2006, 13), attempts for definition imply no explanation for the development of negative emotions in case of anticipation of negative problem scenarios. Siegl adds that the development of worries is depended on constraints of essential objectives

positive sense, worries imply possibilities to cope with future problems in order to take care of oneself.

*Ignored Securities*

Efforts to guarantee security favour innovative technology solutions; one example is the use of biometrical systems for access controls.[31] The organization of administrations and other institutions has also been improved by establishing (digital) network structures. The prime threat is (international) terrorism followed by natural disasters, industrial accidents as well as criminality including organized crime.[32] Based on the aforementioned negative definition, this search for security rests upon the hypothesis that the more security measures that are implemented, the more security will be achieved.

However, the rise of protections is not inevitably related to an actual reduction of the threat potential. As a consequence, the internal security situation according to terrorism would not be stabilized; moreover, the main outcome would be of mere symbolic nature. However, negative (unintended) side effects often accompany counter-strategies in different areas. The comprehensive digital surveillance of citizens to combat terrorism does not only infringe freedom rights, but also supports insecurity about data protection and raises fear about data abuse from the government and especially from its security and intelligence agencies among parts of the population.[33] Hence, the prioritization of specific security interests can lead to the disregard of other distinct securities. As a result, ignored/hidden securities arise from a balance in favour of higher-ranked securities or from forgotten or unknown[34] threats.

Also, the weighing up of the security interests of the public and the parties concerned might bring forth the alleged pursuit of purposes of the majority. Whereas on a societal level security rights of minorities could be unnoticed, on a personal level the assertiveness of one individual or group threatens the security

---

of an individual: existential life areas such as work, relations and financial situation receive the most worries.

31 For example, Ceyhan (2008, 102 ff.); Cole (2001); Graham and Wood (2003, 227 ff.); Introna and Wood (2004, 177 ff.).

32 In particular, civil security research addresses these phenomena; see Fraunhofer IME (ed.) Jahresbericht Annual Report 2014–2015, http://www.ime.fraunhofer.de/content/dam/ime/de/documents/Publikationen/Fraunhofer%20IME%20Jahresbericht%202014_2015.pdf, accessed June 10 2015.

33 See the actual debate on the US and British spy programmes (Prism, Tempora); http://www.guardian.co.uk/uk-news/2013/jul/08/nsa-gchq-spy-programmes-legal-challenge, accessed June 10 2015.

34 Taleb (2007); compare also the famous quote of Donald Rumsfeld, former United States Secretary of Defense, in a statement to the press in February 2002: 'But there are also unknown unknowns – there are things we do not know we don't know'; http://www.defense.gov/transcripts/transcript.aspx?transcriptid=2636, accessed June 10 2015.

of another weaker person or group. One example for a salient structural change in the European Union (EU) and its member states is the governmental dealings with immigrants and asylum seekers in recent years. Fear of terrorism and the continuous economic crisis in the EU has fostered the search for security within the borders of the EU as well as within national borders. In this sense, immigrants and asylum seekers from different cultures are understood as posing a threat to public safety and thereby as a potential security problem (Zedner, 2010, 380 f.). Concerns are related to threats to social cohesion, social security (for example, jobs, welfare system) and to social disorder and crime, especially terrorism, transnational organized crime, trafficking of people, drugs and other illicit goods. The ensuing 'securitization of immigration' relies upon limiting the mobility for the quests of protectionism, selectivity and a more restrictive, control-orientated enforcement of immigrations controls. As a result of this perspective, the concept of citizenship serves as a distinct attribute and gains more weight in governing immigration.[35] This process is accompanied, on the one hand, by a criminalization of immigrants and asylum seekers and, on the other hand, by an immigration policy that seems to be a pioneer for domestic crime control. Finally, recent developments are a threat for immigrants and asylum seekers and imperil their security and, vice versa, their existence is assumed to be a hazard for the wellbeing of the EU and its member states.[36] The management of immigrants and asylum seekers in the EU is only one example of an ignored (in)security. Additional examples of ignored (in)security at the EU or the national level might be found in the context of health, environment or protection of the vulnerable (poor women, children and so on).

## Contrasting Different Types of Security – the 'Security Quadrate'[37]

### *Personal and Societal Security*

An essential differentiation is the separation between personal and societal security. While personal security relates to the personal room of nearness and especially experience, societal security encompasses observable developments on the macro level. This difference has consequences for the estimation of the potential of worries for the individual dependent of the personal involvement.

The personal security feeling is based on a process in which experiences and personal estimations are developed, and seem to be an individual attribution

35 Zedner (2010, 380 ff.), examines the concept of citizenship as one feature of securitization and advocates a 'liberal model of the criminal law and assertion of due process protections as security rights for all individuals against the state' (quote 379); see also Bosworth and Guild (2008, 703 ff.).

36 In detail Knickmeier, Chapter 10 in this edited volume.

37 In the absence of a better term the original German 'Sicherheitsquadrat' was translated literally.

and construction. The perception of the environment is different because human beings do not have a congruent constitution due to their personality and attitudes. Nevertheless, congruencies between people and groups can be observed. These congruencies concern especially the estimation of (in)secure rooms and persons with the help of objective criteria (for example, crime rates), effects of sex and age, as well as the assessment of interrelations between human beings and the environment.

*Ambivalence between Subjective and Objective Security*

According to Zedner the term security is used to connote the 'objective' state of being without, or protected from, threat; it is used to describe the 'subjective' condition of freedom from anxiety; and it is used to refer also to the means or pursuit of these two end states'.[38] In contrast to feelings of safety, objective security refers to the way a real danger or risk can be expressed, for example, a statistical probability of the entrance of damage and a measurable degree of security. The different focal points of objective and subjective security imply incongruity because the subjective estimation of risks can differ from the objective probability of occurrence and vice versa. The promise to guarantee subjective security denotes the social construction of security and not inevitably the prevention of a real hazard or risk. As a result, subjective security cannot be seen as a further development of objective security. Moreover, its ambivalence has to be considered: 'feeling secure' and 'being secure' can coincide but also fall apart. In this context, the Thomas theorem is important concerning the relationship between objective and subjective security: 'If men define situations as real, they are real in their consequences' (Thomas and Thomas, 1928, 571 f.). That means people do not only act on the basis of the objective circumstances, but on the basis of the interpretation of a situation.[39] The given subjective importance of the situation determines the following actual reaction that also influences objective security. The following security quadrate (Figure 3.1) takes this into account and shows the partial inconsistency and respective discrepancy of objective and subjective security.

This differentiation is crucial in understanding ignored securities by means of the exhibited matrix. Especially the two quadrants 'putative secure and putative insecure' are of special interest for the discussed topic because the putative condition expresses misperceptions of the objective situation which can be labelled as ignored security/insecurity. From their subjective perspectives, people can misinterpret given situations and therefore come to a false estimation of the probability of occurrence of a hazard or a risk, for better or for worse.

38 Zedner (2003, 158) [emphasis added].

39 Merton (1968, 477) was inspired by the Thomas theorem and developed the concept of the self-fulfilling prophecy.

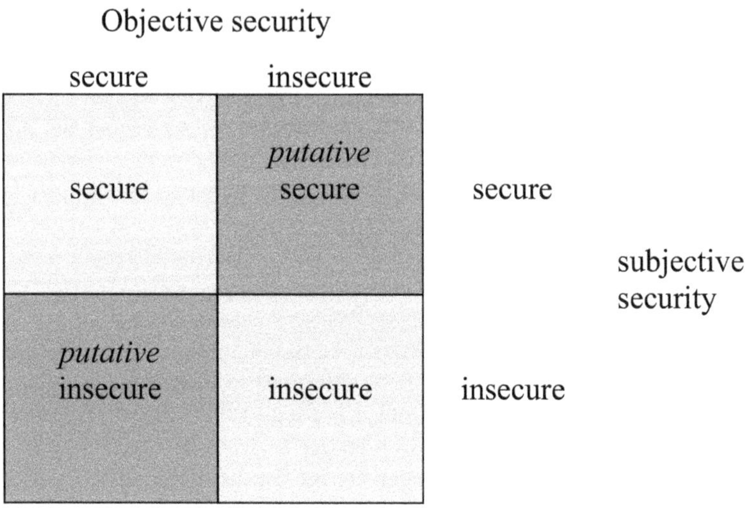


**Figure 3.1 Security quadrate**

On the one hand, when a person feels secure he/she nonetheless actually lives in an insecure condition in certain essential living areas ('putative secure'). One example from the subject area of natural hazards is the perception of vulnerability and risk in flood-affected areas in Germany.[40] Standardized household surveys in chosen districts in Cologne reveal an awareness of the population concerned about the general flood hazard in their city, but more than one-third of the respondents negated living in a flood-affected area, though this area is struck once in a century by a flood (based on statistical calculation) (Fekete et al., 2011, 99 f.). Besides, the impression of having undergone a so-called 'centennial flood' could lead to the false assumption of being secure and therefore feeling safe concerning floods for the next decades.[41] Another striking incident is the long-lasting heat wave in Europe in 2003 (Robine et al., 2007, 2 ff.). In that year, more than 9,000 people died as a consequence of the enormous heat in Germany. Although a heat wave in fact is a threat for the health and the life of human beings, the perception of people is different. The number of fatal incidents (deaths) due to the heat is usually underestimated and the overly hot summer is remembered as a positive exception. In view of the threat potential of a heat wave, the perceived security is deceptive.

40 Using the BBC-Framework developed by the UNU-EHS (United Nations University – Institute for Environment and Human Security), Birkmann (2006, 34).

41 A centennial flood happened in summer 2002 in the Eastern part of Germany, but another centennial flood struck the region in early summer 2013 once again, http://de.euronews.com/2013/06/05/dresden-bangt-schlimmer-als-die-jahrhundertflut/, accessed June 10 2015.

The state of 'putative secure' is characterized by a hidden and unperceived objective hazard or risk and might be called 'ignored insecurity'.

On the other hand, when a person feels unsafe, the surroundings are nevertheless widely secure ('putative insecure'). The governance of security has emphasized the threat of terrorism since 9/11. Therefore, surveillance measures combating terrorism are justified by the promise to ensure objective security (irrespective of its efficacy) in order to satisfy the demand for subjective security. The perception of terrorism as a threat is often distorted according to the actual hazard. While airports and flights are understood as forms of critical infrastructure that need special protection, road traffic is just associated with a manageable everyday risk.[42] Hence, according to statistics, the risk of being killed in a road accident is much higher than in a terrorist attack. From a theoretical psychological perspective these different perceptions and attitudes can be interpreted with reference to attribution theory and more specifically to the concept of the locus of control (Rotter, 1966, 1990, 489 ff.). The internal locus of control relates to the belief of having power over events in one's own life, which can result in a kind of control illusion; in contrast, the external locus of control is related to unswayable outside forces, for example, fate. People are accustomed to traffic as an everyday risk with fatalities, but not to terrorist attacks as extraordinary unusual incidents with a high impact. Whereas terrorism as a dread risk is led by external circumstances and therewith is beyond the internal locus of control, road traffic is believed to be guided by personal decisions and efforts within the internal locus of control. The (apparent) irrational fear of terrorism in the face of its low probability directs to this phenomenon as an ignored security. As a result, ignored securities are interrelated to objective securities and perceived insecurity and, conversely, ignored insecurities to objective insecurities and perceived security.

Subjective security is assumed to be an intricate concept because measures to satisfy feelings of security can be detached of objective security and can satisfy the security interests of the majority of the population or the powerful part of the population without gaining objective security (Zedner, 2009, 19). In the course of the commodification of security, unpopular minorities are banned from mass public property and also public spaces to enhance the subjective security of the consumer and citizen. In this sense, the concept of subjective security could disregard objective securities of marginal people (for example, secure residence due to the political persecution of asylum seekers).

### *Relevance of Personal and Societal Worries for Life Quality*

In contrast to other countries, Germany is widely untroubled by catastrophes, industrial accidents and terrorist attacks. Annually, a single-figure number of

42 Gaissmaier and Gigerenzer (2012, 1449 ff.) about the terrorists' second strike due to increased car traffic fatalities after 9/11; Zedner (2009, 17) referring to plane and road accidents.

cases are recorded: in 2011 in each case one incident was registered concerning the three phenomena.[43] However, the data sources used are partly incomplete and fragmentary according to the damaging incidents and their extent. This result hints to a certain relativity of the registered objective damaging incidents. This relativity is shown all the more in the differentiating definitions, objectives and collection systems as well as the unavoidable intermodulation factors[44] of recorded offences. Nevertheless, relevant reference points arise from these small numbers for the objective security of the four aforementioned phenomena.[45] The remarkable low case numbers of catastrophes, industrial accidents and terrorist attacks allow us to arrive at the cautious conclusion that Germany is a secure country in this regard.

An attribution in terms of the security quadrate and ignored (in)securities can only be achieved on the basis of the surveyed data about subjective security sensitivities and perceptions. Therefore, Figure 3.2 shows the societal worries of the German population. The societal worries are ranked according to the personal relevance of different areas of security.[46]

In a societal regard, the respondents are worried least by terrorist attacks (29 per cent) and even less by natural disasters (18 per cent) in comparison to the other themes. By contrast, nearly half of the respondents consider the development of crime in Germany as worrying which is the third important societal problem of the questioned areas.[47]

Pursuant to first analysis, only one-tenth of the respondents utter great worries of being a victim of a natural disaster or a terrorist attack on a personal level.

43 Information from the open source database EM-DAT (The International Disaster Database; http://www.emdat.be/) and from the report TE-SAT 2011, 36.

44 Just compare the hints in the German Police Crime Statistics, 2011, 7 (Polizeiliche Kriminalstatistik – PKS). 'The unlawful (criminal) acts dealt with by the police, including attempts subject to punishment, are recorded in the Police Crime Statistics. This also includes the drug offenses handled by the customs authorities' (PKS 2011, 6).

45 In this framework crime will not be considered due to its relevance for the living environment and its complexity. 'The unlawful (criminal) acts dealt with by the police, including attempts subject to punishment, are recorded in the Police Crime Statistics. This also includes the drug offenses handled by the customs authorities.' (PKS 2011, 6)

46 Thanks to Dina Hummelsheim for the figures.

47 A comparison to the recorded offences in total (PKS, 2011, 31) is interesting: though there is a small rise of recorded offences in total (0.9 per cent to the previous year), the total number of registered crimes lies closely under six million like in 2010. Since 2005, recorded crime (in total) has decreased apart from 2011. With regard to crime, a divergence between objective and subjective security seems to be seen according to the security quadrate: in a societal regard there is objective security, but subjective insecurity; that would be the case of 'putative insecurity' and crime would be an ignored security. Indeed, it is not possible to give a valid and reliable answer at this point. First of all, the reference points of crime according to the security quadrate have to be explained, as too does the question concerning the criteria for the existence of objective security.

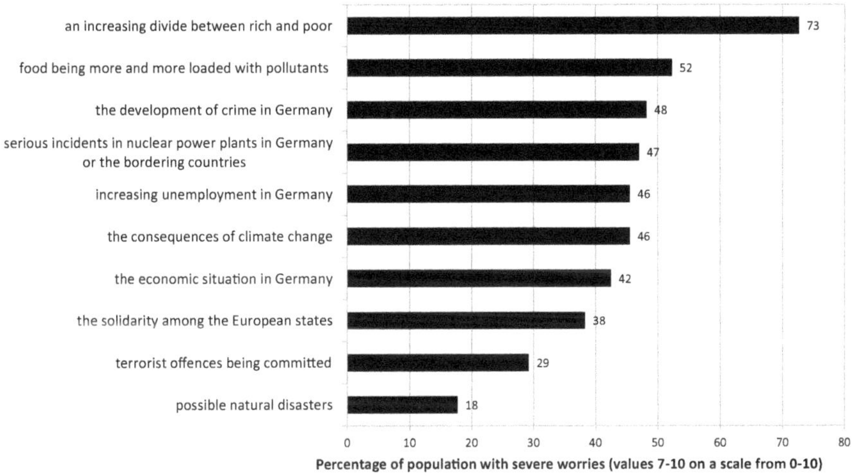


**Figure 3.2 Societal worries among the German population**

*Source:* MPI-Survey 'Security and Quality of Life in Germany' (2012), N=2.525

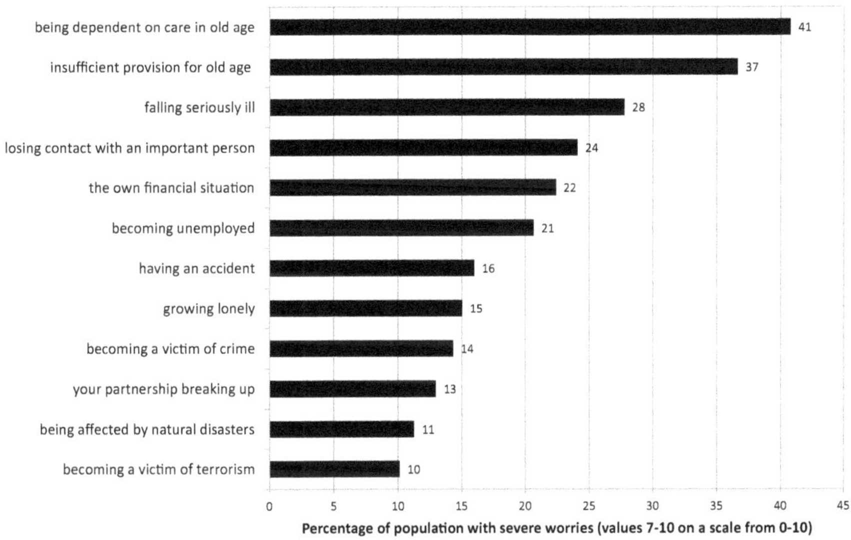


**Figure 3.3 Personal worries among the German population**

*Source:* MPI-Survey, Security and Quality of Life in Germany' (2012), N=2.525

Personal worries are considerably lower than societal worries. The worry of becoming a victim of crime troubles very few people (14 per cent). Instead, vast worries concern existential areas of life like health, aged care or the loss of a loved one.

Referring to the security quadrate, it can be stated that in the personal environment the data on objective security and the present database on subjective estimations of security concerning natural disasters and terrorist attacks agree with each other and thus mean 'secure' in both cases, likewise according to societal society, albeit alleviated. The quadrant 'secure' articulates that no ignored securities/insecurities can be recognized at this early stage.

One should of course bear in mind that the utterances of the respondents are generally expressed as societal and personal worries on diverse phenomena. The term 'natural disasters' especially covers a variety of different catastrophes such as the aforementioned heat wave and centennial flood. Therefore, an analysis in detail could reveal differences within one phenomenon concerning the feeling of security on a societal and personal level as well as towards ignored (in)security. Consequently, an elaborated assessment would consist of multiple security quadrates according to each phenomenon's ramifications and to the typology of individuals.

### *Relevance of the 'Security Quadrate' for Ignored (In)Securities*

In criminology, the security quadrate could be significant to categorize fear of crime in conjunction with the actual objective crime situation. This medium facilitates the assessment concerning crime in general as well as offence-related in particular. The depicted differences and congruencies might be useful to render more precisely theoretical explanations for findings. In such a way, ignored securities/insecurities would be figured out in two steps: the first step refers to the attribution of fear of crime and the objective situation of crime according to the security quadrate and the second step relates to the theoretical background whereupon other ignored securities/insecurities could be discovered and further analysed.

In view of empirical research on unreported crime, the security quadrate contributes to facilitate a comparison between objective data on unreported crime and subjective data on fear of crime. The findings of the German Victimization Survey 2012[48] provide such data on self-reported victimization and perceptions of crime combined with feelings of insecurity; they thereby allow a categorization according to the security quadrate.

Research on victimization offers even more possibilities to find out about ignored (in)securities by means of the security quadrate. One example is

48 Large-scale empirical research on the dark figure of unreported crime began in the 1960s in the US, theoretically even earlier in Europe. The Max Planck Institute, that is, its criminological department, started its own empirical research on hidden delinquency and victimization in the middle of the 1970s.

victimization in the alleged 'safe haven' of marriage and family. In recent decades, domestic violence and child abuse lost their shadowy existence and gained a prominent part in victimology. Increased awareness in public and political spheres can be also observed, though the sanctioning of sexual offenders usually refers to traditional thinking such as electronic tracking. Electronic tracking is based on the idea that sexual offenders are attracted to potential victims at certain places (for example, playgrounds, nursery schools), but empirical results show that sexual offenders mainly approach their victims in familiar surroundings (for example, relatives, friends, neighbourhood) (Duwe, Donnay and Tewkesbury, 2008, 484 ff.). In this sense, electronic tracking seems to fulfil a symbolic function to create subjective safety (Haverkamp, Schwedler and Wößner, 2012, 19 f.).

In conclusion, the security quadrate serves as an analytical tool to assess objective and subjective securities and fosters the further development of explanatory theories of crime. Another achievement is the easier identification of ignored (in)securities.

## Conclusion

Broaching the issue of ignored (in)securities from the perspective of an empirical criminological[49] and social science research project, an attempt was made to bridge the gap between theoretical and factual concepts of security through the use of the analytical security quadrate tool ('Sicherheitsquadrat').

Starting from the two basic and opposed views on security, that is, the subjective one and the objective one, both perceptions were systematically contrasted with each other, resulting in a simple, yet fundamental typology, which allows one to expound the problems of their respective agreements and disagreements. Through this schematization, it could be seen that this analytical view specifies fields of ignored securities as well as insecurities that are addressed by the two quadrants 'putative secure' and 'putative insecure'. Such an assessment supports the identification of the ignored (in)securities of different persons and groups on an individual and societal level. The specific typology allows for a differentiation between the security interests of the majority and the security interests of the minority. In the future, this clear distinction could help contribute to the acknowledgement of minority security concerns, like those of immigrants and asylum seekers, based on a more open-minded approach.

Starting from this initial point, that is, the identification of ignored securities and insecurities and based on the empirical results, it will be possible to not only substantiate theoretical approaches in the study of security in society, but also

49 In the framework of the aforementioned project BaSiD, one module consisted of the German Victimization Survey 2012, which was designed and carried out by the Max Planck Institute together with the Criminalistic–Criminological Research and Advice of the Federal Criminal Police Office in Germany.

to develop recommendations for criminal and social policy. In conclusion, the political weighting of security interests should require, and be influenced by, diverse pluralistic criminal and social policy decisions that harmonize dissimilar security interests.

It is recommended that future research on ignored securities expand its efforts to also include methods and findings from the field of sociology of knowledge, which not only promises experiences in the handling of epistemic problems of how to gain and assure knowledge or make use of tacit knowledge but also on the relevance of ignorance[50] within the state of knowledge (society).[51]

50 Criminological research on assurance might provide insight in two areas: prognoses and crime as an insurance problem.

51 See, for example, Ungar (2009).

# Chapter 4
# The Governance of Crime in the Risk Society: From 'Designing Out Crime' to 'Built-in Resilience'?

Tim Lukas

## Foreword

Resilience has become a fashionable buzzword in the recent security discourses. The concept of urban resilience promises answers to a range of security problems in cities. This chapter discusses to what extent crime prevention through urban planning can benefit from the discourse about urban resilience. The question in this respect is whether the urban resilience perspective can provide the basis for a new paradigm of crime prevention through urban planning. Since the concept of resilience is often used in a narrow technical understanding, the chapter concludes that the concept of urban resilience can primarily be considered promising in the context of crime prevention through urban planning if questions of security are addressed not only in terms of construction and technology, but also within their wider social context.

In recent years, concerns about security have acquired increasing importance in the characterization and development of urban areas. The promise of urban lifestyles, social security and infrastructural supply, as found especially in urban contexts, shows a trend towards re-urbanization, towards a renaissance of the urban, which makes the guarantee of security an essential task for every community. New safety requirements have resulted as a consequence of socio-spatial polarization processes, primarily from different crime and order problems. Additionally, altered public standards as to feelings of security have resulted from local security issues, which are being increasingly dealt with in the scope of municipal competence. The local authorities mainly act in the fields of danger defence (issuing and suspending business licences, casinos, restaurants and so on), in the urban development policies (establishment of utilization structures, avoidance of places of anxiety and so on) and in community crime prevention.

Over the past two decades, the prevention of crime has evolved to a central element of local security policy and crime control. Order and safety partnerships between the police and local authorities, crime prevention committees and public order services are components of a local security architecture in which the cooperation and coordination of the different constituents play an important part. The boom of community-based crime prevention stands in close relation to macro-social transformation processes, which find their expression in the widespread diagnosis of a risk or security society (Beck, 1992, 1998; Groenemeyer, 2010; Legnaro, 1997; Singelnstein and Stolle, 2006). The idea of the ability to shape society as the fundamental principle of handling insecurity is substituted by an increasing orientation towards hazards and risks.

Just like other consequences of the modern age, crime is nowadays no longer seen as a fateful event that cannot be influenced, thus, as a threat, now it is rather considered now as a 'risk of modernization' (Blinkert, 1988; also O'Malley, 2010), which calls for the statistical calculation of the probability of its occurrence and the assessment of possible damages. The local security policy follows this actuarial logic insofar that a risk management for the handling of criminal behaviour is developed in a way that clearly distinguishes itself from repressive strategies of moral ostracism and extermination of this phenomenon.[1] Now preventative efforts focus less and less on the perpetrators, but more on risk-evaluated situations as well as the daily routines of potential victims who generate opportunities for criminal acts as a non-intended side effect. In the centre of security policies are the calculation of risks and the development of situational prevention measures that take form in the local authorities, notably in the condition of urban crime prevention.

Below we will look into the question of how and to what extent urban crime prevention can profit from the current discourse on urban resilience. At present, resilience is a 'fashionable buzzword'[2] in discussions concerning safety that promises readily available answers to different security problems. In question is if a new paradigm of crime prevention in urban development can be derived from a perspective of urban resilience.

## Urban Crime Prevention

The conceptualization of crime preventive principles in town planning is based on strategies that, under the collective term of situational crime prevention, have recently experienced prominence in the explanation and prevention of criminal behaviour. Situational approaches in crime prevention aim at systematically reducing opportunities for criminal acts. The accessibility of instruments and objects of criminal interest should be reduced, whereas the risk of failure of criminal offences should be increased. 'Situational prevention comprises opportunity-reducing measures that (1) are directed at highly specific forms of crime, (2) involve the management, design or manipulation of the immediate environment in as systematic and permanent way as possible, (3) make crime more difficult and risky, or less rewarding and excusable as judged by a wide

1 As to the theoretical model of 'actuarial justice' see Feeley and Simon (1994).

2 Boin, Comfort and Demschak (2010, 1); see also Müller (2010, 1).

range of offenders' (Clarke, 1997, 4). By elaborating and analysing the structure of criminal opportunity, the likelihood of an offence being committed should be quantitatively reduced, qualitatively impaired or at least the direct consequences minimized. In this context, the reduction of excessive fear of crime counts among the negative consequences that are to be minimized.

Situational strategies of crime prevention find their theoretical reasons in criminological approaches that are based upon the idea of crime as an individual utilitarian consideration. Therefore, the causes of crime are located in such situational circumstances that make criminal activities possible, or at least favour them. The routine activities approach supposes that crime always occurs where three things coincide in space and time: motivated offenders, suitable targets and the absence of a capable guardian (Cohen and Felson, 1979). Operationalized as a rational choice, it is expected that the 'reasoning criminal' (Cornish and Clarke, 1986) in preparation of a criminal act, calculates costs and benefits to decide whether to go through with the act or not.

In the practice of city development, it is supposed that by means of a targeted implementation of constructural measures, the local structures of crime opportunities and thus the extent of crime-related feelings of insecurity and the level of criminal and non-criminal deviance can be influenced. The spectrum of promising recommendations reaches from a good nightly illumination, constructive measures for green areas, accesses and routing to the local settlement of infrastructural facilities such as shops or supermarkets.

The basis for including the aspects of urban planning efforts in the analysis of criminal opportunities is the 'defensible space approach' (Newman, 1972; Rolinski, 1980), in which the architectural design from housing and housing developments is interwoven with aspects of security. The main concern of this approach is to increase the social control by promoting responsibility. 'By means of architectural and urban design, natural surveillance – in the sense of informal social control – is to be triggered by inhabitants who by means of signs of appropriation of space symbolically indicate defended and defensible spaces' (Schubert and Veil, 2011, 84). Defensible spaces can be considered as a physical expression of social systems that defend themselves. Thus '"defensible space" is a socio-physical phenomenon' (Newman, 1972, 4). The structural design shall contribute to creating functioning neighbourhoods, representing hereby less an independent recipe for the prevention of criminal behaviour but rather a remedy for the residents' self-help. Stable structures of local self-organization are to decrease the necessity of governmental and municipal interventions.

The model of defensible space constitutes the starting point for a series of further approaches in which, however, the original idea of an increase of informal social control is sometimes reduced one-sidedly in favour of social exclusion and an orientation focusing solely on the architectural and urban design of housing developments. The common basis of approaches such as 'Crime Prevention through Environmental Design'[3] (CPTED), 'Secured by Design'[4] or 'Design out Crime'[5] is the principle of 'Target Hardening', the reduction of the vulnerability of a potential object of crime by active securing behaviour. 'The most obvious way of reducing criminal opportunities is to obstruct the vandal or the thief by physical barriers through the use of locks, safes, screens, or reinforced materials' (Clarke, 1995, 110). Planning and designing should structure city districts in such way that human behaviour therein will be influenced by barriers, access limits and physical hardening of the material environment. So, whereas originally symbolic barriers were to serve the division of different usage and property claims, now real barriers dominate in this strongly tightened perspective, the problematic effects of which on the design of a city impressively become visible by the example of closed housing complexes, the so-called 'gated communities'.

Whereas the mentioned design recommendations exclusively comprise architectural and free-space-planning measures, the so-called 'second generation'[6] of the CPTED concept goes further than the simple design of space, insofar as also strategies of social planning and social management are included. Thus in the focus, there is not only the design of physical space but also the infrastructural work with the users of that space. In return to the original idea of 'defensible space' as a socio-physical system, the 'new' social factor is to promote the inhabitants' integration by offering facilities to the different target groups. 'The social contacts that are so important for the basic principle of "informal social control" cannot be morally enforced but must be obtained as a "side effect" of spatial design' (Stummvoll, 2002, 8). Schools, playgrounds, parks, stores and neighbourhood centres should create an infrastructure that promotes the development of social contact.

In the most recent developments of the CPTED concept, in which the attempt is made to reconcile the guiding concepts of security and sustainability in the implementation of crime preventive strategies, a 'third generation'[7] is now winning more contour. So far, security and sustainability have been considered to be incompatible in the planning process. Often compromises had been made in the environmentally sustainable design of new town quarters in terms of the guarantee of safety or as to the residents' feelings of security. Ecological responsibility and sustainable economic activity should therefore also find their way into urban crime prevention. Now, by using regenerative energies and innovative technologies, not only safety and its perception concerning crime risks are to be increased but

3 See, for example, Crowe (2000); Jefferey and Zahm (1993).

4 Cf. Armitage (2000); Cozens, Pascoe and Hillier (2007).

5 See, for example, Ekblom (2012); Poyner (1983).

6 As to the 'Second-Generation CPTED' see Brassard (2003); Saville and Cleveland (2008).

7 As to the 'Third-Generation CPTED' see UNICRI and MIT Senseable City Lab (2011). As to the relation between security and sustainability in urban crime prevention see also Armitage (2007) and Marzbali et al. (2011). The ecological use of crime preventative measures in town planning is explained by Pease and Farrell (2011).

also the city dwellers' life quality on the whole. In expansion of the conventional security conceptions, where inner and outer security along with physical and economic security are distinguished,[8] now with 'third-generation CPTED', safety problems are addressed that result from the ecological consequences of global warming: 'The premise of third-generation CPTED is that a sustainable, green urbanity is perceived by its members and the outsiders as safe. Third-generation CPTED's focus on sustainable green environmental design strategies insists on practical measures, physically or cybernetically enhanced, that foster the perception of urban space as safe beyond mere concerns about crime' (UNICRI and MIT Senseable City Lab, 2011, 19). In this perspective, crime prevention in urban planning can be understood as an element of a lasting urban development, with the help of which social conflicts as a consequence of socio-economic and demographic processes are to be solved.

## Urban Resilience

While urban crime prevention has only just discovered the topic of sustainability, outside of crime policy, another change of paradigm seems to become apparent insofar that the thought of sustainability is more and more substituted by the concept of resilience. For example, in view of the seemingly inevitable overload of the planet, economist Dennis L. Meadows (contributing author to *Limits of Growth*) demands a change of perspective from the sustainability policy to a policy of resilience (Müller, 2010, 1).

Resilience designates the capability of a system to flexibly absorb, balance and overcome stressful influences in such a way that the proper capacity to act remains preserved. The concept is applied in very different contexts of research. The common basic idea of a capacity of resistance and adaption that is based on fault tolerance seems to stimulate 'an exchange between different scientific and thematic fields' (Felgentreff, Kuhlicke and Westholt, 2012, 70). Resilience can be understood as a 'boundary object' (Brand and Jax, 2007), the basic definition of which only makes transdisciplinary cooperation possible. In technological sciences, for example, resilience characterizes an object's physical ability to go back to its original form after suffering an exterior deformation. In psychology, resilience is defined as the phenomenon of a positive adaption or development in the face of adverse living conditions or events. The resilience approach finds its initial point in ecosystem research, where under the term of capacity, an ecosystem is understood that is capable of tolerating disturbances and changes without transforming into a different system state with new structures, functions and control mechanisms (Holling, 1973; Voss, 2010, 69). Furthermore, in this

8 As to the change in security policy and the extension of the term security see Albrecht (2007).

perspective, characteristics of resilient systems include: the capability of self-organization and the capability to learn and adapt.

In risk and disaster research, resilience is often operationalized as the reverse of vulnerability. The term of resilience is understood as the capability of a society to comparatively quickly cope with a crisis or catastrophe and thus, to re-establish the capacity of the people and technical systems affected to act and function as soon as possible. In this perspective, the concept's popularity can be explained by a paradigm change in the 1970s in the scope of which an almost exclusively natural scientifically characterized hazard analysis was more and more substituted by a more socio-scientific view on social processes in connection with natural and anthropogenic catastrophes. With the resilience approach, a change has happened which shifts the focus away from the 'isolated' natural phenomenon towards society and their coping strategies.[9] Based on the awareness that an all-encompassing guarantee of security is impossible in a highly developed and networked world, it is now less the safeguarding against possible risks that is in the centre of attention, but rather the creation of a general resistance against possible disastrous outcomes despite precautionary measures.

For the guarantee of safety in the specific context of the city, urban resilience means 'the degree to which a city is able to tolerate alteration before reorganizing around a new set of structures and processes' (Resilience Alliance, 2007, 8). The particular vulnerability of the city results from the disastrous consequences that unpredicted events can produce in this scope. Cities are extremely complex systems, the supply infrastructures and traffic junctions of which are considered to be particularly vulnerable. Even minor disturbances can trigger cascading effects and cause damages of a catastrophic extent. In this regard, resilience offers the municipalities the possibility to take measures in their field of responsibility, which anticipatory safeguard resources, provide infrastructures and preserve or create structures that are solid enough to guarantee the future urban development despite critical phenomena (Bürkner, 2010, 23 f.; Hitthaler, 2011, 43).

The concept of urban resilience promises answers to questions as to how disturbances of the social life in cities can be avoided in the future or how their impact on the urban co-existence can be mitigated or coped with. Resilient towns are 'capable of withstanding severe shock without either immediate chaos or permanent deformation or rupture. Designed in advance to anticipate, weather, and recover from the impacts of natural or technological hazards, resilient cities are based on principles derived from past experience with disasters in urban areas' (Godschalk, 2002, 1).

As the view into the past can only provide unsure predictions about the future, urban resilience is characterized by a specific handling of nescience. This especially shows up in urban systems in the face of complex contexts and a lack of predictability of the extent and frequency of critical events. A society's vulnerability to risks and dangers cannot be exactly predicted. Thus the complexity

9 See, for example, Bara (2011, 78 ff.).

of urban structures calls for a holistic concept of urban resilience that integrates constructional as well as social aspects on different levels of urban development.

In the context of coping with structural traumas such as earthquakes or floods, urban resilience mainly focuses on the constructed environment. Resilience, in this context, means 'the capability to reconstruct constructional structures or a city's functional integrity' (Bürkner, 2010, 23, cf. also Kreimer, Arnold and Carlin, 2003; Vale and Campanella, eds. 2005). In the discourse on the prevention of catastrophic events, urban resilience is comprehended as the capability of cities and local authorities to develop effective forms of control and efficient institutions for the handling of supposed or real terror threats (Bürkner, 2010, 23). The consequences of a municipal policy that is so strongly oriented to the 'designing out terrorism-principle' (Coaffee, 2004) can critically become denominated as 'military urbanism' (Graham, 2010). In the end, this leads to the fact that the cities' distinctiveness, their openness and liberality, turn against themselves in such a way that elementary civil rights would be much more restricted due to an increasing technical supervision (Rogers, 2012; Savitch, 2008). Using the city of London as an example, Jon Coaffee shows that through a dichotomous strategy of fortification and technical supervision, risks are shifted and zones of different security levels are created. Particularly controlled high-security zones would be hereby literally disconnected from the rest of the city (Coaffee, 2004).

As opposed to this structurally and technically established, 'built-in resilience'[10]-oriented safety, Martin Innes and Vanessa Jones localize 'the presence of collective efficacy in a community whereby a group of people come together around a shared goal, such as improving feelings of safety and security' as the source of urban resilience (Innes and Jones, 2006, 50). On the basis of the idea of collective efficacy,[11] an urban district is considered to be resilient when its inhabitants stick up for a common goal and if there is a basis of mutual trust and shared norms among neighbours. In this 'soft' perspective, urban resilience represents a bottom-up process in which resistance and flexibility are the result of social cohesion and local expectancies from control and intervention. The population's resilience may be described as the 'capacity of a social entity (For example, a group or community) to bounce back or to respond positively to adversity'.[12] From

10 Thus the subtitle of a publication by Lee Bosher (ed. 2008).

11 As to the concept of collective efficacy, see Sampson (2012); Sampson, Raudenbush and Earls (1997).

12 Maguire and Hagan (2007, 16). As to social resilience, see also Pelling (2003). As a good example for the social resilience of city dwellers, the coping with the so-called 'London riots' in the summer of 2011 can be named. In the face of the circulating violence and the obvious overstrain and reservation of the police, the citizens of the most affected districts took self-help measures and organized cleaning operations and neighbourhood watch patrols in order to protect others and themselves. In the subsequent investigations concerning the causes of the riots, it became clear that the victims of the riots as well as those living in violence-free districts had proven as extremely resilient (cf. Newburn, 2012;

this perspective, resilience is a condition of an urban daily routine, which does not only unfold its effectiveness in the case of damage. In the case of disaster, resilient communities are able to generate capabilities and skills for self-help, the testing of which occurs in the cooperation of social cohabitation. Against this background, the consolidation of the population's resilience pursues the objective 'to return their individual capacity to act to the citizens and to promote the development of civil-social commitment ... The strengthening of social cohesion leads to a stronger informal social control. This is essential for the development of resilience' (Gerhold, 2011, 6).

## Resilience as a Paradigm of Urban Crime Prevention

Urban resilience is primarily seen as an answer to severely damaging events. Correspondingly, the concept is currently cyclically applied in the field of civil defence and disaster management. Critical infrastructures are to be protected from natural hazards or terror attacks and the population is to be prepared for both unknown and unimaginable new risks and insecurities. The thought of applying urban resilience in this context to the phenomena of everyday crime, virtually appearing to be common, calls for an explanation.[13] On the one hand, crime can also be conceptualized as a disaster, which for the affected victims, can easily turn into a 'life disaster' (Hanak, Stehr and Steinert, 1989). On the other hand, though, the measures taken in order to fight terrorism, for example, leave their traces also in the prevention and persecution of everyday criminal acts. Measures are taken before a certain event occurs in order to prevent it, to delay it, or at the very least, in order to mitigate its consequences. Naturally the problems addressed in the fight against terrorism are of a different dimension. The risks that state institutions see themselves confronted with are harder to calculate. Furthermore, the consequences of a possible occurrence of danger are generally more calamitous and the attacks on individual freedom might be more severe and more enduring. And nevertheless, 'they do not differ fundamentally from the problems occurring in everyday criminal prosecution. Here, too, risks need to be assessed ...; here, too, the state's answers – in proportion to the probability of a risk and its severity – must be in an appropriate relation' (Zedner, 2012, 37). The drastically extended scopes of action of the law enforcement authorities in the fight against terrorism do not remain without consequences. If terrorism constitutes a crime, this being the basic idea, then far less extraordinary criminal acts can be fought effectively by the means of criminal persecution and prevention. 'Because new safety measures, the introduction of which is, in the course of extraordinary hazards, considered to

The Riots Communities and Victims Panel, 2012). Davis (2012) reports similar methods from chronically violence-stricken cities and districts in South and Central America.

13 However, see Walklate (2011) who explores the concept of resilience in thinking about the impact of criminal victimization.

be necessary for the protection of the public or even the life of a nation, have the calamitous tendency to be applied and integrated into the everyday methods of criminal prosecution and police work even in the case of minor threats' (Zedner, 2012, 37f.). Thus even simple forms of crime move into the context of threats to national security (Kaufmann, 2013, 1011 f.). But also with a tightened and optimal hazard and risk prevention, safety threats cannot completely be avoided – there is no such thing as absolute security. In case the unavoidable happens, protection and coping strategies must be developed that make up for an effective management for the remedy of consequences of possible damage as well as for the prevention of escalating adversities.

Resilience-based strategies of urban security do not anticipate a concrete event but rather, expect the unexpected. Where the formation of urban resilience implies that cities need to adapt more and more to conditions that are not precisely predictable, the controllability of future events is challenged (Lorenz, 2010). The optimism of a controllability of social threats hereby gives way to an understanding of the fact that safety-relevant problems can just not be eliminated but can only be minimized in their extent and intensity. In crime prevention, more and more advanced methods are conceived in order to make the occurrence of criminal events reliably predictable. By the means of innovative police methods, hopes are raised that crime can be made predictable spatio-temporally in the future (Johnson, 2010). In so-called 'predictive policing', a possibility is seen to contribute to the prevention of criminal acts by analysing information about crimes that have already been committed in combination with highly complex algorithms for the prevention of criminal acts. The preventive assessment of criminal risks, however, is subject to the condition that hazards are calculable and risky situations can be identified through objective criteria based on reliable data. But in risk society, 'the new risks are characterized by the fact that the occurrence of damage is anticipated and that damage is no longer considered to be the result of misplanning or human failure but that they be unavoidably brought up. Then it only comes down to be prepared for the consequences' (Groenemeyer, 2001, 161). This orientation does not only apply to technological or also ecological risks, but is also applicable for the addressing of deviant behaviour. Thus urban resilience can be seen as a logical continuation of crime prevention in the risk society. When crime must be expected – and routine activity theory assumes a motivated offender whose behaviour does not only seem rational but also rewarding – the society and its citizens must be prepared for the normality of crime. 'Resilience in this context does not mean to institutionalise more and more safety measures but to become independent of the thematization boom and strengthen social cohesion' (Gerhold, 2011, 7). A lack of social cohesion weakens informal forms of social control over potential targets.

In a recent article, Stefan Kaufmann and Sabine Blum have traced vulnerability and resilience as a 'wandering of ideas' in and between fields that debate the different risks and safety issues (Kaufmann and Blum, 2013). Taking a look at the development of urban resilience strategies, it can be shown that this development is largely based on principles of urban crime prevention. Anti-terror measures, which nowadays operate under the label 'Design out Terrorism', as well as the idea of 'Design out Crime', can be attributed to situational concepts as they are already generated within the 'defensible space' approach. Physical barriers are to protect against the threat of an attack, the way to the target is to be blocked, and the materials used are to be solid enough to withstand the impact of an explosion. After the terror attack on the World Trade Center in 1993, the *New York Times* wrote: 'Barricades and bollards have become the newest accessory on this country's psychic frontier ... You might call it the architecture of paranoia. They call it "defensible space"' (Brown, 1995). In the disaster discourse, its original idea has been boosted so much with physical and technical control mechanisms that urban resilience can easily be discredited as an element of an extensive surveillance society. In crime prevention, such a one-sided concept would run contrary to those efforts that try to anchor the socio-spatial conditions of urban safety in urban planning as a part of crime prevention and to overcome the tight understanding of situational approaches (Schubert and Veil, 2011). On the other hand, 'perhaps the opposite of a Big Brother State, a logic of resilience would aspire to create a subjective and systematic state to enable each and all to live freely and with confidence in a world of potential risks' (Lentzos and Rose, 2009, 243). Urban resilience can insofar be understood as a community's capability to establish common values and to sustain them by means of an effective social control, the crucial factors being mutual trust among the members of a community who are the ones who make the exercise of collective control action possible.

If urban resilience as a local strategy of establishing safety can adopt so many different shapes that resilience can just as well mean a tightened surveillance and technically supported space and access controls as well as the strengthening of social cohesion and mutual trust, then it is necessary to focus more on the frame conditions under which resilience and prevention interact. Douglas Wildavsky once considered the unit of resilience and anticipation as optimal for dealing with risks (Wildavsky [1988] 2011, 77 ff.). Nowadays attempts are made to put both approaches in a counter-proportion (Perelmann, 2007, 21). Here it shows that the concept of urban resilience, in the scope of urban crime prevention, can especially be considered to be promising when safety issues are addressed not only from a constructional perspective but also in their further social context.

# Chapter 5
# Resilience Discourses and Ignored Insecurities: The Neoliberal Myth of Self-contained Individualism

Monica Colombo and Luigi Ferrari

## Introduction

The concept of resilience has in the recent past rapidly 'infiltrated' (Walker and Cooper, 2011), 'proliferated' (Joseph, 2013) and 'infused' (Coaffee, 2013) vast areas of the social sciences as demonstrated by the impressive increase of publications across a variety of fields – from psychology, sociology, geography, anthropology to public health, ecology, technology and communications.[1] Originated in physics and mathematics, where it refers to the capacity of a material or system to return to equilibrium after a displacement (Norris et al., 2008), the concept of resilience has subsequently been applied to a number of objects, from ecosystems and the built environment to individuals, social systems and communities to describe both the adaptive capacities of individuals (for example, Bonanno, 2004; Butler, Morland and Leskin, 2007; Rutter, 1993; Werner and Smith, 1982) and those of human communities (for example, Brown and Kulig, 1996/1997; Sonn and Fisher, 1998), and of larger societies (for example, Adger, 2000; Godschalk, 2003). Resilience as an operational strategy of risk management has more recently been taken up also in financial, urban and environmental security discourses (for example Coaffee, Murakami Wood and Rogers, 2008; Adey and Anderson, 2011; 2012; Anderson and Adey, 2011).

Several criticisms have been raised over time as resilience discourses have become increasingly ubiquitous not only across a variety of disciplines, but also in politics and policy documents. As Brussett and colleagues (2013, 222) suggest, resilience can be so far regarded as 'the organising principle in contemporary political life'.

In this chapter, we attempt to highlight the importance of a critique of the idea of resilience. In particular, we focus on the proximity between resilience discourses, self-contained and economic individualism, and contemporary neoliberalism.

1 For instance, free-text searching across several databases using the term 'resilience' as the search term sorted 12,568 results in PsycINFO and 13,772 in WOS – Science Citation Index Expanded.

Taking critical discourse analysis as a broad theoretical framework for analysing the way power relations and ideology are (re)produced in discourse and the key role played by the psychological and the social sciences in furthering the (neo) liberal societal regime, we review and discuss the most convincing arguments offered by several scholars in the ongoing sociological and psychological debate about resilience and we attempt to unveil their ideological implications. The starting point of this discussion is the acknowledgement of the '*highly political* – nature of resilience' (Brussett, Croft and Vaughn-Williams, 2013, 221).

Particular attention is paid to the process of individualization which has undergone a dramatic acceleration in the time of economic crisis and which has implied a shift from social (economic) insecurity to individual/personal insecurity. As a consequence of this shift, self-responsibility is emphasized and collective insecurity (as, for instance, in the case of unemployment) is reframed as individual insecurity. From this standpoint, discourses on psychological resilience can be regarded as directly stemming from the neoliberal myth of self-sufficiency and from economic individualism. This shift is evident, for instance, in the case of temporary and 'esodati' workers who are pushed to internalize the causes of their insecurity while the case of workers' recovered companies is discussed as an example of resistance and counter-conduct.

## Psychological Resilience and Self-contained Individualism

Foucault (1977) has provided us with a basis for understanding how knowledge about the individual plays a key role in processes of societal control and the ways any conception of personhood is linked to regimes of social power. Taking a Foucaldian perspective, over the last decades, critical psychology has been steadily challenging not only the epistemological, ontological, theoretical and methodological foundations of mainstream cognitive, clinical and social psychology, but also their covert ideological assumptions. From the beginning, several criticisms have been raised in particular towards the dominant Western understanding of personhood – which psychology has been mirroring over time – for being based in great measure on a liberal individualist framework. According to Sampson's (1977, 1983, 1985, 1988) definition of self-contained individualism, the conception of personhood dominating in Western culture emphasizes firm self-other boundaries, personal mastery, control over the self and the environment, and 'an exclusionary concept of the person' (p. 15). The exclusionary nature of self-contained individualism is as 'if one were to draw a circle marking off the region of self from the region of nonself', and 'the circle would be drawn so as to exclude others from the region defined as belonging to self' (p. 16).

If we look at how resilience is conceptualized within the field of psychology, it is immediately clear that discourses on individual resilience are consistent with Sampson's definition of 'self-contained individualism'. Indeed, within the field of psychology, early research on resilience focused on individual traits to explain people's wellbeing; at the individual level, personal agency and the capacity to act in and on the context are regarded as key factors of resilience. Researchers focused on identifying the characteristics of individuals, particularly children and young people, who thrived while living in difficult circumstances, such as poverty and parental mental illness (Garmezy, 1991; Rutter, 1990; Werner and Smith, 1992). Since the early 1990s, the focus of resilience research has shifted away from identifying protective factors to understanding the process through which individuals overcome the adversities they experience (Luthar, Cicchetti and Becker, 2000) with resilience being examined across a range of contexts, including business organizations (see, for example, Riolli and Savicki, 2003), education (see, for example, Gu and Day, 2007), and communities (see, for example, Brennan, 2008). Definitional debates have characterized the development of research in this field; for instance, the construct of resilience has variously been defined as a trait, process or outcome (see Table 5.1); despite these differences, all definitions share a focus on individual 'privatized' personal resources and adaptive capacities as a core element. From this respect, according to Joseph (2013) 'the idea of resilience does two things. It supports the organisational structure of the advanced liberal societies through the assumptions it makes about social relations … And it supports the idea of an individuated neoliberal subject … It uses the idea of an external threat to encourage individual adaptation. People are encouraged to engage actively with a wider situation that is deemed beyond their control. Hence the emphasis is on adapting to and exploiting a situation, rather than trying to change the wider social condition' (p. 262).

Bottrell (2013) has recently suggested that in advanced capitalist democracies, 'the policy framing of resilience encapsulates the principle of individualisation that has been characteristic in the neoliberal realignment of social policy to economic aims. Resilience is typically represented in social policy as a matter of individual attributes that are integral to autonomy, self-invention and choice biographies' (p. 2). In *Inventing Our Selves: Psychology, Power, and Personhood* (1998) Rose proposes that psychology is now employed as a technology that allows humans to buy into an invented and arguably false sense of self. In this way, Foucault's theories of self have been extensively developed by Rose to explore techniques of governance via self-formation. In this vein, resilience discourses can be regarded as a technique used to responsibilize people's behaviour. The main effect of the resilience literature is to place the burden of responsibility on the individual while ignoring the causes of macro change.

## Community Resilience and the Ecological Metaphor

As mentioned above, the concept of resilience has subsequently been applied to describe both the adaptive capacities of individuals and those of human communities, and of larger societies.

In commenting the table, Norris and colleagues (2008) point out that both at the individual and community or social level, most definitions emphasize 'a

**Table 5.1 Definitions of resilience (adapted from Norris et al., 2008 and Fletcher and Sarkar, 2013)**

| *Citation first author, year* | *Level of Analysis* | *Definition* |
|---|---|---|
| Gordon, 1978 | Physical | The ability to store strain energy and deflect elastically under a load without breaking or being deformed |
| Bodin and Winman, 2004 | Physical | The speed with which a system returns to equilibrium after displacement, irrespective of how many oscillations are required |
| Holling, 1973 | Ecological System | The persistence of relationships within a system; a measure of the ability of systems to absorb changes of state variables, driving variables, and parameters, and still persist |
| Waller, 2001 | Ecological System | Positive adaptation in response to adversity; it is not the absence of vulnerability, not an inherent characteristic, and not static |
| Klein et al., 2003 | Ecological System | The ability of a system that has undergone stress to recover and return to its original state; more precisely (i) the amount of disturbance a system can absorb and still remain within the same state or domain of attraction and (ii) the degree to which the system is capable of self-organization (see also Carpenter et al. 2001) |
| Longstaff, 2005 | Ecological System | The ability by an individual, group, or organization to continue its existence (or remain more or less stable) in the face of some sort of surprise … Resilience is found in systems that are highly adaptable (not locked into specific strategies) and have diverse resource |
| Adger, 2000 | Social | The ability of communities to withstand external shocks to their social infrastructure |
| Bruneau, 2003 | Social | The ability of social units to mitigate hazards, contain the effects of disasters when they occur, and carry out recovery activities in ways that minimize social disruption and mitigate the effects of future earthquakes |
| Godschalk, 2003 | City | A sustainable network of physical systems and human communities, capable of managing extreme events; during disaster, both must be able to survive and function under extreme stress |
| Brown and Kulig, 1996 | Community | The ability to recover from or adjust easily to misfortune or sustained life stress |

**Table 5.1 Definitions of resilience (adapted from Norris et al., 2008 and Fletcher and Sarkar, 2013) (*concluded*)**

| *Citation first author, year* | *Level of Analysis* | *Definition* |
|---|---|---|
| Sonn and Fisher, 1998 | Community | The process through which mediating structures (schools, peer groups, family) and activity settings moderate the impact of oppressive systems |
| Paton and Johnston, 2001 | Community | The capability to bounce back and to use physical and economic resources effectively to aid recovery following exposure to hazards |
| Ganor and Ben-Lavy, 2003 | Community | The ability of individuals and communities to deal with a state of continuous, long-term stress; the ability to find unknown inner strengths and resources in order to cope effectively; the measure of adaptation and flexibility |
| Ahmed et al., 2004 | Community | The development of material, physical, socio-political, socio-cultural and psychological resources that promote safety of residents and buffer adversity |
| Kimhi, 2004 | Community | Individuals' sense of the ability of their own community to deal successfully with the ongoing political violence |
| Coles and Buckle, 2004 | Community | A community's capacities, skill, and knowledge that allow it to participate fully in recovery from disasters |
| Rutter, 1987 | Individual | Protective factors which modify, ameliorate or alter a person's response to some environmental hazard that predisposes to a maladaptive outcome |
| Masten, 1990 | Individual | The process of, capacity for, or outcome of successful adaptation despite challenging or threatening circumstances |
| Egeland, 1993 | Individual | The capacity for successful adaptation, positive functioning or competence … despite high-risk status, chronic stress, or following prolonged or severe trauma |
| Luthar et al., 2000 | Individual | A dynamic process encompassing positive adaptation within the context of significant adversity |
| Connor et al., 2003 | Individual | The personal qualities that enables one to thrive in the face of adversity |
| Butler et al., 2007 | Individual | Good adaptation under extenuating circumstances; a recovery trajectory that returns to baseline functioning following a challenge |

capacity for successful adaptation in the face of disturbance, stress, or adversity … Across these definitions, there is general consensus on two important points: first, resilience is better conceptualized as an ability or process than as an outcome (Brown and Kulig, 1996/1997; Pfefferbaum et al., 2005); and second, resilience is better conceptualized as adaptability than as stability (for example, Handmer and Dovers, 1996; Waller, 2001). Adaptability takes different forms' (p. 129). Norris and colleagues make a distinction between 'Engineering resilience', which makes a system return to one pre-designed state or function after a disturbance, and 'ecological resilience' which allows for many possible desirable states that match the environment (Gunderson, 2000). They regard the second type of resilience as the relevant one for human communities, organizations and societies. In their theoretical model of community resilience, resilience is regarded as emerging from a set of 'networked adaptive capacities' (p. 135). They identify four primary sets of networked resources: Economic Development, Social Capital, Information and Communication, and Community Competence. Despite the recognition of the fact that individual and community resilience are subject to larger sociological and economic forces (for instance, economic growth, stability of livelihoods, equitable distribution of income and assets within populations), resilience is basically defined as 'a process that leads to adaptation, not an outcome, not stability. We emphasized the likelihood that stress and crises induce transient periods of dysfunction, but adopted wellness as the eventual outcome of interest, the manifestation of adaptation to an altered environment' (p. 144). In Norris' model, the consistency between the idea of resilience itself and new forms of governance implying network forms of organization, individual responsibility and policy dealing with risk is not taken into account. In this vein, Norris and colleagues can claim that the strategic value of the idea of resilience 'lies not, or not only, in its inspirational message but in whether it leads to effective interventions and policies that increase the probability of adaptation by enhancing adaptive capacities' (p. 146). From this quote, it is clear that the idea of resilience, with its background in ecological systems, 'fits with new philosophies of the social order that emphasise a shift away from a focus on things like social structure, class, the state and stable identity in favour of a belief in contingency, complexity and uncertainty' (Joseph, 2013, 254).

## The Resilient Self, Security and Economic Individualism

As for the concept of resilience, the concept of security has also increasingly gained attention in the social and psychological sciences. In Chapter 3, Harald Arnold and Rita Haverkamp report a study on security and quality of life conducted at the Max Planck Institute for Criminal Law in Germany in 2012 which can be regarded as an example of a research tradition which aims at identifying and classifying the fears and worries shared by a given population. If we look at Figure 5.1, which shows the percentage of the population with severe worries concerning a variety of domains, it is possible to identify three main areas of perceived sources of threat.

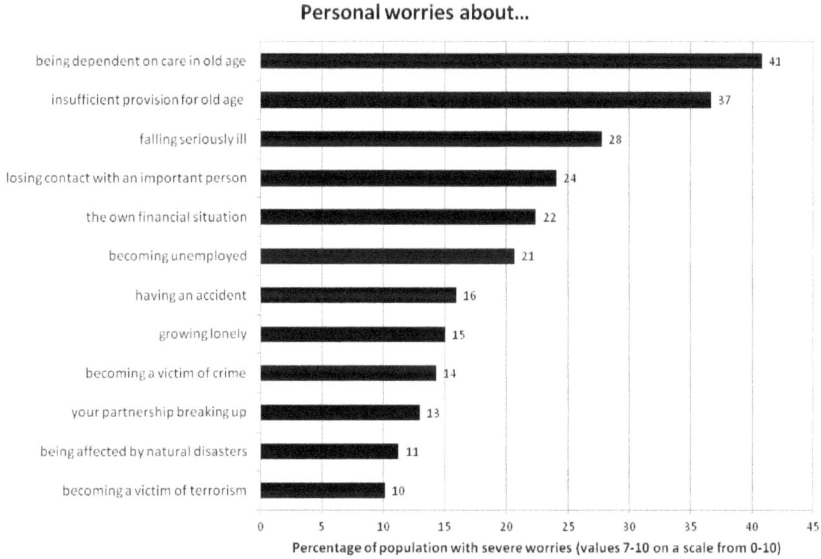


**Figure 5.1 Personal worries among the German population**

*Notes: 1)* external physical threats, such as natural disasters, accidents, crime, and so on; *2)* relational and social-economic threats, such as unemployment, financial difficulties, becoming poor, losing a partner, and so on; *3)* individual, such as disability and physical impairment.

*Source:* MPI-Survey, Security and Quality of Life in Germany' (2012), N=2.525

From a psychological point of view, each category represents a peculiar set of threats to the self, defined as the personal sense of physical and mental unity. However, if we look closer at these three categories, they all imply forms of self-dissolution and fragmentation which are more similar to one another than it may appear at a first glance. Indeed, the above-mentioned threats to the sense of self are strictly intertwined and often imply one another. Physical impairment, relational desolation, social 'superfluity' in the relationship with others, the incapability to act for oneself: all these experiences represent a form of negation, if not of dissolution, of the individual self.

This profound sense of threat to personal unity and this lack of all forms of security is symbolically represented in an extraordinary way in Kafka's novel *Die Verwandlung* (1915). As it is well known, the protagonist, Gregor Samsa, wakes up one morning and finds himself transformed – for unknown, enigmatic reasons – into an insect: he is unable to work and to support his family as he used to do before, unable to establish any relationships and to communicate, unable to take care of himself. Samsa is physically assaulted by his father and, in the end, he is psychologically attacked to death by the all family and Samsa's self, symbolically represented by the insect's corpse, is thrown away like rubbish by the servant. In his oniric apologue, Kafka has represented extremely well the anxiety associated to superfluity, which in contemporary society constitutes a synthesis of many basic insecurities. In his novel, anticipating the later development of the neoliberal conception of personhood, Kafka represents the isolated individual – this isolation is the core of the novel – who is constantly at risk to become superfluous due to uncontrollable and uncomprehensible reasons (Ferrari, 2014). In our view, the novel is very powerful in virtue of the *reductio ad unum* which operates of all the anxieties and worries which characterize the neoliberal subject and which are well documented in the study reported by Arnold and Haverkamp.

If we examine the MPI-Survey more carefully, a contradiction clearly emerges as well: respondents who are worried about isolation at the same time fear dependence from others: 41 per cent of the respondents are afraid of becoming dependent on care in old age. The results show that dependence is the most widespread worry among the respondents and here we will regard it as a more general fear going far beyond the issue of aging. It is noteworthy (see Ehrenreich and Hochschild, 2002) that care givers who are able to cover up their actual role adopting a set of concealing rituals and present themselves as if they were just keeping mere company to the care-dependent elderly on social occasions, are very much appreciated and better paid in the USA. Of special interest is the significance of the invisibility of careworkers: as Rivas (2002) highlighted in her study of the relationship between personal attendants and their charges, the disabled person needs the assistant to be invisible in order to maintain a sense of his or her independence. The need to offer the image of a resilient, autonomous and intact self is essential to such an extent that to this purpose individuals are willing to manipulate and to deceive themselves. The refusal to be labelled dependent is based first on the refusal to be infantilized objects of paternalistic concern, and second, on the supposition that the source of dependence is internal to the individual.

The respondents to the MPI-Survey seem to be wrapped by two opposing needs for security: they perceive both isolation and dependence as a risk. This contradiction has been grasped very well by Schopenauer who exemplified it with the myth of the porcupines (*Parerga und Paralipomena: kleine philosophische Schriften*, 1851) which was later taken on by Freud in *Massenpsychologie und Ich-Analyse*, 1921. When porcupines are feeling cold they cannot get closer to one another because they would sting themselves; they avoid contact, but isolated, they freeze. We can then draw an early preliminary conclusion: if the individual is atomized then the self is simultaneously threatened by depletion and devourment, because of the lack or the excess (surplus) of relationships respectively.

As mentioned above, the issue of security is strictly interconnected to the complex dialectic relationship between individualism and collectivism which characterizes our time. From this standpoint, we deem it necessary to address it, taking an historical approach in order to grasp present ongoing (occurring) changes. The porcupines' dilemma takes on different meanings in different historical periods and the importance attached to it varies accordingly. For instance, in traditional agricultural/pastoral civilization, dependence from others was accepted as a normal condition in any circumstances. Moreover, up to the modern age, dependence has being regarded as a value; at the same time, the self was conceptualized in a completely different way (see Ferrari, 2010).

During the Middle Ages, the individual feared such threats as starvation, physical aggression and disease (type i); consequently, what has been defined as 'medieval nervousness' didn't deal as well as today with relational and socio-economic worries (type ii) and with internal anxieties (type iii). Changes in the perception of security are well documented as well as the persistence of the agro-pastoral mentality in our age. For instance, from interviews conducted with elderly people who used to live in farmsteads in Pianura Padana (Lombardy), it emerges that, up to the 1950s, dependence was perceived as less threatening and distressing than it is today to the extent that workers accepted that they were to be locked up at night by the farmstead owner. It is noteworthy that elderly peasants, physically impaired and unable to work, living in extremely poor conditions, were sustained by the community without being disregarded by it. They experienced very harsh conditions of life, but they did not perceive any insecurity. This example, as well as many others, highlights the distance from the present situation. In Sardinia, up until recent times, poor families were allowed by the community to collect some head of cattle from others. Incidentally, medieval law codes include plenty of limitations to private property in order to avoid starvation. These 'thieveries' were not humiliating for the individual, on the contrary, they made the individual look like a *balente* – the Sardinian dialectal term for *valente-capace*, in English, 'smart'. The *balente* was the poor shepherd who rescued his family from starvation thanks to his dexterity, but limited the collection of cattle from others' property at what was strictly necessary to him to survive. The Sardinian *balente* has no affinity with the modern resilient subject: the abilities and capabilities he performed were part of a social, and thus collective, device aimed at safeguarding survival and security. While these 'thieveries' were socially allowed, the robbed was morally obliged to kill the 'thief' if he had exceeded the rule of the right measure, which was socially watched over in order to avoid any violation. At the same time, the head of a family who was not a *balente* and let his family suffer in poverty, was an object of collective scorn (contempt) and social dishonour (see Pigliaru, 1955).

## The Myth of Individual Resilience

Historically, the modern self has emerged in different times in different contexts. In Italy, it is a recent phenomenon which has accelerated since the Second World War and, even more, after the fall of the Berlin Wall in 1989 and the decline of any form of collectivism and socialism which in turn has allowed the liberal model of the individual to become hegemonic. At the same time, social capital as well as the importance of primary groups, starting from the family, has been constantly decreasing over time. All these phenomenon are connected to the economic growth which has characterized Western societies after the Second War World where traditional relationships have been substituted by market relations. Indeed, only if income constantly grows is it possible to buy services appropriate for reassuring the self (as in the case of the elderly Americans cited above).

This model has been seriously challenged by the economic crisis of 2008. We cannot say what will occur in the future, but we can suppose that the economic recession will significantly change the perception of security. What about resilience in this scenario?

Psychological resilience can be regarded as a typically individualistic construct rooted in the individualization of the subject which characterizes capitalism. The idea of resilience is also the product of a culture which does not conceive any long-lasting (endurable) standstill or, even, a degrowth. However, economic crises put in the foreground (emphasize) the fragility of security and resilience. Again, Kafka was prophetical. In his novel *Das Schloss* (1926), the protagonist, known only as K, is a land surveyor who, for an unclear mistake, has been summoned by the castle authorities. Nobody informed K about the mistake and his job for the organization never begins. He spends most of the novel doggedly trying to overcome the bureaucracy of the village and to contact the castle official Klamm, however he is continually thwarted and frustrated. The castle authorities keep silent. They will never deny the possibility of a future success and, even more important, that success will depend on K's capabilities. Success will simply never arrive.

Since the mid 1900s, the novel has been traditionally interpreted as a metaphor of the despotic and almighty power of institutions and bureaucracy. These interpretations belong to the period of maximum growth (expansion, development) of individualism and of the powerful economic institutions that emerged after the Second World War. Today, the novel can be regarded as a metaphor of the impossibility to enter into organizations. More generally, the novel can be regarded as the symbolic representation of the self-inherent inadequacy to cope with a competitive world. On the one hand, the protagonist lives in a condition of chronic insecurity and uncertainty because he constantly feels inadequate, he experiences the anxiety of not matching up, of not doing enough, of not having those necessary qualities which Kafka describes as shifty and never clear. On the other hand, K also represents the condition of the subject when, by exclusively self-referring his/her condition, he/she is unable to recognize the objective, and thus beyond his/her control, boundaries of his/her experience.

Beyond metaphor, in our view, the current economic crisis implies that individuals are placed at a crossroad:

1. they can do as the land surveyor K and rely on individual resilience;
2. they can acknowledge the socially determined origin of their insecurity and look for forms of social resistance and counter-conduct.

In the following section, we will attempt to exemplify these two options.

The first kind of approach can be led back to such well-known historical examples as the growth of neurasthenia in the USA in the nineteenth century, which has been defined as a reaction to the social discomfort provoked by the diffusion of the individualist model. This syndrome – also defined as the 'American nervousness' – can be regarded as a particular form of psychological rejection towards the economic growth, which is characterized by corporate concentration and a loss of importance of the role of local communities in the labour market (Bellah et al., 1985). Brown (1992) has examined the medical observations concerning the changes that occurred at that time in the mentality (attitudes) of Americans. The 'torment', later included as a symptom in the neurasthenic syndrome, stemmed from the lack of mediation between the individual and the labour market, which was by now nationally based (see Ferrari, 2010). In this market, self-esteem relied on self-promotion and on the capacity to positively impress employers in order to defeat competitors. According to Brown, competition has contributed to make individuals more adaptable, easy-fitting and flexible, but it has also increased both a painful perception of being inadequate and a need to take care of the self.

Other latest examples deal with the various forms of temporary work which have been increasing more and more as a consequence of the economic recession and the reactions to it. In Italy, temporary work is less legally protected than other forms of employment and implies a condition of objective weakness for the worker. Abuses and exploitation on behalf of the employer are feasible and can overtly manifest themselves. However, in this situation it can happen that the worker does not perceive (or neglects) the evident objective cause of his/her discomfort, focusing on his/her non-existent faults in his/her work relationships, thus self-referring and internalizing the causes of his/her condition (Ferrari and Veglio, 2006).

But if the social origin of discomfort (distress) is neglected, then insecurity becomes boundless. The results of research conducted on temporary workers (Ferrari and Veglio, 2006) show to which extent this condition is characterized by ambivalence. Temporary workers should be expected to show a low identification with the organization due to the fact that the work time in the company is too limited to allow for identification processes or for developing emotional attachment to the workplace. On the contrary, the results show that temporary workers are deeply psychologically tied to the organization even if their work satisfaction is low. The authors conclude that such a paradoxical attitude can be explained only by the fact that temporary workers are overwhelmed by the anxiety of a superfluous self and, just like the land surveyor K, they have to fight against the twin anxiety of being 'washed up' because of their own faults. Even a temporary job attenuates (diminishes) *pro tempore* the distress of becoming like a social insect and fills them with gratitude and positive feelings towards those whom can anesthetize this distress.

Temporary workers have to face a profound insecurity which did not exist in periods of economic growth. This does not refer to unemployment – which is inherent to capitalism and its way of production. We refer here to a specific kind of insecurity which stems from the internalization process mentioned above and which can be regarded as a secular (laical) version of the Puritan ethics of hard work as a sign of grace. Therefore, the reaction to social insecurity consists in emphasizing the individual's faults and in emending a 'guilty' self. Resilience can be regarded as part of this process, as a way to react to what is perceived as an individual failure.

The so-called 'esodati' workers – this term, coined by the Italian media in 2012, refers to the thousands of workers who have lost or left their employment as a consequence of corporate restructuring plans, trade union agreements or individual economic agreements and are not yet entitled to a pension, due to the increase in retirement age and to the tightening of the retirement requirements provided for by the Pension Law Reform – represent another dramatic example. From a set of interviews conducted with these workers it emerged that they tend to hide their condition and to avoid others' solidarity; they feel ashamed and blame themselves. In brief, while they have been victim of a social and political 'trap', they have internalized guilt. They report that these feelings and attitudes are shared by the people living in their social milieau, who tend to blame them and their families (Cigna et al., 2014).

In these cases resilience is not possible. In our view, the idea of resilience is an individual myth bond to liberal culture. The only way to face a condition like that of 'esodati' or temporary workers is to abandon the individualist framework informing liberalism. As documented by several emblematic examples, recognizing the social origin of insecurity allows for a shift from individual resilience to collective forms of action. We consider here the case of worker-recovered enterprises, which we deem one of the most relevant.

Worker-recovered enterprises are self-managed companies. A well-known example was the LIP factory, a watch and clock French company based in Besançon which bankrupted in 1973. More than 1,000 fired workers managed the company till March 1974 when they were rehired and mobilized the area, receiving the solidarity of the local communities. This collective mobilization supported the financial resurgence of the company until the firm was liquidated again in the spring of 1976. A more recent case is that of the *Fabbrica Brukman*, a textile factory based in Buenos Aires established at the beginning of the 1900s (Ferrari et al., 2006). As in the case of LIP, Brukman workers self-managed the company after the owners had left as a result of the effects of the Argentine economic crisis, the so-called *corralito/corralon*, at the beginning of 2000. In March 2002 the first attempt of evacuation by the police occurred. It is noteworthy that a neighbourhood *cacerolazo* (*cacerolazo*: citizens' mobilization protesting with saucepans and lids) forced the police to withdraw. The customers' support and the neighbourhood solidarity gave rise to a new relational network which allowed the female workers of the Bruckman to resist five additional attempts to remove them from the factory. Since then – and especially after the 2008 crisis – the number of self-managed companies has also been increasing in Europe, for instance in Greece, France and

Italy. A common element is the search for establishing relationships with residents in the areas where the companies are based. This is happening nowadays for Ri-Maflow, a company which, before becoming bankrupt, produced air conditioners for BMW, based in Trezzano sul Naviglio (Milan). The company has been recovered by its workers and it has now become a reference point for unemployed people in the area.

The seamstresses working at the Brukman told about their initial astonishment and desolation when confronted with the closure of the factory. Their astonishment and a sense of being abandoned have initially caused the weakening, if not the dissolution, of previous social relationships in a process of atomization of individuals. After the company became bankrupt, for a period of time, each seamstress experienced a sense of being trapped in her impoverished and helpless self. However, a resocialization of their experience quickly followed. They reacted as most Argentinian citizens were doing, gathering in the *Asambleas Barriales* (neighbourhood assemblies) as a way to contrast the collective and personal devaluation and the uncertainty which characterized those dramatic days. Indeed, the seamstresses soon began sharing their worries and fears and they formed small groups in which they could experience new forms of relationships. They successfully broke the shell of individual guilt and of individual shame.

As a conclusion, we would like to report the words of an Argentinian psychoanalyst, Esteban Gomez, who – like everybody in Argentina – was overwhelmed by the economic crisis. He admits that he rescued himself not thanks to his professional individualistic approach, but by participating in the *Asambleas Barriales*:

> The positive side of this experience resides in language and in the possibility it offers to give new meanings to ourselves and to what happens to us. It is not the same to lay on the sofa insulting politicians and to attend a meeting with your neighbors in order to plan the repainting of a square ... You still don't have money in your wallet, but you feel better, you regain dignity and self confidence ... If to this first level of subjective, personal and individual production we add other people living in the same condition, we reach the highest level to which an individual can locate, that is social participation, no longer as an individual, but as a citizen (cited in Ferrari et al. 2006).

## Resilience, Security and the Neoliberal Governance

In a recent special issue of *Politics*, Joseph (2013) argues that the idea of resilience is 'an anglo-saxon import' and that it has more to do with particular forms of governance than with security. For instance, in comparing current British and French security discussions and in pointing out the significant differences between the two approaches, he suggests that 'these differences are largely historical and have less to do with security ... than with an understanding of the governance of populations and the strategies and techniques used to responsibilise people's behaviour, then we can see why it is that the term is seen as having an Anglo-Saxon bias' (p. 253).[2] According to Joseph, the idea of resilience has become so popular in the English-speaking world because 'it fits easily into the current political vocabulary. It is consistent with new ideas about forms of governance, network forms of organisation, individual responsibility and ways of dealing with risk. More specifically, the idea of resilience, with its background in ecological systems, fits with new philosophies of the social order that emphasise a shift away from a focus on things like social structure, class, the state and stable identity in favour of a belief in contingency, complexity and uncertainty' (p. 254). In a society organized in response to risk (Beck, 1992) and preoccupied by safety (Giddens, 1999), the idea of networks promotes the individualization thesis by claiming that collective forms of organization are outdated and being replaced by new, more dynamic and more flexible social forms where individuals are 'nodes' in an interconnected system. Risk awareness and the risk management of both vulnerable and 'dangerous' citizens are central features of the contemporary ideology of welfare. As Palidda (Chapter 2, this volume) points out, insecurity and security has taken on increasing importance and political weight since the 1980s and more so since the 1990s, first of all, in the United States and then in Europe and the rest of the world. Even more so after 11 September 2001, the concept and the practices of security governance in all countries have given priority to the fight against insecurity understood as urban delinquency, irregular immigration and terrorism. According to Palidda, this approach has been legitimated thanks to a 'mass distraction' incited by the 'governance through fear'.

In his analysis of the impact of neoliberal capitalism on society, Di Palma (2013), problematizes neoliberalism as the transformation of social collective solidarity into risk management and individual responsibility. He shows how the 'ownership society' that characterizes neoliberalism is based on possessive individualism and commodification, and on the decline of the notion of collectivity. These economic, social, cultural and political transformations seem to be hardly consistent with the emergence of new forms of community organizing foretold by liberal–democratic thinkers such as Schumpeter, Keynes, Polanyi, Mills, Galbraith, Elias, Castel, Bauman, Bourdieu and others.

Indeed, market ideology subverts the modern state and social solidarity while at the same time, as Joseph (2013) suggests, strategies and techniques of governance operate though the encouragement of free conduct: 'Hence governmentality is most meaningfully understood, certainly in today's world, as a liberal form of rule.

2 Joseph (2013) suggests that 'as an indication of how influential the idea of resilience has become in the UK, the Cabinet Office has its own website devoted to its promotion. Here we find resilience defined as, 'The capacity of an individual, community or system to adapt in order to sustain an acceptable level of function, structure and identity' (Cabinet Office, 2011, 4). This definition is taken from a pamphlet by Demos (Edwards, 2009), the UK think tank once close to New Labour.

By liberal we mean that it works through encouraging free action and individual responsibility and decision-making. We might further distinguish between liberal and neoliberal forms of rule and distinguish the latter as a more assertive strategy of introducing the norms and values of the market as the means for assessing social conduct. Society is subject to the "dynamic of competition" (Foucault, 2008, 147) where people are expected to apply the logic of enterprise to their actions. Neoliberal governmentality promotes this as an exercise of freedom, where individuals are freed from social constraints and empowered to act' (p. 259).

This is clearly evident in the way resilient discourses encourage people to become active citizens who, rather than relying on the state, take responsibility for their own social and economic wellbeing. Institutionally, this takes the form of devolved state powers and belief in the need to empower local actors and the private sector.

The historical context for this neoliberal approach is the belief that the institutional settlement of the post-war period is no longer able to deal with a new set of social problems. This justifies the dismantling of the welfare system and the programmes of privatization and deregulation characteristic of the last three decades (see Palidda, Chapter 2, this volume). Contemporary governmentality is driven by the idea that one must not govern too much (Foucault, 2008, 319) and that government should be aware of the need to limit its own actions in order to encourage free conduct in its 'natural' (that is, artificially constructed) environment. Freedom might be understood as the independence of the governed (Foucault, 2008, 42) based on the belief of encouraging rational and autonomous behaviour free from state interference.

The idea of resilience can be regarded as part of this neoliberal rationality of governance; of particular importance is its promotion of the idea of responsibility which runs from the state down to the community right down to the micro level of individual conduct.

A highlighting example is presented and discussed by Bulley (2013) who argues that the UK Government's Community Resilience Programme is less about responding to disasters and more a matter of producing community and governing its behaviour: 'The passing over of responsibility to local volunteers and organisations is not only about empowerment, but also about forming identities and relationships that can be more efficiently managed and directed. However, this attempt is hamstrung by its basis in a nostalgic, romantic view of community and the effacement of poverty and inequality as central to the vulnerability/resilience binary. The effect may be a more intense government of communities rather than their empowerment through resilience' (p. 265). In his paper, he discusses a key aspect of the programme, that is 'the ignorance of poverty and inequality as hallmarks of vulnerability and its treatment of community as an unproblematic "good"' (p. 270). Vulnerability is in large part the 'product of social inequalities (gender, racial and socio-economic inequalities). Why are poverty and inequality entirely absent from the UK community resilience agenda? Bully's answer is that 'because this would require local and central government spending and policies targeting "equity in hazard vulnerability, focusing on poorer areas" of the community' (Godschaulk, 2003, 141). But being part of the coalition's 'Big Society', which follows on from New Labour's 'Third Way', the aim of the agenda is rather to make local people themselves responsible for their own resilience (p. 271). As Rose (2000, 1,400) has pointed out, in this 'double movement of autonomization and responsibilization' communities and individuals are granted freedom within certain limits, but made morally and financially responsible for their own success and failure. The idea of resilience works to manage expectations and behaviour in a neoliberal political economy of a government which aims 'at accomplishing more through a lesser exertion of force and authority' (Gordon, 1991, 24) by 'governing at a distance' (Rose, 2000, 48–49).

According to Chandler (2013):

> Resilience understands the problems of conflict, underdevelopment and a lack of rights and democracy as being self-generated products of communities or societies themselves. For this reason, the solution is not seen to lie merely in transforming policy governance at the level of elites or installing new formal institutional frameworks. The ideological power of resilience is driven by the understanding that 'we' cannot fix 'their' problems but, equally, that they cannot be expected to break out of the reproduction of these problems or 'traps' without external assistance. Changing or adapting behaviour and understandings needs to come from within; resilience cannot be 'given' or 'produced' by outside actors, only facilitated or inculcated through understanding the mechanisms through which problematic social practices are reproduced (p. 277).

# PART II
# Specific Aspects in Security and Insecurities Governance

Chapter 6

# Neoliberal Decriminalization of Business Law: The French Case Study and Similarities with Other Democratic Countries[1]

Mikael Kazgandjian

This chapter provides an analysis of changes in French business law as a case study, revealing a shift in what is considered legal and illegal in the current frame, which coincides with the turn towards neoliberalism in almost every country and at world level.

In recent decades there has been a considerable decriminalization regarding the rules of business law and at the same time a strengthening of the penalties for crimes such as offences relating to drugs, attempted theft, 'urban incivilities', illegal immigration and ecologist mobilization considered as a form of terrorism, and so on.[2] This has resulted in a sharp increase in arrests and a prison population increasingly made up of marginal people, while there are fewer sanctions for 'white collar' crimes. The 'production' of police and the penalties decided by the courts have also been affected by the mainstream discourse that classifies crimes attributed to 'undesirable' people as the most intolerable or dangerous for the 'good of society', that is, for neoliberal law and order. I aim to show that both the neoliberal discourse and power strategies that have been devised as a rational government are based on the principle of '*laissez faire*' and arguments such as 'dangerousness' and 'zero tolerance', concepts that reinforce each other and create the need for security technology that is inseparable from them.

However, crimes regarding the area of business law often have particularly serious economic and social effects on the lives and work of a large part of the population. Frequently they also have adverse effects on public health and on the environment, even if these are not immediately direct: in particular disasters provoked by property speculation, activities and products that increase pollution, accidents or diseases and so on (see Chapter 1).

1 English revised by Deborah Kay.

2 See Chapter 7 by Quassoli and Colombo and Chapter 2 by Palidda.

## The World of Business Illegalities

Among the authors of the so-called 'Chicago School', particular importance can be attributed to Edwin Sutherland (1883–1950) and to his work on white-collar crime. According to this author the many studies on illegality which refer to the 'social disorganization' concept focus their attention on the crimes committed by the popular classes, easily visible both statistically and socially. However, more often there are other ignored forms of illegality that are more difficult to grasp. Thus, he proposes we rethink how to approach these phenomena and further extend the field of study by incorporating financial illegality, and become interested in all breaches of the law whether or not they result in criminal prosecution. Apropos of empirical research, it is interesting to recall that Sutherland uses a biography (the masterful example is *The Thief*, 1937), to trace the course of a professional criminal career, in order to understand the rules and codes of conduct and to explain the learning and internalizing it can generate. In his most famous book, *White Collar Crime* (1949),[3] which was censored[4] at the time, he coined the concept of 'differential association' in some way as an alternative to 'social disorganization', as it seeks to describe the individual process of the interpretation of legal provisions which may favour transgression of the law.

It is interesting to recall here the conceptual neologism of 'illegalism' (Lascoumes, 1996, 78–84) that Foucault adopts some decades later, especially in *Discipline and Punish* (1975); this reference allows both for the overcoming of the criminological notion of delinquency as well as the legal categories of crimes (infraction, offence, crime and so on) and of criminal proceedings; it also helps our understanding of what power does in terms of the differential management of illegalities (Fischer and Spire, 2009).

According to Sutherland, the process of interpreting rules of law is the result of learning and therefore of the internalization produced by proximity with illegal worlds, but also of porosity with legitimate professions (bankers, politicians, lawyers, judges, police and so on). So this type of deviancy is not an ideological and moral one, thought out in terms of social disorganization and attributed to an individual by a psychologizing analysis of this phenomenon. To this end, Sutherland also analyses the special treatment of 'acts committed by individuals of high social status commensurate with their economic and professional activities'. If they are offences 'they are treated as if they were not crimes, with the effect – and perhaps the goal – of eliminating all stigma referring to crime' (Sutherland, 1945, 10).

If this argument is seen as paradigmatic it can help uncover a fragment of ignored insecurities in contemporary France, although there is a strong disinterest

---

3 Despite its importance and the fact that it is often cited, this book has not yet been translated into French, but it continues to influence some French researchers, including Thierry Godefroy and Pierre Lascoumes.

4 See Lascoumes (2013): Only in 1983, 47 years after the first edition was published, was the real version of the now famous 'White Collar Crime' finally made accessible.

in social science vis-a-vis white-collar illegalities, and a certain unwillingness to make it the main object of an essential case study.

It is: 'An object (argument) still surprisingly neglected in France and, to a lesser extent, by the French academic world, including those areas specialised in the analysis of rule violations of all kinds' (Chantraine and Salle, 2013).

And yet this is a subject that is well-covered by the media, although it is often obscured or veiled by sensationalistic description, and an approach that is perhaps overly justicial, populist and sometimes psychologizing, but it is also deprived of a politico-sociological reading and therefore an understanding of the observable phenomena. The challenge is to show that illegality within the business world[5] is just as common as illegality in the popular world as a sociological phenomenon.

A retrospective look, starting from the 1970s and concentrating on the early twenty-first century, reveals three highlights in the French case: (1) the decriminalization process of the rules of law concerning business activities; (2) the stability of penal control in business law despite very strong growth in trade and finance; (3) the demolition of specific law devices in business affairs.

## An Outline and Fragments of a Genealogy of Business Law

I do not aim to deconstruct and reconstruct the history of business law, but to show the utility of a genealogical approach (Foucault, 1977, 163–177).

The genealogy is not opposed to history but rather to the meta-historical deployment of ideal meanings and undefined teleologies. It opposes the search for the 'origins'. (Foucault, 1971). Through the study of three emblematic cases it is possible to go back in time to find moments of criminalization in French business law which led to strategies. That is to say, times when the standard was discussed and disputed before being incorporated into the law, and therefore by the state, regardless of the reasons for its origins.

We can agree on the fact that criminal business law appears as a discipline under this name in 1960, but it actually dates further back in history, especially when it refers to the issue of trade money, and many ancient civilizations applied sanctions against those who violated the rules and codes in this area. However, criminal action was limited to the money sector, and there were no real criminal courts. Previously, there had been little need to create firm business laws, but

5 I refer to the concept of world according to Becker (Becker, 1982 and 2006, 163–180), that is, a metaphor that opens multiple possibilities for the interpretation of reality: the discoveries during immersion in social life that contains and consists of real people, the interactions and actions that are useful to their goals; the results of negotiations between them (that is, their agreement); everyone interacts and reacts – consciously and unconsciously (thought and un-thought co-exist). They gradually develop their own reactions, trying to adjust these (this can lead to adequate or inadequate results in terms of their convenience – as Goffman (1974, 9–14) suggests).

during the eleventh century, when a number of the European nobility left for the Holy Land to take part in the Crusades, they began to understand the importance of trade to supply their campaigns. The proof lies in the institution of the trade fairs in Champagne, Flanders and Antwerp (Le Goff, 1956 and Braudel, 1979). These fairs usually lasted six weeks, during which there were only two weeks of exchanges, and four weeks to complete post-regulation exchanges. These fairs created the need for a range of commercial and financial institutions such as the Bill of Exchange and consequently, a series of sanctions.

Certain events, which should be outlined, helped to develop the various areas of criminal law. The first major case regarding financial crime to take place on French soil was the case known as the 'Templars', that is, the trial of the Order of the Temple in 1307 (Demurger, 2005). The Templars were both a religious and military order that had been established during the Crusades, and its members were involved in the movement of goods and money necessary to their purposes. This allowed the order to accumulate a large fortune and considerable power, but the King of France, Philip IV, or Philip the Fair, resented the importance of the order and the wealth and power they had accumulated, so he set out to destroy them. He ordered the arrest of a hundred important members and also the grand master of the order, Jacques de Molay (Demurger, 2002).

The order was accused of committing fraud to avoid paying taxes, and various other criminal offences. The conclusion of this case was the execution of Jacques de Molay, who was burned at the stake on the island of the Jews in Paris.

In terms of tax evasion, and from a genealogical point of view, the Jacques Coeur case (Heers, 2008) is significant. This case was named after the merchant and supplier of currency who was appointed master of the mint by King Charles VII of France in 1436 (Ruffin, 2012), and who began trading, making vast amounts of money for himself, thus provoking the king's envy. Charles ordered his arrest for having transferred a portion of the funds he had raised to foreign countries, notably to the cities and ports of the Ottoman Empire, without royal permission, and also for issuing securities without authorization. Jacques Coeur was thus accused of diverting abroad a portion of the funds that would normally reach the royal coffers. Prosecution was inevitable and Jacques Coeur was imprisoned, although he later escaped and spent the rest of his life in exile.

However, he was rehabilitated *post-mortem*, under Louis XI, who did not hesitate to condemn the greed of his predecessor. Balzac, who was only too willing to state that the judge is 'the most powerful man in France' in *Splendeur et misère des courtisanes* (1838–1847), paid homage to Jacques Coeur in his book (Balzac, 1845, vol. XI, 488).

Finally, it is worth mentioning the story of John Law of Lauriston, as recorded by Adolphe Thiers.[6] Law was a rich Scottish heir who had been sentenced to death in Scotland, but managed to escape to France, and whose name was subsequently

6 See Thiers (1858). Historian and President of the Republic, Adolphe Thiers was interested in the Law affair and his finance system; he wrote the story when he was 29 years

Frenchified to 'Lasse". However, he gave his name to the original 'Law System' which became known as Case Law in 1720.

Mr Law was a keen observer of gambling and these observations enabled him to accumulate knowledge and considerable wealth. He was also appointed Controller General of Finances in France, and the French Regent gave him the opportunity to introduce paper money into France through a bank linked to the General East India Company, although the transactions were frantically speculative.

The codes and rituals of traditional trade were upset by the principle of exchanging precious metals for paper money. The issue of securities was initially very successful, but Law was called by the King and forced to give back some of the money for military operations. Securities were guaranteed for a quarter of the value, but for the remaining three-quarters, the value depended on Law's business prosperity; this led to a insider trading generating a rumour that reversed the frenzy in the opposite direction, causing the temporary end of system. Despite being successful for several years, the experience ended with the first bankruptcy in French history. Afraid of being put on trial, Law took voluntary exile in Venice, but he had led a significant portion of the French population to poverty and ruin.

This historical fact, together with the failure of the *Assignats* experiment during the revolutionary period, was one of the principal causes of the French people's distrust of paper money. This country was also the last in Europe to establish a banking system, in the nineteenth century.

## From the State Penalties of the Nineteenth to Twentieth Centuries to Neoliberal Decriminalization

In the Napoleonic Penal Code of 1810, there are very few provisions under the Penal Business Law – only articles 402–418 relate to this field. Article 402 concerns bankruptcy, while other items relate to sanctions against opening a pawnshop without permission, violating a trade secret, illegal price fixing, and finally a provision regarding auctions. To return to the concept of bankruptcy, the origin of the term comes from the Italian *banca rotta*, in reference to the bankers who, following a bankruptcy, were made to break the public bench where they did their business and were forbidden to practice their profession. However, the process of the criminalization of economic life was more intense in the early twentieth century, and primarily concerned offences in relation to consumption (Law on fraud in 1905), and diversions committed by corporate leaders (with the misuse of corporate assets), and finally related to the life of businesses. During the years from 1960–1970, the field of financial markets and the stock market were added. This process concerning sanctions took place during a period of growth in the role of the state, which had been growing since the English and French

old and was not yet an academic or minister, www.academie-francaise.fr/les-immortels/adolphe-thiers, accessed on 17 September, 2015..

revolutions, both under less authoritarian regimes as well as more absolutist and dictatorial governance. To this end, Stenson and Sullivan (2001) wrote this about the UK, although the words might apply anywhere:

> For more than three centuries, or since John Locke, liberal states have practiced a schizophrenic form of criminal justice. One form of such justice is for the rich, enlightened, and occasionally blessed. Quite another is for the 'noxious' and 'degenerate' underclass that have declared themselves quit of the principles of human nature.

This process was halted or even reversed with the advent of the neoliberal revolution that began in the 1970s (in the name of freedom of action for businesses, to promote economic development). This reversal of movement, that is to say the decriminalization of business law in France, began in the mid-1970s and intensified between the late 1990s and the early 2010s. Laws regulating new economic or financial security are proof of this. These laws repeal dozens of offences, representing more than half of the penalties of company law. The idea was that the inflation of criminal law into business law could lead to an inconsistent and unpredictable criminal risk, and threaten economic activities.

Therefore, the life of businesses was decriminalized in favour of corporate governance. Today, convictions relating to economic and financial legislation represent less than 1 per cent of the crimes punished by the courts. In this perspective, the neoliberal frame that has been imposed since the 1980s, and especially since the 1990s, is legitimized by the discourse that is summed up in the maxim 'less government, more market', in other words more freedom of action for economic actors and of course for the stronger players.

The Italian example is paradigmatic of the evolving rules described above.[7] The collapse caused by the current economic crisis is just one aspect of the most serious consequences of this process. It is important to note that the Obama administration gave the impression of wanting to stop or limit the damage primarily in response to the Madoff scandal and the subprime scandal and so on. But we can also note that the so-called output measures of the crisis have almost always tried to save the banks and finance companies responsible for the crisis and have only increased the adverse consequences for most of the population of the planet (Lascoumes, Godefroy and Favarel-Garrigues, 2009). Proof of this is the continuous increase in the distance between rich and poor and the freedom of action for financial speculators; freedom

7 [Note by the editor] In fact, the Italian case can be considered as a good example of how it is almost impossible to impose a rigorous democratic regulation of business laws on the most powerful actors. Laws are made by parliamentarians who are elected with the decisive support of these actors: thus democracy is a decoy; economic, social and political regulation always depends on the relationships between these forces. The economic and political power, and then the law, are always in the hands of those who also have the military and police power (see Joxe, 2012 and Palidda, 2010).

in each country for the greatest excesses of the financial sector through transnational games, among others, the use of tax havens, even within the European area (Roux, 2010; della Porta and Vannucci, 2012; Ruggiero, 2013).

## The Impact of Dismantling the Repressive Features of Criminal Law Cases

According to Foucault any dispositif is registered in a relationship of power, both explicitly or implicity, and it implements and builds knowledge (Foucault, 1977). The regulation of current illegalities in the various segments of society, as among the spheres of power, reproduces itself firstly through the force relationships. Legitimacy does not always correspond to legality; on the contrary there are informal rules that are considered absolutely legitimate in certain social circles or segments (as Simmel, 1999/1894, suggests – see also Palidda, Chapter 2).

These will function as a circle of social and moral legitimation and as a coalition of economic interests. Within the circles of the dominant or stronger players, illegalities and crimes are considered legitimate, as in any social circle which uses an illegal and also a legal economy to sustain its gains. Therefore non-codified rules exist that are hidden but socially shared, and that can be quite contrary to penal, civil or administrative rules. In this regard Foucault said that:

> Only fiction can believe that laws are made to be respected and that the police and the courts exist for their enforcement. Only a theoretical fiction can believe that we agreed once and for all the laws of society to which we belong. Everyone also knows that laws are made and enforced by each other. But it seems that we can take a step further. Illegality is not an accident, but a more or less inevitable imperfection. This is an absolutely positive social functioning, whose role is expected in the general strategy of the society. Any legislative device provided protected and profitable areas where the law may be violated, others where it can be ignored, others again where offenses are punished. Ultimately, I would gladly say that the law is not made to prevent a particular type of behaviour, but to differentiate ways of circumventing the law itself' (Foucault interviewed by Roger-Pol Droit, 21 February1975, 16).

We are thus dealing with a game of 'anamorphosis of the rule of law', that is, the possibility that what is illegal becomes legal (Palidda, 1992). This is especially typical of the illegalities carried out by the strongest actors, but also of crimes that are considered legitimate and justified and seen as absolutely acceptable in the eyes of much of the population. To this end, we can draw four general conclusions (Godefroy, 2010, 31–34):

- 25 years ago the police dealt with 100,000 annual cases regarding economic and financial legislation, but they have dealt with approximately 20,000 per year over the last decade.

- The number of convictions by the courts in economic and financial matters has remained steady. It fell from 28,467 convictions in 1990 to 27,152 in 2008. There has been no distinct rise in the number of sanctions or the severity.
- Despite the introduction, in 1994, of the criminal liability of legal persons and its generalization since 2006, less than 1,000 companies are criminally prosecuted each year and when they are, a quarter of them are acquitted, a rate six times higher than that observed for individuals.
- Half of the fines imposed on legal persons do not exceed 3,000 euro, while the Penal Code provides five times the amount of fines for natural persons.

Parallel to the dismantling of specific law devices, control is exercised increasingly by prosecutors under their minister. This control over prosecutors is enshrined in the Perben II Act of 9 March 2004, and reinforced by the recent law of 13 December 2011; it concerns the allocation of litigation and the relief of certain court proceedings and extends to all corruption offences, fraud and breaches of trust and also the opportunity to have an arraignment proceedings on recognition of prior conviction (CRPC), that is to say, to 'plead guilty'.

However, more broadly in France (Delmas-Marty, 2004/1986, 55), there are currently nearly 15,000 types of criminal offence (Jean, 2008, 32). Knowing that the Criminal Code of 1810 cited around 500 types of criminal offences, there has been a frenzied proliferation of rules over the last two centuries. Nevertheless, crime statistics suggest that only 200 offences are regularly committed; among them a small group of 60 offences make up 90 per cent of convictions (theft, drunk driving, drug use, violence and so on; criminal law case offences are certainly excluded) (Ibidem, 32).

According to Kitsuse and Cicourel (1963) it is possible to say that police, legal and prison statistics only measure what these institutions (the state) want and what can be measured (that is, according to their discretion or free will).

### A Synthetic and Technical Description of the Coulon Report and Reactions to the Judicial Doctrine That He Raised

In a letter to Jean-Marie Coulon, former President of the Paris Court of Appeal, who chaired the commission in charge of a working group on decriminalization in 2007, the Minister of Justice, Rachida Dati, underlined the need for the decriminalization of business law. She stated and therefore conditioned the objectives of this working group around three main points before they began debating them:

- It was to examine all crimes that affect businesses.
- From there the proposal to remove unnecessary sanctions, or propose their adaptation or replacement by civil or commercial proceedings, or

administrative, disciplinary sanctions or fines was made. As such, it was estimated that some sanctions were 'inadequate', citing the fines that punish threshold crossing without declaring inequity, or increasing contributions in kind for SA or SARL.
- Finally, when criminal sanctions and civil or administrative sanctions co-exist, the working group was to question the justification for this combination, and the coordination of these different types of sanctions.

During their work, the working group cited deception alongside embezzlement and fraud as the main offences that form the basis of criminal law cases. They made 30 proposals, some of which were the object of criticism by the doctrine. The most important criticism was focused on the 25th proposal of the Coulon report which stated the intention of:

> Changing the rules of the requirement for public action by posing as a starting point intangible date facts and extending the statute of limitations (Dechenaud, 2010).

This proposal created difficulties regarding the offence of the 'misuse of corporate assets' where the limitation period could start running from the date on which the allegations were proven, or known about, if there had been a cover-up. That also holds jurisprudence. This proposal is not without precedent, it aimed to contradict the case law developed in the mid-1930s concerning occult offences. The Supreme Court declared, with regard to offences including the misuse of corporate assets, that the limitation period for public action was to run only from the date on which the hiding of facts ceased.

Such reform was thus to substantially weaken business law. With the Coulon report the starting point of limitation would be from the constitution of the offence. For doctrine, the risk to the conclusions of this report was that if business leaders concealed an offence, such as the abuse of corporate assets, they could remain unpunished, even though the working group recommended an increase in the limitation period.

## The Discourse of Power as a Strategy for Getting around the Coulon Report

Foucault said that the aim is to show discourse as a strategic field because it is common and it may become both a place and a confrontation instrument:

> This is primarily because speech is a weapon of power, control, subjugation, qualification and disqualification; that is the issue of fundamental struggle.[8]

8 Foucault (1977). Here there are some aspects similar to Goffman's suggestions.

In the balance of force, discourse is not just a surface registration but an operator. In other words, discourse is the same content that is configured as knowledge and thinking power and thus represents a 'science'. In this case, it is a legal science directly related to financial and economic science, which is then translated into the practices of the police and the judiciary on the one hand, and on the other, into the practices of businessmen in the business world.

Shortly after his election, on 30 August 2007, during his speech at the MEDEF University, the then President of the Republic, Sarkozy said:

> The criminalization of our business law is a serious mistake, I want to stop it. It must give the French people the taste for risk and entrepreneurial spirit, but how can it be done if financial risk is systematically added as criminal risk? If an error of management can land you in jail?

However, this speech expresses the absolute opposite of what Sarkozy says about so-called 'urban incivility' or 'juvenile delinquency': see 'scum'.

On 12 September 2007, the government spokesman, Laurent Wauquiez, said: 'There will be no decriminalization of misuse of corporate assets' (*Le Monde*, 13 November 2007) but the question about the decriminalization of business law is not 'new' because it was 'triggered' by the law of 15 May 2001, regarding new economic regulations, adopted under the left-wing government of Jospin.

On 4 October 2007, the then Minister of Justice, Dati was setting up the working group to draft the report and Coulon estimated that:

> Excessive criminalization of risk … hinders economic activity. This is an obstacle to its development, it is a diffuse and difficult risk to identify, which paralyzes economic initiative, instead of being just a frame to prevent violations.

For its part, the Union of Magistrates (Syndicat de la Magistrature) chose not to participate in the relevant commission, preferring to recommend comprehensive reform aimed at establishing 'an ethic of economic responsibility' (Mediapart.fr, 17 February 2008). As an anecdote, the day the working group on the decriminalization of business law began its work, the EADS case made headlines in the press (*Le Monde*, 4 October 2007).

## The Result of These Proposals and the Current Situation

Due to the economic crisis of 2008, but also the political and/or financial scandals that have erupted – for example, EADS, Clearstream, Kerviel, Bettancourt Woerth, Sarkozy-Gaddafi Libyan money, Strauss-Kahn, Tapie-Lagarde, Karachi-Takieddine, Dassault system, Amazon, Cahuzac, Bygmalion, or the recent revelations about Switzerland HSBC customers *(Le Monde*, 4 March 2015)[9] – the the main proposals for the transformation of the bill made by the Coulon report could not be put into practice.

However, some proposals in this report, such as the 25th proposal for the limitation period, were considered in 2010, in front of the future Code of Criminal Procedure project. The Coulon report is thus in the same spirit as another report by the Léger committee (2009), which called for the abolition of the judge, without giving the prosecutor independence vis-à-vis the Ministry of Justice.

Today, however, the legislative trend rather leans towards a return to the criminalization of business law. The most recent example is reflected in the nomination of a National Financial Prosecutor Eliane Houlette, previously General Counsel at the Court of Appeal in Paris (she will be the face of this new institution), requested by the President of the Republic, François Hollande, after the intense excitement caused by the Cahuzac case (involving the President himself). This initiative was supported by the current Minister of Justice, Christiane Taubira, and was finally adopted by an organic law dated 6 December 2013, strengthening the evidence given by 'a true story, a law' and could become an adage. The draft organic law further specified its role, stating that the financial prosecutor is:

> Competent to prosecute offenses of breach of probity and tax fraud in cases of great complexity, mainly because of the large number of perpetrators, accomplices or victims, or the geographical jurisdictions in which they extend (Long et al., 2015/Conseil Constitutionnel, DC n° 2013–679).

But, at the European level some groups and federations have intensively lobbied the management of the internal market to promote a European directive on business confidentiality which must be seen as a complement to the transatlantic free trade treaty. In this respect the draft European Directive has retained the demands made in a joint letter addressed to the US and European negotiators, Michael Froman and Karel De Gucht, from the European employers' federation and the National Association of American companies.[10]

---

9 [Note by the editor] A similar list for the Italian case would be much longer; also in other countries the illegal affairs of important businessmen, political personalities and high bureaucrats of state are frequent and considerable; the neoliberal turn has provoked an important growth of these illegalities thanks to the logic 'less state more market', that is, more private interests over the *res publica*. Among other aspects, it is interesting to note that in Switzerland there is a veritable institutionalization of conflict of interests: the elected at all levels of administration and state have the right to continue to work for private companies even when they have to write and vote a law concerning the interests of these companies.

10 See European Commission (2015) and the criticism by Médiapart of 28 April 2015.

## Conclusion

Literature is often a useful way of seeing that critical and sceptical postures are not new regarding the relationship between power and the law. In his book *Ocean* (1864) Victor Hugo wrote: 'Some men, legislators, judges, philosophers, honest people, say we walk in righteousness. / – Yes, like thieves on the highway' (Hugo, 2002/1864).

Furthermore, the character of the *Azzeccagarbugli*,[11] that is, the lawyer/shyster, comes into the first modern novel in popular Italian literature *The Betrothed* (1827) by Manzoni (grandson of Beccaria). He says: 'for any law passed we can find the loophole'; it is emblematic in ironically unveiling that legal knowledge is on the side of the strongest players.

In France, after the opening of a judicial inquiry (26 February 2014) involving the former President of the Republic, Nicolas Sarkozy, in a recent case of influence peddling and breach of confidentiality during investigations, phonetaps revealed that an informant was responsible for telling Sarkozy that the Court of Cassation had warned that the Advocate General's requisitions against him would be carried out as late as possible, and they would seek an annulment for the seizure of presidential diaries. According to these recordings, Sarkozy's lawyer, Mr T. Herzog, told the former President of the Republic that he welcomed the dedication of its informant because they had worked well, and said that the Supreme Court should follow requisitions, commenting in an almost anamorphic 'unless the law eventually prevails' (Mediapart, 18 March 2014).

We later learned that the Supreme Court informant was Azibert Gilbert (Mediapart, 30 June 2014), *avocat général,* a high magistrate under the orders of the public prosecutor of the Supreme Court and also former director of the National School of Magistrates. In a twist of fate, Mr Gilbert is the editor of the Code of Criminal Procedure 2015, published by Lexis Nexis, and mentioned in numerous bibliographies suggested to law students and a valuable tool for lawyers.

It can be argued that such a possibility is certainly not given to weaker players, as testified by the enemies of the moment and the sort of people filling the prisons (Simon, 2007).

In the activity of the French state observed here, as in the actual reality of the moral idea that spreads, there is an amazing correlation between the development of criminal law cases (*de facto* the impunity of dominant actors) and the evolution of the prison population due to the rise in the imprisonment of the real or presumed authors of pettier illegalities, that is, for less serious offences, compared to the harmful consequences of the actions of white-collar criminals against the population and *res publica.*

---

11 Authorship is attributed to Machiavelli by Cortelazzo and Zolli (1988). The term was used for the first time in 'N. Machiavelli' (1971–1985).

This has occurred thanks to a government consensus of 'fear' and 'zero tolerance' vis-à-vis those designated as being responsible for this fear, in other words, the people who are criminalized for social issues.

Throughout the second half of the nineteenth century a considerable amount of literature was written concerning crime (Foucault, 2012, 221). The glorification of the criminal and the assertion that crime is everywhere is a threat that creates ominous signs throughout the whole social body, producing a fear of crime that is sown in the conscience of the whole population. This vision became scientifically legitimate in the Italian positivist school (Lombroso's disciples). In his book *La criminologia*, Garofalo (1905/1887) theorizes the law of the third element or principle, or the 'mysterious enemy, unknown in history; his name is the CRIMINAL!' This author successively expresses the same anguish of the dominant people when faced with the revolutions of 1848 and the colonized 'refractory to European civilization'; he then theorizes the definition of any criminal as a political enemy (Sennelart, 2007; Palidda, 2015a).

In a different tradition, one notes that the current criminological doctrine of social defence, but also many scientific disciplines, tends to use the legal system as a system of truth:

> Long time, and still to a large extent, medicine, psychiatry, criminal justice, criminology remains at the confines of the truth of the events in the standards of knowledge and of the production of truth in the shape of the event, which always tends to hide under it and be justified by it (Foucault, 1999, 341).

We can see then that we are facing a form of 'mass distraction' that channels popular consensus against the chosen enemy of the moment, and also one of 'police distraction' leading them to target the most vulnerable (see Palidda, Chapter 2).

It appears evident therefore, that the reduction of penalties for crimes in the business sector has been accompanied by the strengthening of punishments vis-à-vis lesser crimes. This neoliberal evolution of standards was made possible by the fact that neoliberal discourse has become dominant and has been internalized by the public as much as by the players in the public administration.

In his *Sociology of Law*, Max Weber wanted to bring lawyers and non-lawyers to reflect on the state of the law and the law of his time. He noted that lawyers still had an ideological relationship with different social groups, which made them fall much more easily on the side of 'order', that is, on the side of the political power that was considered 'legitimate' in a given place at a given time. In this regard, these jurists were different, noted Weber (2013, 282), from the jurists of the French and English Revolution and the Enlightenment, but also different from the jurists of the despotic and patrimonial reigns of the princes in Germany, as well as the jurists sitting in city councils and parliaments up until the mid-nineteenth century.

The contemporary French case shows that most of the new rules of law, their uses and applications, have allowed the dominant actors more leeway to circumvent the law to their own advantage. At the same time the evolution of rules concerning criminal law cases is leaning more towards the intensification of penalties for less serious illegalities, creating an increase in the prison population. This seems to be the result of neoliberal change, that is, the free will of the stronger actors to the detriment of the underclass.

## Chapter 7

# The Governance of Urban Security: Ignored Insecurities and the Securitization of Urban Space (Milan's Case Study)

Fabio Quassoli and Monica Colombo

### Neoliberalism and Urban Security

Security can be regarded as a set of governmental devices (*dispositifs* according to Foucault) which in modern states have ensured the management of the population and allowed the ordinary and regular functioning of liberal societies (Foucault, 2008). In this framework, 'security devices' are a set of governance techniques that emerged in the eighteenth century in order to act on the population, this being defined as a phenomenon which can be regulated according to a criterion of statistical normality and not according to a norm (Foucault, 2004b/2007). The series including 'population' – understood as the object of government in the place of the former unit constituted by the 'territory' – and 'security devices' – understood as a set of technologies with which to manage the population – is integrated with the notions of threat and risk, which become key elements of governance in liberal societies (Foucault, 2008).[1]

Governmentality studies and related work from across the disciplines, including political science, geography and sociology, have documented complex shifts in the governance of cities in the past three decades, often using 'urban neoliberalism' or 'neoliberalization' to describe these changes (Lippert, 2014). Security discourses and practices can be regarded as a key element in the neoliberal governance of cities and Foucault's notion of 'police' is crucial for understanding contemporary urban governance.

---

1 Subsequent studies have focused on the controversial relationship between security and freedom which characterizes Foucault's notion of security. Two main theoretical approaches have emerged: *governmentality studies* (Rose, 2000) and *surveillance studies* (Marx, 1988; Lyon, 2001). The former approach has carried forward the Foucauldian sociological and genealogic research project focusing on the different forms of governance in the so-called post-neoliberal societies. The latter approach has focused on the complex set of procedures which have the population as their object, such as monitoring techniques, personal data filing, surveillance and profiling practices which involve increasingly sophisticated techniques based on an obsessive risk prevention logic.

The most important consequence of the neoliberal governance of security has been the proliferation of a multiplicity of ignored insecurities in a variety of different domains, which range from health and environmental disaster to shadow economy and the exploitation of immigrant workers (see Introduction). In particular, since 1990, the governance of urban security has revolved around two main axes: delinquency and crime prevention and control of irregular immigration. This approach has been effective thanks to a form of 'mass distraction' (see Chapter 1) informed by 'governance through fear' (Simon, 2007) and supported by widespread media alarmist campaigns (Furedi, Altheide, Maneri and others quoted in bibliography and so on). Fourier suggests:

> At once an expression of governmental standards and a collection of affects (Isin 2004), the contemporary citizen is both expected to take responsibility for their own security and expectant of a certain level of physical security. On the one hand, we have the neoliberal citizen, best described by Nikolas Rose (1999) as the dual subject of rationality and responsibility, and on the other hand, we have what Engin Isin calls the 'neurotic citizen', which is the subjective corollary to the discursive generation of insecurity in the last few decades. Contemporary citizens are then being governed through their rationality and through their anxiety (Fourier, 2014, 309).

Studies on urban security confirm that a transnational approach to neoliberal urban governance has emerged and consolidated over time. It is characterized by the creation or expansion of municipal by-laws targeting the urban poor and ethnic minorities, the aggressive enforcement of these by-laws via order maintenance policing (broken windows theory and zero tolerance approach), the privatization of security, the literal or *de facto* privatization of public spaces, and the emergence or re-emergence of an often racialized discourse on the poor (often from ethnic minority backgrounds) as dangerous and criminal (Amster, 2003; Beckett and Herbert, 2006; Caldeira, 2000; Herbert and Brown, 2006; Palidda, ed. 2011; Robins, 2002; Wacquant, 2002). According to Samara (2010, 199):

> The process of neoliberalisation (Peck and Tickell, 2002) thus has two prongs built into its disciplinary structures of governance. One prong is a process for policing, of adherence to the economic logic of market-led growth by, for example, municipal bond rating agencies (Hackworth, 2007), the other for securing that growth and defending it from disorder and disruption.

In regard to the relationships between neoliberal governance and security policies, a useful insight is provided by David Garland, who has highlighted the spread of a 'culture of control' (Garland, 2001) defined by a shift in criminal justice from penal welfarism to an orientation centred on crime control and criminal justice, and which implies decline of the rehabilitative ideal and the renewal of punitive segregation. As a consequence, there has also been a spread in 'criminology of everyday life' centred on situational prevention intended to control urban spaces instead of pursuing the offender's social reintegration (ibid.). The culture of control emphasizes harm-reduction and risk-management intended to minimize criminal opportunities while indiscriminately addressing any kind of crime or deviant behaviour. In this framework, attention to the interests of the victims increases proportionally. In the case of crimes where an individual victim cannot be identified, or of petty crimes where the damage to victims is of little importance, the existence of a collective victim, that is, 'the (local) community' (together with its quality of life), is often invoked.

Thus, the 'new culture of control' relies upon a conception of a community whose identity is defined in purely negative terms, because social bonds are supposedly created through collective victimization, and whose borders are defined by referring to the threats raised by constant waves of incivilities and deviant/criminal behaviour.

A negative definition of a common belonging is a sort of logical precondition for both the spread of actuarial crime policies and the legitimization of racial/ethnic profiling (Harcourt, 2007; Palidda, ed. 2011). Because a sense of community does not rely on a process of social and cultural integration but on a collective shock, those who commit crimes or behave improperly perform the crucial function of reconnecting people and places through the defence of a territory (urban public spaces). Thus security becomes urban security, and it deploys itself as a panoply of surveillance devices used to track people's behaviours (Lyon, 2001; *Conflitti globali*, 2007). As security and crime prevention policies target specific groups such as foreign immigrants, they become crucial tools with which to govern people through fear (Simon, 2007). A fear-centred media discourse becomes a powerful device for the manufacture of consensus and legitimation of bio-power control practices (Altheide, 2007).

Taking the Italian case as a starting point, this chapter aims to provide some understanding of the many socio-political, historical and discursive processes involved in the emergence of the neoliberal governance of urban security. Drawing on archive data, official statistics and research reports, we describe the general policy framework which has emerged and consolidated in urban security policies and discourses in Milan since the 1990s. Special attention is paid to the adoption of Local Pacts for Urban Security and to the use of ordinances as the major devices employed to cope with urban insecurity and to implement urban security policies centred on situational prevention and the criminalization of urban conflicts and marginal people/groups. Although the analysis focuses on the Italian context, and on the city of Milan in particular, the chapter might be of potential interest for all who are concerned with governmental devices and governance techniques which have allowed the ordinary functioning of liberal societies over the last decades, at the same time contributing to the proliferation of ignored insecurities. The chapter also aims to offer evidence that the neoliberal governance of security has been informed by exclusionary discourses and practices which have been increasingly permeating the management of everyday urban spaces.

As many other European cities, the resident population of Milan has diminished by almost 500,000 inhabitants (a neoliberal gentrification effect) in the last 40 years. From 1993 until 2011 the government of the town has been in the hands of a coalition of right-wing parties, included the racist Northern League, sometimes settling for a compromise with the left, often marked by a generalized corruption in several segments of the local private and public administration – after '*mani pulite*' (clean hands) there has been an increase in corruption and mafia affairs. Milan has become Berlusconi's most important fief. In this period, the episodes of violent racism against Roma, immigrants and marginals have been frequent (Chapter 9 this volume; Naga, web; Palidda, 2000).[2]

## The Spread of (Urban) Security Concerns in Italy over Recent Decades

We may begin by recalling that, during the 1990s, concerns about urban insecurity proliferated in Southern European metropolitan areas. They started from cities in Central and Northern Italy where an increasingly widespread 'feeling of insecurity' became one of the key issues in political and media debates (Maneri, 2011): newspapers, newscasts and opinion polls began to describe citizens as fearing for their safety.

As in most EU countries, in Italy numerous alarmist campaigns have been mounted at different times in the last 20 years. They have taken up a wide range of issues, such as illegal immigration, drug dealing and street prostitution. More broadly, there have been repeated protests over the deterioration of local neighbourhoods and widespread micro-criminality. These campaigns have continued in recent years, culminating at the end of the last decade, when they strongly influenced the political agenda (ordinances against windscreen washing, illegal street hawking and begging, the clearance of gypsy encampments, as well as a plethora of injunctions approved in many towns and cities of Lombardy, Piedmont and Veneto to reduce the rights and liberties of immigrants).

Demos research reports demonstrate the significant increase in hardline ethno-nationalist sentiment in many European countries over the last two decades (Bartlett, Birdwell and Littler, 2011; also see Bartlett et al. 2012 on Hungary; Birdwell, Bartlett and McDonnell, 2012; Richardson and Colombo, 2014). Electoral and media campaigns have addressed generalized fears regarding national and cultural identity, the supposed threat represented by immigration and minority ethnic and

2 Interestingly, though, the number of regular and irregular immigrants has augmented almost four times in the same period and consequently there has been a strong development of all sorts of activities (artisans, merchants, entrepreneurs and so on, a useful ethnicization of a part of local economy). The same phenomenon is known in all Northern Italy despite the racism incited by the Northern League, which has been and still is very useful in denying immigrants equal rights and forcing them towards inferiorization or neo-slavery [Editor's note].

religious communities – particularly a perceived growing power of Islam – and have been demanding intervention at all levels: from an EU and national level to a local (urban) one.

The sociological and criminological literature on security policies refers to two ideal-typical models (Baratta, 2001): the typically North American insistence on protecting the citizen's right to safety versus the French perspective that affirms the safety of rights. Until the beginning of the 1990s, actuarial security policies based on both situational prevention and zero tolerance (De Giorgi, 2000; Garland, 2001) differed from the more inclusive French model aimed at developing a 'policy for the city' and which was based on a logic of social prevention (Heilmann, 2012). Over time, the distance between the two models has diminished, and the security policies of many French cities have shifted to a mix of social and situational prevention, with the increasing prevalence of the latter (Robert, 2009).

As in other European countries, the history of Italian security policies started at the beginning of the 1990s in the Central–Northern regions, where a mix of social and situational approaches called 'new prevention policies' – was launched (Pitch, 2008; Stefanizzi, 2013). However, at a time when worries about the consequences of crime outweighed those about its (social) causes, 'securitarian' policies quickly took over from welfare ones, and situational prevention replaced social prevention (Garapon and Salas, 1996). The safeguarding of both neighbourhood decorum and the safety of residents justified increasingly muscular crime policies under the aegis of the zero tolerance ideology, as if any problem that could be framed in terms of safety could be magically resolved by having more police officers on the streets and by introducing more severe sanctions.

Simultaneously, the powers of local administrations on security issues were increased, and agreements were signed between central and local administrations so that the latter could start extraordinary programmes to protect citizens' safety.

As Menichelli (2013) has recently suggested, in Italy the notion of urban security has been used to shift sovereignty from national to local authorities. This has been made possible by the political and legislative processes of the past decade, which have opened up a space of autonomy where local authorities can assume a proactive role in the provision of security to citizens. This is exemplified by the flows of public funding made available for security-related projects throughout the country:

> As a result, most regions passed laws aimed at promoting integrated systems of security in their cities, and an explicit right to security appeared for the first time in Italian legislation (Menichelli, 2013, 2).

A usual corollary of such policies has been an appeal to local communities. The focus on local communities and on the importance of keeping the neighbourhood in order is meant to have a double effect: on the one hand, to stimulate participation; on the other, to discourage breaches of the law and the spread of incivilities.

However, revitalizing local communities and stimulating participation has had to reckon with the weakening of the traditional support networks (families and neighbourhood communities) in urban areas affected by widespread consumerism, greater geographical mobility and fragmentation of social relations (a sort of 'solidarity deficit–feeling unsafe–fear of crime' nexus).

The greater geographical mobility associated with professional careers has rendered both social networks less dense/wide and neighbourhoods more anonymous. The urban landscape has changed in regard to both the expanding importance of city users and the growing social and cultural complexity of urban experience due especially to changes caused by the increasing presence of immigrants.

Highly diversified and mobile urban populations may impact on the capacity of neighbourhood networks to provide both (feel of) social protection and informal social control.

Those new urban populations have provoked in the old residents the perception that they are a besieged minority (like pandas at risk of extinction). This perception, intentionally fuelled by political and media discourses, has triggered defensive reactions towards the presumed invaders (the '*ante portas* stranger' described by Bauman, 1995). When shrewdly exacerbated by moral entrepreneurs, cynical politicians and the media, this perception may trigger action based on group conflict and a non-negotiable opposition between 'us' and 'them' (Maneri, 2013).

Moreover, to be borne in mind is the decreasing level of trust in public powers and institutions among citizens: a crucial problem for the functioning of democracy through social and political participation. Protests concerning the absence/distance of public powers from everyday problems in many neighbourhoods and a feeling of abandonment have been the focal points of many political mobilizations led by citizens associations (groups of active citizens). And citizens associations have been the key players in alarmist campaigns concerning urban insecurity (della Porta, 2004).

## Urban Security Policies in Milan

### *Local Agreements (Pacts) for Urban Security*

In 2005, the 'challenges' raised by the 'new culture of control' and by growing safety concerns induced public administrations in the Centre–North of Italy to inaugurate a new frame for urban policies: 'local pacts for urban security'. These pacts consisted of a formal agreement between the prefect (the local representative of the Ministry of the Interior) and the city mayor, who was held responsible for the citizens' security at the local level. Their main purpose was to strengthen the relations between the (national and local) police forces, third-sector associations and private actors by promoting a variety of actions intended to make city neighbourhoods safer, reassure the citizenry, enhance citizens' quality of life and encourage civic participation. In the case of the province of Milan, about 40 projects were launched under a common regional strategy (Stefanizzi, 2013).

These projects shared four basic characteristics:

1. They involved, together with city councils and local police forces (crucial referents for the pacts), both third-sector associations and private actors.
2. They were aimed at making inhabitants of a town/neighbourhood responsible for combating urban decay.
3. They were intended to mix situational prevention and social prevention within a common framework.
4. They were the most important attempts to promote the governance of urban security.

**Table 7.1 Types of actor involved**

| Actor | Number of cases | % |
|---|---|---|
| City council | 37 | 93% |
| Association | 26 | 65% |
| Religious institution | 12 | 30% |
| Social cooperative | 12 | 30% |
| Province | 9 | 23% |
| Citizens association | 7 | 18% |
| School | 5 | 13% |
| Public firm | 5 | 13% |
| Consortium | 5 | 13% |
| Trade union | 5 | 13% |
| Private firm | 5 | 13% |
| Other | 8 | 19% |

*Source:* IReR 2009

The analysis of local pacts for urban security shows that, as in many EU countries and also in Italy, the trend is towards regional safety and security management, and crime prevention seems to be replaced by a broader concept of the co-production of security by state and municipal authorities.

However, the analysis of empirical data concerning the content of those projects shows that the most important issues addressed were 'perceived insecurity' and 'crime', whilst the social dimension of urban problems was regarded as much less significant.

The data on the general purposes of local pacts are even clearer.

In two out of three cases, they concerned situational prevention (police patrols and garrisons) plus CCTV systems (48 per cent of the cases) and crime suppression

(40 per cent of the cases). It is true that purposes inspired by a more socially concerned approach were also pursued; however, they occupied a sort of niche within projects targeted on enhancing both the efficiency and efficacy of crime prevention and suppression by police forces.

**Table 7.2 Issues addressed**

| Issues | Number of cases | % |
|---|---|---|
| Perceived un-safety | 27 | 68% |
| Crime | 23 | 58% |
| Social disease | 16 | 40% |
| Urban decay | 14 | 35% |
| Social conflict | 13 | 33% |
| Antisocial behaviour | 11 | 28% |
| Road safety | 10 | 25% |
| Squatting | 7 | 18% |
| Dangerous people | 6 | 15% |
| Insufficient public services | 4 | 10% |
| Dangerous people – immigrants | 3 | 8% |
| Safety at workplaces | 3 | 8% |
| Dangerous people – young people | 2 | 5% |
| Dangerous people – gypsies | 2 | 5% |
| School dropout | 1 | 3% |
| Dangerous people – drug addicts | 1 | 3% |

*Source:* IReR 2009

The activation of situational control devices (neighbourhood patrols, the greater and more visible presence of policemen on the streets) were the main purposes pursued by projects in seven out of ten cases, whilst social prevention was the main purpose of only one out of five projects.

The types of actions implemented further confirm what we have just highlighted, because patrolling and surveillance were the main activities foreseen (nine out of ten cases) and their frequency was more than twice that of more socially-oriented actions.

In short, local pacts for urban security, far from being examples of urban governance focused on the social dimension of insecurity, show a clear tendency

Table 7.3 General aims

| General aim | Number of cases | % |
|---|---|---|
| Situational prevention: patrolling and/or strongholds | 27 | 68% |
| Social prevention: social assistance | 19 | 48% |
| Organizational efficiency | 19 | 48% |
| Situational prevention: video-surveillance | 17 | 43% |
| Education and information | 17 | 43% |
| Crime repression | 16 | 40% |
| Social prevention: community building | 15 | 38% |
| Enhancing street circulation | 12 | 30% |
| Urban requalification | 12 | 30% |
| Training of actors involved in the agreement | 3 | 8% |

*Source:* IReR 2009

Table 7.4 Types of action

| Types of actions | Number of cases | % |
|---|---|---|
| Patrolling and surveillance | 26 | 93% |
| Coordination and evaluation | 20 | 71% |
| Education and training | 15 | 54% |
| Problems mapping | 15 | 54% |
| Communication and dissemination | 11 | 39% |
| Support to social marginality | 11 | 39% |
| Creation of shelter | 9 | 32% |
| Video-surveillance | 8 | 29% |
| Urban renewal | 8 | 29% |
| Fighting crime | 8 | 29% |
| Territorial recovery | 5 | 18% |
| Disease prevention | 5 | 18% |
| Support to victims of crime | 1 | 4% |

*Source:* IReR 2009

towards situational prevention and the criminalization of urban problems, conflicts and groups. From this standpoint, the case of Milan confirms how urban governance has been increasingly driven by security concerns, narrowly defined as protecting public order and economic growth under conditions of neoliberalism (Hörnqvist, 2004).

Yet it is evident that only focusing on the social dimension of insecurity would make it possible to understand and respond to crucial ongoing urban changes (the weakening of support networks, the fragmentation of social relations and the 'solidarity deficit–fear of crime' nexus mentioned above).

### *The Ordinances*

However, the case that best clarifies the basic urban governance strategy pursued by right-wing politicians and administrators in Milan over the last decade is the use made of 'ordinances'.[3]

After 2008, enactment of law 125/2008 gave Italian mayors greater injunction powers so that they could take prompt action in regard to 'serious dangers that jeopardize public safety and urban security'. In the province of Milan, the ordinance became the main device employed to cope with urban insecurity and decay.

Between 2008 and 2010, the majority of the mayors in the province of Milan issued at least one ordinance concerning urban security.

But what problems were targeted by the ordinances? They were 'consumption and sale of alcoholic beverages' (25 per cent), 'street prostitution' (23 per cent), 'vandalism' (13 per cent), 'urban decay' (11 per cent) 'aggressive begging' (9 per cent); almost 8 per cent of them concerned illegal encampments, bivouacs and public gatherings, 6 per cent the purchase, consumption and sale of drugs, and, less frequently, late-night rowdyism and street hawking.

These ordinances perfectly embody the basic features of the dominant discourse on urban security highlighted above. The approach is substantially situational insofar as:

1. a set of behaviours are defined unacceptable from the point of view of urban decorum or because of the perception of insecurity that they generate among a minority of citizens;
2. the prevention/suppression of those behaviours by the local police (the city council's chief partner) is the main purpose of the ordinances issued;
3. prevention/suppression activities indiscriminately address (1) crime (for example, drug dealing), adding nothing to the provisions already contained in the criminal code; (2) deviant and undignified behaviour (for example, drinking alcohol in public places, street prostitution, late-night rowdyism); (3) legal and normal activities (for example, reduced opening hours of shops and bars, time limits on the sale of alcohol);

3 Data on ordinances are taken from Cittalia–ANCI (2012); see also Stefanizzi (2013).

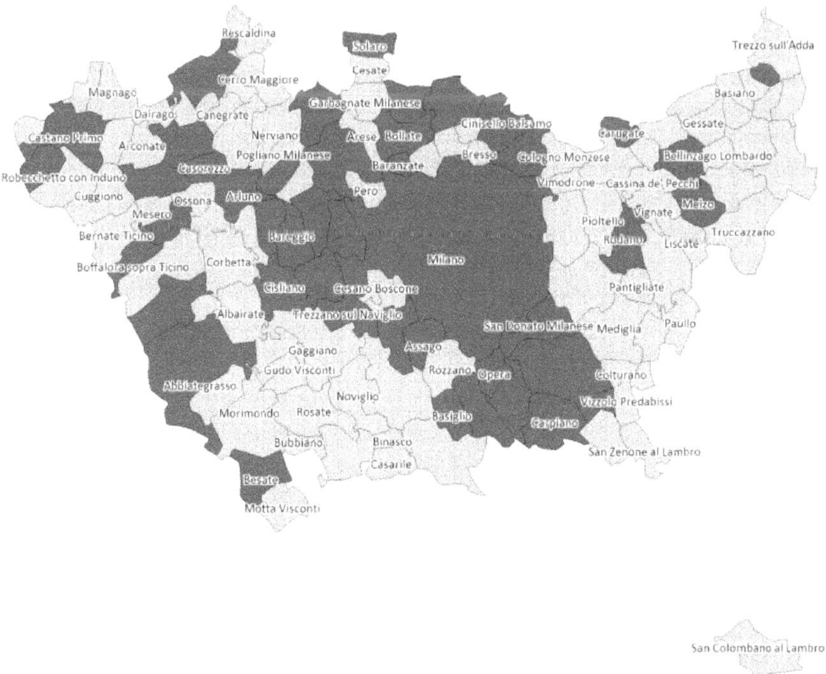


**Figure 7.1 Municipalities in the province of Milan where at least one ordinance was issued between 2008 and 2010**

*Source:* Cittalia – Fondazioni ANCI Ricerche, 2012, 'Le ordinanze: azioni amministrative per la sicurezza urbana. Indagine pilota sulle ricadute delle ordinanze sindacali sulla sicurezza urbana', Unpublished Research Report

4. whenever possible, administrative and penal sanctions are provided in order to control and discourage such activities.

The case of the city of Milan is emblematic in regard to the use of ordinances: between 2008 and 2009, eight of them were issued. They concerned 'phenomena and behaviours' such as street prostitution, damage to public and private property, public drunkenness, begging and the consumption/sale/purchase of drugs. Thereafter, in 2010, seven more injunctions were issued in regard to the management of shops and the decay of certain urban areas and neighbourhoods. Finally, the mayor of Milan issued an *ad personam* ordinance concerning just one block considered to be the residence and gathering place of irregular immigrants and the headquarters of drug dealers operating in the area.

The main targets of ordinances are immigrants, both regular and irregular. In some cases, they are cited directly, in that they are believed to be engaged in drug dealing, prostitution, begging and other kinds of illegal behaviour. In other cases, they are the implicit target, as when limits on opening hours affect areas with numerous 'ethnic' businesses (kebab shops, phone centres, ethnic cafés and restaurants) or that are gathering places for foreigners; or when the local police systematically check commercial licences, toilet facilities and the management of certain types of businesses (again kebab shops, phone centres, ethnic cafés and restaurants). From this standpoint, the use of ordinances as a security device can be regarded as strictly connected with two pervasive trends which have been consolidating in urban governance in most Western countries: the xenophobic discourses on the rise in many countries and the growing socio-economic disparities.

Together with immigrants, young people are also targeted, either as protagonists of the Milanese *movida* or as would-be artists cutting their teeth as graffiti writers.

A final point to be stressed is that an ordinance is an act of *imperium* by the mayor, who issues an injunction by virtue of an unquestionable power recognized by the law. As soon as ordinances are promulgated, they become *ipso facto* laws, at least until an administrative court responds to an appeal by declaring their illegitimacy (the usual conclusion in the case of Milan).

Therefore, ordinances are unilateral acts by the public powers, and they lack any aspect that connects them with a logic of urban governance because they are based on a hierarchical relation between mayors and local police forces. Upon issue of an ordinance, the local police is expected to modify its agenda of priorities, immediately starting the checks and controls by which injunctions must be implemented.

It is clear that if ordinances are evaluated in terms of efficiency and rationality, they can only be judged very negatively. On the other hand, they are strategic devices for a communicative strategy aimed at reassuring 'legitimate' citizens from a symbolic point of view. An ordinance is incontrovertible evidence that the city council is doing everything in its power to respond to a widespread sense of insecurity, and that it is using an extraordinary device to quickly remedy a situation that can no longer be tolerated.

## Security and Immigration Policies

Further exemplification of our main arguments is provided by policies and projects promoted/sponsored by the city council of Milan between by the early 1990s, which targeted the growing presence of immigrants in the city. Indeed, if we consider what has been happening in other European cities (Caponio and Borkert, 2010), we can identify, in principle, three scenarios concerning the kind of policies triggered by analogous urban transformations.

Firstly, there are nationally or locally framed 'integration policies' (Zincone, Penninx and Borkert, eds. 2011) intended to enhance the socio-economic integration of immigrants into the local context (policies to facilitate either entry by immigrants into the local labour market or their access to real estate, public services and welfare provisions, or the '*mixité*' policy launched in some French and German cities in order to prevent residential segregation and the formation of 'urban ghettos').

Secondly, there are 'multicultural policies' aimed at supporting linguistic and cultural diversity in order to promote multicultural cohabitation, new uses of public spaces, cultural associations, multi/intercultural projects in education and so on (Martiniello, 1999; Modood, 2007; Vertovec and Wassendorf, eds. 2010).

Thirdly, we can find policies aimed at 'exploiting' the economic potential of immigrants – in this case classified as 'ethnic communities' – deriving from the economic importance of 'ethnic'/transnational networks (Kloosterman, 2010; Morawska, 2003); or through the 'commodification of cultural diversity': the promotion of ethnic neighbourhoods and precincts, the sponsorship of 'ethnic' festivals and events, and so on (Aytar and Rath, eds. 2012).

These three kinds of policies entail different kinds of governance of urban changes which, in their turn, involve different types of immigrant actors (groups, communities, leaders and so on), as well as institutional players (for example, local police forces, chambers of commerce), third-sector associations and, last but not least, private actors.

Yet, as far as the city of Milan and its provinces are concerned (but similar considerations apply to many cities and towns in the Centre-North of Italy), a fourth type of policy predominated in regard to projects concerning foreign immigrants: *security*, or put better, *safety policies*, meaning policies to protect citizens (usually native residents and city users) from the threats and dangers raised by various social groups, principally immigrants (Petrillo, 2003; Bellinvia, 2013).

Very often, urban conflict – arising, in all its various forms, from the competition between different groups for the use of public spaces/resources – has been overtly redefined as a security issue. Of course, the sector of the citizenry to which the dominant discourse on urban security both explicitly or implicitly refers and seeks to reassure consists of elderly autochthonous people (Semprebon, 2011).

Through the distinction between a legitimized interest (the 'victim') and an illegitimate one (the 'threat'), the expected resolution of urban conflicts – reframed as an urban security issue by those actors socially advantaged in terms of their legitimization, political credibility, access to the media and so on – matches the demand for safety addressed to the local political powers by (autochthonous) residents. Those in an advantaged position – for instance, the long-time residents of a neighbourhood – call for external intervention by public institutions in order to re-establish social order from outside. They do so because they are unable to impose or informally negotiate from inside (through a spontaneous, endogenous process) the conditions (rules and regulations) of cohabitation with other groups.

The public institutions are expected to engage in conflict resolution through the police force (local and criminal) in order to regain apparent control over the urban habitat by taking repressive or preventive action against those (street prostitutes, graffiti writers, young city users, immigrants) accused of threatening public safety.

This opposition between groups in highly asymmetrical positions with regard to the social legitimization of their interests and their respective resources (in terms of economic, social and cultural capital) has manifold consequences: an increase in the relative disadvantage of the weak, an increase in social exclusion and the risk of a perverse securitarian loop (short circuit) generating not only a lack of integration but also the segregation and the exclusion of those groups identified as the preferential targets of repressive action.

Moreover, such policies threaten the sphere of rights and liberties in two ways: on the one hand, by limiting access to public spaces and 'militarizing' them (to the point of creating gated/fortified communities); on the other hand, by increasing social inequalities with regard to the risk of victimization and criminalization to which the members of a minority group are subject.

In conclusion, we may say that, in the considered period, political parties and civil servants in Milan (and throughout the Centre-North of Italy) have done nothing but mitigate the growing worries and fears caused by complex changes in contemporary societies and that cannot be alleviated by stronger crime prevention and suppression measures. When safety concerns have come to the fore, political institutions have seemed unable to appraise social problems in all their complexity. On the contrary, politicians and public administrators went into a sort of hysteria that even induced them to take to the streets and lead street protests against urban insecurity (as if it were not their duty to reassure citizens) and evoke draconian policies often contrary to the basic principles of Italy's Constitution, European law and declarations of human rights.

## Conclusion

Our analysis has illustrated how the securitization of the Milanese urban fabric has been accomplished by shifting from a conception of urban governance centred on the social dimension of urban problems to a different one centred on situational prevention and the criminalization of urban conflicts and marginal people/groups.

Clearly, this shift has been brought about by deploying a number of discursive resources. A crucial figure of speech informing public discourses on urban safety at all levels (political, institutional, media) refers to a clear-cut divide between objective risks and the subjective perception of them.[4] This distinction has been very useful in legitimizing citizens' demands for intervention and promoting local police forces as the most legitimate actors in the field of urban security. In the absence of objective risks, discourse and practices of urban security enable a shift from the subjective perception of a potential risk to the construction of actual risks. Consistently with this shift, it is through a process of foregrounding/

4 On the discrepancy between the crime rate and perceived insecurity see Pitch (2007), Palidda (2000, ed. 2011, 2015b); Maneri (2010, 2011, 2013).

backgrounding that some phenomena are constructed as in/security and others are ignored.

Moreover, the divide between objective risks and their subjective perception involves a psychologization process whereby urban in/security is framed as a problem which concerns individual subjective perceptions, fears and anxieties. This in turn allows the inclusion under the urban security rubric of a wide range of phenomena besides crime (Roché, 1993; Maneri, 2013). As a consequence of the construction of urban security in terms of subjective perceptions, the management of urban security is matched with the management of citizens' fears and unease. It allows for 'a form of "governing through affect" that draws on and targets the affective subject for certain strategies and regulations aimed at designing people's behaviours and attitudes in the public domain' (Fortier, 2010, 17). Moreover, this is a discourse that, besides emphasizing certain dimensions (crime and perceived insecurity) and privileging certain policies (specifically those based on situational prevention) over others (social prevention and the relative safety of numerous urban areas as testified by objective indicators), entirely neglects areas with severe and widespread risks resistant to inclusion in the vocabulary and grammar of urban in/security: from the various aspects of the shadow economy. These take the form of undeclared employment, workplace accidents and occupational diseases in unprotected sectors such as the construction industry, tax frauds, irregularities in leasing contracts, contemporary forms of the gangmaster system in urban areas (Saitta, Shapland and Verhage, 2013), poverty, marginalization and social exclusion. Further instances are road safety (suffice it to say that every year dozens of pedestrians and cyclists are killed in Milan, and hundreds are injured), threats to public health and safety caused by various forms of environmental pollution (car emissions, contaminated brownfield sites) or hydrogeological instability with its attendant risks.

Another interesting question concerns the relation between the vocabulary of risks and security and politics. The construction of the city as a place exposed to a multiplicity of risks, which urban governments are expected to manage with different instruments, standards and approaches, has been interpreted by some authors ' … as an effort to depoliticize public problems, to suggest technical solutions to often complex issues, to redefine rules of accountability, and finally to promote a modern approach to the resolution of urban problems – in line with a general trend that characterizes organizations, public and private alike' (Borraz and Le Galès, 2010, 26). However, the case of Milan, as well those of other cities and towns of Northern and Central Italy, seems to challenge this idea, because the issue of safety has been increasingly politicized and used as a means to cover up other problems (the ignored insecurities) on the agenda, and as an effective tool of ideological manipulation.

Furthermore, in/security discourses and security policies, insofar as they reinforce residents' claims for control and social order, can be regarded as discursive practices used to 'naturalize' exclusionary practices. Indeed, the diffusion and legitimization of an 'ideology of safety' may turn the legitimate demand to live in a safe community into an attempt to legitimize racism and xenophobia: for the discourse on security is closely interconnected with the discourse on cultural identity, the us/them contraposition, the (imagined) community repertoire and so on. The presence of an out-group implies a breakdown of community boundaries, and is associated with feelings of fear, hostility, insecurity, strangeness and distance.

In conclusion, it should be clear that security policies based on the crime-repression strategy which seems to have underpinned actions undertaken throughout Italy by both right and left-wing administrations may be effective at a symbolic level (their actual impact cannot be fully discussed here). Nevertheless, such policies prove able to cope solely with certain 'surface' manifestations of increasing urban insecurity, whilst they are unable to deal with the more profound motives/reasons for the unease expressed by citizens. On the contrary, their fate is to go on nourishing a sentiment of insecurity. If (1) insecurity arises from processes not directly connected with the crime rate but with changes in the labour market, the dismantling of the welfare state, the depletion of relational resources and social capital, and the crisis of traditional political actors (mass-based parties) in terms of their capacity both to be locally rooted and promote bottom-up social/political participation; and if (2) the intensity (force) with which insecurity is perceived is affected by the role played by the media in the social construction of insecurity; then crime policies can be regarded as means to confirm the perception of cities as unsafe (thus fostering the sense of insecurity rather than helping to reduce it). Moreover, the adoption of these policies contributes to security being shaped in terms of a private (individual, personal) good. As a consequence, the social/collective dimensions of insecurity are backgrounded, if not expunged, thereby discouraging any form of collective mobilization and introducing into the public discourse, either overtly or covertly, a notion of *cives* which excludes a large part of citizens – the poor and the homeless, drug addicts, prostitutes, immigrants – who are regarded as threats and, as such, must be neutralized (and rendered harmless): as annoyances or, sometimes, infections, which must be removed/eliminated.

# Chapter 8
# The Choices of the Court with Respect to Which Crimes to Pursue: A Case Study

Cecilia Blengino and Giovanni Torrente[1]

## Trends of Criminal Justice in Italy

What links the condemnation of a homeless immigrant to six months in jail for stealing a four-euro piece of cheese with the goal to promote the efficiency of Italian judicial organization?

The first testifies the rise of 'zero tolerance' policies aimed – also in Italy[2] – to criminalize not 'facts' that are detrimental to citizens, but also 'people' who scare and worry them (Prina, 2003, 131); it is inspired by criminological paradigms oriented to the 'enemy's penal law'[3] based on perceived insecurities rather than on real criminality (Simon, 2007).

Curiously, the early trend that promotes efficiency and productivity criteria as indicators of good administration of justice – applying an 'economic thinking' to criminal justice (Garland, 2001; Garapon, 2012) – may be crucial in assigning a crucial role for the construction of the 'opportune enemies' to the judicial activity.[4]

The logic of efficiency touched the judiciary system in 1999 with the introduction of the 'reasonable lasting of the judicial action' as a new constitutional principle,[5] aiming to respond to the condemnations by the Court of Strasbourg for

---

1 Cecilia Blengino wrote the first three sections and Giovanni Torrente wrote the following two sections. The final section reports common conclusions by the authors.

2 As already observed in the USA, Britain (Garland, 2001) and other European countries (Wacquant, 1999a; Garapon, 2012).

3 This paradigm, theorized by the Dutch philosopher Guenter Jacobs, has been explained in Italy mainly by Pavarini (2006), Tondini (2007), Ferrajoli (2008) and Viganò (2008).

4 With this expression, Pavarini (2006, 17) refers to the impact of Italian laws that, since the last decade, have identified those who are socially excluded as dangerous criminals: drug addicted, homeless people, repeat offenders, immigrants. For example, Law n. 251, 5 December 2005 (so called 'ex Cirielli law') has increased punishment against repeat offenders, also limiting them to access to some benefits or to alternative measures to detention, while Law n. 94, 15 July 2009 has introduced the crime of illegal immigration.

5 Art. 111 of the Italian Constitution, reformed by the Constitutional Law n. 2, 23 November 1999.

delays in Italian trials.[6] As the main result of this reform, the 'fair trial' is no longer guaranteed by the cross-examination principle and the equality of the parties, but it is also connected to its quickness in responding to the claim for justice.

Already consolidated in the Anglo-Saxon area and in other European countries, the rise of 'managerial justice' in Italy follows the crisis of the paradigm of judicial 'ritual' (Garapon, 2001) and the definition of new criteria for accountability of justice and judicial decisions. The increase in the demand for justice and the claims against persistent slowness led judicial system justice to think 'in terms of flow'. So far, the crucial question is not to know whether justice judges fairly, but whether it disposes an efficient flow of the proceedings (Garapon, 2012, 38).

Efficiency objectives introduce a new element of complexity into the implementation processes of the criminal law.

Even if these goals can be either achieved through strategies of defining deviance down or actuarial policies (Garland, 2001, 215), the trend of the Italian judiciary is undeniably oriented to be swayed by a growing 'penal populism' (Simon, 2007). Magistrates have started to play a crucial role in receiving and returning people's fears in the juridical field, having radically changed since last decade both the ways they represent themselves, to account their activities and to ask for social legitimation.[7]

Prosecution clearly shows the relationship between public management criteria and criminalization processes (Hester–Eglin, 1992), since the prosecutor decides whether to request the initiation of the criminal trial. Pre-trial inquiries are crucial in the process of secondary criminalization, representing a key phase for the reception and interpretation of people's fears and security concerns.

## Prosecution and New Public Management

It is necessary to analyse the role played by the various subjects who operate in the 'juridical field' (Bourdieu, 1986) in order to find the logic and the social interests motivating them.

Observing how prosecutor's offices work in order to concretely implement criminal law is fundamental to understand the processes of criminalization.

Besides the efficiency standards and the contraction of economic and human resources, the complexity of prosecution in Italy is increased by the 'mandatory prosecution principle'. In the perspective of the law in action, this constitutional

6 The constitutional reform has been followed by Law n. 89, 4 March 2001 (the so-called 'Pinto law') that has introduced compensation for the huge delay in trials.

7 Evidence of the importance given to citizens' concerns and fears is offered by Bogna and Maddalena (2003). The authors – who are two well-known Italian prosecutors – argue the reasons behind the creation of teamwork aimed to protect citizens from immigrants' vexations and urban disorders. The perception of immigration as a criminal emergency rises also from the annual General Prosecutors' opening speeches (Sarzotti, 2004).

principle in fact presents many conflicting aspects with the 'reasonable lasting one' and the reasonable lasting principle.

Formally, the prosecutor doesn't have a discretionary choice in deciding to prosecute, but he should make enquiries every time he learns about an alleged crime. The need to respect both these specific features (the mandatory prosecution principle and the reasonable lasing principle) characterizes the Italian debate on the so-called organizational question of justice (Fabri, 1997; Sarzotti, 2007; Montana, 2009; Zanier, 2009). The latest judiciary reform laws charge the head prosecutors with a strategic role to pursue the efficiency of public prosecutor's offices. They are both magistrates and managers. They must pursue any crime and solve them quickly with their limited resources: this obliges them to apply some priorities in their work, without violating the mandatory prosecution.

Recognized as being exclusively responsible for prosecution, the prosecutor is charged, by law, with some strategic tasks – such as defining the organization of the office; delegating his collaborators to manage homogeneous types of crimes; and identifying types of offences that could be dealt with by automatic mechanisms.[8]

These new roles have been interpreted in very different ways (Sarzotti, 2007; Blengino 2011), depending on the characterization of the local legal cultures (Cotterrell, 1997) characterizing every local judicial office.

With respect to leadership, organizational differences between local offices are largely influenced by the prosecutors' ability to disclaim the accountability of their actions to the judgment of target audience perception, since the main standard of reference in their decisions' orientation may be the legitimacy of the justice administration, or the account for the performance of their office (Piana, 2010, 124).

Referring to that specific segment of the judicial action, the strategic manager's role becomes particularly evident. The way tasks are assigned underlines the local dimension of the criminal response: management procedures can be crucial for the fate of crime reports and prosecution itself (Fabri, 1997; Blengino, 2007; Zanier, 2009).

Public prosecutor's offices may be considered to be open organizations (Crozier and Friedberg, 1995, 1996), or 'local criminal fields' (Sarzotti, 2007, 52), characterized by specific relationships with 'institutional' and 'non-institutional' actors. Open organizations are: the Ministry of Justice, the Superior Council of Magistrates (hereinafter CSM), local courts, the Bar and police forces. Judicial activity may also be influenced by other social actors, such as local governments, social services, neighbourhood committees, associations and media. Inevitably, prosecutors' managing strategies take into account not only juridical rules, but also the environment in which they operate. Therefore, they are inevitably influenced by the perception of criminality emergency and the local claim for justice.

8 Art. 1 Legislative Decree n. 106/2006.

## Organization of Work and Procedures

The managerial logic reorganizes criminal justice in a 'reactive' way. The same idea of 'criminal response' turns the classic idea of judicial action into the perspective of 'problem solving' (Garapon, 2012, 35).

This phenomenon seems to be linked to the whole redefinition of the success of the judicial system. Judicial organizations have faced up to the claims of inefficiency and have begun to represent themselves in a more realistic way. In this context, judicial organizations may find they can increase their productivity (without employing additional resources) by establishing procedures that reduce the importance placed on some deviances while focusing more on others.

These solutions underline the 'flexibility' of the mandatory prosecution principle, highlighting the local dimension of the implementation of the rules and emphasizing the new trend which means not only must there be an institutional response to any offence, but the answer must also be 'tailored in the right way' (Garapon, 2012, 38). The most evident connection between criminal policies implementation and the external area of the local criminal field is represented by the prosecutor's choices regarding the definition of priority criteria in the *notitiae criminis* procedure.

If a particular crime is considered more serious than others, it will be assigned to a 'specialized group' and pursued carefully. Crime perception influences the definition of the internal proceedings with the construction of different ways to deal with different types of *notitiae criminis*. The assignation of competences and resources to specialized teams that are in charge of the inquiries for specific crimes clearly shows the link between productivity requirements and social alarm perception.

The creation of specialized groups – firstly trialled by the prosecutor's office of Turin in the 1970s – is now a common practice for the main public prosecutor's offices, even if the spur from the CSM for the creation of groups has been implemented in very different ways in the Italian territory.

The connotation of specialist groups is strongly indicative of the ways in which prosecutors play an organizational relay (Friedberg, 1996) with the external environment. The definition of the groups seems to be a concrete expression of the change that has occurred in the relationship between criminal justice and political context (Garland, 2004, 285–286).

The creation of specialized groups translates the criminal policy intentions – more or less binding – from national actors such as the Parliament and the Ministry of Justice, but it also reflects the characteristics of the local claim for justice. To that end, it is interesting to note the changes that have occurred over the last few years in the perception of the main criminal threats and, consequently, in the resulting idea of security. Far from the political and social context dominated in the 1990s by the alarm produced by Mafia's crimes and corruption, a phenomena known as Tangentopoli (Bribesville), today the focus of the specialist groups is on a 'misdemeanours emergency' that, despite the absence of statistical data, is considered as increasingly intrusive and brazen by the general prosecutors (Sarzotti, 2004, 242). In fact, in order to achieve the goal of efficiency in judicial offices, pursued by a recent law that has introduced penal impunity for a list of petty crimes, the discretion of the prosecutor is required since, in concrete cases, it is for him to decide that a crime is 'insignificant'.

The attribution of a *notitiae criminis* to the competence of a specialized group tends to attribute the tools for the realization of the 'due process', envisioned by the Italian Procedure Code (Sarzotti, 2007, Blengino, 2011), to the inquiries and then to the trial.

Nevertheless, new public management procedures tend to reserve the Procedure Code's model inquiries to a few selected types of crimes. The same procedural code suggests prosecutors should define separate procedures in relation to some characteristics of *notitiae criminis*.

Prosecutor's offices normally tend to differentiate the management of the *notitiae criminis* regarding known, unknown or arrested offenders (Di Federico et al., 1994; Fabri, 1997, 176; Blengino, 2007, 152 ss.; Zanier 2009: 93 ss.).

Prosecutor's offices front these classified 'typical' *notitiae criminis* with standardized procedures that are designed to accelerate the productivity of the offices and promote the immediate outcome of the proceedings, whenever it is possible, for all cases marked by the repetition of the conduct, and also to achieve economies of scale and reduce waste in human resources and time (Sarzotti, 2007; Zanier, 2009; Piana, 2010; Blengino, 2011).

To this purpose, several Italian prosecutor's offices have now simplified procedures and offices dedicated to the management of 'simple cases' (so-called SDAS). Normally, this 'simplicity' is attributed to misdemeanour cases and illegal immigration offences. Considering a *notitiae criminis* as a 'simple' affair has very relevant consequences in terms of criminalization processes. The construction of strategies to speed up the criminal process, classifying certain crimes as 'simple affairs' may in fact represent the basis for the development of standardized procedures that are allowed to escape from the constitutional principles that constitute the basic requirements of due criminal process (Garapon, 2012). Even if these procedures are different, they share the use of standardization of implementation processes (Lippi and Morisi, 2005, 114) as a method to prevent the rise of back issues. Their management is, therefore, entrusted mainly to administrative organizational units, with total autonomy in the area of the judiciary.

*Notitiae criminis* with an unknown – and unknowable – author represents here an emblematic downsizing of deviance aimed at increasing productivity. Inquiries relating to such crimes should require a higher investment of resources and energies than those required when the identity of the author of the offence is already known. Such *notitiae criminis* force public prosecutors to face the dilemma represented by the need to divert their scarce financial and human resources to investigations whose fruitful result is considered as possible or reasonable. Official data reveal that the problem of prescription involves a marginal number of *notitiae criminis* with unknown authors (Sarzotti, 2007, 364), also because a typical procedure for these *notitiae criminis* has been constructed to lead them to a quick dismissal request (Blengino, 2011).

Furthermore, the categorization – in the wake of the EU enlargement of 2007 – of some individuals as *Unpersonen* (non-persons, Pavarini, 2006) is the basis for the development of standardized procedures. The exhortation to assess those crime reports 'immediately' and 'as many as possible' may lead the prosecutor's office to consider the possibility of applying a measure of condemnation as a priority. The strategies dictated by the need to implement the principles of fair trial efficiency can find here a disturbing and paradoxical convergence with the theoretical assumptions of the doctrine of the enemy's law.

As the public prosecutor's office in Turin has chosen to apply new public management principles to its organization and proceedings, it may represent a good example to reflect on the actual new public management influence on the criminalization processes.

## The Case of the Public Prosecutor's Office in Turin

The Turin model shows an approach that is characterized by a constant attempt to keep up with the prosecution files and provide a judicial answer to the largest number of crime reports. This approach can be traced back to the organizational strategies adopted by the public prosecutor Vladimiro Zagrebelski, and later confirmed and refined by other prosecutors that have taken turns in covering the role of head prosecutor.

The organizational choices made to deal with the excess of *notitiae criminis* concern are divided into two groups, A and B. The A-band consists of the most serious offences, considered as more important and are thus given priority. B-band crimes are considered less relevant from a social and criminal process point of view, so can be dealt with later. As a result, the latter are left aside when their number rises above the operative capacity of the prosecutor's office. Following this organizational approach, the efficient-culture model has produced 'alternative circuits' in which B-band crime reports can be processed more swiftly. Such an effective approach, which is quite unique in the Italian penal judicial scenario, is probably the most significant accomplishment of the new public management criteria: the adoption of organizational choices geared towards the rationalization of available resources.

## Risk Management: Efficiency and Urban Security

The public prosecutor's office in Turin has shown a very particular approach in the repression of crime with a known perpetrator in the last few years. This was characterized on the one hand by a spirit of efficiency in finding feasible solutions for getting through the backlog of criminal files, and on the other hand by a judicial culture founded on the value of providing effective action, in order to tackle the various types of crime present in its territory. Despite dealing with criminal files in various ways throughout the years, the Turin office seems to have finally opted for the 'double track' system. When faced with limited resources, the prosecutor's strategy is to focus mainly on A-band crimes, as they are considered more serious and socially urgent. As opposed to other Italian offices, however, the organizational scheme adopted in Turin also focuses on the processing of other types of offences that reach the prosecutor's office. This is done by means of the double track system, which provides B-band crime reports with a swift way through to trial. In particular, this last stage of the process requires the criminal file to be submitted immediately to the debate, without any real preliminary investigation taking place, but through an exclusive use of the information contained in the criminal report.

It is important to note that, despite its effectiveness from a process-efficiency point of view, this method presents some flaws concerning fair trial guarantees at the pre-trial stage.

The effectiveness of the classification of crime reports, in fact, seems to be characterized by a high level of administration of the service, which is carried out through semi-automated routine procedures. As a result of such automation, subjects that are frequently highlighted should be assessed by way of common stereotyping as part of the decision-making process.

The 'flagging up' of 'normal' or 'typical' crimes committed by certain categories of people seems especially to prompt the immediate sending for trial of a *notitiae criminis*. It is by means of these conceptual tools that the jurists observed in this study appear to be able to get through such a high number of cases every day. As already highlighted by ethnomethodology and symbolic interactionism researches (Sudnow, 1965; Blumberg, 1967; Quassoli 1999, 2002), the automated implementation of such work-classification criteria is, on the one hand, very profitable in the management of tribunal organization, as it allows decisions to be made on cases about which there is little information; however, on the other hand, it raises issues regarding fair trial guarantees. As the latter constitutes one of the basic principles on which the penal process is founded, there is great concern about the factual knowledge of the historic event being substituted for a swift probabilistic reconstruction by means of a 'mundane reason' (Pollner, 1987), which allows one to find a simple explanation to otherwise highly complex situations. Besides, such semi-automated practices need to be based on the fact that certain subjects, especially those who are underprivileged, do not make complete use of the tools provided by the penal code. Thus, the absence of an investigation into B-band crimes seems to rely on the fact that the evidence already at the office's disposal is sufficient to allow condemnation at the trial.

Numerous examples of the consequences of such semi-automated procedures have been found during this research. The following interview appears to be emblematic of this:

> I remember this episode where one night a man was hit by a stone thrown through his window by someone outside in the dark. Thus, he reported the fact to the police and wrote that he went out, turned on the lights in the park and couldn't

> see anybody, nobody at all. Then, he hinted at the fact that he had had a small argument with one of his neighbours and so on. This neighbour was a non-EU national, by the way. Such a kind of report is written as if deliberately wanting to accuse the Albanian neighbour, who is also said to have been threatening him, despite the man's initial declaration that he hadn't seen anybody at all. It could have been a meteorite ... It became a penal decree ... that is the nature of ... there certainly are some investigations like, for goodness sake, the one on Telekom Serbia, where one rightly goes deeply into things ... and other investigations that seem like a big joke. I mean not real enquiries but just messing about to keep up the investigation figures.[9]

In this case, the file is dealt with very swiftly by imposing a pecuniary sanction on the neighbour. It is evident that in this the procedure results are unjustified from a legal point of view, as it does not meet the criteria of completeness of evidence. So, on what basis could the prosecutor issue the penal sentence? The alleged culprit is one of the victim's lodgers with whom he had an argument the previous day. What's more, he is a non-EU national of Albanian origin. Despite being the only elements known to the prosecutor's office, these are sufficient to sentence the subject. It is obvious that the penal sentence is based on probabilistic reasoning, which stems from prejudice and common stereotypes. However, the sentence is carried out notwithstanding the awareness that the accused foreigner is very unlikely to file an appeal for having been unfairly sentenced. Living on the margins, lack of knowledge of judicial tools and ways to access them are factors that favour the prosecution, which can dispose of a criminal file of scarce importance by issuing a sentence that satisfies the victim's requests.

At times, these speedy procedures stand out for the rigour applied to certain types of crime. A clear example of this is the 'urban security team', which was founded between 1999 and 2000. It had the specific aim to contrast 'street' crime committed by immigrants. The particular toughness shown by the public prosecutor's office in Turin when dealing with such crime is explained by the fact that the task of the urban security team is to ensure that alleged criminals are not released after being arrested or sentenced. This is supposed to be achieved also by filing an appeal for the sentences that allow release on probation. What seems to be evident in the light of this research is how the repressive force of the police is followed by the concept of negative prognosis carried out by the penal jurists towards immigrant street criminals. As a magistrate said in an interview:

> Urban Security started as a reaction against the behaviour of the judges who considered the crimes committed by non-EU subjects as non-existent, in the way that non-EU citizens always had their sentences waived as first-time offenders, because they gave a different name every time they were arrested. Moreover, we were of the opinion that a non-EU subject without a stable residence could

9 From an interview with a Turin administrative officer.

> not be treated in the same way as a citizen who had a house and a job … We believed that, having to make a prognostic judgement, the two situations could be treated differently, in the sense that you can control resident citizens because you know where they live … so you can find them, even if you don't keep them in prison, you can send the police to check on them. You can tell them to go into a residential community and then check whether they did. There are several technicalities that do not work for non-EU members. It isn't their fault; it's just that it doesn't work for them. So, as far as those who don't work are concerned, they are to be considered illegal immigrants, as they have no chance of being regularised. They have no family. It's not surprising that the easiest thing that can happen is that they carry on offending if they are released.[10]

As already observed in other studies carried out in Italy (Mosconi and Padovan, 2005), when facing the same crime committed by an Italian subject, the lack of regular stay permit or a documented job and social-family ties, determines a negative prognosis of the defendant's future. From the perspective of the criminal policies adopted, the case of the public prosecutor's office in Turin could be interpreted as an opportunity to experiment in practice by implementing the 'enemy's penal law'. In particular, the connection between urban security and such penal theories can be traced back to the choice of implementing special organizational tools in order to crackdown on crime linked to immigration. From this perspective, it becomes imperative to notice how the prosecutor's office staff seems to consider the traditional juridical tools as ineffective, thus falling back on prison sentences as the best system to neutralize dangerous people. Evidently, individuating such a repressive tool shows a worrying connection with what has happened in North America in the last few years (Wacquant, 2008; Simon 2007) and the increasing expansion of the area of social security entrusted to incarceration.

## Two Different Interpretations of Security

Reflecting on the connections between the organizational strategies of Turin's prosecutor's office and the general discourses about security and insecurity in the Italian criminal field, we think that the double track system adopted by Turin's office is a mirror of two different interpretations of security and insecurity.

Regarding the A-band crime, we have to consider that the Turin's public prosecutor's office has led some of the main investigations in Italy in the field of white-collar crime over the last 20 years. In particular, the 'Safety at work crimes' specialized group, coordinated by the prosecutor Raffaele Guariniello, carried out several investigations concerning accidents at work. As known, the death of many people due to work accidents is a huge problem in Italy. Nevertheless, it is an often-ignored issue in discussions about safety. Furthermore, Italian prosecutors

10 From an interview with Turin's public prosecutor.

rarely considered the question with the tools of criminal law: few investigations have been launched and even fewer have led to the conviction of the managers responsible. Turin's specialized group has partly reversed this trend. During these years, in fact, the 'Safety at work' group has been the protagonist of the main proceedings in Italy against companies found guilty of serious crimes against the safety of workers. With the cooperation of health services, the 'Safety at work' group contributed to create an 'Observatory on occupational diseases'. These activities have been a first important step in the social construction of the image of workplace safety as a social problem. They also played a crucial role in the Italian process of criminalization of the 'safety at work crimes'. It is not accidental that the most serious condemnation of a manager in Europe has been pronounced by the Court of Turin (Blengino and Torrente, 2013). In fact, both the recent sentences pronounced by the Court of Turin in the Thyssenkrupp[11] and the Eternit[12] cases would not have been possible without the efforts made by the prosecutor's office. Regarding this specific area, it has played a crucial role in the emergence of a type of insecurity traditionally ignored by the dominant discourses on safety and crime.

The approach regarding B-band crimes is very different. In this case, we have to point out a close connection between the moral panic against urban crime, the law and order policies adopted in Italy[13] and the organizational choices implemented by Turin's prosecutor's office. In particular, in choosing to adopt practices that ensure a very swift incarceration of the author of less serious urban crimes, we can find the connection with a rhetoric established in Italy in the last 20 years. This rhetoric identified prison as the only way to neutralize the new dangerous classes excluded from the neoliberal competition. Even in carrying out this task, the public prosecutor's office in Turin has been fully efficient.

11 The sentence pronounced on the 15 April 2011 by the court of first instance in the Thyssenkrupp case condemned the managers for intentional homicide. On the 28 February 2013, this sentence was reformed by the Court of Appeal to manslaughter.

12 The sentence to ten years of prison inflicted to Erald Espenhahn for manslaughter is until now the most serious sentence ever pronounced in Europe against a manager for 'accidents' at work.

13 See the other chapters in this book.

Chapter 9

# Who Is Afraid of Whom? Turning Security Threats Upside-down in the Governance of Roma People in Today's Italy[1]

Barbara Giovanna Bello

> In the final analysis, human security is a child who did not die, a disease that did not spread, a job that was not cut, an ethnic tension that did not explode in violence, a dissident who was not silenced. Human security is not a concern with weapons – it is a concern with human life and dignity.
>
> *UNDP, Human Development Report, 1994.*

## Introduction

It is largely documented that after 1990 overtly intolerant discourses and practices have targeted those perceived and constructed as dangerous 'Others' in most European countries, be they ex-colonized peoples, Roma,[2] migrants, refugees or minorities who belong to the most marginal or poor people. Independently of their national origin, the Roma communities belong to those groups most concerned by discrimination (including institutional discrimination). The set of legislations and policies concerning the governance of Roma people has hardly been responsive to their genuine needs (Bello, 2014a). The old and new biopolitics adopted in particular by the Italian governments have prepared the way for today's situation of Roma marginalization and discrimination (Clough Marinaro, 2009). The ongoing precarious living conditions of many Roma families from Eastern European countries and their spatial segregation in ghetto-like and ethno-based camps, have gone hand in hand with the strategy of 'bio-political control and the gradual stripping of the Roma's rights which has been evolving in Italy,

1 I am indebted to Pietro Saitta and Nora Markard for their helpful comments on this chapter.

2 In this chapter, I use the term 'Roma', in the same way as the EU's documents, as 'an umbrella term which includes groups of people who have more or less similar cultural characteristics, such as Sinti, Travellers, Kalé, Gens du voyage, etc. whether sedentary or not' (see footnote n. 1 in EC, 2011). I use sometimes the acronym RSC (Roma, Sinti and Caminanti) when I refer to the Italian 'National Strategy'.

and particularly in Rome, for nearly thirty years'(ibid.) and with the anti-Roma rhetoric used by both left-wing and right-wing leading politicians during electoral campaigns to gain more votes. For now, the legacy of securitization and anti-Roma policies developed during the 'state of Roma emergency' (declared in 2008 and officially ended in 2013, infra) co-exists with the implementation of the Italian National Strategy for the Inclusion of Roma, Sinti and Caminanti Communities (in the following 'National Strategy')[3] adopted on 28 February 2012, in response to the European Commission's 2011 (EC, 2011) 'Framework for National Roma Integration Strategies up to 2020' (hereafter 'Roma Framework'). Despite their well-intended aim to stop discrimination against the Roma communities, such targeted policies like the 'National Strategy' also include their own risks for the Roma populations (Rostas, 2012; Sobotka and Veermersch, 2012; van Baar, 2012) because, for instance, they do not challenge the *status quo* of the commoditization of social inclusion managed by public–private partnerships, nor the 'policy-centric and institution-centric approaches' (van Baar, ibid.) to Roma activation and participation at the local level (Kapuy, 2004).

In this scenario, the issue of Roma people's human (in)security has been barely explored and obscured by the dominant discourse of Roma-related issues as a matter of 'public security'. Drawing inspiration from Newman's still embryonic Critical Human Security Studies (CHSS in the following, Newman, 2010), which offers interesting inputs in order to investigate Roma people's human security in today's Italy,[4] and from Tomovska's work (2010), I reverse the question and look diachronically at the human security of Roma in Italy, in order to shed light on the pitfalls of old and new policies concerning them.

---

3 The National Office against Racial Discrimination (UNAR), which issued the 'National Strategy' and is entrusted as National Focal Point (NFP) for implementing it, benefitted highly from the outcomes of researchers and activists and upheld most of their findings and stances. Among others, the International Conference on 'The legal status of Roma and Sinti in Italy' (Milan, 16–18 June 2010, at Bicocca University and ASGI – Association for Legal Studies on Migration) was a unique event that gathered together about 900 participants (researchers, policy-makers, representatives from several local, national and international institutions and NGOs, including Roma representatives; see Bonetti, Simoni and Vitale, eds, 2001).

4 To my knowledge, Tomosvska's (2010) work is the first thorough attempt to explicitly look at the situation of Roma communities moving from a 'critical security studies' approach. More precisely, she relies on the concept of 'security as emancipation' elaborated, within the Welsh School, by Booth (1991) and, later on, by Neufeld (2004). Tomosvska's contribution shows the centrality of Roma people's security, but it misses a broader reflection on the relations and mutual nurturing between human security and critical security studies. According to Newman (p. 92), the normative strands of critical security studies – such as the 'Welsh School' – which claims to seek to change the world for the better, can help bridging between human security and critical security studies, since they share 'a more positive view on security' (ibid., 86).

### The Neoliberal Governance of the Roma in Italy: From the Construction of the Roma as a 'Public Security' Threat to the New Rhetoric of Roma Integration

In the lack of ethnically disaggregated census data, Roma people in Italy are estimated to number between 150,000 and 180,000 inhabitants (about 0.25 per cent of the total Italian population, Associazione 21 Luglio, 2014). These low numbers do not justify the paranoid fear of the 'invasion' of Roma newcomers from the new EU countries that has spread since the 2007 enlargement of the European Union.[5]

Italian public discourse has deployed many labels to describe the Roma people that have reinforced stereotypes about them through harsh and inadequate legislations, policies and practices. Among them are 'fortune tellers', 'child kidnappers',[6] 'thieves', 'deviant' and 'criminal' people, which echo positivist criminologist Lombroso's (1911) theory that the Roma are the 'living example of a whole race of criminals and have all the passions and all the vices of criminals'. Alternatively, heteronyms like 'zingari' (Gypsies) and 'popolazione zingara' (Gypsy people) (Piasere, 2004; Vitale, 2010) are used instead. The term 'nomads' has served the purpose of fuelling the non-Roma Italian people's perception of the Roma communities as just living temporarily in Italy and as better regulated by short-term and provisory measures (ERRC, 2000). The 'nomad' theory (Sigona, 2003) has survived for decades, disregarding the fact that approximately half of the Roma people living in Italy are Italian nationals and are no longer engaged in traditional jobs (horse traders, carousel owners, and so on) requiring a nomadic or semi-nomadic lifestyle (with very few exceptions, for example, among the Camminanti from Noto in Sicily). The 'nomadization' of Roma has deprived them of the 'responsibility for decision-making normally afforded adult human beings' (ibid., 12.) and of the recognition as 'citizens' and 'peers' in society, legitimizing authorities to forcibly evict them.

While many regional legislations entail fragmented provisions aimed at 'accommodating' the 'Gypsy' or 'nomadic' cultures within their territories, mainly in the field of housing and education, the first ever national law (the Nomad Emergency Decree of 2008) about Roma communities in the last decades addresses the Roma people as 'nomads'.

---

5 To be sure, the fear about 'poverty migration' concerned other countries as well (for example, France, Spain, Germany, UK) and made Germany and the UK apply transitional restrictions on Bulgarian and Romanian workers' free movement rights for up to seven years after accession (until 31 December 2013), as allowed by accession treaties, in order to protect their social security systems and labour markets.

6 On the other hand, Roma people and some scholars call attention to the cases of Roma children removed from home and also given up for adoption (Associazione 21 Luglio, 2013).

In the wake of the EU Enlargement in 2007, the fear of massive Westward mobility by Roma EU citizens from the new member states (Romania and Bulgaria) formally entitled with the same rights and freedom of mobility as all the rest of EU citizens, quickly spread the rhetoric of public order and zero tolerance from the local to the national authorities. Even if the concern referred to poverty migration from new EU countries in general, the 'Romanian Roma' were particularly scapegoated and quickly became the new 'public enemy' (Sigona, 2008). Interestingly this binomial sounded even tautological for many Italians. Borrowing from linguistics, Romanian Roma people became a sort of 'living synecdoche', since a part meant the whole or, conversely, the whole a part.

National and local politicians, as well as media, used isolated incidents to frame the Roma-related issues as a 'public security problem' or a 'problem of criminality', as in the often-cited case concerning the rape and murder of Giovanna Reggiani in 2007, by a young Romanian Roma man (Maneri, 2011). This crime provoked a massive reaction against both the Roma people and the larger Romanian community legally residing and working in Italy (amounting to about one million at the beginning of 2008) and culminated in the adoption of the Decree Law n. 181/2007, introducing rules for the expulsion of EU citizens for reasons of public security. While not denying the brutality of the crime, pro-Roma activists mobilized against it (the Decree was never converted into Law by the Italian Parliament).[7]

In May 2008, the Italian Government adopted new discriminatory measures against Roma, namely the aforesaid Nomad Emergency Decree, declaring a state of emergency in the Campania, Lazio and Lombardy regions until 31 May 2009, then extended to the Piedmont and Veneto regions.[8] For these five regions, the emergency was further prolonged to December 2010 and then for one more year.

Under this provision, Italian local authorities were empowered to order, *inter alia*, the eviction, expulsion of the Roma people, a census of the Roma on the basis of ethnicity, fingerprinting and data retention, and the replacement of non-authorized Roma camps by 'equipped villages'. Advertised by local authorities as places which should guarantee the security both inside and outside their perimeters,

7 This attitude of criminalizing whole populations, rather than adopting a case-by-case approach, has created a backlash against Roma in other countries as well. The Romanian Government, for instance, took a strict stance against big groups of seasonal Roma workers leaving the country, deemed responsible for damaging the reputation of Romanian citizens. Besides, in 2010 a draft law was presented in the Romanian Parliament (rejected by the Senate in 2011) in order to rename the Roma minority 'Tigan' to avoid confusion between similar terms Roma and Romanian in Europe. Romanian Roma (and people perceived as such) were doubly discriminated against, because they were considered as undesirable citizens in Romania and unwelcome newcomers in European countries.

8 The state of emergency was extended through two ordinances of the President of 1 June 2009, Nos 3776 and 3777 relating to the regions of Piedmont and Veneto.

these new villages were nothing more than ghetto-like settlements controlled by private guards and, in many cases, video surveillance, aimed at preserving public order and controlling the Roma people.

Moving from Foucault, van Baar (2011) explains that the Italian scenario illustrates:

> how the Roma's problematization – be it in terms of nomadism, illegality, or public or social security – is mobilized to create a legal state of exception and legitimate unorthodox policy interventions.

On 4 November 2011 the Council of State (no. 06050/2011) ruled that the declaration of the nomad emergency by the Berlusconi Government had been 'unfounded and unsubstantiated' (Amnesty International, 2012, 8).

The post-Berlusconi Government, led by M. Monti, however, decided to appeal the sentence in Italy's Court of Cassation in February 2012, even if many NGOs (Association 21, ASGI, Amnesty International, Human Rights Watch and Open Society Justice Initiative) tried to call on the government to drop the appeal (HRW 2012).[9] It is striking that, at the same time, this very government adopted the 'National Strategy' that clearly condemns emergency solutions and stresses the need to endorse long-term policies allowing Roma people to find durable solutions for their housing, legal status, access to employment and to their children's access to education.

At the national level, the discursive shift (from emergency and public security to social inclusion) (Bello 2014a, 9) of this unprecedented national policy is marked by the terms 'Rom, Sinti and Caminanti' (rather than 'nomadi') used to address these communities. This is highlighted in bold type in the text of the 'Roma Strategy':

> The old conception, which associated these communities with the sole connotation of 'nomadism' has been overcome: this term is outdated both linguistically and culturally, since it does not correctly portray the current situation (National Office on Anti-Racial Discriminations 2012, 4).

In a nutshell, the 'National Strategy' plans interventions in four macro-areas, namely employment, education, housing and health. It also stresses the need for robust monitoring of its implementation, which should involve Roma communities. Among the compelling initiatives urged by the 'National Strategy', are: the elaboration of an *ad hoc* bill concerning the recognition of the Roma communities as a national minority, a specific task to help undocumented Roma

9 The Supreme Court eventually put an end to the nomad emergency in Italy on 2 May 2013, ruling that 'there is no evidence of any unfruitful attempt to use the ordinary means' (Cassation Court, Italy, 2013) which justified the declaration of the state of emergency in 2008.

people gain access to *legal status* (ibid., 4) and the Remembrance of *Porrajmos* (the extermination of the Roma during the Second World War). Apparently, the 'National Strategy' aims to raise awareness on the heterogeneity among various Roma groups (see also Piasere, 1999) in terms of nationality, legal status, religion, migration trajectories and lifestyles. Furthermore, it acknowledges the need to take into account the claims of non-Roma people living near the Roma settlements and to set up bilateral bodies, including organizations representing both the RSC (acronym for Rom, Sinti and Caminanti, used in the 'National Strategy') people and the local community.

The 'National Strategy' upholds the remark made by activists and researchers, that the emergency responses of institutions are ineffective in either the short or long term, 'costly in economic terms and damaging to the social cohesion of the country' (National Office on Anti-Racial Discriminations 2012, 79; also Lunaria, 2013).

At present, it is too early to appraise the work of the 'National Strategy' on a local level, the follow up by local authorities (Regione Emilia-Romagna-Il Difensore Civico, 2013), and the involvement of Roma communities as well as of non-Roma neighbours beyond tokenism. For this reason my analysis can only focus on some of the more compelling aspects of the Roma people's human security in this new phase of Roma governance.

## Critical Human Security Studies as an Approach for Investigating the Situation of Roma People in Italy

'Human security' has flourished in the United Nations where it is defined as 'the protection of the vital core of all human lives from critical and pervasive threats' (Alkire 2002, 2). The interest in a concept of security focused on people rather than on nation-states, can be traced back to the United Nations Development Programme (UNDP) aimed at designing a Human Development Index. According to the second chapter (entitled 'New dimensions of human security') of the 1994 Human Development Report, human security is necessary for enabling people to exercise 'safely and freely' (UNDP, 1994, 23) their choices and to make them feel 'confident that the opportunities they have today are not totally lost tomorrow' (ibid.). At its early stage, human security was considered substantiated by two major elements, 'freedom from fear' and 'freedom from want', but in 2003 the Commission on Human Security extended it beyond the dimension of people's protection in order to encompass the 'freedom to take action on one's own behalf' (ibid., 10).

The threats to human security were summarized in a list encompassing seven categories: economic security; food security; health security; environmental security; personal security; community security; political security (ibid., 24).

The issue of human security needs to be addressed not only in developing countries, but also with regard to people moving to and living in European

**Table 9.1 Types of human security threats**

| Type of security | Examples of main threats |
|---|---|
| Economic security | Persistent poverty, unemployment |
| Food security | Hunger, famine |
| Health security | Deadly infectious diseases, unsafe food, malnutrition, lack of access to basic health care |
| Environmental security | Environmental degradation, resource depletion, natural disasters, pollution |
| Personal security | Physical violence, crime, terrorism, domestic violence, child labour |
| Community security | Inter-ethnic, religious and other identity-based tensions |
| Political security | Political repression, human rights abuses |

*Source:* United Nations Trust Fund for Human Security, 2009

countries[10] (UNESCO, 2009, 100). Dangers that happen 'close to home' (Kapuy, 2004, 4.) to native, migrant or refugee Roma people in Italy and in other European countries (in particular in France, Commission nationale consultative des droits de l'homme, 2014; Barbulescu, 2012) recall attention to the fact that:

> for many people in the world the greatest threats to security come from internal conflicts, disease, hunger, environmental contamination or criminal violence. And for others, a greater threat may come from their own state itself, rather than from an external adversary (Newman, 2010, 79).

The concept of human security has been rebuffed by 'critical security studies', an umbrella term encompassing critical and non-traditional approaches to security that question, on the one hand, the centrality of state security and, on the other hand, the engagement of human security with policy as well as its willingness to cooperate with the state to ensure security. This attitude is judged by critical security studies as 'legitimizing the business-as-usual of hegemony and privation' (ibid., 87) through the supine acceptance of hegemonic discourse, based on the assumption that governments genuinely commit to dealing with people's human security. In fact, the state remains 'the main provider of individual security' (ibid., 81), downgrading the potential of grassroots and bottom-up initiatives in which individuals are involved.

---

10 For instance, no mention of 'human security' is made in documents such as European Commission (2010). For the concept of 'human security', in relation to external relations, see Study Group on Europe's Security Capabilities (2004); Kaldor, Martin and Selchow (2007); Kaldor and Martin (eds. 2010); and Chapter 2 by Palidda.

Even if both human security and critical security studies involve 'de-essentializing and deconstructing prevailing claims about security' (ibid., 86), the former lacks the 'conceptual sophistication' (ibid., 86) and transformative power of the latter. Therefore it tends to be problem-solving, rather than embarking on 'epistemological, ontological or methodological debates' (ibid.) and challenging the structures that perpetuate insecurity.

Newman explores what critical security studies and human security could learn from each other and paves the way to an approach, called Critical Human Security Studies (hereafter CHSS), which merges contributions from the two different approaches to security into one strand, in a way that is critical- (like the 'critical security studies') *and* policy-relevant (like human security). More in detail, CHSS enriches human security with a critical and theoretical framework that makes it more explanatory and critical rather than normative and rhetoric, while, at the same time, providing critical security studies with 'a policy-oriented intellectual agenda' (ibid., 92) that allows it to have an impact on reality.

## Reversing the Question of Security: Is the Roma People's Human Security Threatened at the Local Level?

The Roma communities in Italy have been confronted with almost every type of human security threat listed in Table 9.1. Even if this contribution is confined to the analysis of the Roma people's human security in Italy, to the blatant threats to personal security and community security because of their impact on Roma individual and collective life and identity (UNESCO, 2009, 67ff.), we should bear in mind that the various dimensions and menaces to human security are not sealed off from each other (Tomovska 2010, 67). In fact:

> [T]hreats to human security are mutually reinforcing and interconnected in two ways. First, they are interlinked in a domino effect in the sense that each threat feeds on the other … Second, threats within a given country or area can spread into a wider region and have negative externalities for regional and international security (UNTFHS 2009, 7).

### *Personal Security*

Personal security can be defined as security from physical violence and threats, ranging from direct and indirect discrimination, harassment and assaults to inter-ethnic tension, abuse and torture. Human beings can be threatened by other individuals, groups, institutions or states (UNDP, 1994, 30). Violence can occur both in out-group relations and in in-group relations and can be based on one discrimination ground (for example, ethnic origin, religion and so on) or more than one ground (for example, gender and ethnic origin) as happens to 'minorities within minorities' (Eisenberg and Spinner, 2005).

The critical security scholars find that human security does not often question 'existing structures of power, gender and distribution in relation to economic and political organization' (Newman, 2010, 89). Threats to personal security need to be analysed both when they come from outside the Roma communities and when they originate from within them, in a way that takes into account how gender, age, sexual orientation and other grounds impact the individual experience of the human (in)security of Roma people (Bello, 2014b; Fremlova and Georgescu, 2014). Concerning in-group relations, the personal security of Roma women is particularly at risk in the case of gender inequalities and domestic violence (ERRC, 2011, 11). In the same way as non-Roma women, Roma women often do not seek to escape from an abusive relationship either because they lack economic independence, or because of the emotional attachment to the perpetrator, or because they are expected to tolerate male dominance in the family in order to comply with the expectations of the traditional role of women (ibid.). However, there are specific structural obstacles both within and outside the communities that prevent Roma women from reporting violence or asking for help. Illiteracy among Roma women and their marginalization from the rest of society surely hinders them from being able to access the relevant services (hotlines, shelters) set up for victims of violence. Other causes concern the undocumented status and unregistered marriage of many of them, as well as the fear of losing parental rights over their children (ibid., 9). Often they distrust the police and authorities and fear being victimized by them because of ethnic discrimination and tense out-of-group relationships. Examples of daily threats to the human security of Roma people, including young people and women, span from verbal and physical assault to humiliating treatment (ibid.), as in the well-known case of a young Roma girl who was strip searched by policemen in front of the Station in Padua (in Northeastern Italy) in broad daylight in 2005, because she was suspected of hiding drugs on her body (Bonatelli, 2005). This persistent racism means that Roma women have to face the well-known dilemma of 'further stigmatising the group by exposing intra-group hierarchies' (Kocze, 2009, 21 quoted in Bello, 2014b; Crenshaw 1991). The 'National Strategy' mentions that women are doubly discriminated against and places great emphasis on the 'empowerment' of Roma women, 'which is instrumental to the improvement of the condition of the person and also of the family structure, as a whole' (National Office on Anti-Racial Discriminations 2012, 21; Istituto per gli studi sui servizi sociali onlus, 2013). Scholars warn that such 'progressive' tools as community participation, empowerment and social capital have been depoliticized and have dismissed their original interest for power relations. Their transformative and emancipatory objectives are 'replaced with objectives designed to preserve the status quo' (Miraftab, 2004b, 241; Freire, 1970, 131) in a way that leaves power relations unchallenged.

The issue of dis/empowerment within the Roma communities is a relevant one, because the stigma attached to the members of Roma communities is a threat to their personal security from their childhood. The 'National Strategy' stresses the importance of 'supporting interventions designed to deconstruct social labelling'

(UNAR, 2012, 52), in order to promote the social inclusion of Roma children. When dealing with Roma children and young people who break the law, however, this document simply stresses the need to pay specific attention to them (ibid.), but it does not provide any data for the application of measures as an alternative to detention penalties for Roma children. Even though juvenile justice authorities usually apply these measures in order to allow the compulsory attendance of school, or participation in training courses or work activities (ibid., 13), not so long ago, the Juvenile Court in Naples did not apply them in the case of a 16-year-old Romanian Roma girl (Angelica), who was suspected of child abduction in a flat in the Ponticelli district of Naples in May 2008, because:

> [she was] fully integrated in the traditions typical of Roma culture which create a real danger of recidivism, in the absence of realistic processes for the analysis of her life experiences (HRW, 2012, 48).

According to the court, no alternative measures would be adequate because of the minor's 'adherence to Roma traditions which by common experience determine in their adherents a lack of respect for the rules' (ibid.). The decision was eventually annulled by the Court of Cassation in 2010 because it was *contra ius* and discriminatory, but Angelica, who was immediately labelled as 'zingara rapitrice' (Gypsy kidnapper), spent about four years of her life in the juvenile institute in Nisida (near Naples).

Another case is useful in helping to understand how the stereotype of 'Gypsy kidnappers' threatens the Roma people's personal security within and across national borders, legitimizing the immediate removal of children from their families. On 20 October 2013, the media reported the news about Maria, a little blond-haired girl found in a Roma Camp in Larissa, Greece, suspecting that the child could have been abducted or bought by the Roma family she was living with. Some days after, the biological mother, a Bulgarian Roma woman, admitted she was forced to leave her daughter in Greece before returning to Bulgaria, because she could not afford to feed her. Before Maria's mother clarified the situation, the case had already created alarm across the borders about 'Gypsies' kidnapping non-Roma children. At around the same time, in fact, another blonde girl was removed from her Roma parents in Ireland and then was returned to them, after DNA tests confirmed their relationship. On 22 October 2013, Italy's Northern League party urgently called on the Italian Ministry of the Interior to investigate similar cases of alleged child abduction by Roma people in all settlements in the country.[11]

While public discourse has focused on the non-Roma children 'kidnapped' by the Roma people, Saletti Salza (2010, 2012) sheds light on the disproportionate number of Roma children made available for adoption in comparison to non-Roma ones. According to this scholar, Roma children are often removed from their

11 The information is available on the Northern League's webpage, accessed on 11 November 2013.

families because of material poverty, rather than parental inadequacy, even though the Italian law forbids this as a justification for removal (Bello, 2014a).

Another compelling issue which must be considered within the 'personal security' stream, concerns the barriers that Roma people experience in accessing legal documents and the recognition of their legal status. The 'National Strategy' places a strong emphasis on the need to work on the legal status of many undocumented Roma people who arrived in Italy in the aftermath of the Balkan war. Scholars and NGOs have raised awareness on the obstacles which hinder the access of many of these people to either the Italian nationality or the nationality of their country of origin (once belonging to Yugoslavia), or even their stateless status (Consiglio Italiano per i Rifugiati, 2013; Perin, 2011). Such studies encompass Roma people who lost their nationality as a consequence of the dissolution of the Yugoslavian state, people who were already undocumented before the national collapse, as well as children born in Italy from undocumented parents. It is worth mentioning that the Italian Government has not yet adopted long-term solutions for supporting undocumented Roma people's transition to a legal status. As a consequence, hundreds of Roma people in Italy have been denied the possibility to exercise basic rights like property rights, mobility across the country and abroad, access to regular employment, to quality education, to healthcare and to political participation.

The 'National Strategy' also acknowledges the role played by the mass media in spreading hate speeches against Roma communities (National Office on Anti-Racial Discriminations, 2012, 20) and it identifies, as one of the priorities necessary to tackle it, the launch of a computer-based monitoring system of discriminations committed by the media within the 'Italian Equality Body', with a specific thematic area dedicated to Roma-related discriminations, stereotypes and prejudices (ibid., 90). Although forms of formal and informal social control try to prevent harmful attacks or to limit the damage of a resurgent hatred against Roma people, it cannot be denied that the persistent rhetoric centred on public security and public order has detrimental effects on their basic human needs. On social networks and on national TV it is possible to find several supporters of the 'Salvini[12] method', as 'the deliberate targeting of highly stigmatised and voiceless communities to grab media attention and feed the induced moral panic with banal but highly effective sound bites' (Osservazione, 2015). On 3 March 2015, Mr Buonanno (European deputy of Northern League), called Roma people the 'dregs' of society while speaking with the Roma-activist Dijana Pavlović during the talkshow 'Piazza Pulita'. Anti-Roma discourses derive from issues of public security, from the assumed economic loss that comes from investing in the inclusion of the 'undeserving Roma' people imposed by the EU, rather than in the initiatives to support older or unemployed non-Roma Italian nationals. It is worth mentioning that, according to the 2012 Special Eurobarometer survey, Italy features among the EU member states where most respondents (51 per cent)

12 Mr Matteo Salvini is the new leader of Italy's right-wing Northern League.

assume that society could not benefit from better integration of the Roma (EC, 2012b, 116).

The proliferation of intolerance on the web does not affect only Roma people, but also some politicians and artists or intellectuals sympathetic to their cause to refugees and migrants.[13] All these people were exposed to the verbal assaults of the same right-wing formations who spread xenophobic sentiments among voters.

### *Community Security*

Community security consists of security from inter-ethnic, religious and other identity-based tensions (UNESCO, 2009). Community security entails protection of ethnic groups and community identity and protection from oppressive traditional practices towards women, or discrimination against ethnic/indigenous/refugee groups (UNTFHS, 2009, 18).

As previously in the case of personal security, I also analyse the issues related to community security both with regard to in-group and out-group relations. Among in-group practices, early marriages and virginity testing still occur in some Roma communities. European Roma Rights Centre (ERRC) (2011) field research in Italy found that 47 out of 74 female Roma respondents married below the age of 18 and five of these declared that they were forced to marry (see European Roma and Travellers Forum & Romani Women Informal Platform 'Phenjalipe', 2014).

Certainly, early marriages are not inherently typical of the Roma culture. They are rather seen as the legacy of patriarchal structures in which the Roma girls' agency interplays (Marcu, 2014) and, furthermore, they should be discussed in light of the socio-economic conditions and spatial segregation of various Roma communities, as well as of the racism Roma children have to face at school and their consequent drop-out. Oprea explains that:

> the state-sponsored structural marginalization contributes to the vulnerability of under-age Romani women to arranged marriages. The educational level of Romani parents affects their decisions regarding when and whether or not to marry off their daughters. In addition, the lack of educational and employment opportunities for Romani women in Romania also contributes to many Romani parents' decisions – or as they say 'she's not doing anything with her life anyway, might as well'. In addition, better educational and employment opportunities help to empower Romani women within their relative communities by increasing their options and social mobility' (quoted in European Roma And Travellers Forum & Romani Women Informal Platform 'Phenjalipe' 2014, 10).

13 Mrs Cécile Kyenge, the Congolese-born Italian former Minister of Integration, and the Lower House President, Laura Boldrini, the President of the Tuscany Region, Enrico Rossi, after he posted on his Facebook page a photo of himself with his Romanian Roma neighbours.

The discrepancy between 'social age' and 'biological age' within the Roma communities and the need to reconcile family duties and school attendance/work should also be taken into account when planning measures to support Roma children, something which the 'National Strategy' seems to be aware of, as the following statement shows:

> in the communities, sometimes a boy of only 12 years is already an adult, able to work to produce wealth, eventually marry, and procreate, as a child of the same age can be given in marriage (National Office on Anti-Racial Discriminations 2012, 51).

As far as out-of-group relations are concerned, the most blatant institutional discrimination and violation of community security is residential segregation in mono-ethnic camps. These settlements, which were planned to just temporarily accommodate Roma refugees from Former Yugoslavia and Romania, have become permanent solutions for those among them who were unable to access public or private dwellings. The policy of Roma segregated settlements has severe consequences on almost all dimensions of human security. In the case of Angelica, described in the previous section, as a reaction against the alleged kidnapping communicated by the national mass media, a group of about 400 non-Roma inhabitants from Ponticelli assaulted the Roma settlement where the Roma girl used to live with her family. This camp was one of the biggest in the area and hosted about 50 Roma families. Stones were thrown against the shanty town and Roma inhabitants were threatened and insulted. Other attacks against nomad camps in the same area took place in the following days and fires started by the assailants destroyed shacks. By the end of the 15 May 2008, all Roma residents were forced to move from Ponticelli to different districts of Naples and were hosted in other camps and in a school. Whereas the Roma were in the spotlight, there was silence about the allegations made by some local administrators, that the Camorra (the local criminal mafia organization located in the Campania Region) may have encouraged the attacks in order to clear the Roma settlements and to speed the kick-off of construction work before the expiry of the deadline set by the municipal council for the allocation of funds for an ambitious requalification project for the area.[14]

With regards to economic security, Roma segregation puts Roma people at risk from discrimination in their access to jobs, because prospective employers can easily discover the ethnic identity of candidates living in segregated settlements through their address. Additionally, forced evictions may compel Roma people to flee to areas of the city far away from their workplace or to spend days without shelter.

14 See the webpage of the Municipality of Napoli, accessed on 25 November 2013. See also Presa Diretta RAI (2009) devoted attention to these suspects.

Three kinds of camps have developed across the years: authorized or 'equipped' camps (established by local authorities and supplied with basic facilities); semi-authorized or tolerated camps (spontaneous settlements recognized *ex post* by municipalities, or former authorized); and unauthorized/illegal camps (mostly without access to drinking water, or electricity, heating or gas for cooking).[15] The police and local authorities have particularly targeted this last kind of camp and they still forcibly evict people living there without procedural safeguards and without offering them any housing options (ECRI, 2014). This leads to what Merton (1948) calls the 'self-fulfilling prophecy': evicted Roma people rebuild new illegal encampments in other parts of the city and local authorities constantly evict and circulate (Sigona, 2003; van Baar, 2011). The policy of forced evictions carried out during the school year has also dramatically increased the statistics concerning Roma children's drop out from education and damaged their aspirations to improve the quality of their lives.

The 'National Strategy' acknowledges the inadequacy of Roma evictions and emphasizes the need to adopt:

> an integrated and sustainable medium-long term approach (and therefore the exclusion of on-the-spot or short-term projects) (National Office on Anti-Racial Discriminations 2012, 34).

While the season of 'Nomad emergency' and Roma fingerprinting[16] came to an end in 2013, the vicious circle of Roma ghettoization and eviction still takes place.

15 Life in non-authorized camps is particularly at risk, due to the absolute lack of heating, water, electricity or other essential services. In February 2011 four Roma children, aged 4–11 years, died in a non-authorized camp in Rome (Redazione online, 2011), when fire swept through their shack while they were sleeping and their parents were out looking for food. Just four years before, in 2007, another four Roma children, aged 4–12, died in a similar tragedy in Livorno (Redazione online, 2007). To no avail, the European Committee of Social Rights ruled a decision in 2005, in response to a collective complaint lodged by the European Roma Rights Center against Italy, stating that Article 31§1 of the Revised European Social Charter (COE 1996) 'guarantees access to adequate housing, which means a dwelling which is structurally secure; safe from a sanitary and health point of view, i.e. it possesses all basic amenities, such as water, heating, waste disposal, sanitation facilities, electricity; not overcrowded and with secure tenure supported by law … The temporary supply of shelter cannot be considered as adequate and individuals should be provided with adequate housing within a reasonable period'.

16 Procedures were slightly different in the various cities, Roma people were usually identified, fingerprinted and, in some cases, mug shots were also taken. Operations also included private house searches by the police without showing any search warrant. They usually started early in the morning and could last the whole day, in spite of low temperatures during winter and with little consideration for the needs of children, women and old people. In Rome, Naples and Milan, the census took place from June to October 2008. The final data, delivered by the prefects of these cities assess that 167 camps were found, of which

Today, just like yesterday, a substantial budget has been allocated to 'solve the Roma problem', even if the rhetoric guiding it has changed. In fact, in the past, the mobilization of resources and the 'public–private' partnerships had the purpose of 'controlling' the Roma through the implementation of the newly adopted 'Piani Nomadi' (as a consequence of the Nomad Emergency Decree).[17] These measures were described by politicians as innovative tools for integrating 'the part of the Roma people who truly showed respect for Italian (European) rules' (Redazione online, 2010), but they never stepped up any effort to support Roma people's transition from the camps to permanent housing at least for the Roma people who aspired to it (Amnesty International, 2010) and they rather aimed at the relocation of Roma families from non-authorized camps to authorized ones.[18]

Pro-Roma NGOs provide data to explore 'the nature of the institutions which provide security, and the interests served by them' (Newman, 2010, 81) at the local and national level. For instance, Associazione 21 luglio writes:

> As the inquiry of Associazione 21 Luglio shows, 35 public and private bodies operate in the "Nomad Camps Inc" business. Estimating that each of these bodies necessitates an average of 12 workers, there are over 400 individuals employed in the mechanisms of the "Roma question". It is worrying that the percentage of funds assigned directly to these bodies, without going through the regular process of public calls for competitive bids, is in some cases of 100%. (Associazione 21 Luglio, 2014).

By contrast, today, the 'National Strategy' provides the mobilization of resources and relies on public–private partnerships (which belong to the tenets of neoliberal governance) with the aim of 'including' the Roma people (rather than controlling them). It is still far from clear, though, whether power relations, dynamics of local justice and the structure–agency binary can be challenged and transformed by its implementation, in such a way that change is truly beneficial for the Roma people, rather than for 'Roma business'.[19] Besides, the economic crisis and high

---

124 illegal – 12,346 persons were found living there, of whom 5,436 were minors. In March 2009 the Veneto Region started the census in almost all its provinces (Osservazione, 2009).

17 The prefect of Rome estimated that about 32,445,000 euro was invested in Rome within the framework of the 'Roma emergency' only in 2009 (see Berenice et al., 2013).

18 The eligibility criteria set by the local authorities for a place in the authorized camps are strict and encompass the commitment to send children to school and to attend inclusion programmes, as well as a clean criminal record. The local authorities in Milan deliberately infringed the agreements promising public housing for Roma families who accepted to leave the camp situated in Triboniano Street. In 2011, the Milan Court (24 January 2011) upheld the complaint lodged by 10 Roma people, eventually finding that the non-allocation of public housing was discriminatory on the basis of the applicants' ethnic origin.

19 The 'Capital Mafia' scandals in March 2015 (www.ansa.it) unveiled the risk that criminal organizations seek to control the 'market' of Roma housing emergency and the reception of refugees.

unemployment rates can easily be used as an argument to justify anti-immigration and anti-Roma resentment, as well as Eurosceptic discourses.

## Conclusion

The so-called insecurity attributed to Roma communities in Italy and in most European countries must be focusing on Roma people's ignored human security, namely on personal and community security. Despite the recent shift in the Italian Government's rhetoric concerning the Roma, from a 'public security' to an 'inclusion discourse', Roma are the people at most risk of victimization of abuses, violence and sometimes homicide. From Newman's studies and Tomovska's research on Roma, the Roma people's insecurity in the neoliberal frame of securitarian governance can be understood. There are three main concerns about the Roma people's human security in today's Italy and European countries. Firstly, by calling on the old and new EU member states to take Roma rights more seriously through the adoption of national strategies, the European Union runs the risk of generating major tension between Roma communities and other citizens (Sigona and Nidhi, 2011), reinvigorating populist rhetoric against the Roma people, above all at a time of high unemployment rates. To this purpose, Vermeersch gives evidence that the:

> EU institutions will need to find ways to deal with the ambivalence inherent in their 'European' calls for tackling the problems (my note: of the Roma) at hand. Such calls may indeed push member-states into adopting new policies that will benefit their Romani citizens. But the new tendency to single out the Roma as a European priority and a special European concern has also, rather paradoxically, opened up new opportunities for nationalist politicians to plead against new national measures to help the Roma (Vermeersch, 2012, 1197).

In the multilevel EU architecture, once policies such as the 'National Strategy' are adopted, they need to be implemented by local authorities. The Italian 'National Strategy' explicitly states that the implementation of social inclusion policies 'is, and remains, a specific responsibility of local Authorities' (p. 31).

As a possible solution, Sobotka and Veermersch (2012) suggest that:

> If EU mechanisms want to truly be a catalyst for social change … they will have to find ways to address the local community as a whole and pursue strategies that can persuade local authorities to become more committed to the socioeconomic integration of all their citizens. In other words, there is a need to find a language and a strategy of social policy and human rights protection that promotes the EU initiatives as beneficial for the Roma and non-Roma alike.

Secondly, measures such as the 'National Strategy' are deeply rooted in neoliberal premises (Sigona and Nidhi, eds 2011, 121 and 133), based on 'public–private partnerships' and activation. The 'Roma Framework' (2011, *supra*), for instance, argues that the intergenerational vicious cycle of Roma poverty and exclusion will be broken by putting emphasis on the 'activation' of unemployed Roma people (especially the younger ones) and integrating them into the labour market.[20] Besides, the measures envisioned by the Roma Framework:

> Rely on a neoliberal trend for governing social and minority-related affairs through processes of decentralisation, the outsourcing of public services to private stakeholders, and the correlated mobilisation of civil societal agencies (van Baar 2012, 1291).

The political aim of such targeted integration and inclusion measures implies the need to evaluate and measure their achievements with the support of experts, researchers and various stakeholders, and through various sets of indicators. Evaluation does not necessarily imply transformative aims (Mertens, 2009) and may easily fall into the trap of the widely criticized instrumentalist approach to law and policy (Griffiths, 1999 and 2003; Vermeule, 2007). van Baar (2012) points out that copious pieces of research and reports have explored the challenges of enhancing the Roma people's situation:

> In terms of the discrepancies between inadequate, largely lacking, well- or not-well-meant policies on the one hand, and the ways in which they have been implemented at a variety of levels (grassroots, community, national, etc.) on the other. Following this binary opposition, supranational agencies and NGOs have often been understood as emancipating agencies that 'only' need to exercise their powers more effectively to challenge Roma marginalization more productively.

It is precisely this role of 'emancipating agencies' (Miraftab, 2004a), though, that leads to the third and last concern with regard to this new phase of Roma governance in Italy, that is the risk of depoliticizing key components of these policies and depriving them of their transformative potential.

---

20 The 'Roma Framework' states: 'The Roma population is young: 35.7% are under 15 compared to 15.7% of the EU population overall. The average age is 25 among Roma, compared with 40 across the EU. The vast majority of working-age Roma lack the education needed to find good jobs. It is therefore of crucial importance to invest in the education of Roma children to allow them later on to successfully enter the labor market. In Member States with significant Roma populations, this already has an economic impact. According to estimates, in Bulgaria, about 23% of new labor entrants are Roma, in Romania, about 21%.' European Commission (2011, 2). See also World Bank (2010); Sigona and Nidhi (eds. 2009).

# Chapter 10
# Refugees: Stranded between the Priorities of European Security[1]

Susanne Knickmeier

## Introduction

Since the beginning of the Syrian civil war in March 2011, nearly 140,000 fatalities have been estimated.[2] Despite international efforts like the UN resolutions on Syria (for example, in May and September 2013 as well in February 2014), the bloody conflict and serious human rights violations continue. Around 3.9 million people have fled the fighting, mostly into neighbouring countries like Jordan, Iraq, Lebanon or Turkey. Some have also tried to reach the European Union (EU), the first challenge of which is to enter 'Fortress Europe'. Refugees coming from Turkey must now go by boat, after the Greek Government built a fence between Turkey and Greece to stop land border crossings.

In recent years, thousands of people from conflict-torn states like Syria, Iran, Afghanistan, Somalia, Sudan or the Democratic Republic of the Congo have drowned trying to reach the European Union (UNHCR, 2014b: 16*)*. In October 2013, a boat with more than 500 refugees from Eritrea and Somalia attempting to sail from Libya to Italy capsized in the Mediterranean near the Italian island of Lampedusa. More than 300 people died (UNHCR, 2014a). Another incident happened in January 2014: the Greek coastguard is accused of turning a group of 28 Afghani and Syrian refugees illegally back to Turkey, when the refugee boat towed by a Greek coastguard vessel capsized at high speed and 12 people died (UNHCR, 2014a; European Parliament, 2014a). According to Amnesty International, from August 2012 to May 2013 in at least 39 cases refugees were stopped trying to cross the Aegean Sea or the Northern land border along the river Evros and unlawfully returned to Turkey by Greek coast and border guards (Amnesty International, 2013a, 9).

It is beyond dispute that the European member states, as sovereign nations, have the right to control who crosses their borders and who resides in their territory

1 I am grateful to Dr Christopher Murphy for his invaluable comments as well as translation and editing support.

2 In 2014, the United Nations announced that 'the death toll may now be well over 150,000 – the UN had stopped releasing figures because it became "impossible to count all the bodies" ... ' (United Nations, 2014, 1). See also German News Agency (dpa) 2014.

(Angenendt, 2003, 182). But, as signatories to the United Nations Convention Relating to the Status of Refugees (1951) and referring to other conventions, for example, the European Convention on Human Rights (1953), or the Charter of Fundamental Rights of the EU (2009) the EU member states also have to grant asylum to refugees or can resettle them for humanitarian reasons. Though only authorities in the destination country can review, through due process, whether somebody is a refugee or not, the EU has aggravated physical access to its territory over the last years, thereby worsening the conditions for refugees and diminishing their chances of resettlement.

Migrants[3] try to reach the EU due to political problems, insecurities or bloody conflicts in their home countries, maybe without papers, perhaps with the help of facilitators, who could be members of Mafia groups exploiting migrants' needs. Instead of offering them security, migrants are increasingly connected to crime and seen as a risk factor and threat to Europe's internal security and welfare. Over the last decades, the challenging topics of migration and security have become an indivisible unit high on the political agenda of the EU;[4] this has resulted in several restrictive migration and asylum laws (Huysmans, 2000, 757). Legal, administrative and practical measures, focusing on resistance, deterrence and detection, have been implemented to strengthen control measures at the EU's external borders. In 2010, the European Commission emphasized the importance of appropriate measures to prevent and combat cross-border crime (for example, illegal entry) at the European external border (European Commission, 2010b, 4).

The European internal security framework identifies (illegal) migration as a security risk (for example, European Commission, 2010b, 11) that has to be controlled, ignoring the fact that most migrants aim to attain protection from insecurities in their home countries. This chapter, based on theoretical insights and legal analysis, discusses: (1) the dichotomy of national and human security; (2) European security policy; (3) border security measures; and (4) the consequences of these policies. It is, for example, questionable, whether migration constitutes a real or perceived risk/threat for Europe's security, or whether the chosen measures are an effective prevention strategy to reduce (irregular) migration or just an act of securitization and criminalization. A massive security operation is supposed to keep migrants out of the EU; yet who protects their security?

## Dichotomy of National and Human Security

What exactly the term security entails has been debated. According to Zedner, security can be used to describe an objective state of being protected from threats

3 In the following text, 'migrant' is used as a general term, because the European measures of border control are related to every migrant irrespective of the legal status.

4 For example: the Stockholm Programme of 2010 or the Internal Security Strategy of 2010.

or danger, a subjective feeling of safety, or the means of pursuit of either of these (Zedner, 2003, 158). In his social contract model, Hobbes argued that the sovereign must guarantee his citizens′ security (Hobbes, 1881, 133). States regularly focus on the 'traditional term' of security that is based on national security and therewith the requirement to maintain the integrity of the nation state through military power (Daase, 2011, 143). A certain lack of conceptual clarity is necessary for a state to govern unanticipated challenges (Dalferth, 2008, 180). Over the years, the term of traditional security was broadened and threats like economic and environmental threats, terrorism or inner-state conflicts were included (Daase, 2010, 10).

With regard to migration, European states fear an alleged 'lost control' over their national borders (Vollmer, 2011, 317). Irregular migration can be considered to undermine the state's sovereignty and right to control border-crossings and residency. However, it has to be noted that irregular migration is often a direct consequence of missing legal access to the EU for people fleeing persecution or (armed) conflicts. Additionally, Europeans fear insecurities for their social and political security as well as for their internal market through an assumed invasion of migrants and organized crime (Huysmans, 2000, 757, House of Lords, 2008, 50). Referring to a representative survey about political attitudes, around 48 per cent of the respondents agreed (at least partially) with the statement that migrants come to Germany to exploit the social system (Decker, Kieess and Brähler, 2015, 32).

The traditional term of security does not include human being's interest in security and the question, how safe and free people are as individuals. In 1994, the Commission on Human Security developed the term 'human security', thereby introducing the individual's need of security and human rights. The Commission pointed out that security is 'freedom from want, freedom from fear, freedom to take action on one's own behalf' and listed seven values of human rights: economic, food, health, environmental, personal, community and political security (United Nations Commission on Human Security, 2003, 4; United Nations Development Programme 1994, 24). In the same vein, Zedner emphasizes 'security in its objective sense promises freedom from risk and the liberty of citizens to pursue their individual and collective goals free from injury, harm, or loss' (Zedner, 2003, 168).

The human security framework offers migrants protection against several insecurities. But the clashing interests between the European perception and fear of migration as a security threat on the one hand, and the wish of migrants to attain protection from insecurities (like war, persecution or mistreatment by police authorities) in their home countries on the other hand, seem to create an intractable struggle.[5]

5 See also Chapter 3, Haverkamp and Arnold in this edited volume.

## European Security Policy

What are the European strategies designed to govern these assumed potential threats, risks and colliding requests? Does the area of freedom, security and justice, mentioned in several European policy papers, apply only for Europeans?

The Treaty of Amsterdam (1997) postulates in Article 1 of its substantive amendments the will 'to maintain and develop the Union as an area of freedom, security and justice, in which the free movement of persons is assured in conjunction with appropriate measures with respect to external border controls, asylum, immigration and the prevention and combating of crime'. Security should, referring to the Internal Strategy for the European Union 2010, be driven by crime prevention, which has, in the case of border control measures, to concentrate on detection and deterrence to reduce irregular migration. Threats (for example, irregular migration) should be prevented or at least their consequences minimized (for example, by push-back actions) (Daase, 2010, 14). Therefore, risk analyses (strategic and operational analyses) have been implemented and are conducted to identify and control risks as soon as possible, for example, by the European border agency Frontex (see: Frontex, 2014).

### *Legal Measures*

In the area of migration, discussed within the framework of security, the European member states have prepared a Common European Asylum System (CEAS). CEAS should lead to harmonized asylum laws and standards within the EU and tackle and prevent illegal migration. But, as argued in the following examples, the Eurodac regulation and reception conditions directive affects not only illegal migrants,[6] but for example asylum seekers, too.

Eurodac is a large database of fingerprints to identify applicants for asylum and illegal migrants within the EU. The legitimacy of the compulsory obligation to provide fingerprints is controversial, due to the acute possibility of surveillance and restrictions on the right of informational self-determination.[7] With the passing of the revised Eurodac Regulation in June 2013, access for police authorities was simplified. Although asylum seekers are not suspected of a criminal offence, their data are accessible for police authorities. It could be argued that contrary to European citizens, asylum seekers are under general suspicion and treated like potential criminals.

The Reception Conditions Directive regulates common standards of living conditions of asylum seekers. It seeks to ensure the maintenance of fundamental rights and expresses that detention is a measure of last resort. In the context

6 The term 'illegal migrants' is discussed below.

7 The informational self-determination is, for example, guaranteed by the German Constitution, in Art. 2 sec (1) (general personal liberty) combined with Art. 1 sec (1) (guarantee of human dignity).

of the revision in 2013, the member states agreed that no one shall be held in detention for the sole reason that he/she is an asylum applicant; however, detention is allowed when authorities need to determine the identity or nationality of an applicant or when protection of national security or public order so requires (Art. 8 Reception Conditions Directive). Contrary to the simplified conditions of detention of migrants, investigative custody in Germany is only allowed in narrow circumstances (§ 112 StPO[8]). The deprivation of liberty as a last resort has to be proportional. Nevertheless, in Greece migrants and asylum seekers, even minors, vulnerable or disabled people, can be detained for up to 18 months by a police administrative decision (Greek Presidential Decree 116/2012). It is deemed legally permissible based on the risks of absconding and the threat to public order. However, it could also be argued that this is an act of criminalization that runs contrary to human rights that grant asylum seekers certain liberties and due processes while their applications for asylum are examined.

*Border Control Measures*

According to the Schengen Agreement (1985), national borders within the Schengen area were abolished and the free movement of European citizens (not migrants) was assured. Movements of persons are no longer monitored at national borders, but rather at the external borders of the Schengen area. These external controls require the intensification of cross-border police cooperation, which led to the establishment of the European border agency Frontex in 2005 and the Schengen Information System (SIS).[9] The Internal Security Strategy (2010) refers to the strengthening of security through border management and emphasizes: 'The EU can treat migration management and the fight against crime as twin objectives of the integrated border management strategy' (European Commission, 2010, 11).

One section of European migration control is the externalization to third countries (like visa requirements for selected third-country nationals or offshore border control measures of the EU, for example, in Northern Africa) or the switch to private organizations (for example, airlines that have to control visas). In April 2014, Malta and Nigeria signed an agreement for the repatriation of migrants arriving irregularly in Malta (Government of Malta, 2014). After the transition of power in Libya, the Italian Government renewed its Memorandum of Understanding on migration control and border security in April 2012. Libya, a state that has not ratified the UN Refugee Convention yet, committed itself to stop migrants within its territorial waters and to take them back (Vassallo Paleologo, 2012). Migrants travelling through Libya on their way to Europe are sheltered in detention centres or live without any legal status in the street. Amnesty International visited

8 German Code of Criminal Procedure.

9 The SIS is a shared governmental database with over 46 million records (as of 2013) on individuals, lost and stolen property or alerts on persons (for example, entry bans) (Council of the EU 7389/13SIS-TECH 36 COMIX 156).

seven detention centres for migrants in spring 2013 and found evidence of ill-treatment, in some cases torture and inhumane conditions of detention (Amnesty International, 2013b). Repelled people have no chance to apply for asylum, nor receive security from their home-countries' insecurities. Using the argumentation of internal security, the EU thus shies away from its liability towards migrants and enables inhuman treatment that would indeed be utterly forbidden within the EU.

To support border control, several technical measures have been developed, for example, the abovementioned SIS and Eurodac or planned 'smart border' measures like Eurosur,[10] Entry/Exit Systems and the Registered Traveller Programme. The European Commission underlines the advantages of 'smart borders' as enhancing mobility and security (European Commission, 2013b). Conversely, data storage and data exchange restricts the right to informational self-determination and produces its own set of new insecurities.

A move towards a common European border agency was the foundation of Frontex in 2005. Frontex is a specialized, independent body with legal personality, operational and budgetary autonomy. The agency, covering all aspects of border policy, was set up to coordinate intelligence-driven cooperation between the member states and to strengthen security at the EU's external borders (sea, land and air). To control migration, Frontex coordinates joint operations planned on the basis of risk analysis (Frontex, 2013).

As mentioned, border police have been accused of illegal push-back refoulements of refugees at the Greek–Turkish sea border by several human rights organizations. The push-back actions took place in the area of Frontex's operation. The involvement of the agency is questionable. In April 2014, the European Parliament approved the establishment of rules for the surveillance of the external sea borders in the context of an operational cooperation coordinated by Frontex that is scheduled for implementation in mid-2014. The regulation emphasizes in its amendments that 'saving of lives and the protection of fundamental rights should be a priority', but also notes that 'measures against those persons who have crossed the border in an irregular manner' should be taken (European Parliament, 2014b) and that border guards may 'warn and order' a vessel not to enter the territorial waters of a member state (European Parliament, 2014c). But who can control whether a migrant boat is diverted or pushed back? Is there a difference between diverting against one's will and pushing back? What happens if people in ineligible boats are warned and sent back to their destination country? Do they have a realistic chance to return safely? How can it be guaranteed and proven at high sea that boats are not being sent to states where there is a risk of persecution, torture or other serious harm?

As a consequence of a tragic incident in October 2013, when over 300 migrants drowned off the coast of Italy, the Italian Government decided to strengthen controls at its maritime border in the Mediterranean Sea with the military and humanitarian

10 European border surveillance system, started in December 2013.

operation Mare Nostrum (Ghelli, 2013). The operation ran over one year and was discussed controversially. Although special emphasis was placed on its humanitarian impact and more than 100,000 people were saved from drowning, funding also went into the upgrading of military and police operations and equipment. Human rights organizations are concerned that illegal push-back actions could be camouflaged as humanitarian rescue missions. Although asylum applications could be only reviewed through due process, it was reported that authorities identified and questioned rescued migrants on military vessels without the support of translators (Andres, 2014, 51).

After several discussions in the EU about funding the operation Mare Nostrum, it was stopped one year later and replaced by the Frontex joint operation Triton. Triton was developed to manage migration, to 'ensure effective border control … and at the same time to provide assistance to persons or vessels in distress' (European Commission, 2014a). Frontex's focal point is still border control and crime prevention, for example, illegal entry, but not humanitarian search and rescue.

In the context of controlling Frontex, it has to be noted that the member states have not delegated their sovereign rights. Thus, it is not the European Parliament controlling Frontex's operations, but the national parliaments. They are informed by way of reporting through Frontex (which decides about the information given) (Deutsche Bundesregierung, 2009, 5). As such, it is questionable how far national parliaments are able to control what happens at the external border.

## Consequences of European Border Security

Police authorities or Frontex operate on the basis of risk analysis to identify risks and 'dangerous' people and to deprive them of their liberty, before they can be delinquent (for example, illegal entry). Migrants can be objects of police measures (for example, push-back actions against the Syrian refugees), although they have not committed a crime. Border police usually assume an attempt of illegal entry irrespective of the migrants' actual legal status (some migrants have refugee documents from the United Nations High Commissioner for Refugees (UNHCR) and their possible intention to seek asylum.

Although precise figures are difficult to obtain, numbers of (irregular) migrants and the assumed influx play an important role in security discourses, for the justification of repressive policies and the governing of risks. Vollmer has pointed out that 'higher numbers are used as a justification for control and enforcement policies; lower numbers calm the political landscape' (Vollmer, 2011, 330). The UNHCR estimated that there are more than 51.2 million refugees worldwide in 2013 (UNHCR, 2014b, 2). More than 430,000 people sought asylum in the EU with its 500 million inhabitants in 2013 (Eurostat, 2014, 1) – the highest number in years. This suggests that migration movements do not depend on border security measures but on political problems worldwide. In line with this, Frontex counted

107,365 illegal entries in 2013 (Frontex, 2014, 14). But still the number of asylum seekers cannot be labelled as a flow of refugees.

The abovementioned policies lead to a securitization and criminalization process of migrants. The Copenhagen School considers security as a figurative construction of contemporary politics (Waever, 1995, 76). Every problem is considered as to whether it is security related and, if so, classified as part of a broader security policy (Berndt, 2007, 110). If a problem is securitized within society, it is irrelevant whether there is actually a real threat as below mentioned political and legal measures[11] are automatically accepted by citizens. Huysmans explains that securitization of migration is dependent on national identity and the differentiation between 'us' and 'them' (Huysmans, 2000, 757, 758). This discussion was particularly apparent concerning the housing of asylum seekers in Berlin in July 2013, where people demonstrated against a house for asylum seekers in their neighbourhood. Within the group of demonstrators, people that usually had nothing in common were unified against the group of asylum seekers (Bullion, 2013, 10).

Securitization processes can lead to criminalization of refugees and migrants through legislation,[12] legal practice[13] or rhetoric (Pickering, 2005, 141). In common language usage, various groups of migrants are not differentiated between. Refugees are regularly labelled as 'illegal migrants'[14] and therefore automatically criminalized. Though even if a refugee has crossed the border illegally, it is not a criminal action, but rather justified by the UN Refugee Convention, so long as the refugee seeks asylum immediately, Art. 31 UN Refugee Convention. Nevertheless, police authorities initialize, from time to time, preliminary investigations that appear in crime statistics and imply illegal behaviour by migrants.[15]

The criminalization in the language use fosters the differentiation between citizens and non-citizens. According to Bosworth/Guild 'non-citizens are always and already identified as a source of potential risks in contemporary political discourses' (Bosworth and Guild, 2008, 705, 711). Citizenship determines whether individual rights and their protection by the state are ensured and about social and political participation or exclusion (Zedner, 2010, 382).

11 For example: foundation of Frontex, pushback actions in Greece.

12 For example: Eurodac database. In Germany, asylum seekers are banned from working, are subject to residential obligations that restrict their free movement and have a duty to cooperate. Infringement of these rules is considered a criminal offence.

13 For example: Detention of migrants in Greece.

14 Political programmes, police authorities and the European border agency Frontex usually concentrate on the term 'illegal migration'. Illegal migration is usually defined as a movement across borders or a stay within a country by violating the nation's immigration law.

15 In 2012, the German police registered nearly 85,000 offences against the Residence Act, but just about 7,000 convictions.

However, migrants are seen as illegal and left to their own on their perilous crossing to Europe. It was the civil society who installed an alarm phone[16] which can be called by migrants in distress. The approach is simple and clear. People answering the phone inform search and rescue services about the incident and compel authorities to act. The alleged security of Europe seems to be a justification for the violation of international laws. Can law-breaking concerns about non-Europeans be accepted to secure the justice and freedom of Europeans? The following examples focus on border control operations by Frontex and describe debatable actions at the Southern sea border.

The first deals with the escorting back of migrants to a third state (push-backs), as happened in the abovementioned example at the Greek–Turkish border. This can violate the UN Refugee Convention, the European Convention on Human Rights, the order of non-refoulement and their right to be heard. National criminal laws may also be violated if migrants are physically beaten or if their vessel is damaged. Everybody has the right to apply for asylum, though to make a claim one has to be on land within European territory. In 2012, the European Court of Human Rights ruled in the case Hirsi Jamaa and Others vs. Italy that spurning migrants and asylum seekers on the high seas violates human rights principles and the international obligation not to return individuals to countries where they could be at risk of human rights abuses and ill-treatment (European Court of Human Rights, Judgement 23.02.2012, Application no. 27765/09).

Another question is the handling of castaways. The UN Convention on the Law of the Sea (1994) states that it is a legal obligation to rescue anybody who is distressed while being at sea. This obligation will be incorporated in the abovementioned regulation for Frontex. Nobody knows how Frontex vessels divert migrant boats. What happens to diverted boat people in international waters who – caused by their bad equipment – have no chance to reach a third country? In 2011, NATO, operating in Libya, failed to rescue 72 migrants and caused, according to a report of the Council of Europe, the death of 63 people (Council of Europe, 2012). At the same time, local lifesavers were accused and criminalized. In 2007, seven Tunisian fishermen rescued 44 Africans. They were charged as facilitators for illegal entry and their boats, their basis of existence, were confiscated in Lampedusa. Four years later, in September 2011, the captains were discharged by the Sicilian High Court (Migrants at Sea, 2011).

## Conclusion

Although the EU underlines the importance of human rights and the granting of asylum, various measures have led to the establishment of 'Fortress Europe' through pre-emptive criminalization of migrants, erosion of due process and militarization of border security (Frontex). Political measures and legislation

16 See: Watch the Med. Online: http://watchthemed.net.

focus on the European right to security and consider migrants as a risk for Europe's internal security and welfare system, ignoring the migrant's wish to (human) security. Border management and control have been enlarged through securitization processes, even though the UNHCR pointed out that measures of control and deterrence infrequently make an impact (UNHCR, 2001, 4). They are not effective strategies to regulate migration, which seems to be instead dependent on political developments in the migrants' home countries (referring to the numbers of asylum seekers). Thus, the security measures in place at Europe's borders cannot and will not solve the problem of migration.

# PART III
# Aporias or Bad Governance of the Disasters in the Twenty-First Century

Chapter 11

# Industrial Disasters: The Italian Case[1]

Pietro Saitta

## Introduction

This chapter focuses on the environmental crisis in the *Bel Paese,* as Italy is often referred to with some irony. In fact, in spite of narratives and self-representations based on ideas of its landscape, heritage and slow food, for decades Italy has witnessed a systematic erosion of its territory and a dramatic increase in pollution, stemming from a substantially unregulated industrial presence. From North to South, at different stages of post-Second World War development, both large process industries (oil, petrochemical- and chemical-related) employing thousands of workers, and smaller family-based firms, operating in highly polluting sectors, have become increasingly central to the local economy. For example, Sardinia (Sarroch, Porto Torres), Sicily (Milazzo, Gela, Augusta, Siracusa, Melilli, Priolo) and Apulia (Manfredonia, Taranto) are home to some of the largest petrochemical and steel works in Europe. In the Central and Northern regions of the country, other similar industries are also present (Falconara, Livorno, Marghera); but these areas have also been at the epicentre of another type of industrial development, and one which has been much studied in international economic literature, the 'Italy of industrial districts'. In fact, Veneto, Lombardy, Le Marche and Tuscany are regions of Italy that have based their wealth on the concept of 'small is beautiful': that is, on the concentration of a series of small factories in given areas, which specialize in different phases of the same production process, run and operated by members of a single family with the help of a few workers (Becattini, 2002).[2]

In addition, since the 1950s, the construction industry has become a leading sector within the national economy – an occurrence that has lead to massive speculation processes, and has contributed towards making Italy the third European country for human activities-related agricultural deficit (Rondinone, 2012).

Yet, over the years, Italy has undergone a severe process of industrial reduction. A number of dangerous operations (related to mining, steelworks,

---

1 English revised by Deborah Kay.

2 Beyond their serious environmental impact, these 'districts' also played an important role with regard to labour conditions and health. Work accidents, poor levels of security, exploitation and black employment were, and still are, quite common features of such development models. Probably, the Italian encounter with neoliberalism and its quintessential forms of production began in such milieus. See: Palidda (2013).

refuse dumping and so on) are no longer active, but their legacy is relevant in terms of soil contamination. Since 1998, different decrees and acts issued by the Ministry of the Environment and the national government[3] have identified 57 sites of national interest (SIN) in need of remediation – roughly corresponding to an area equal to 3 per cent of the national territory. In addition, about 13,000 sites of regional interest should be added, which still need accurate monitoring with regard to pollution levels.

Furthermore, the presence of military bases and shooting ranges in the South of Italy – especially in Sardinia (Salto di Quirra, Perdasdefogu, Capo Teulada, Capo Frasca, S. Stefano, La Maddalena and so on) – have frequently been at the centre of controversies and alarms over the radioactive nature of their wastes and emissions.

Finally, still in the South, but more especially in Campania, for decades now large parts of the region have been grappling with an unprecedented waste crisis, which has also produced vast social turmoil as well as effects on public health (Chapter 12 Petrillo in this volume). Not by chance, a huge area surrounding Naples, including the villages of Acerra, Nola and Marigliano, has been renamed the 'triangle of death' (Senior and Mazza, 2004). But, in spite of its notoriety, the Neapolitans' revolt over the management of waste sites and projects for the construction of new incinerators is, so to say, just the tip of the iceberg. However, the 2007–2008 Naples 'trash crisis' and the related riots made clear to many mainstream international observers that Italy faces serious environmental, health and social questions (Petrillo, 2009; De Biase, 2015).

Certainly, this is not an exhaustive list of the cases and problems that exist in Italy, but it should be enough to suggest that the environmental issues at hand are complex, and affect local communities in different ways.

## Issues

### *Epidemiological Issues*

Epidemiological rates in industrial and formerly industrial areas are the most obvious sign of the magnitude of the problems faced by regions exposed to toxic products. In this regard, the 'Sentieri' study (Pirastu et al., 2011) is a portrait of the sanitary conditions of 44 of the 57 sites of national interest identified by the Italian Government in the past few years. Such a review shows that, during the years 1995–2002, 'for a total of 403,692 deaths (both men and women), an excess of 9,969 deaths is observed, with an average of about 1,200 extra deaths per year.

3 Ministerial decree, Environment 18 September 2001, n. 468 (National Program for the environmental restoration of polluted sites); Law n. 388/2000; Ministerial decree 468/01; Law 179/02; Law 266/05; Legislative decree 152/06; Ministerial decree 11/4/08; Ministerial decree 25/5/08; Order of the President of the Ministerial Council 3716/2008.

Most of these excesses are observed in Italian polluted sites located in Southern and Central Italy' (Pirastu et al., 2011, 30).

By going into more detail, one sees, for example, that in Gela (Sicily) malformation rates – especially, hypospadias – are among the highest ever recorded in medical literature. But pollution-related death rates are also significant and in excess for all sexes and age groups. Such rates are compatible with other data indicating that, historically, in Gela's spring water, arsenic, mercury, benzene, vinyl chloride, 1.2 dichloroethane were found in unprecedented quantities: for example, arsenic: 250,000 mg/l (legal limit: 10 mg/L); mercury: 6,600 mg/l (l.l: 1 mg/L); benzene: 34,000 mg/L (l.l: 1 mg/L), and so forth (Mudu, 2009, 11).

Taranto (Apulia) hosts a large steelworks, and here, mortality excesses for all types of cancer (between 10 and 30 per cent for different pathologies) and respiratory diseases (40 and 50 per cent for men and women, respectively) have been found (Pirastu et al., 2011b, 136). Some of these rates can be correlated to other data, and show that the food chain is severely affected by emissions from local industries: for example, a study of firms operating in the foodstuff sector, and located in an area within 10 km from the industrial centre, indicated that 26 per cent of samples collected presented an excess of dioxins (Iavarone et al., 2009).

Manfredonia (Apulia) is home, among other things, to a petrochemical plant that exploded in 1976, releasing enormous quantities of arsenic into the environment. The effects of that incident became visible only in the early years of the new millennium, when a mortality study showed increasing rates of pulmonary and cholestatic tumors (Portaluri, 2006, 218).

Between 1948 and 1997, Pieve Vergonte (Piedmont) was the site of a chemical industry specializing in the production of DDT. Particularly affected by this production were rivers, trees and fish (Bettinetti, Croce and Galassi, 2005). Several years after production ended, epidemiological rates still show an excess in mortality rates for all types of tumour (Pirastu et al., eds. 2011, 118–120).

In the Litorale Domizio Flegreo and Agro Versano (the 'triangle of death', in Campania) there are several landfill sites, rubbish incinerators and, above all, a number of illegal sites used for dumping waste and toxic substances from all over Italy. Excesses in mortality rates for all kinds of tumors, and for respiratory, circulatory and digestive apparatus diseases have been registered in the area (Pirastu et al., eds. 2011, 86).

The sites mentioned above are, of course, only a small selection of cases (five out of the 44 presented by the Sentieri Project). However, they are all of particular interest because they are paradigmatic of the issues under discussion here. In particular, these events suggest that Italy has had to face a vast number of 'disasters' at different moments in its past – that is, events that have been the cause of thousands of deaths, malformations and diseases, and that have affected the environment, the food chain, and, ultimately, the safety of entire communities and regions. Ultimately, they are events that have failed to be acknowledged as such, since the remediation of these 'sites of national interest' has never taken place, in spite of the fact that some of these accidents happened almost 40 years ago, and certain discoveries concerning epidemiological rates were made a couple of decades ago.

On this subject, Molotch (1970, 139) distinguishes between a 'creeping event' and 'routinization of evil'. The former is an event that occurs at a gradual and piecemeal pace; the latter is an occurrence that takes on the quality of an expected event and, thus, loses its urgency. The Italian experience seems to provide many of these cases, and suggests that in this country 'evil' is largely routinized – to the point that the definition of urgency can be considered merely a formal legal process that does not imply prompt institutional reactions or solutions.[4] Certainly, in opposition to what Molotch claims (1970, 143), in the history of Italy an 'accident' is not 'an occasion in which miscalculation leads to the breakdown of customary order'. Perhaps the realms of non-industrial disaster are a better demonstration of this, in that the earthquakes and landslides that have hit Italy in past years have not produced countermeasures aimed at reducing the impact of such events, where responsibility is both natural and human. On the contrary, these have become occasions for setting in motion, sometimes on a larger scale, the sort of classic speculative dynamics that national capitalism is based on (Nimis, 2009; Saitta, 2013).

The judicial history of disasters can help us to understand this point. In the Italian case, courts are important players, and unlike other countries, many investigations into environmental issues are set up by judges, following indications by citizens or associations who provide clues that contaminations are taking place. At the beginning of an official investigation experts are appointed, and levels and sources of contamination are quickly ascertained, as in the cases of Gela or Taranto, for instance. Frequently, this process is followed by a decision – typically, that of shutting down a plant or an industrial complex. But, contrary to what one could expect, such resolutions are not welcomed and generate a backlash both at the top and the bottom of communities. The result is that legal decisions are, *de facto*, bypassed by the virulence of political and social reactions (see *infra*). As a consequence, the social and technical structure that caused the contamination remains substantially intact, and operations start up again (Saitta, 2009).

Over the course of the following pages, each of these points will be discussed. For now, what is important is to highlight that the nature of epidemiological rates is trivial, and that the salience of an environmental crisis does not derive from its magnitude, but from indirect mediating social factors (Beamish, 2002, 11).

### *Social Issues*

An analysis of the most typical patterns in matters concerning the environment and human activities would show that, usually, critical considerations of current development models become part of the public and political agenda when disasters

4 Similar tendencies were recorded in other countries. Weale (1992), Szasz (1994) and Beamish (2002) observed that remedial policies and structures are discretional, and this nullifies the intended benefits of regulatory legislation.

have already taken place, and in many cases, only at a very local level. More often than not, with some noticeable exceptions, catastrophic events do not produce changes in the general organization of the economic forms of a country, and debates develop mostly at the level of the affected communities – although such debates are usually not immune from contradictions, and requests for dramatic changes in organization are opposed by many members of the same communities, who favour current models rather than u-turns, whose effects are unpredictable in terms of occupation, salary and future prospects.

This peculiar situation is due to the fact that, in most cases, and prior to the public definition of a state of crisis, the industrial presence has had all the necessary time to become part of the 'landscape'. Over the years, industrial plants permeate the life of places, even their aesthetics, and produce what one could call 'dependence', by virtue of industrial salaries, the rise of linked industries, the development of industry-related facilities (hotels, restaurants and other similar services), and the extension of urbanized areas, so that the life of places tends to become integrated into the 'moral' and 'material' space of the industry. Industries, then, become part of the inhabitants' mental and economic horizons, and their dreams, life projects and aspirations are consequently related to this industrial presence. Frequently, because of their lights, the emanating fumes and tubes, 'monster plants' are also integrated into the inhabitants' everyday landscape, in the same way as the mountains or the sea.

Interestingly, when, for a number of different reasons, such dreams come to an end, or change into a nightmare as the effects to health become apparent, people's reactions are unpredictable. Strikes, violent demonstrations in favour of the industry and so on are likely to occur, witnessing people's attachment to the 'industrial dream'. The industry, then, appears to be a particular kind of disciplinary organization, capable of producing apparently 'illogical' responses in the 'victims' of its activities. A discipline, moreover, made possible by the social conditions within the communities: in fact, according to the repeatedly mentioned Sentieri study, 60 per cent of the population of the sites of national interest belongs to the two more disadvantaged quintiles.

In the case of Italy, this type of scenario mostly corresponds to situations in places such as Gela and Taranto, where the activities of local process industries belonging to giant groups, such as Eni and Italsider, are suspected of having caused, over the course of 60 years, a wide array of health problems among workers and citizens (Comba et al., 2012). But, in reality, Italy provides dozens of examples relating to different production and activities, including dangerous manufacturing (asbestos – see chapter 12), decommissioned technologies (nuclear plants), food industry (agriculture) and speculations (waste disposal).

We might argue, then, that disasters are not only human or natural events, nor should they be considered merely the consequences of the cynical conduct of speculators. Rather, many of these catastrophic events should be deemed socially constructed facts that involve entire communities, and whose origins lie in an 'agreement' between the 'perpetrators' and the 'communities' themselves.

To describe this dynamic it is worth mentioning what might more simplistically be called the 'pet-coke riot', which took place in Gela in 2002 (and similarly, in Taranto in 2012).[5] This story exemplifies the nature of the social structure and order present in the territory as well as the concept of hegemony, understood as being a type of cultural direction operated by a dominant group over a subordinate group (Gramsci, 1975). It is also possible to argue that this event brings together many of the issues that can easily be summarized in the concept of 'dependency'. Here, dependency is intended as the internalization of the idea that the factory plays a central role in the lives of the individuals and the community, and that the city's survival depends on it. Communities, then, must support the factory or industry, and should not impede those actions considered necessary for its survival and expansion. Within this ideological framework, the fact that the stability of the industry involves high costs is of no matter, and the resulting social costs are seen as the necessary price to be paid for maintaining the social and industrial body. It is an almost metaphysical body which combines the interests of local society in having an income, and the capital needed to generate profits.

In more detail, the 'pet-coke riot' was a large demonstration that took place the day after an industrial plant was seized on the orders of a local court. After an investigation, and following the provisions of a national mandate (D.L. 22/1997), the court declared the use of coal oil (or pet-coke) in that plant illegal, and put a ban on it.

Pet-coke is the solid residue obtained from a refinery process called 'coking'. Despite having a high calorific value, the use of this substance as a fuel is not common in Europe due to its environmental impact. Environmental analyses have shown an abundant concentration of elements related to the combustion of pet-coke, which are likely to generate toxicity, cancer and malformations (Bosco, Varrica and Dongarrà, 2005). But pet-coke is also an inexpensive and effective way of producing power for the refinery.

The residents' reactions, then, took place within a framework characterized by an actual health and environmental crisis, and by a legal decision issued to defend public health. Under the motto 'better sick than unemployed', about 20,000 people took to the streets in defence of the refinery and against the seizure

5 In 2012, at the end of an investigation into a case of environmental disaster, a Taranto magistrate ordered the seizure of a local steel mill, and the arrest of eight corporate executives. Experts appointed by the court had ascertained the role of the company in the contamination of the area, a pollution process, moreover, compatible with the extremely concerning causes and rates of mortality recorded in the city. Before the company announced mass dismissals, 8,000 workers, in agreement with the main trade unions (Cgil-Cisl-Uil), occupied parts of the city for days (streets, highways and other strategic infrastructures were taken). Soon after, the national government launched an investment programme aimed at assisting the company in the amelioration of the industrial plant. Interestingly, a consultative referendum aimed at expelling the company was abandoned, and at the moment, parts of the industrial plant are closed; however, from time to time, new alarms are raised.

order, erected barricades, closed all access to the city and engaged in clashes with the police. This should be enough evidence to conclude that, in 2002, what took place can be summed up in terms of resistance against misery, and the struggle for employment – any kind of employment. Or we can conclude that the day was the culmination of a disciplinary process aimed at persuading the local masses to internalize the ethics of production, labour and profit at the expense of life. However, it would be too hasty to describe the matter solely in these terms.

Indeed, we might also claim that the residents were misinformed, and generally unable to understand the nuances of the topic, and the saliency of the violations committed by the refinery. However, it is difficult to believe that members of the community were completely naïve and insensitive. The effects of industrialization, in fact, were clear to the majority of the population due to the epidemiological rates and their familiarity with disease, malformations and suffering. It is more likely that residents were simply ambivalent, and experienced mixed feelings of ecological concern and loyalty to the establishment. This hypothesis is backed up, *inter alia*, by the apparent tranquility of the same population after the seizure of 90 tanks the following year.

According to a classic study by Hytten and Marchioni (1970) on the Gela case, the way the oil company running the operations was perceived was essentially due to the fact that it was a 'state company', that is, a company historically entrusted to opposing the oligopoly of private companies in the sector; a company, moreover, that also pursued capitalist objectives, but was requested to do so in order to produce 'social profit': by providing jobs and welfare. In other words, the petrochemical plant was perceived as public property, a friendly presence and certainly not as an alien entity.

To this psychological analysis, we should also add an endemic deficit in political representation. Political parties and trade unions represent collective and social demands only in a very partial way, because many trade unionists and politicians inherit their job at the factory from their fathers (this is called the 'passage from father to son', and used to be a very common practice), or expect to pass on their positions to their children. In so doing, they experience the dominion exercised by the company directly and in its most elementary forms. One could claim that people contract debts with the industry from their very first contact with it, so it should not be surprising that, according to many labour leaders, modern trade unions ought not to be aggressive, but should always aim to negotiate and bargain for solutions (Saitta, 2009).

However, by helping to organize a protest against the seizure of a company identified as the only real, big employer in the area, the trade unions also felt they were fulfilling their task, by trying to help save jobs, salaries and welfare. And if this is so, how is it possible to escape the contradiction of devotion to a corporate ideology that is also a defence of welfare? And how to make the view of these trade unionists less corporate-centric? And above all, is it possible to avoid the production of toxic substances and pollutants in milieus of that sort?

A paradoxical response to some of these questions was provided by one local senator – who worked for a new characterization of pet-coke, and its inclusion in the listing of fuels. Using a simple sleight of hand, and in total opposition to scientific evidence and the principle of caution, the politician asked his political community to transform a highly toxic substance into an almost innocuous one.[6]

Such a manoeuvre, conducted on the edge of legal formalism, did not protect the corporation from harsh criticism, and this new measure was accompanied by a communication plan aimed at showing the company's renewed attention towards the environment and health. The company announced that 200 million euros were to be spent on the environment – more specifically, for the adoption of so-called SNOx technology, which helps to clean residual fumes (a 'Best Available Technology', as required by the EU).

In this way, the age of disagreement came to an end (or until very recently, when a new oil-spill provoked the reaction of a younger generation of activists interested in the 'commons'. But it is still too early to comment on this). The extraordinary communication strategy of the company, in admitting past mistakes and announcing changes in policy based on large investments and social responsibility, have actually helped to build 'new' legitimacy, thus making it possible to rely once again on legal formalism, job blackmail, asymmetry and fear. Or, in other words, to acknowledge the limitations of the available options.

### *Forms of Opposition*

It would be wrong to assume that these processes are without opposition. From this point of view, it may be useful to consider the concept of popular epidemiology (Brown and Mikkelsen, 1990), which can be broadly tracked in many cases and, certainly, in Gela and Taranto.

Popular epidemiology is a bottom-up epidemiology that arises from uncertainty, from anecdotal and unsystematic observations on the abnormal spread of certain diseases, and from the dissemination of a kind of popular medicine, conveyed by the media and circulated in flurries of information. It may be argued that popular epidemiology is an extraordinary catalyst for insecurity, and a vehicle for spreading often unsustainable assumptions, whose only effect is to multiply the fears and anxieties of vulnerable people. However, if, from an academic and political perspective, these characteristics sound negative, we should bear in mind that popular epidemiology is also a form of resistance that can help bring issues to the surface and fill in relevant information gaps, often providing real insights.

Interviews conducted with former leaders of local environmental movements in Gela are, in this sense, exemplary. An analysis of the individual and collective processes that guided their political involvement shows how certain practices arose from the (sometimes extreme) need to make sense of their own pain, and from the difficulty of placing the loss of loved ones in the natural order of things. This suggests that there are strong elements of irrationality behind this sort of

6 The result of this intervention was the following decree: D.L. n. 22 of 7 March 2002 – *Urgent instructions concerning the use of pet-coke in combustion plants.*

practice, and yet, this type of thinking also has the extraordinary capacity to perceive trends, since the sensitivity of these witnesses to an abnormal cancer rate was confirmed by subsequent research, carried out also on the basis of their reports (Pasetto, Comba and Pirastu, 2008). This becomes apparent when one considers the method used to highlight the number of deaths attributable to inaccuracies in death certificates, which consisted of analysing the number of 'exemptions' – that is, the number of people who are exempted from paying for the drugs needed to treat certain diseases. Certainly, it reflects epidemiological imagination.

Our analysis would not be complete without reflecting on the lack of trust displayed by many witnesses. In particular, we need to address the widespread view that doctors were corrupted by the local oil company. Some witnesses had misgivings about the fact that data referring to deaths from cancer and other diseases attributable to environmental pollution had long been unavailable because of deliberate omission. The view expressed by some was that for years doctors had intentionally filed false death certificates to hide the effects of pollution, and to protect the petrochemical establishment.

The plausibility of this hypothesis will not be discussed here, but it is worthwhile reflecting on the impact produced by the bad management of information on similar topics, and certain micro-practices in the health field. In line with this, the construction of reality by social actors seems to consist of apparently tiny pieces that have the power to create huge distance between the institutions and the public, thus enhancing the culture of suspicion. Also, some medical practices that have been in use for decades, have not only caused a substantial loss of objective information for studying pathological phenomena, but have also discredited health institutions and increased distrust in them. These occurrences demonstrate the strategic role of communications in improving the relationship between public institutions and the general public, and they also provide important indications of local distrust. Anyone interested in improving the relationship between the institutions and the general public in the Gela area, or similar places, should consider the micro-dynamics that go into determining a serious crisis in trust.

## Discussion

The word 'disaster' should not only indicate catastrophic events that take place at a given time, resulting in a relevant number of victims within a few short moments, and marking the life of the affected communities (Smith, 1999). Rather, there are any number of disasters taking place in the industrial areas of the world, events that are constant and invisible, and which last over many decades – and accompany the daily life of the people living in the areas that host industrial plants.

Moreover, one should be cautious when distinguishing between natural- and human-related disasters. In fact, the consequences of many natural events – earthquakes, landslides or floods, for example – should be read in connection to human activities. In many cases, the erosion of the territory, urban development in hazardous areas, the low quality of materials deployed for erecting infrastructures, and speculations of all sorts, are elements that help explain the dimension of disasters in both material and human terms. In other words, at this point in human technological and social development, any analysis of the notion of disaster should also involve the connections and relationships between 'human' and 'natural' – namely, technology and economics, nature and human activities.

Such connections and relationships can be considered as parts of a discourse that takes place within specific frames and sets of narrations, based on the concepts of development, trust in technology and risk control. Such a frame has been constituted over the course of the past two or three centuries, but the process has been going through 'acceleration phases' over the past 30 or so years – in spite of the opposition that has developed during the same period.

The relationship between production, wealth, risk and safety has raised the concern of both academicians and lay actors. Often, visions of these problems by 'experts' and 'non-experts' have been very different (Sowby, 1965; Starr, 1969), and the former have not omitted to blame the latter for their irrationality (for example, in the field of nuclear energy). In a similar vein, others have advocated a more subjective approach, that might explain fear and distrust either in psychological (Fishoff et al., 1978; Slovic, Fishoff and Lichtenstein, 1980) or in institutional and cultural terms (Sjöberg, Moen and Rundmo, 2004).

In spite of different aims, perspectives and ideologies, these visions all share the idea that risk, fear, safety and the like, are social constructions; that is, events and concepts whose meaning does not exist *per se* or objectively. However, constructivism is often professed for different reasons. In fact, some proponents believe that reality exists, and that the main problem is the perception of it – a perception that needs to be understood in order to be changed. For these authors and many stakeholders, the problem of science (and economy) today, is being understood by those who resist technological changes, innovation and the development of infrastructures.

For other authors, instead, the adoption of a constructivist perspective is tantamount to revealing the interests hidden by dominant discourse within societies. Rationality, safety, security and insecurity, then, can be seen as different ways of bypassing objections, and arousing subsequent emotions in the public. After all, among many other things, government is precisely the art of managing collective emotions. Clearly, in most cases, the elites and the people do not share priorities. The art of management to which I referred earlier, then, consists of bringing together very different needs, and helping the people to accept what the elites deem to be important. This is, ultimately, the complex 'process of processes' that begins with 'issue installation' and ends with the 'conversion' of the public (Cohen and Hamman, 2005, 142).

In a nutshell, such an operation requires that those in charge of setting the public agenda select certain problems and neglect others. Folk devils and street crime are good examples of this multipurpose selection process, aimed at imposing inconsistent priorities that have no relation to 'facts' (for example, the statistic relevance of a phenomenon) (Cohen, 2002). In other words, there is solid evidence to show that it is the definitional activity of the state, rather than the reported incidence of, say, crime or drug abuse, that has shaped public concern regarding crime and security in Europe and the US from the 1990s to the present time (Beckett, 1994; Furedi, 2002; Palidda, ed. 2011). On the contrary, there is other very solid evidence to show that many communities within the same continent face very serious environmental problems, related to their economic activities, technologies and industries. Notwithstanding, these issues – certainly not in Italy – have not received the same attention during the same years, and have not been on the agenda.

In this regard, Simon (2007) noticed that 'government through crime' belongs intimately to the history of the liberal political order. In this perspective, the word 'liberal' indicates that the state and its elites operate 'on' – rather than only 'against' – the liberties of its subjects. It is not difficult, then, to relate this statement to the lessons of authors like Rusche and Kirchheimer (2009), Williams (1961), Foucault (1979) and Ignatieff (1978) for whom the history of the liberal states and their pedagogy, discourse and penal institutions is, after all, that of a long and complex disciplinary process aimed at making subjects accept the unnatural world of production, with its fumes, smells and slums. In the light of these observations, like a closing circle, it is possible to maintain that a line connects this history of the West (and its colonies) and the events narrated in the previous section; that is, the 'pet-coke riot' and a 'job-at-any-cost'.

The plausibility of such a hypothesis seems to be confirmed by the fact that the instalment of massive industrial plants, especially in the *Mezzogiorno*, or South of Italy, has often been the result of a 'social contract' between the populations in the area and the giant companies willing to establish there. When, back in the 1950s, the well-known figure, Enrico Mattei (president of ENI, the national oil company), visited the poor lands of the South, he met crowds of peasants looking for a better and more modern life. In his speeches, he promised them wealth, wellbeing and modernity, and the peasants, especially in Sicily, became his allies in the war that placed him in opposition to a national government that did not support his projects. At this point, Mattei's dream was not only his; an entire population wanted the oil wells that would make their dream of becoming the Texas of Italy come true. The Italian 'development theoreticians', who were active in the same years, did likewise: their analysis, in fact, focused on the notion of economic dualism, and portrayed Southern society in general as backward. The South, in other words, found itself couched in a discourse that propelled it towards 'modernity'; that is, to the material conditions and development models based on the sudden erosion of the land, land speculations and a wave of investments that changed the features of their society, but not the exploitative structure of its organization (Sapelli, 1989; Saitta, 2009; Petrillo, Chapter 12 in this volume).

As should now be clear, this process, which started in the 1950s, is at the basis of the epidemiological and environmental question that already had its current features some 20 years ago (Di Luzio, 2003). Nevertheless, the problem of these ailing areas, which are more often than not situated in Southern Italy, has never been one of 'unhealthy development', but rather, it has always been a problem of maladjustment, poor social capital and, finally, the mafia.

## Conclusion

This chapter has focused on some aspects of what can be defined as the environmental question in Italy. The idea behind it, is that this country is witnessing an emergency that strives to be acknowledged as such. From the North to the South, the results of epidemiological investigations are clear: environment-related diseases are, in a large part of Italy, one of the most important causes of mortality. Beyond that, the decommissioning of plants is not followed by remediation, and only a very small number of the areas identified as being in need of remediation have been treated. Moreover, in addition to the health crisis, the closure or resizing of many operations has determined a serious social crisis. The industrialization model present in many parts of Italy has not been directed, and it has reproduced and broadened biases that were already typical of the social organization of the country. Such capitalism without direction has produced some serious effects, the magnitude of which is only now becoming apparent.

On a larger scale, however, pollution, land speculation and extreme urbanization are, in fact, very common ills that affect the country in a fairly uniform way.

Over the course of the past 20 or so years, no serious policies have been undertaken; on the contrary, the contemporary management of society has contributed to aggravating the situation through a set of measures aimed at broadening speculation, eroding land and reducing controls. Moreover it has, *de facto*, denied the urgency of such environmental questions by focusing on more virtual enemies (immigrants and crime, above all) and imposing an agenda that refuses to deal with more pressing problems, in other words, with those issues related to the quality of the environment and public health, but also to safety at work, and the quality of housing and infrastructures.

More than by crime and immigrants, Italy seems to be threatened by earthquakes, landslides and floods, whose social and economic costs are difficult to bear (Ance-Cresme, 2012).

Moreover, as is so often the case, the Italian courts have tried to compensate for the lack of political leadership in these matters, although these attempts fail to produce the desired effects, as these sorts of issues cannot be solved by means of judicial decisions. The environment is too strictly intertwined with development and the economy, and the solution to problems must be, *in primis*, political.

The lesson that can be drawn, then, is that Italy needs to reprioritize its most pressing problems, and that it needs to do so with some urgency.

## Chapter 12

# The Use of Natural Disasters to Turn the Rule of Law into Corrupt Governance, Favouring Private Companies[1]

Antonello Petrillo

### Introduction

This chapter describes and analyses a very significant case study on the economic and political profitable use of a natural disaster with devastating consequences on the health of local population, on corruption and on the growth of organized criminality connected to all sort of affairs.

The research presented here took place over about two years and was aimed to examine a fairly small and less notorious area of the Campania region, which was one of the reasons why it was designated as the location for such a delicate project.[2]

Over the course of many years, the Campania region of Italy has been exposed to all sorts of environmental devastation by both national and international powers. This total disregard for any type of biopolitical objective regarding the health and safety of the resident population, which constitutes the basis of neoliberal 'governmentality' (Foucault 2004b/2009, 2004a/2010), has been achieved by developing a specific order of discourse aimed at the inferiorization of the local population by constant reference to their state of underdevelopment, and by means of very structured narrative patterns (from the theories of the positivist school of anthropology which accompanied the period after national unity in 1861, up to Banfield's 'amoral familism' of the post-Second World War period and to the 'progressive' paradigms of the following decades – Petrillo, ed. 2009, 2011, 2014).

As well as general indexes regarding the quality of life (the lowest in Europe), health data reveals some really worrying statistics: with the exception of other diseases, the National Cancer Institute 'Fondazione G. Pascale' in Naples reports a disturbing increase (over the last two decades) in the cancer mortality rate: up to

1 English revised by Deborah Kay.

2 The following material was collected through a URiT (Research Unit into the Social Topographies by the Suor Orsola Benincasa University of Naples) for a research project, coordinated by Petrillo and led by A. D'Ascenzio, M. De Biase, G. Della Cerra, G.P. Di Costanzo, S. Ferraro and A. Petrozziello. See Petrillo (ed. 2015).

47 per cent in men and 40 per cent in women, in comparison to general data on a national scale (Caporale, 2013).

## 'Green' Irpinia

The province of Avellino, also called Irpinia, lies beyond the Neapolitan city boundaries and rises above the Campania plains, becoming an area thickly wooded with chestnut trees and where the land is intensively cultivated; the Southern interior, an 'area of bones rather than flesh' (Rossi-Doria, 1982). After the Second World War, TV opinion polls and the collective imagination very often associated the area with bleak images of poverty and primitive farming methods, mixed with archaic customs and ancient magical rites, and historically, the source of endless migrations towards the Americas at first, and later to Central Europe and Northern Italy.[3] Later on, the post-industrial/neoliberal redefinition of consumption and lifestyles, which change according to the particular behavioural trend in vogue at the time, from 'hedonistic', to 'cultural' or 'health fanatic' and so on, has led – as it has with other rural contexts – to a reassessment of these areas, so that they are seen as urban decompression areas, suitable for weekend trips, and for walking or horse-riding – as an escape to the country, as well as being associated with healthy food and a fascination for old villages dominated by ancient Norman castles, crisscrossed by winding, narrow stone streets. The institutional adverts refer to it as 'green Irpinia', the 'green space' of Campania and so on. However, apart from any rhetoric in favour of developing tourism, the reality of the situation, as far as accommodation and facilities are concerned, is that things have actually improved very little so far, and it is quite obvious that there needs to be a very different kind of development model.

After the devastating earthquake in 1980, which left about 280,000 people homeless, 8,848 injured and 2,914 dead,[4] Irpinia became, in the name of reconstruction, an uncontrolled testing ground for the realization of a totally free development model, resulting in the construction of nine 'Industrial Parks', four ASI (Areas of Industrial Development) and two Industrial Districts. Here, in this peripheral fraction of the national territory, it is possible to see the policies that were supposed to help develop this part of the South, but which have actually failed to do so (a case with many analogies to many other in all the world). Most of

3 We are in the land wonderfully recalled by Ettore Scola in what is considered the iconic movie of the great neorealist season of Italian filmmaking: *Trevico-Torino – Viaggio nel Fiat-Nam* (1973).

4 The 6.9 magnitude earthquake struck an area of 17,000 km$^2$, inhabited by six million people; 687 villages were affected (36 of which were razed to the ground; of the others, between 50 per cent and 75 per cent of buildings were destroyed): it was one of the most disastrous seismic events in Europe in recent years . The province of Avellino was the epicentre.

the companies, without any real industrial plans and motivated by the opportunity to grab the huge state aid funds available, and by the prevailing lack of supervision by local institutions, as well as their compliance, have suffered from a chaotic and disastrous 'turn over', causing most of them to close down. What remains is a landscape dominated by buildings that can only be described as 'white elephants', as they are no longer in use but continue to be highly polluting. The same can be said for big businesses, whose reasonable presence is due to sizeable state aid funds: the IVECO/IRISBUS is closed and the FIAT/FMA has been working for some time now at a reduced rate. In other words, there has been no development at all. Nowadays, the province of Avellino occupies 93rd position in the list of Italian provinces that generate wealth, and industry represents 14.3 per cent of the added value produced *in loco*,[5] and 19.1 per cent of the total of 138,600 people employed.[6] Also, unemployment continues to grow: according to the CISL Observatory (end of 2013) the 'declared' number of unemployed was 17.3 per cent of the whole population (7.4 per cent in 2009) and this data becomes even more tragic when we consider the underemployment in the agricultural, building and commercial sectors, also the huge increase in demands for social security support within the business industry,[7] the ageing population and, last but not least, the data concerning youth unemployment, which is as high as 58.5 per cent.

Against this landscape of waste dumps and closed factories, it is also possible to see an uncontrolled and illegal urbanization of the area. The 2011 census of the province recorded 430,214 inhabitants in an area of 2,806,07 km$^2$, with quite a low average density (153.32 inhabitants per km$^2$). Nevertheless, the populations living in the 119 towns are distributed quite unevenly: in some, there are just a few hundred inhabitants,[8] in others there is a more significant urban density. More than a quarter of the population of Irpinia lives in Avellino, the main provincial town; if its population (55,448 inhabitants) has been quite stable over the last decade, the average growth in the towns in the region, between 2001 and 2011, has exceeded 20 per cent, and has sometimes been as high as 40 per cent. As far as the history of the development of local housing is concerned, there is one simple and brutal truth regarding the 'land' and the 'families' that possess it; its value has multiplied thanks to the only business the Southern interior has in abundance, the only business that does not require any initial investments or advanced technologies: the building industry, the most *labour intensive* and low cost of all industries.

---

5 Against 3.3 per cent represented by agriculture, 5.7 per cent by the building industry and 76.6 per cent by the service sector (Source: Camera di Commercio di Avellino, Rapporto Irpinia 2013.

6 Work is guaranteed by agriculture 4.1 per cent, building industry 8.6 per cent and service sector 67.96 per cent.

7 Between 2012 and 2013, 329,613 hours of redundancy payments were approved, 268,516 of which were within the industry sector.

8 Moreover, many of them continue to lose inhabitants – with an average of 15–20 per cent in the last decade – due to new migration flows.

The earthquake of 23 November 1980 had a decisive role in the definition of economic, urban, social and political structures. An *ad hoc* law (219/1980) allowed reconstruction to take place regardless of the main standard regulations concerning the use of the land, tender notices and contracts, public expenditure supervision, city planning, commerce, environmental impact, occupational safety and so on. This resulted in an inversion of the public–private relationship with regard to building costs, an excessive increase in the mayor's power, at the expense of council bodies (administrative powers regarding 'safety and emergency' were increased) and there was harsh repression – backed up by a strong military presence in the territory – of any attempts to be locally involved in the reconstruction, or to resist or disagree with any of the choices coming from above. To sum up, national government, the local political class, the ministerial bureaucracy and the economic actors concerned, all took advantage of the natural catastrophe, to establish a state of exception parallel to the normal democratic rule of law: a mixture of normality and exceptional circumstances that gave the stronger actors absolute freedom to operate within the usual areas of healthcare, territory and environmental protection, and to manipulate the boundary between legality and organized crime. This is an ideal situation for the diffusion of the new *laissezfaire* politics which, at that time, were spreading all over the country. The 1980 earthquake can be seen as the place where Naomi Klein's 'shock economy' (2008) meets Foucault's biopolitics.

It was not by chance then, that the Special Commission appointed to deal with the emergency situation in Irpinia soon became the framework for the current Civil Protection Department. In other words, a powerful device was created and in the name of the 'exceptional state of emergency', was quickly extended to very different fields: waste disposal, important events such as the G8, the Pope's funeral, sporting events and major public works. In fact, during those years, Irpinia became a huge laboratory for testing out – for the first time on such a large scale – new management technologies that perfectly coordinated the territory and the population. What resulted was a series of scandals over excessive wastefulness, as well as scandals over corruption and fraud (these and the subsequent investigations by magistrates were referred to as 'Irpiniagate'), which all came to the unbelievable sum of 66 billion euros, at current value. About 280,000 displaced people were forced to move in just a few days and were scattered around the territory on the basis of speculation and land ownership. Because of this emergency, in the decades that followed, it was possible to completely rearrange the territory and the economic, social, cultural (lifestyle and consumption) and political setup. Along with illegal urban speculation and a local political class that had ways of bypassing the regulations, the '*camorra*' turned its hand to the construction equipment industry, to the building and management of huge housing complexes and commercial parks and, later on, to the big 'business' of waste. Low-quality housing complexes, impersonal and ugly shopping malls, mega cinemas, amusement arcades, car dealerships and factories that were opened rapidly and just as quickly closed, garbage dumps and dumping grounds began to spring up all over the landscape.

## 'Light Blue' Irpinia

A year and a half after the earthquake, in the spring of 1982:

> That afternoon I was there, together with some other young people … . I desperately needed a job, but what happened that afternoon upset me for a very long time afterward. Graziano's car arrived while we were in the square and one of his lackeys was staring at us. He got out of the car, came up to us and said: 'All those aged 18 to 20, take a step forward.' I was 20 and moved forward. Graziano went on: 'You're all hired and you can start working now; as for the others you can go home. We need young people.'[9]

This is not a United Fruits plantation, we are not in a colony and the story-teller is not Fanon. The 'wretched of the earth' have become a specific generational cohort, the earthquake survivors who were here, together, one sunny afternoon, in a square in a suburb near Avellino. Graziano – the 'Engineer', as he is known because of a supposed degree in Chemistry from the Sorbonne University – is somewhat of a legendary businessman hereabouts: people talk about incredibly generous tips to restaurants or hotel waiters, some tell of a worker who was paid 100,000 Lire one day, just to follow the shadow of the only tree in the square; he has been seen getting out of a helicopter at the stadium to watch a match played by the team that will soon be his property. Graziano is, among other things, the owner of Isochimica Inc., a new company in need of healthy young employees; he cannot risk hiring unhealthy people because they would not be well enough to do what is required of them: it is asbestos. Nicola, one of the workers who became ill while working there, is telling his story now. He was just a young man then, and did not have the faintest idea that his green Irpinia would soon become light blue Irpinia, due to the colour of the deadly mineral.

The voices of the men and women involved in this story have never been properly heard, they have held many meetings but their plight has not been recognized, all of which goes to show how difficult it has been for them to have their rights, hopes and failures acknowledged. These are the stories of young workers, some of whom had only just turned 18, and who are envied by their peers for finding a 'good job' amidst the dearth of employment in this post-earthquake period in Irpinia. They have grown older fully aware of the meaning of death, together with the partners who washed their clothes when they came home from work. They are the stories of rail-workers and the inhabitants of a neighbourhood known as *Ferrovia*, a working class part of town where the concept of 'fringe' is ubiquitous (city planning, services, quality of life) and which has become the frontier between sickness and death: an allusion to the hundreds and hundreds of cubic metres of asbestos that was removed in the former factory, and that are still here, in the schools, the football grounds and the houses. These are stories that open up other

9 Nicola, 51, former Isochimica worker, 27 December 2011.

stories: stories that concern political, institutional and social responsibilities. They are stories which call into question local mayors, local and national politicians and administrators, and the absurd and destructive development models planned for this area after the 1980 earthquake. They are stories that involve the state and its departments: the health system, occupational medicine, supervisory bodies, the social security system, legal powers and the trade unions that so often seem to be suffering from selective mutism. The unheard voices that have gathered together are at least equal to the tons of asbestos that have been removed: for more than 30 years, silence and asbestos have covered the 'bare lives' of hundreds of people. Men and women who have been left alone to fight a desperate battle which does not seem to have touched local society.

It is the story of a small factory (with no more than 300 workers at the height of its production) in a small town, and it is a little story, but one which is inextricably linked to a bigger historical picture. For many years, the Isochimica industry, based in Avellino, played an important part in a serious nationwide problem: the disposal of the asbestos that had been an integral part of many railway carriages. Asbestos had long been used as a heat-acoustic insulator and fire retardant in the inner panels of train carriages on railways all over the world; in Italy, the decision to introduce it was made during the 1950s following a serious fire, and the commission to have it systematically 'sprayed on' was given to a firm in Genoa (Davidson Rhode). At the end of the 1960s, however, people became increasingly aware of the great dangers related to asbestos,[10] and in 1975, the State Railway (it was a state company at that time) conclusively decided to stop using asbestos for the new wagons and to gradually introduce fibreglass, also because European and National regulations began to impose the total removal and reclamation of asbestos. In several Italian towns, the OGR workers (from the State Railway's Major Repairs Workshop, the *Officine Grandi Riparazioni*) and their trade unions attempted to oppose this dangerous new procedure to remove the insulation panels from the carriages. In 1972, the National Trade Unions developed the first 'Bargaining Platform' against the dangers of toxic substances in the workplace, and used it to demand that a 'facility doctor', and 'qualified technicians trained in environmental surveying' be taken on, in addition to a pollution monitoring system. In 1979, Medicina Democratica (Democratic Medical Institution) disclosed an epidemiological

10 Particularly, Irving Selikoff's survey on a wide sample of American workers clearly showed the strong connection between asbestos and cases of cancer; thus inducing, in 1970, the Occupational Safety and Health Administration to strictly minimize the workers' exposure to asbestos; in 1989, the Environmental Protection Administration issued new regulations to definitely interrupt any processing of products containing asbestos. The Epidemiological Research Centre (CRE) of the European Cancer and Environmental Science Foundation 'B. Ramazzini', with the support of the Bologna Section of the Italian Anti-Cancer Association, furnished proof of the devastating effects of using asbestos in the Italian carriages on workers, travellers and their relatives, publishing the results from 1986 on (Maltoni et al., 1986; Maltoni et al., 2002).

survey highlighting the increase in cancer cases among the OGR workers involved in the handling of asbestos in a Northern Italian town (i.e. Casalmonferrato – see chapter of Blengino and Torrente). This meant that the workers became increasingly alarmed, that more knowledge was gradually and widely spread, and that the first demands were made at the beginning of the struggle; as a result, doctors, journalists and magistrates began to take an interest.

The State Railway was forced to appoint an *ad hoc* inquiry committee (a government committee) and to accept – albeit partially – the workers' requests, including rules intended to make changes to the designated locations in which the reclamation was carried out, and to improve individual and collective protection, workplace hygiene, safety and processing methods. The personnel exposed to asbestos were to have their working hours reduced from 40 to 36 hours per week, periodic health inspections and a special wage allowance. Obviously, the introduction of these measures implied high costs and it all took a very long time; so, between 1981 and 1983, the workers and the company agreed on an 'experimental phase' that was supposed to keep any actions involving asbestos to a minimum, while testing was carried out on more suitable methods, infrastructures and work tools. From here, they prepared a 'Ten-year Plan' for the removal of the asbestos from the train carriages, which was to be carried out in 10 different industrial premises (OGR), and involved six external firms.[11]

The magistrates' investigations and battles were seen as a form of obstruction to almost all the State Railway's plans, which were to remove asbestos from: 516 locomotives, 1,002 light vehicles, 271 light thermal vehicles and 4,044 carriages. In spring 1982, the Isochimica company removed asbestos from at least 1,740 carriages and 499 electric locomotives.[12] It is fair to say that the 300 workers from Irpinia removed asbestos from practically the entire fleet of electric locomotives belonging to the State Railways and almost half of the carriages. Basically,

11 An exhaustive reconstruction – using original documents and direct evidence – of the workers' struggles against the use of asbestos at the State Railways industrial premises is in the dossier *Amianto – Piano di intervento di prevenzione del rischio amianto del Consiglio dei Delegati dell'O.G.R. – F.S.*, published by Regione Piemonte (Health Department) in 1987, as well as in the report *Bonifiche dell'amianto presente sul materiale rotabile F.S. Cronistoria di una costante crescita della consapevolezza del rischio attraverso il concreto contributo dei lavoratori*, presented by Alessandro Alberelli during the First National Conference on Asbestos by the Department of Health (Rome, 1–5 March 1999).

12 Actually, the figures are rounded down. They were given to Magistrates on 31 October 1988 by the Avellino Tax Police (Romano, Volpicelli, Mansi, Relazione di perizia tecnica nel Procedimento penale n. 878/86 contro Graziano Elio + 2, Naples, Court, 15 October 1989), but they refer to 1983, that is the official launch date of the work process – which was actually in 1982 – and they obviously don't include the activities that followed and continued up to 1991 on the same site but under a different company name (El.Sid, Co. Ltd).

(also because of transport privatization and the subsequent heavy staff cuts)[13] the State Railways 'outsourced' the national disposal of the asbestos to Isochimica in Avellino. As in many other cases concerning the outsourcing of dangerous work processes (which is a peculiarity of globalization), it was the work processes which moved and not the workers (in Italy they moved from the countryside to the industrialized North).

The biography of the businessman Elio Graziano states that he comes from Irpinia, is the son of a rail worker, and that he was a State Railways chemical manager for 28 years, mostly at the OGR in Bologna. In 1968, taking advantage of the early retirement offered by the company, Graziano came back to the South and established two factories in Fisciano[14] which produced special chemical products almost entirely designed for the same State Railways (detergents for industrial use, paint removal agents, fire retardants, herbicides for the railway lines and so on). Sometime later, his activities diversified, and included the production of detergents for domestic use and sodium-, lithium-, barium- and chlorine-based products designed for different kinds of industries (from battery factories to fireworks factories). But the 'real' business started in 1979, with a 140-billion-Lire supply contract for TNT sheets ('Non-Woven-Fabric', a synthetic polyamide-based fibre) designed for use on train bunks in sleeping carriages. Later on, in 1988, this supply contract would be the main reason for the disintegration of Graziano's empire, due to the so-called 'Golden Sheets' Scandal'. This scandal unveiled the all-pervading corruption in the contract system (and outsourcing as well) which later became what we know as *Tangentopoli* (Bribesville). After several convictions, the whole State Railways' Board of Directors was forced to resign. However, in spite of all this, in the early 1980s Graziano was at the height of his powers: thanks to the TNT (NWF) supply and the Isochimica disposal of asbestos, his two companies received two of the most lucrative orders from the State Railway. Graziano's fame grew, and in 1982 the President of the Republic decorated him *Cavaliere del Lavoro* for his service to industry,[15] but in Avellino the businessman became a real legend. The neo-*Cavaliere*, almost officially affiliated to the Socialist Party, was everybody's friend in his hometown and generously sponsored events and initiatives without making any distinctions. Even the local press and local political

13 Between 1980 and 2000 the train transport sector lost 46 per cent of its workers all over Europe, and the Italian State Railways was in line with the rest of Europe, suppressing104,292 positions. The privatization of the State Railway was realized through the so-called 'stew' method, which consists of splitting a company into several smaller firms, thus preventing any collective action by the employees or negotiations with these.

14 Between Salerno and Avellino, near an area that would soon be occupied by the new University of Salerno *campus*.

15 'After the scientific merits received', the businessman points out in his lively tale: *La vita dell'Ing. Graziano*, autobiography collected by URiT and signed by the author, page 2. This title was also bestowed on Berlusconi who has not retired from politics as a leader of the right, despite being condemned by the courts for various criminal charges.

parties received his attentions, and there is talk of a brand new sports car donated to a very popular local anchorman. And then a truly emblematic episode took place: the camorra was raising its game and the local public prosecutor was shot in the legs during an aggression; also, along with several other people, one of the biggest local building entrepreneurs, Antonio Sibilia, was arrested. At the time he was also chairman of the local soccer team, Avellino (then in the First Division – Serie A), and was already infamous for having obliged one of his players – Juary – to donate a gold medal to the camorra boss Raffaele Cutolo, with a dedication. This is what Graziano had to say:

> It was the end of the 1985/86 Championship and I was called by the then Mayor Venezia and by the solicitor Basile, who told me I was urgently required by the Chief Prosecutor of Avellino. We went towards the City Court where the Prosecutor was waiting for us. When I was standing in front of him he said: 'Graziano I have risked my life to fight against crime in this town, now I want you to take over the soccer team because I know for sure that you're an honest, clean hard worker and I know that you possess the financial means to keep the team in the top division' (Serie A).[16]

And the team really did fall into 'his hands' – even though its stay in Serie A was short-lived – which made Graziano something of a 'popular hero'.[17]

## Young Workers

Besides the obvious web of corruption and the lack of supervision, Avellino was a suitable place for the realization of the State Railway's dangerous productive division, due to a *habitus* that had been created through a social process over the course of time (Bourdieu, 2005, 85ff.), a rational system of individual and collective sanctions, including the creation of politicians, institutions, local trade unions and families who took on the role of the 'custodians' of a development and social welfare 'contract', and were ready to call potentially dissenting workers to 'order'.

Dressed simply in jeans and trainers, with no overalls or protective masks, the youths start working. Initially, the work is carried out in an area where there are no buildings or constructions to shelter the work and for a long time after, there are no facilities, or other types of building: just workers and asbestos, right in the middle of a residential area. At a much later date, the local railway police chief and

---

16 *Ivi*, p. 5. Syntax, grammar and punctuation belong to the original text.

17 Some years later – when Graziano was under arrest with a mandate from the Rome public prosecutor's office for the 'Golden Sheets' episode – the team went back to Sibilia, as he had been acquitted of the charges against him.

several workers would tell the investigating magistrates[18] that for a year, between spring 1982 and 1983, a huge number of State Railway carriages were dismantled, the asbestos was removed and they were reassembled by the Isochimica workers – supervised by a retired rail worker, who was called in specifically to train them (*sic!*) – on a blind track only 70 metres from where travellers, commuters and students were waiting for their trains. The asbestos was scraped out, gathered into ordinary plastic bags and stored inside the station, to be periodically loaded onto lorries whose destination no witness has ever been able to disclose. The State Railway Inspection Office in Naples has always denied any knowledge of the operations conducted on that blind track, but the dates on the communications between the State Railway and the contractors are evident: even though there was no evidence of direct authorization, the State Railway managers knew that the disposal of asbestos in Avellino dated back to a year before the official opening of the facility and even before the two 'partners' had formally signed an order contract. The same can be said of the 'Servizio Materiale e Trazione' in Florence: the removal of the asbestos from the carriages implied that someone must have ordered to be send them there.

Franco, hired 'thanks to a recommendation', remembers his 'first day of work':

> I'll never forget my first working day at Isochimica. I entered a tiny wagon and I found four young workers next to each other, completely dust-covered, scraping the asbestos out using a putty knife with their bare hands.

Graziano himself, 'especially at the beginning', asked him 'to bury the removed asbestos in a hole that had been dug inside the facility'.[19] However, none of them paid any attention to that sort of thing; they were happy and so were their families: the young men had finally got their 'secure job' and, for once, they had not been forced to emigrate or to slog under the sun in the building industry or in the fields, as their forefathers had.

At that time, these young workers knew nothing about asbestos: if any of them showed even the slightest interest in it, they were quickly told it was a material once used as an insulating material to help keep the travellers warm and that it needed to be replaced with a more 'modern' material, such as fibreglass.

> The work was hard, but during the breaks we joked around; we were young, we talked about girlfriends and marriage which seemed to be possible thanks to the job we had. Sitting on the steps of the train carriages – covered with asbestos – we ate our sandwiches, had a cup of coffee and soon we were back scraping the

18 The enquiry is that of 1989, started by the City Court Examining Magistrate, Modestino Roca, following a charge brought by Giovanni Maraia, Provincial Secretary of the Italian political party Democrazia Proletaria.

19 Franco, 56, former Isochimica worker, 27 December 2011.

> wagons. Hard work seemed normal to us, after all we were not office workers, just workmen.[20]
>
> During the first two years we worked without having the faintest idea about what asbestos was. When we couldn't breathe the dust anymore, we made paper masks to protect our noses and mouths and prevent ourselves from inhaling it. But, as well as the asbestos, we used several different kinds of acid to clean the wagons, the fumes were very strong and we didn't have anything to protect ourselves. Also, the wagons coming from Florence were really dirty and, after having inhaled thousands of asbestos particles, I was taken to the Infectious Diseases Ward at the Hospital, because I had contracted brucellosis while cleaning a carriage. It's a miracle I'm still alive!

One day – in 1984 when they had been working for two years – one of the workers brings in a news item all about the dangers of asbestos, about workers becoming ill and about the protests in the Northern Italy factories. Some of them read it and eagerly hand it to each other in secret, as if it were a porn magazine, in the factory toilets. Some begin talking about it at home and just get slapped for it. The brightest among them try to involve the trade unions, they've heard that it's the place to go, because it's the union representatives' job to defend the workers. Mario still has painful memories of the deep sense of frustration and powerlessness they felt at those first attempts at protest:

> The first thing we did when we realized how truly dangerous the situation we were being exposed to was, was to call the Factory Council to try and work out a shared approach to protest. But, there has never been any sense of unity in the history of Isochimica. Most of the Council delegates were paid off by Graziano and it was the same for the trade unionists. Most of them regularly got Graziano's 'bribes', and he, in turn, paid for their silence.[21]

In the group organizing the protest, there are never more than 30 people (about a tenth of the workforce); the others are held to ransom or accept the boss's 'generosity', or they let the union reps persuade them to act cautiously so that they don't lose their jobs.

From the very beginning, any attempts to fight simply meet with a wall of silence. But it is quite a conquest when the workers manage to get an agreement for a technical survey; it is at the suggestion of the unions, in order to calm the situation, and is carried out by the Catholic University of the Sacred Heart of Rome. The experts come for the first time on 10 June 1985: and their report is terrifying. There is no need to make complex instrumental collections of the data on dust

20 Antonio, 50, former Isochimica worker, 3 July 2013.

21 Mario, 53, former Isochimica worker, 5 April 2012.

concentration: the asbestos is everywhere, visible to the naked eye, scraped out of the wagons using simple painting tools, crumbled by the hands of the workers during processing, collected into huge piles on the ground, put into plastic bags, pushed around by brooms and flushed away with water into manholes that lead directly to the sewage pipes; asbestos is not only inside the work-sheds (one of which has no ventilation system); asbestos has permeated the storage unit, the offices, the changing rooms, the canteen, the infirmary and it obviously invades the outside service area from where it has spread all over the town. The workers, who had only a vague idea of the danger, can see the terror in the experts' eyes, even before they read the report.

> There were two of them. As soon as they entered the plant they stopped and couldn't walk any further. Their faces were white with fear. Even though they had the proper protection equipment, they already knew that it was a mess and stared at us as if to say, 'how come you're still alive?' I remember their expressions when they were looking at the filter masks we used. When one of them saw the flimsy paper masks some colleagues had, he looked as if he was about to faint ... [22]

At the end of the *tour*, the experts provided an unequivocal report:

> According to what we have personally witnessed at the plant, we can assert that there are not, at the moment at least, any occupational health and safety protection measures whatsoever in place for the workers (most of whom are young), particularly in relation to the extreme level of danger of the material they are handling ... Most of the asbestos handled belongs to the crocidolite family, which is the most dangerous kind of asbestos and for which the American Conference of Governmental Industrial Hygienists[23] has established the most restrictive 'threshold limit value'.

Half the measurements taken showed a crocidolite concentration exceeding at least 50 times the TLV, [24] and the inspectors obviously called for the immediate introduction of a certain number of safety measures, initially some form of stop-gap measure, and a total reorganization of the processing methods together with a reform of the whole work cycle.

22 Carlo, 51, former Isochimica worker, 7 June 2013.

23 Altieri A., Paoletti A., *Relazione tecnica sullo stato igienico-ambientale dello stabilimento Isochimica*, Istituto di Medicina del Lavoro dell'Università Sacro Cuore di Roma, prot. 79/V, 22 April 1985.

24 Threshold Limit Value, set by the EU Directive at 0.2 fibre/cc.

The press – generously bribed by Graziano – ignored everything[25] up to 1988, when the factory was closed. The young workers were not understood, nor were they even listened to, either by their fellow citizens or by their relatives: 'Everybody thought we were just lazy or that we didn't want to work'.[26] And the neighbourhood where the factory was located did not react positively either:

> After the occupation of the factory and the continued incriminations about our working conditions and the precarious clean-up interventions, we decided to organize some demonstrations, hoping to sensitize the people of Borgo Ferrovia to our issue. During a rally in Borgo Ferrovia, we were badly insulted by the people and some of them spat on us from their balconies.[27]

Like anywhere else, the inhabitants prefer not to hear about pollution that causes cancer; mainly because they cannot afford to move to a more healthy environment, especially after having worked hard and made sacrifices in order to buy the house they are living in, but also because they refuse to consider the fact that their lives may be at risk.

## 'Asbestos is Less Harmful Than Coca Cola'

As quoted by numerous newspapers, this infamous sentence was attributed to the businessman Graziano, making the humiliation of being spat at certainly not the worst injury that the workers had to endure. In the wording of the civil action to sue for damages that was brought by some of the workers in 2011 (addressed to the President of the Republic, among others) it is possible to read:

> The complicity was such that these criminals broke every possible law without any kind of contestation or any attempts to restore legality … [28]

The post-earthquake emergency legitimized this complicity on a social level, too; besides the massive subsidies given to whoever wanted to invest in the

25 There were a couple of exceptions – humanly remarkable but statistically irrelevant; some court documents report that in an interview, a young journalist, who had been charged with libel by a colleague who was very close to the company, gave an account of a 'punitive raid' against a journalist formerly known for his unwelcome articles.

26 Carlo, 51, former Isochimica worker, 27 December 2011.

27 Michele and Gino, former Isochimica workers, respectively 53 and 54, 5 October 2011.

28 *Atto di denuncia-querela con richiesta di misure cautelari personali e reali*, presented on 12 November 2011 at Legione Carabinieri 'Lazio', Latina Station, p. 4.

area,[29] the main health and safety regulations for the protection of the workers were actually suspended. Isochimica never obtained the reconstruction subsidies[30] or any particular exemption from regulations, nevertheless it benefitted from a widespread mental attitude that legitimized malpractice. Law no. 219 did not require, of course, the revocation of the regulation which stated that asbestos 'must be used in the countryside, far away from residential areas.'[31] In February 1983, six months after the start of the work to remove the asbestos from the carriages in the local railway station, the Town Council of Avellino, in possession of a legal building licence and a positive health-sanitation assessment, authorized the construction of a facility on the edge of a very crowded neighbourhood, 'for the installation of chemical thermo-acoustic insulation'.[32] The certificates necessary to start work were presented by the company in 1985, but according to Town Council records, no Certificate of Use and Occupancy had been either granted or required for the entire premises. Only in 1986, when the processing activities had already been underway for four years, was Plant B given a certificate of conformity to standards: needless to say, the asbestos was treated in Plant A, where not even one inspection was carried out by a municipal technician.[33]

In 1986, a courageous magistrate launched an investigation and opened a first criminal proceeding; but it was dismissed by the public prosecutor's office.[34] The same magistrate launched a second proceeding in 1989, but it stalled when a reform to the Code of Criminal Procedure withdrew the investigating magistrate and he was moved to another role and to a different location. The documents gathered for that investigation (a number of surveys) clearly displayed the evidence of an environmental-health disaster, but nobody decided to use them. Another investigation in Florence showed that after the removal of the asbestos, the wagons were sent back to the local OGR Railway Stations to be refitted

29 Art. 32, Law no. 219/80 provided for an advance and non-repayable aid package of 75 per cent to all those who would invest in the area.

30 The government aid, granted on 16 March 1984, was repealed by the Council of Ministers on 21 June 1989.

31 Art. 216, R.D. 7/27/1934 no 1265, M.D. 2/12/1971, 12/23/1976 and 3/2/1987.

32 Town Council of Avellino, Construction Permit no. 2856 of 2/26/1983.

33 Dr Giuseppe Pianese, at the time health officer for the Municipality of Avellino and then in charge of ecological services for the Local Health Authority (USL in Italian), declared to magistrates that 'the projects presented by Isochimica never mentioned the use of asbestos during processing phases, on the contrary, they denied any use of toxic substances'. He also declared he could not attest to 'the existence of any Certificate of Use and Occupancy concerning the entire plant, presumably built in 1983, in spite of the surveys carried out'. Similar declarations were made by Mr Luigi Masi, Chief of the Municipal Technical Bureau, and by other officials in the judicial proceedings.

34 The old Chief Prosecutor, Antonio Gagliardi – the same man who, in Graziano's opinion (v. *supra*) had wanted him to keep the local soccer team in Serie A – has recently publicly declared that he does not remember and that, anyway, at the time, not much was known about environmental matters.

before being introduced back into the rail network. The Tuscan workers – given the peculiarly primitive processing methods used in Avellino – found particles of asbestos everywhere, so, after sending the wagons back to the South,[35] they decided to act to protect their health and turned to the magistrates. The news of this new investigation began to spread all over Avellino, and as a consequence, the Local Health Authority or ASL (formerly USL) finally certified the Isochimica plant as an 'industry detrimental to health of the highest degree'. During a local town council session, on 9 April 1988, at the request of the workers to put forward an order to close down the industry, the council discussed the matter as a mere 'item on the agenda' and voted to 'take note of the situation'; but on 8 September 1988 it was forced to impose a 'temporary suspension' of the company's activities, officially noting, for the first time, that it was operating without a permit and that, as it was located 'near residential areas and public buildings (nursery and primary schools), represented an imminent danger for the surrounding residential area'.[36] On 13 December, an order from the Florence Public Prosecutor, Beniamino Deidda, imposed the definite closure of Isochimica, but after only a few months and a number of appeals, the asbestos removal activities were started up again under the new name of El.Sid srl, and continued for more than three years.

According to official records, it is evident that Isochimica has treated thousands of cubic metres of asbestos; in other words, thousands and thousands of tons. From an economic point of view, asbestos is an extraordinary 'capital investment multiplier' (Bonanni, 2009, 63) – it slowly kills those who extract, transport, process or use it and those who are exposed to it.[37] Nevertheless, as is clearly shown by Mr La Verde's declarations to the Parliamentary Committee of Enquiry on Occupational Injuries:

> It was not until 2003 that the Avellino Local Health Authority stated its intention to start monitoring the Isochimica workers: but, in spite of the requests and insistence of the press and the people involved, the results relating both to the precise number of people examined, let alone the outcomes of these

35 The Irpinia workers would find unfriendly messages from their Tuscan colleagues in the carriages, as Nicola remembers: ' … inside the wagons we found handouts written by the OGR workers in Florence. Our Northern colleagues called us 'grazianini asbestos-eaters' and they also wrote: 'the asbestos is yours, keep it yourselves!' (Nicola, 51, former Isochimica worker, 27 December 2011).

36 Town Council, *Committee resolution no 2746*, 8 September 1988.

37 Kagan and Jacobson (1983); Waxweiler et al. (1983); Battista et al. (1999). For a general report on the medical literature about asbestos pollution, see Dodson and Hammar (eds. 2006) and Hallenbeck and Hesse (1977). The work by Battista et al. (1999) contains interesting research material specifically about the asbestos-related diseases among the train factory workers.

> examinations, are still unknown. Likewise, we do not even know the precise number of workers who worked in the Isochimica plants over the years.[38]

To add insult to injury, and in this case a 'biological one':

> Of the 61 workers examined by INAIL (The National Institute for Insurance against Occupational Injury), 27 have been officially recognized as having biological damage caused by the effects of asbestos at a level between 0 to 4%, and the others between 5 to 25%. Just one person has been officially recognized as having 50% damage. Finally, other subjects have proved to be positive for different kinds of cancer (Ibid.).

Even today, many of the workers have still not been called for examination by the Local Health Authority; following the Local Health Authority's check-ups, others with more 'negative' results that can be directly connected to asbestos-related diseases, have decided to go for further examinations in other Italian medical centres, and have been found to be extremely ill (with asbestosis, mesothelioma and other diseases). Because there has been no official recognition of their diseases, no compensation has been granted, nor have any of them been entitled to early retirement: the luckiest of them get by, working in other factories in the area. In the above-mentioned Enquiry Committee deposition, the workers were reproached by Senator Donaggio (Democratic Party) for not having supported the adoption of the National Center of Research (CNR) project, during the 1990s, which provided for the automated removal of asbestos, preventing any human contact with the material. In Senator Donaggio's opinion, the workers refused to accept the project so as not to lose their status as people damaged by asbestos exposure and, as a consequence, the right to statutory and particularly Social security benefits (Ibid.). And, as the President of the Enquiry Committee clearly says (Tofani, Party of Liberty/Berlusconi):

> Asking to further extend Social Security benefits to other asbestos-exposed workers is unacceptable, also because we run the risk, as has occurred in the past, of reducing the benefits in order to contain spending (Ibid.).

On 30 May 2013, the new public prosecutor for Avellino, Rosario Cantelmo, ordered the seizure of the former Isochimica plant alleging the offence of environmental disaster. The investigations are still covered in the greatest secrecy and the Senate Hygiene and Health Committee Prosecutor's declarations, made in March 2014, are also classified. Just a few days after this, on 25 March, during a citizens assembly at the parish hall in Borgo Ferrovia, Antonio Giordano, a well-known scholar

38 Council Committee Enquiry relating the phenomenon of occupational injuries, and mainly concerned with the so-called 'deaths at work ', 16th Legislature, *Summary Report no 96, 02/22/2012*, p. 1.

and expert in Cancer Research and Molecular Medicine in Philadelphia, and who has devoted a great deal of his research to the situation in the 'Campania Land of Poisons' (Giordano and Tarro, 2012), efficiently summed up the tragedy by saying that the case was far from solved: the Isochimica case is as serious as the 'Triangle of Death', and is 'one of the biggest cases of asbestos dumping ever to take place in Europe, thanks to the connivance between ruthless businessmen, complicit technicians and corrupt politicians'. It is very difficult to quantify the damages: 80 per cent of the former workers have been officially declared ill, affected by asbestos-related diseases, meanwhile the 'official' number of dead is at least 15,[39] to which we should also add Pasquale Soricelli, who committed suicide in 2011 (the Local Health Authority had refused to recognize his illness, and consequently he was not given the right to early retirement, which resulted in a form of serious depression). In any case, according to the experts, we are likely to see the peak of the diseases in 2020. Of course we must not forget the workers' wives, too, who were responsible for washing their husbands' work clothes; some of them are already ill, others have died affected by cancer, but in these cases any correlation is both scientifically and legally extremely hard to demonstrate. There are also the 'spin-off' workers, the staff in charge of the cleaning and the disposal of factory waste and there are, finally, those who were residents or attended schools in the area. Nevertheless, no adult 'screening' has started yet, nor is it likely to: the site has not been officially included in the Minister of Health's map of areas at risk from asbestos-related diseases; and there are no funds; so, for the moment at least, those who need to carry out any further diagnostic testing will have to do so at their own expense.

The remediation of the site is a much more complex question. To sum up, in 2004, following the infringement procedures brought about by the European Union, the Avellino Town Council and the Special Government Commissioner for Waste Management were designated to carry out the remediation of the site. In reality, only superficial 'safety operations' were carried out. Since then, there has been a tragic and systematic 'passing the buck', and there is little likelihood that any remediation will ever actually take place: there are problems relating to the ownership of the land (Isochimica went bankrupt in 1990), uncertainties about the jurisdiction (municipality, province, region, national government, Local Health Authority and so on) and, more particularly, a lack of funds (or rather an underlying unwillingness to find them, due to an evident lack of interest in the lives of humble people).[40] In the same year, the Isochimica premises (42,000 sq.m.) was rented to Eurokomet Ltd, with a pre-emption right to buy the property once it had definitely

39 Mr Carmelo Padula's Press Conference, Director of the Prevention Department of the Local Health Authority (4 February 2014).

40 In April 2013, the President of the Regional Committee for the monitoring of the Remediation, Antonio Amato, declared to the investigating magistrates that: 'To the best of my knowledge, and with reference to any information in my possession, the delay in the remediation of the Isochimica area is due to an attitude of "passing the buck" between the Municipality, curatorship, other bodies and society» (City Court of Avellino, Validation

been remediated; in the meantime, the company would work to completely recover the site within two years. The company, which had been founded six months earlier by a local business consultant with the purpose of 'planning promotional events, entertainment, hotel and restaurant management, distribution of food' (Fierro, 2008), hoped to create *in loco* an exhibition centre, and perhaps a shopping centre. It was an extremely complex network of contracts and subcontracts with a series of different companies,[41] which makes it very difficult to reconstruct what really happened. What is certain is that the first remediation work started in July 2009 (that is, three years after the fixed deadline for the completion of the works) but the work has never been finished. There is still a huge amount of debris[42] on the premises, and what is concealed underground is even worse: tons of asbestos that have been randomly buried. But the worst situation is to be found outside the plant, in the surrounding areas, where, in May 2014, ARPAC (Regional Agency for Environmental Protection) certified the existence of 27 fibres of asbestos per litre of air. But the asbestos is everywhere, in the streams, in the fields, in the woods of Irpinia, where the thousands of tons of crocidolite that was removed from the carriages, is probably buried, still unaccounted for. It is a well-known fact that small quantities of asbestos have been hidden in different locations all over the territory: sometimes the monitoring agencies find it, but it is very difficult to identify where it comes from. The most concrete information comes from the workers and the inhabitants of the area and from the Criminal Proceeding (no. 878/86) Administrator's declarations.

Isochimica, with its tons of asbestos, is still there, opposite a school and a football ground, in a popular neighbourhood built in the 1960s and which boomed after the earthquake in 1980. Lots of families moved there in order to find jobs and better living conditions, and work was to have come from the different factories in the area. Currently, after the continual turnover and many closures, the number of officially operating factories is over 60, employing less than 6,000 workers. These factories have a very reduced occupational impact, an extremely low survival rate and, furthermore, most of them have a very high environmental impact, which has frequently been denounced by residents, wine producers and environmental associations. As far as fire in this area is concerned, the experts have defined it as one

Order for the urgent Preventive Seizure issued by the Public Prosecutor and Preventive Seizure Decree by the, Examining Magistrate, Giuseppe Riccardi, 15 June 2013).

41 Incidentally, the HGE is the name of the old IRM, that was involved in the 2005 fire, and whose activities were authorized by the Region by means of the *Executive Decree no 554 of 11 May 2010*, signed by the then Coordinator for Environmental Protection and Pollution Abatement Department, Luigi Rauci (Rauci is the same manager who was ordered, in May 2014, by the Court of Auditors, to pay damages in the sum of four million euros for allowing the 1,023 legal statements against the managers of several wastewater treatment plants and the owners of illegal dumping sites in Campania) to become statue-barred.

42 City Court of Avellino, *Validation Order of the Preventive Seizure etc.*

'of the biggest fires of its kind to occur in Italy'.[43] But this is not a typical situation, taking place in just any suburb anywhere in the world: here, politics and groups of interest have been allowed to play without any respect for the rules or for public health. Nevertheless, the ASI triumphantly attributes itself the merit of having succeeded in 'transforming a traditional mountain economy into an economic and modern industrial reality'. The balance between costs and benefits, though difficult to calculate, seems to be decidedly negative. In the towns belonging to the Local Health Authority District, the death rate exceeds the regional average for several types of cancer (*in primis* leukaemias). Over the last two years it has been shown that about 80 per cent of deaths are due to cancer (Scannella, 2013), and it is not by chance that journalists, with all due caution, have recently used the expression 'Cancer Valley', to identify the area.

## Heterotopia and Thanatopolitics in the Italian South (*Mezzogiorno*)

Without being particularly pessimistic, and considering the extent of the disaster, it seems objectively difficult to imagine that there will be a realistic remediation of the Isochimica premises in the near future. Who could possibly foot the bill? The local authorities, who are up to their necks in debts? The region, struggling with a general environmental disaster (the 'Land of Fires', 'Triangle of Death' and so on)? This region is one of the most polluted in Europe, and the state is trapped by its need to make spending cuts. The EU seeks to balance the budgets, and people are consumed by one of the most serious economic crises in history. In these circumstances, it is possible that, sooner or later, there will be a hasty 'burial' of the site: this action will be considered in public speeches by technicians and politicians as the only way to guarantee total and definite safety; once the struggling people have come back home and the media spotlights have been turned off, all traces of the asbestos will be lost forever. Perhaps the victims will finally receive limited compensation (the former workers, at least). The courts may even hand down some sentences.

What lessons can be learned, from the ILVA in Taranto and the 'Land of Fires', about the development model that has been imposed on the Italian *Mezzogiorno* in recent years, and about its terrible costs in terms of health and environment? In Foucault's thoughts about biopolitical governmentality – meaning 'state care for individual life', and its 'protection' and 'improvement' (Foucault, 2009, 2010) – environment is a fundamental variable with which the population interacts in a reciprocally conditioned relationship. Traditionally, in political terms, rivers, forests, fields, different territories and so on were seen as 'resources', at most, as

43 Paper presented in 2008 by Bellizzi, D'angelo and Vassallo to the V International Conference on the Risk Assessment and Management in the Civil and Industrial Settlement of Pisa University (http://conference.ing.unipi.it/vgr2006/archivio/Archivio/2008/Articoli/23.pdf, accessed on 16 September, 2015.

means for obtaining taxes and revenues; but everything was destined to change with the advent of the *Polizeiwissenschaft*. For Von Justi 'the population and the environment are perceived as being in a perpetual living interrelationship which falls within the domain of the state to manage' (Foucault, 1994, 160). So, what meaning can we give to the kind of situation described above, where the government seems to be so openly disengaged in its objectives of protection, care and improvement of the population, until it ironically becomes death itself? Actually – although we do not always remember – (Mbembe, 2003) – Foucault's idea of biopolitics does not exclude the existence of a completely political dimension of death: 'Since the population is nothing more than what the state takes care of for its own sake, of course, the state is entitled to slaughter it, if necessary. So the reverse of biopolitics is thanatopolitics' (Ibid.). The transition from the ancient sovereign power with 'the right of death' (*potestas vitae ac necis*) to the modern state's biopolitical power, marks, instead, a redrawing of the political value of death within another theological system:

> There has been a parallel shift in the right of death, or at least a tendency to align itself with the exigencies of a life-administering power and to define itself accordingly. This death that was based on the right of the sovereign is now manifested as simply the reverse of the right of the social body to ensure, maintain or develop its life (Foucault, 1988b, 120).

So, death is no longer the extreme form of 'deduction' imposed by the sovereign on his subjects (the appropriation of goods, wealth, services, time), it is no longer a simply 'negative' gesture of compensation for *lèse-majesté*, but a way to attain a higher state (divine) of wellbeing, a sacrificial means for the conservation and protection of life itself, 'a modality – or a tonality – of its same conservation' (Esposito, 2002, 162). In modern states, the overturning of the historical right of the sovereign 'to take life or let live' into 'the right to let live or make die' results in a powerful structural tension: 'The coexistence, in political structures, of large-scale destroying systems and institutions concerned about individuals' lives is a strange fact that requires further investigation. It is one of the central antinomies of our political raison (Foucault, 1988a, 137). This antinomy is clearly readable in Foucault's vision of the terrible wars of the twentieth century: wars are now 'waged on behalf of the existence of everyone; entire populations are mobilized for the purposes of wholesale destruction in the name of life necessity' (Foucault, 1988b, 121). The redefinition of power as the government of life, implies the activation of different kinds of justification mechanisms: the horrors of war can be tolerated only if interpreted as a continuation of the same caring for life ('massacres have become vital', Ibid.), in turn, this implies a clear perception of life as something that deserves to be protected: and it is not by chance that the spectre of race has reappeared so violently during the last century, in the shape of openly declared genocide, as the elimination of any type of biological threat. The close connection between these racialization systems and the concept of biological threat had already allowed Foucault to show its absolute flexibility, and to clarify how it can disregard the political boundaries of the nation in order to turn in on itself, so that its real target becomes the 'sub races' who are recognized as the 'biological enemy' of the same national population (Foucault, 2003).[44] The disguised reintroduction of the ancient sovereign right over life and death, is, first of all, based on wordplay: it is thanks to this that it is possible 'to establish a break between the biological and political body – the population –, between what has to live and what has to die, these two utterly dependent categories. In other words, the growth of what needs to be strengthened – the healthy race – depends on the elimination of the negative, or sub race' (Chiantera-Stutte, 2006, 256). 'Racism without race' can therefore appear as a form of racism in which categories such as 'ethnicity', but also 'culture', have the same discriminatory function played by 'race' in the past. In liberal theory the supposed 'naturalness' of market development has always been tempered by the idea there are certain cultures capable of blindly standing up to the endogenous matrix. New importance is then given 'to the cultural struggle against the socio-symbolic instances of tradition, seen as being primarily responsible for the failure of developmentalism' (Romano, 2006, 311). According to a well-established narrative plot of 'internal orientalism' (Schneider, 1998), this type of discourse reinterprets, from a cultural point of view, the old mechanism of hierarchical attitudes of differentiation towards the Southern people as a physical race, that was introduced by the school of positive anthropology after the Unification of Italy (Teti, 2011). The ethical – political categories, such as Banfield's 'amoral familism' (1967) or Putnam et al.'s defect of 'civicness' (1993), as well as the numerous and detailed moral descriptions of the way the local populations live that fill the bookshelves with their cold journalistic style (Saviano, 2006), owe most of their considerable fortune to their functionality regarding this order of discourse. Behind the developmentalist theories there is, actually, a theoretical game which results in the identification of:

1. the features of those traditions which prevent bio-economic substance from growing freely;
2. (*anti*)-cultural imaginary architecture, able to easily tolerate growth and the difficulties of unlimited enhancement;
3. the factors that most efficiently contribute to spreading and strengthening this architecture (Romano, 2006, 311).

It is precisely the inferiorization, the cultural de-legitimization and, particularly, the moral de-qualification of the local population which are at the basis of the whole public intervention system in the *Mezzogiorno*. This means that it has not received the biopolitical protection owed to it by the modern state, while the

44 Nowadays, the tragic evidence of this is represented by the ability to bring back to life as 'ethnic' some historically forgotten or utterly false identities, within many contemporary conflicts, from the Balkans to Sri Lanka, to Africa …

state has been allowed to govern the territory according to the dynamics of rapid, aggressive and chaotic consumption of the land that, over the last few decades, has globally characterized most of the so-called 'peripheral' economies. It is necessary to connect the issue of the Italian South to those types of differential government of territories which characterize current globalization, thus breaking with the tradition of 'geographical exceptionalism' in which the subjects are seen as a mere sequence of special cases (Harvey, 2000).

Just like Foucault's 'heterotopia', it is possible to think of these *espaces autres*[45] of the capital as sites which possess 'the curious property of being in relation with all the other sites, but in such a way as to suspect, neutralize, or invert the set of relations that they happen to designate, mirror, or reflect'. In 1972 Ferrari Bravo had already observed that 'the passage of the economic and social Southern status from a 'divisional' sort of relationship to an 'integrated' one, compared to the whole National (and International) economy' does not mean 'establishing an economic structure in the South which is the same as, or similar to, that in the North', it means, instead, re-establishing a subordinate relationship according to the new concept of 'backwardness as a developmental resource', including any adjustments (for example, consumption models), but also the reproduction of the differences (particularly as far as working conditions are concerned), so that, 'as regards the growing internationalization of economic life (and not only economic)', 'the sort of economic crisis present in the south' can no longer be represented 'as one "side" of a dual relationship, but, if anything, only as the backward side, together with others, of a larger and better integrated economic system' (Ferrari Bravo 2007, 25, *passim*). Hypermodernity and backwardness can co-exist, primarily, in the form of work organization, which means that a large share of the work can be given to a local workforce rather than being destined, as usual, to foreign manpower (Moulier Boutang, 2002, 85ff.). Global reunification in the name of 'market laws' does not imply the standardization of production organization: particularly, the world of work regulations is a landscape made up of 'variegated areas' which endlessly combine different kinds of work regulations (Ong, 2006), a real 'mixed constitution' (Hardt and Negri, 2002, 285 ff.) that does not follow the boundaries set out by geo-political maps, but which can indifferently represent pieces of 'North' in the 'South' (as decribed by Ong when talking about the technological *enclaves* in China and South-East Asia) or pieces of 'South' in the 'North' (which is what seems to occur with 'labour-intensive' production and the black economies so firmly rooted in the richest areas of the world). The distinction – which was still possible up to the 1960s, and the last of the struggles for decolonization – between 'colonial economies' and 'metropolitan economies', seems to be unfeasible in modern and global capitalism, characterized by numerous and complex relationships which are very closely intermingled.

45 'Des espaces autres' is the title of a Conference given by Foucault on 14 March 1967 at the 'Cercle d'études architecturales' in Tunis. Here quoted from the first Italian edition (Foucault, 1994).

So the 'colonial exception', observed by Fanon, with its ability to combine modernity and backwardness ('slavery, serfdom, barter, a skilled working class, and high finance' Fanon, 2007, 60), seems to be tempered in a well-structured 'normality' of local 'micro-exceptions'.

The 'heterotopic' and 'heterochronic' nature of some parts of the world, both on the outskirts as well as in the beating heart of the West, allows capital enhancement to take place even in structurally negative conditions, for example through the integration and the use of 'historical' and 'local' forms of work organization and regulation. In the case of Isochimica, these forms can be, at the same time, 'feudal' and 'very modern'; in the national border spaces of globalization, such as the Italian *Mezzogiorno*, the old nepotistic relationships can easily merge with the refined mechanisms of judicial exceptions, which are more and more present in the contemporary management of territories and populations (Giroux, 2006). 'Corruption' itself, which is apparently typical of Southern Italian political and business relationships and the constant 'overstepping' of local economies into illegal actions, can be read – from this point of view – as real government mechanisms, rather than as random system 'deviations'. In our story, the 'young workers' and the inhabitants of a popular neighbourhood in a Southern Italian suburb have saved, for once, the revenues of the State Railway by absorbing as much asbestos as they could into their lungs. In this, as in other cases, the harshness of global economic competition has transformed the presence of a widely available workforce, from which circumstance has removed any form of conflictuality, into a full and limitless 'supply' of expendable, or rather, 'disposable' bodies.

# Annex
# The Governance of the Anthropocene and Health, Environmental and Economic Crimes Against Humanity

We have added this annex because the debate that has arisen mainly in France following the COP21 summit, deserves attention[1]. In fact, the debate on climate change, and in particular, on the consequences of the Anthropocene has been ongoing since 2000, but has been confined to narrow circles of experts, especially those involved in the so-called earth- or nature-, or environmental-sciences.

As suggested by recent studies, such as those published in the journal *Nature*[2], the current geological era, the Holocene[3], is considered as having come to an end after the Second World War (if not at the end of the nineteenth century) because of human activities that have created dramatic changes in the ecological balance of planet Earth. The new era is called the Anthropocene, because it is marked by the particularly destructive impact of human activities[4].

1 For the first time in over 20 years of UN negotiations, the summit COP21 (November December, 2015), also known as the 2015 Paris Climate Conference, aims at achieving a legally binding and universal agreement, whose purpose is to keep global warming under 2°C. See the petition 'Stop to climatic crimes' promoted by numerous authors: http://350.org/climate-crimes-it/ accessed on 9 November 2015, and some articles by Jade Lindgaard, Joseph Confavreux and Lucie Delaporte, Michel de Pracontal, in http://www.mediapart.fr, accessed on November 9, 2015. See International Conference 'How to think the Anthropocene? Anthropologists, philosophers and sociologists face to climate change', http://www.fondationecolo.org/l-anthropocene; abstracts: http://www.fondationecolo.org/l-anthropocene/resumes, accessed on 9 November 2015

2 See http://www.nature.com/news/anthropocene-the-human-age-1.17085, accessed on 9 November 2015.

3 See the famous graph of geological eras https://it.wikipedia.org/wiki/Scala_dei_tempi_geologici, accessed on 9 November 2015. The age of the Earth is estimated to be about 4.57 billion years and the Holocene period dates back to 11, 700 years.

4 See *Nature*, quoted. Anthropocene is the term coined in the eighties by biologist Eugene Stoermer; it was adopted in 2000 by the Nobel Prize for chemistry Paul Crutzen in the book *Welcome to the Anthropocene*. The term derives from *anthropos* (human) and from Holocene (https://en.wikipedia.org/wiki/Anthropocene) accessed on 9 November 2015.

This change, and the obviously huge concern that it arouses, has become the main theme of reflection for many researchers, not only in the field of natural sciences (geologys, biology, physics and so on) but also among some academics in the humanities, political and social sciences sectors. Among others who signed the petition "STOP climate crimes" (at the COP21 summit in Paris) were Bruno Latour and Clive Hamilton[5].

In fact, until a few years ago, the main focus and passion for the study of different aspects concerning the changes in the field of knowledge about planet Earth, was generally restricted to circles of archaeologists, biologists, geologists and anthropologists, and did not arouse much interest among political and social scientists. From Plato and Aristotle up to the more political philosophers (Saint Augustine, Ibn Khaldoun, Thomas More, Machiavelli, Thomas Campanella, Hobbes and Locke), and up to Durkheim and some of our contemporaries, these 'sciences' almost always contained a particularly imperative 'prescriptivism'; that is, they claimed to provide 'recipes' to 'solve' the problems of the political organization of society – often in the name of prosperity and posterity, peace and even happiness 'for all' (e.g. 'equal justice for all'). However, more recent history and the current situation show that, on the contrary, there is increasing inequality, and a rise in atrocities and genocide. In fact, there have been more wars than periods of peace. The claims of prescriptivism have therefore proved unsuccessful, or perhaps worse, since these sciences have produced useful knowledge for the dominant class, who have been primarily responsible for the worst atrocities. From Socrates to Foucault, Parrhesia has always been confined to a niche granted by power, in which it is possible to be criticized or desecrated because there is an increasing asymmetry between the dominant and the dominated that has now been reduced to a few attempts at resistance, which are often desperate or simply impotent because of the erosion of the possibilities for political action (the case of Greece is very eloquent).

In the face of the triumph of liberal think tanks, the opposing perspective has proved to be impracticable, in spite of the obvious crisis of the liberal model, which manages to survive without major obstacles due to the lack of viable alternatives. Considerable support has been offered by many political and social scientists who

5 Bruno Latour, 2015. *Face à Gaïa. Huit Conférences sur le nouveau régime climatique.* Paris: La Découverte. See the article and related video http://www.mediapart.fr/journal/culture-idees/141015/bruno-latour-sur-le-climat-nous-devons-comprendre-qui-estlennemi-de-qui and also the conference: https://www.youtube.com/watch?v=kPxbuluOrgE, accessed on 14 October 2015. See also Clive Hamilton, *Requiem for a Species. Why we Resist the Truth about Climate Change*. London: Earthscan 2010. Jason W. Moore: http://www.jasonwmoore.com/About.html, accessed on 9 November 2015; *La mondialisation des risques*, edited by Soraya Boudia and Emmanuel Herny, Presses Universitaires de Rennes, 2015; J-B. Comby, 2015. *La question climatique. Genèse et dépolitisation d'un problème public.* Paris: Raison d'Agir; Laura Centemeri and Xavier Daumalin, eds, 2015. *Pollutions industrielles et espaces méditerranéens XVIIIe-XXIe siècle*. Paris: Karthala.

are converts to the liberal cause. Consider, for example, the legitimacy granted to 'humanitarian wars', 'zero tolerance', or the shadow economies and the lack of attention to the related 'side effects' such as neo-slavery, the blurring of the formal and informal economy, corruption, and the victims of this interrelatedness.

Now, just when it seems that we are 'close to the abyss', faced with rampant involution in all fields, a 'revolutionary' reaction has emerged (see the petition at the COP21 summit in Paris mentioned above). This reaction is, perhaps, the result of research that seeks to go beyond disciplinary boundaries, in a universe of knowledge comparable to that experienced by Leonardo da Vinci, who was interested in empirical knowledge in all fields (anatomy, geology, physics, art and architecture), but not in abstract philosophical reflections.

On presenting his latest book, Latour observes that the 'ecological stakes and climate disruption are likely to turn what could have been a passing crisis into a profound alteration of our relationship with the world. . . . we could have acted thirty or forty years ago and we did nothing or very little . . . Strange situation, having passed a series of thresholds we have crossed into total war and have noticed practically nothing at all'[6]. In fact, it may not be surprising that the economic and political authorities, along with some political and social scientists, have not noticed this war either, or else they have deliberately ignored it in favour of other wars.

As the article in *Nature* shows, several recent studies show that the Anthropocene period began after the Second World War[7]. This would be evidenced by facts of enormous importance: nuclear testing and the general increase in radioactivity; the large-scale production and abuse of fossil fuels and the chemical industry (also a factor in local wars); large dams that have destroyed the ecological balance of huge territories particularly in India, Africa and Latin America; extensive deforestation and unscrupulous overbuilding; abuse of water resources, fertilizer and pesticides; criminal treatment of toxic waste, and so on. In short, the air, the land, the waters and the seas have been mercilessly polluted.

According to Latour, it is correct to say that 'we are in a situation in which war has been declared'. Since the Second World War, authors have made constant pleas to our attention so as to avoid a Requiem for the species (the title of Clive Hamilton's book[8]), and now many experts fear that the COP21 summit will be yet another failure.

Obviously this is not a question of a trivial dispute between pessimists and optimists, but if we want to see some useful and effective outcomes from acting publically and politically, it is necessary to thoroughly understand the reality of the stakes and of the dominant actors. Today's lobbies are more and more focused on realizing maximum profits at any cost, even to the detriment of humanity and the

6 See footnote 5.

7 See footnote 2.

8 See also the interview with Hamilton: Médiapart, http://www.mediapart.fr/journal/culture-idees/051115/clive-hamilton-lanthropocene-est-l-evenement-le-plus-fondamental de-l-histoire-humaine, accessed on 9 November 2015.

Earth. This is due, above all, to the link between financial games, military lobbies and other businesses (particularly, the oil, car and chemical industries, among others).

As we have noted in the research presented in this book, ignored insecurities, like all the other consequences of the Anthropocene era, have come about, more particularly, due to the neoliberal turn in the governance of security. This is where we find the mass distraction mentioned by Latour, and also the idea of the 'capitalocene' as proposed by Jason W. Moore (although we agree with Hamilton that it is useless to make this into a terminological controversy).

The frequency and severity of health and environmental disasters increases as a result of the perpetuation of war, of the distance between wealth and poverty (the wealth in Africa, India and Latin America is only about 5 per cent of global wealth[9]) and the increase in neo-slavery and irregular migration caused by the prohibitionism of rich countries in providing undocumented workers for their shadow economies. According to Jason W. Moore, we can observe that all disasters are the result of economic and socio-ecological projects. Factories and financial centres are just as important as the choices made by multinational companies operating in the agriculture and forest management sectors. The modern world system emerges through the relationship between economy, society and nature.

Human history is marked by disasters of all kinds and types, but the disaster *par excellence* – quantified by the number of victims and from the point of view of the damage caused – is war. It is therefore no accident that the lexicon of war also pervades the observations made by geologists, biologists, researchers in Earth sciences and, of course, journalists in search of an scoop (for example 'water bombs' or 'war of floods'). So why not consider health and environmental disasters as crimes against humanity?

Predictably, the logic of liberalism does not give up so easily in the face of the growing awareness surrounding a possible 'requiem for humanity'. Research programmes in the field of risk prevention, both in Europe and elsewhere, are recycling what is already being used by the military and the police (such as the adoption of drones in Mare Nostrum operations, in the name of rescuing migrants who are at risk of drowning). Sophisticated satellite systems, radars and sensors, new 'postmodern' mega-projects included in different disciplines, are often based on research areas controlled by financial-military-police lobbies (in NATO countries in joint venture with US companies). In parallel, a bevy of psychologists preach resilience which is limited only to individual abilities to adapt to situations of risk, disaster or terrorist attack. This approach does not favour any ability for collective action; in fact, it pushes people more and more towards atomization, to impotence and to the hope that they will always be protected by a military-policing form of governance (supported by psychologists and embedded opinion leaders). All the wealthier countries turn to new security businesses in the name of disaster prevention,

9 See Global Wealth Report of 2015, available at https://www.worldwealthreport.com/ accessed on 9 November 2015.

and enslave their Civil Protection units especially during speculative response work (the Hurricane Katrina case, for example). However, no country has promoted a real recovery programme for situations of risk; although it is a well-known fact that without it and the necessary reclamation, proper prevention will never take place. And yet, a serious reclamation programme could create millions of jobs, as well as a sustainable future. Unfortunately, a part of the population is complicit, as they are ensnared by the dominant discourse, lured by small concessions and privileges or the illusion of obtaining them. The victims have no protection, they endure their plight passively and are sometimes even accomplices to their tormentors. In the meantime, health and environmental disasters take place again and again, and worsen, not simply because the dominant players have no intention of renouncing their profits, but because they increase speculation thus making the weaker players defenceless and exposed to risk. Inevitably, the scandal surrounding Volkswagen will end with the repair of the 'infected' cars, but without any sanctions for the damage caused by the pollution and to its victims. In order to aid the refugee crisis, funds will be dished out to NGOs without stopping the production and trade in the arms that fuel permanent wars and fund ISIS[10], nor will a single thought be given to the economic and environmental disaster areas from where the exodus began.

It is a shocking fact that even after the massacre in Paris on 13 November 2015, all countries seem to have adopted a solemn and hard-line commitment against pseudo-Islamic terrorism, while all these countries are complicit in the trade of arms and oil with this horrible new enemy. In the name of the war against this pseudo-Islamic terrorism, all new security expenditure will be considered legitimate by the European Commission; on the other hand, no government has made proposals for the urgent funding of a serious reclamation program for the sites affected by serious risks to health and environmental disaster. The relatives of the millions of those who have died due to cancer or neo-slavery do not claim vengeance, but they do note that the governments of almost all countries have little or no respect for those who have died.

There seems to be a very high risk that humanity will succumb to the neoliberal hand. The governance of military and police affairs that pervades public health and civil protection is likely to lead to a new version of thanatopolitics. But, as long as life continues there will always be a fight for survival, and therefore resistance against the 'non-creative' destruction of neoliberalism. As we have learnt from those who resisted fascism and nazism: even when all seems lost and you have the impression that you will not be able to resist, even for a short time, it is still possible to win.

---

10 See: Isis Inc: How oil fuels the jihadi terrorists, http://www.ft.com/intl/cms/s/2/b8234932-719b-11e5-ad6d-f4ed76f0900a.html#axzz3rezLOwry, accessed on 16 November 2015; and: http://www.mintpressnews.com/how-the-us-its-allies-and-syriaunwittingly-corporatized-isis/201748/, accessed on 16 November 2015; and The escalation of permanent war, http://www.alfabeta2.it/2015/11/16/guerra-permanente, accessed on 16 November 2015.

# References


Adey, P. and Anderson, B. 2011. Event and Anticipation: UK Civil Contingencies and the Space Times of Decision. *Environment and Planning A.*, 43, 12, 2878–2899.

Adger, N. 2000. Social and Ecological Resilience: Are They Related? *Progress in Human Geography*, 24, 347–364.

Ahmed, R., Seedat, M., van Niekerk, A. and Bulbulia, S. 2004. Discerning Community Resilience in Disadvantaged Communities in the Context of Violence and Injury Prevention. *South African Journal of Psychology*, 34, 386–408.

Albrecht, H.-J. 2007. Perspektiven kriminologischer Forschung. Der Wandel im Konzept der Sicherheit und neue Aufgabenfelder der Kriminologie, in *Kriminologie im 21. Jahrhundert,* edited by K. Liebl. Wiesbaden: VS Verlag, 177–201.

Albrecht, H.-J. 2009. *Sicherheitswahrnehmungen im 21. Jahrhundert*, Vortrag. https://www.mpicc.de/shared/data/pdf/workshop_sicherheit09_albrecht.pdf, accessed on 3 September, 2015.

Alkire, S. 2002. A Conceptual Framework for Human Security, Working Paper 2. http://www3.qeh.ox.ac.uk/pdf/crisewps/workingpaper2.pdf, accessed on 17 September, 2015.

Altheide D. 2002. *Creating Fear. News and the Construction of Crisis*. Piscataway: Aldine Transaction.

Altheide, D. 2007. The Mass Media and Terrorism. *Discourse & Communication*, 1, 3, 287–308.

Amster, R. 2003. Patterns of Exclusion: Sanitizing Space, Criminalizing Homelessness. *Social Justice*, 301, 95–221.

Anderson, B. and Adey, P. 2011. Affect and Security: Exercising Emergency in UK Civil Contingencies. *Environment and Planning D: Society and Space*, 29, 1092–1109.

Andreatta, F. 2014. Prefazione, in *Sicurezza globale*, edited by P. Foradori and G.P. Giacomello. Bologna: il Mulino, pp. 7–10

Andres, J. 2014. Mare Nostrum – Humanitäre Operation oder Deckmantel militarisierter Migrationspolitik? http://edoc.vifapol.de/opus/volltexte /2014/5565/, accessed on 3 September, 2015.

Angenendt, S. 2003. Regelung und Vermittlung: Die Rolle internationaler Migrationsorganisationen, in *Migration im Spannungsfeld von Globalisierung und Nationalstaat*, edited by D. Thränhardt and U. Hunger. Wiesbaden: Westdeutscher Verlag.

Arendt. H. 1958. *The Origins of Totalitarianism*. New York: Meridian (2nd edn).

Arendt, H. 1963. *Eichmann in Jerusalem: a Report on the Banality of Evil*. New York: Penguin.

Armitage, R. 2000. *An Evaluation of Secured by Design Housing wthin West Yorkshire*. London: Home Office Briefing Note 7/00.

Armitage, R. 2007. Sustainability versus Safety. Confusion, Conflict and Contradiction in Designing Out Crime, in *Imagination for Crime Prevention. Essays in Honour of Ken Pease*, edited by G. Farrell, K. J. Bowers and S. D. Johnson. Monsey/NY, S. 81–110. Collompton, UK: Willan

Arnaud, A. J. and Farinas Dulce, M.J. 1998. *Introduction à l'analyse sociologique des systèmes juridiques*. Bruxelles: Bruylant.

Aytar, V. and Rath, J. (eds) 2012. *Selling Ethnic Neighborhoods*. London: Routledge.

Bagnasco, A. 1977. *Tre Italie. La problematica territoriale dello sviluppo italiano*. Bologna: il Mulino

Bagnasco, A. 1981. La questione dell'economia informale. *Stato & Mercato*, 1, 173–196.

Ball, K., Haggerty K. D. and Lyon D. (eds) 2012. *Routledge Handbook of Surveillance Studies*. New York: Routledge.

Balzac, de H. 1845. *Splendeur et misère des courtisanes*. Paris: Furne, XI.

Banfield, E. C. 1967. *The Moral Basis of a Backward Society*. New York: Free Press.

Bara, C. 2011. Resilienz und Verwundbarkeit in der Schweiz: Die Unwetter-Katastrophe von 2005 in *Bulletin zur Schweizerischen Sicherheitspolitik*, edited by A. Wenger and D. Trachsler. Zürich: Center for Security Studies (CSS), 77–105.

Baratta, A. 2001. Diritto alla sicurezza o sicurezza dei diritti?, in *La bilancia e la misura*, edited by S. Anastasia and M. Palma. Milano: FrancoAngeli, 19–36 (also in www.ristretti.org).

Barbulescu, H. 2012. Constructing the Roma People as a Societal Threat. The Roma Expulsions from France. *European Journal of Science and Theology*, 8,1, 279–289.

Barkun, M. 1977. Disaster in History, in *Mass Emergencies*, 2, 219–231. http://massemergencies.org/v2n4/Barkun_v2n4.pdfm, accessed on 3 September, 2015.

Barthelemy, P. 1997. Économie souterraine et structures industrielles dans les pays de la Communauté européenne. *Région et développement*, 69, 5.

Bartlett, J., Birdwell, J., Krekó, P., Benfield, J. And Gyori, G. 2012. *Populism in Europe: Hungary*. DEMOS Report. http://www.demos.co.uk/files/Demos_Hungary_Book_web-1.pdf?1327923915, accessed on 3 September, 2015.

Bartlett, J., Birdwell, J. and Benfield, J. 2012. *Populism in Europe: Hungary*. Demos Research Reports.

Bartlett, J., Birdwell, J. and Littler, M. 2011. *The New Face of Digital Populism*. Demos, http://www.demos.co.uk/files/Demos_OSIPOP_Book-web_03.pdf?1320601634, accessed on 3 September, 2015.

Barton, A. H. 1969. *Communities in Disaster: A Sociological Analysis of Collective Stress*. Garden City, NY: Doubleday.

Battista, G., Belli, S., Combat, P., Fiumalbi, C., Grignoli, M., Loi, F., Orsi, D. and Paredes, I. 1999. Mortality Due to Asbestos-related Causes among Railway Carriage Construction and Repair Workers. *Occup Med (Lond)*, 49, 8, 536–539.

Bauman, Z. 1995. Making and Unmaking of Strangers. *Thesis Eleven*, 43, 1–16.

Bauman, Z. 1997. *Postmodernity and its Discontents*. New York: New York University Press.

Bauman, Z. 1998. *Globalization. The Human Consequences*. Cambridge: Polity Press.

Bauman, Z. 1999. *In Search of Politics*. Redwood City, CA: Stanford University Press.

Bauman, Z. 2000. *Liquid Modernity*. Cambridge: Polity Press.

Bauman, Z. 2004. *Wasted Lives. Modernity and its Outcasts*. Cambridge: Polity Press.

Baylis, J. and Smith, S. (eds) 1997. *The Globalization of World Politics*. Oxford/New York: Oxford University Press.

Beamish, T. D. 2002. *Silent Spill. The Organization of an Industrial Crisis*. Boston, MA: MIT Press.

Becattini, G. 2002. From Marshall's to the Italian 'Industrial Districts'. A Brief Critical Reconstruction, in *Complexity and Industrial Clusters*, edited by C. A. Curzio and M. Fortis. Heidelberg: Verlag, 83–105.

Beck, U. 1992. *Risk Society. Towards a New Modernity*. London: Sage.

Beck, U. 1998/2009. *World Risk Society*. Cambridge: Polity Press.

Becker, H.S. 1982. *Art Worlds*. Los Angeles, CA: The University of California Press.

Becker, H.S. 2006. Howard. S. Becker et Alain Pessin: Dialogue sur les notions de onde et de champ. *Sociologie de l'art*, 1, 8, 163–180.

Beckett, K. and Herbert, S. 2006. *Dealing with Disorder: Social Control in the Post-industrial City*. Paper presented at the conference Place Matters: Seeking Equity in Diversity, Diversity Research Institute, University of Washington.

Beckett, K. 1994. Setting the Public Agenda: 'Street Crime' and Drug Use in American Politics. *Social Problems*, 3, 425–447.

Bellah R. N., Madsen R., Sullivan W. M., Swidler A. and Tipton S. M. 1985. *Habits of the Heart. Individualism and Commitment in American Life*. Berkeley, CA: University of California Press.

Bellinvia, T. 2013. *Xenofobia, sicurezza, resistenze. L'ordine pubblico in una città 'rossa' il caso Pisa*. Milano: Mimesis.

Bello, B. G. 2014a. *Roma Matrix. Country Report*. https://romamatrix.eu/italy-country-report, accessed on 17 September, 2015.

Bello, B. G. 2014b. *General Rapporteur's Reflections on the Conference: How to Navigate the Archipelagos of Discrimination? United for Dignity*. Conference on the specific situation of Roma young people affected by multiple discrimination. http://enter.coe.int/roma/Media/Files/United-for-Dignity-Conference-final-report, accessed on 3 September, 2015.

Bender, J., Rosen, A. and Gould, S. 2015. *This Map Explains Why The F-35 Has Turned Into A Trillion-Dollar Fiasco*. http://uk.businessinsider.com/this-map-explains-the-f-35-fiasco-2014-8?r=US&IR=T, accessed on 3 September, 2015.

Berenice and Al. 2013. *Segregare costa. La spesa per i campi nomadi a Napoli, Milano, Roma*. http://www.lunaria.org/wp-content/uploads/2013/10/Segregare -costa_sintesicorretta_4ottobre2.pdf, accessed on 3 September, 2015.

Berger, P. L. and Luckmann, T. 1966. *The Social Construction of Reality: A Treatise in the Sociology of Knowledge*. New York: Garden City.

Bernaciak, M. 2014. *Social Dumping and Market Expansion in Europe*. Brussels: ETUI

Berndt, M. 2007. *Die Neue Europäische Sicherheitsarchitektur*. Wiesbaden: Verlag für Sozialwissenschaften.

Bettinetti, R., Croce, V. and Galassi, S. 2005. Ecological Risk Assessment for the Recent Case of DDT Pollution in Lake Maggiore (Northern Italy). *Water, Air and Soil Pollution*, 162, 1, 385–399.

Bigo, D., Bonelli, L. and Deltombe, T. (eds.) 2008. *Au nom du 11 Septembre*. Paris: la Découverte.

Bigo, D. and Tsoukala, A. (eds) 2008. *Terror, Insecurity and Liberty: Illiberal Practices of Liberal Regimes after 9/11*. London: Routledge.

Birdwell, J., Bartlett, J. and McDonnell, D. 2012. *Populism in Europe: Lega Nord*. DEMOS.

Birkmann, J. 2006. Measuring Vulnerability to Promote Disaster-Resilient Societies: Conceptual Frameworks and Definitions, in *Measuring Vulnerability to Natural Hazards: Towards Disaster Resilient Societies*, edited by J. Birkmann. Tokyo: United Nations – University Press, 9–54.

Bittner, E. 1990. *Aspects of Police Work*. Boston, MA: Northeastern University Press.

Blengino, C. 2007. Esercizio dell'azione penale e processi organizzativi: la selezione del crimine come output della procura in *Processi dei selezione del crimine. Procure della Repubblica e organizzazione giudiziaria*, edited by S. C. Milano: Giuffrè, 117–226.

Blengino, C. 2011. *Percorsi di giustizia. Efficienza del sistema penale e implementazione delle politiche criminali*. Torino: L'Harmattan Italia.

Blengino, C. and Torrente G. 2013. Nuovo Management pubblico e crimini d'impresa: il caso della procura di Torino. *Studi sulla questione criminale*, 1, 75–94.

Blumberg, A. S. 1967. The Practice of Law as Confidence Game: Organizational Cooptation of a Profession. *Law and Society Review*, 1, 15–39.

Blinkert, B. 1988. Kriminalität als Modernisierungsrisiko? Das ‚Hermes-Syndrom' der entwickelten Industriegesellschaften. *Soziale Welt*, 39, 4, 397–412.

Blok, A. 1973. *The Mafia of a Sicilian Village, 1860-1960: A Study of Violent Peasant Entrepreneurs*. New York: Harper & Row.

Bogna, P. and Maddalena, M. 2003. *Il giudice e i suoi limiti. Cittadini, magistrati e politica*. Rome, Bari: Laterza.

Bodin, P. and Wiman, B. 2004. Resilience and Other Stability Concepts in Ecology: Notes on Their Origin, Validity, and Usefulness. *ESS Bulletin*, 2, 33–43.

Boin, A., Comfort, L. K. and Demchak, C. C. 2010. The Rise of Resilience, in *Designing Resilience. Preparing for Extreme Events*, edited by L. K. Comfort, A. Boin and C. C. Demschak. Pittsburgh, University of Pittsburgh Press, 1–12.

Boltanski, L. 2008. Institutions et critique sociale. Une approche pragmatique de la domination. *Tracés. Revue de Sciences humaines*, Special issue. Présent et futurs de la critique, 17–43, http://traces.revues.org/2333, accessed on 3 September, 2015.

Boltanski, L. 2012. *Enigmes et Complots. Une enquête à propos d'enquêtes*. Paris: Gallimard.

Bolton, M. 1997. *Recovery for Whom? Social Conflict after the San Francisco Earthquake and Fire*, 1906–1915. PhD thesis. University of California, Davis.

Bonanni, E. 2009. *Lo Stato dimentica l'amianto killer. Diritti negati – Ricorso a Strasburgo*. Sesto san Giovanni: Comitato per la Difesa della Salute nei Luoghi di Lavoro e nel Territorio.

Bonanno, G. 2004. Loss, Trauma, and Human Resilience: Have We Underestimated the Human Capacity to Thrive after Extremely Aversive Events? *American Psychologist*, 59, 1, 20–28.

Bonatelli, B. 2005. *Percossa e denudata in pubblico*. http://www.meltingpot.org/articolo5364.html, accessed on 13 July, 2013.

Bonetti, P., Simoni, A. and Vitale, T. (eds) 2001. *La condizione giuridica di Rom e Sinti in Italia*. Milan: Giuffrè.

Bonß, W. 2011. Un-Sicherheit in der Moderne, in *Zivile Sicherheit*, edited by Zoche, P., Kaufmann, S. and R. Haverkamp. Bielefeld, transcript Verlag, 43–69.

Booth, K. 1991. Security and Emancipation. *Review of International Studies*, 17, 4, 313–326.

Borraz, O. and Le Galès, P. 2010. Urban Governance in Europe: The Government of What? *Metropoles*, 7.

Bosco, M. L., Varrica, D. and Dongarrà, G. 2005. Case Study: Inorganic Pollutants Associated with Particulate Matter from an Area Near a Petrochemical Plant. *Environmental Research*, 99, 1, 18–30.

Bosher, L. (ed.) 2008. *Hazards and the Built Environment. Attaining Built-In Resilience*. London: Routledge.

Bosworth, M. and Guild, M. 2008. Governing through Migration Control: Security and Citizenship in Britain. *The British Journal of Criminology*, 48, 6, 703–719.

Bottero, M. 2016. Klimt e la Giurisprudenza, in press.

Bottrell, D. 2013. Responsibilised Resilience? Reworking Neoliberal Social Policy Texts. *M/C Journal,* (online), 16, 5, http://journal.media-culture.org.au/index.php/mcjournal/article/viewArticle/708, accessed on 3 September, 2015.

Boudia, S. & Herny, E. eds. 2015. *La mondialisation des risques*, edited by, Presses Universitaires de Rennes

Bourdieu, P. 1986. La force du droit. Eléments pour une sociologie du champ juridique. *Actes de la recherche en Sciences Sociales*, 64, 3–19.

Bourdieu, P. (ed.) 1993. *La misère du monde*. Paris: Seuil.

Bourdieu, P. 1976. Le sens pratique. *Actes de la recherche en sciences sociales*. 2, 1, 43–86.

Bourdieu, P. and Wacquant, L. 1998. Sur les ruses de la raison impérialiste. *Actes de la Recherche en Sciences Sociales*, 121, 1, 109–118.

Brand, F. S. and Jax, K. 2007. Focusing the Meanings of resilience: Resilience as a Descriptive Concept and a Boundary Object. *Ecology and Society*, 12, 1, 23.

Brassard, A. 2003. Integrating the Planning Process and Second-Generation CPTED. *The CPTED-Journal*, 2, 1, 46–53.

Braudel, F. 1979. *Civilisation matérielle, économie et capitalisme XVe –XVIIIe siècle* (Tome 3. Le temps du Monde). Paris: Armand Colin.

Brennan, M. A. 2008. Conceptualizing Resiliency: An Interactional Perspective for Community and Youth Development. *Child Care in Practice*, 14, 1, 55–64.

Brodeur, J. P. 2003. *Les visages de la police. Pratiques et perceptions*. Montréal: Les Presses de l'Université de Montréal.

Broeders, D. and Engbersen, G. 2007. The Fight Against Illegal Migration Identification Policies and Immigrants' Counterstrategies. *American Behavioral Scientist*, 50, 1592–1609.

Brown, D. and Kulig, J. 1996/1997. The Concept of Resiliency: Theoretical Lessons from Community Research. *Health and Canadian Society*, 4, 1, 29–52.

Brown, P. L. 1995. Designs for a Land Of Bombs and Guns. *New York Times*, 28 May.

Brown, P. and Mikkelsen, E. 1990. *No Safe Place. Toxic Waste, Leukemia, and Community Action*. Berkeley, CA: University of California Press.

Bruneau, M., Chang, S., Eguchi, R., Lee, G., O'Rourke, T., Reinhorn, A., et al. 2003. A Framework to Quantitatively Assess and Enhance the Seismic Resilience of Communities. *Earthquake Spectra*, 19, 733–752

Brussett J., Croft, S. and Vaughan-Williams, N. 2013. Introduction: An Agenda for Resilience Research in Politics and International Relations. *Politics*, 33, 4, 221–228.

Bulley, D. 2013. Producing and Governing Community through Resilience. *Politics*, 33, 4, 265–275.

Bullinger, H.-J. and Röthlein, B. 2012. *Morgenstadt. Wie wir morgen leben*. München: Carl Hanser Verlag.

Bullion, C. 2013. Willkommen in der nächsten Runde der Asyldebatte. *Süddeutsche Zeitung*, 12 July.

Bürkner, H.-J. 2010. *Vulnerabilität und Resilienz – Forschungsstand und sozialwissenschaftliche Untersuchungsperspektiven*. Erkner: IRS- Institut für Regionalentwicklung und Strukturplanung.

Butler L., Morland, L. and Leskin, G. 2007. Psychological Resilience in the Face of Terrorism, in *Psychology of Terrorism*, edited by B. Bongar, L. Brown, L. Beutler, J. Breckenridge and P. Zimbardo. New York: Oxford University Press, 400–417.

Buzan, B. and Hansen, L. 2009. *The Evolution of International Security Studies*. Cambridge: Cambridge University Press.

Buzan, B., Wæver, O. and de Wilde, J. de 1998. *Security. A New Framework for Analysis*. Boulder, CO: Lynne Rienner Pub.

Caldeira, T. 2000. *City of Walls: Crime, Segregation and Citizenship in São Paolo*. Berkeley, CA: University of California Press.

Calvino, I. 1974. *Invisible Cities*. Orlando, FL: Harcourt Brace & Company.

Caponio, T. and Borkert, M. (eds) 2010. *The Local Dimension of Migration Policymaking*. Amsterdam: Amsterdam University Press.

Caporale, A. 2013. Napoli, la terra dei fuochi: aria infetta, rifiuti e tumori attorno al Vesuvio. *Il fatto quotidiano*, 9 November.

Carbonnier, G. 2015. Reason, Emotion, Compassion: Can Altruism Survive Professionalisation in the Humanitarian Sector? *Disasters*, 39, 2, 189–207.

Carpenter, S., Walker, B., Anderies, M. and Abel, N. 2001. From Metaphor to Measurement: Resilience of What to What? *Ecosystems*, 4, 765–781.

Cartuyvels, Y. and Mary, Ph. (eds) 1999. *L'Etat face à l'insécurité. Dérives politiques des années 90*. Bruxelles: Labor.

Centemeri, L. & Daumalin, X. eds. 2015 *Pollutions industrielles et espaces méditerranéens XVIIIe–XXIe siècle*, Paris: Karthala

Ceyhan, A. 2008. Technologization of Security: Management of Uncertainty and Risk in the Age of Biometrics. *Surveillance & Society*, 5, 2, 102–123.

Chandler, D. 2013. International Statebuilding and the Ideology of Resilience. *Politics,* 33, 4, 276–286.

Chantraine, G. and Salle, G. 2013. Pourquoi un dossier sur la délinquance en col blanc? *Champ pénal/Penal field,* (online journal), http://champpenal.revues.org/8555, accessed on 3 September, 2015.

Chevalier, L. 1958/1984. *Classes laborieuses et classes dangereuses*, Paris: Hachette.

Chiantera-Stutte, P. 2006. Razzismo, in *Lessico di biopolitica*, edited by R. Brandimarte, P. Chiantera-Stutte, P. Di Vittorio, O. Marzocca, O. Romano, A. Russo and A. Simone. Rome: Manifestolibri, 256.

Chivers, C. J. 2014. ISIS' Ammunition Is Shown to Have Origins in US and China, *New York Times*, 5 October. http://www.nytimes.com/2014/10/06/world/isis-ammunition-is-shown-to-have-origins-in-us-and-china.html, accessed on 3 September, 2015.

Christie, N. 1986. Suitable Enemies, in *Abolitionism. Towards a Non-repressive Approach to Crime*, edited by A. Bianchi and R. Van Svaaningen. Amsterdam: Free University Press.

Cigna, E., Ferrari, L., Gabrielli, G., Laini, M. and Suriano, M. 2014. *Esodati. Anche un disagio psicologico*. Roma: Ediesse.

Clarke, R. V. 1995. Situational Crime Prevention, in *Building a Safer Society. Strategic Approaches to Crime Prevention*, edited by M. Tonry and D. P. Farrington, Chicago: University of Chicago Press, 91–150.

Clarke, R. V. 1997. Introduction, in *Situational Crime Prevention. Successful Case Studies*, edited by R. V. Clarke. Guilderland, NY: Harrow and Heston, 2–43.

Clough Marinaro, I. 2009. Between Surveillance and Exile: Biopolitics and the Roma in Italy. *Bulletin of Italian Politics*, 1, 2, 265–287.

Coaffee, J. 2004. Rings of Steel, Rings of Concrete and Rings of Confidence: Designing out Terrorism in Central London pre and post September 11th. *International Journal of Urban and Regional Research*, 28, 1, 201–211.

Coaffee, J. 2013. Rescaling and Responsibilising the Politics of Urban Resilience: From National Security to Local Place-Making. *Politics*, 334, 7, 240–252.

Coaffee, J., Murkami-Wood, D. and Rogers, P. 2008. *The Everyday Resilience of the City: How Cities Respond to Terrorism and Disaster*. London: Palgrave/ Macmillian

Cohen, J. E. and Hamman, J. A. 2005. Presidential Ideology and the Public Mood: 1956–1994, in *In the Public Domain: Presidents And the Challenges of Public Leadership*, edited by L. Cox Han and D. J. Heith. Albany: SUNY University Press, 141–162.

Cohen, L. E. and Felson, M. 1979. Social Change and Crime Rate Trends: A Routine Activity Approach. *American Sociological Review,* 44, 4, 588–607.

Cohen, S. 1972. *Folk Devils and Moral Panics: The Creation of the Mods and Rockers*. London: McGibbon and Kee (3rd edn 2002. Routledge).

Cole, S. A. 2001. *Suspect Identities. A History of Fingerprinting and Criminal Identification*. Cambridge, MA: Harvard University Press.

Coles, E. and Buckle, P. 2004. Developing Community Resilience as a Foundation for Effective Disaster Recovery. *The Australian Journal of Emergency Management*, 19, 6–15.

Collective. 2006. CASE Critical Approaches to Security in Europe: A Networked Manifesto, *Security Dialogue*, 37, December, 443–487. http://www.eui.eu/Personal/Researchers/bueger/Documents/Case%20manifesto%20as%20published.pdf, accessed on 3 September, 2015.

Comba, P., Pirastu, R., Conti, S., De Santis, M., Iavarone, I., Marsili, G., Mincuzzi, A., Minelli,G., Manno, V., Minerba, S., Musumeci, L., Rashid, I., Soggiu, E. and Zona, A. 2012. Environment and Health in Taranto. Southern-Italy: Epidemiological Studies and Public Health Recommendations. *Epidemiologia & Prevenzione*, 36, 6, 305–320.

Combes, M. 2014 Sommet climat de l'ONU : vers une agriculture sous l'emprise de la finance et des multinationales? http://blogs.mediapart.fr/blog/maxime-combes/100914/sommet-climat-de-lonu-vers-une-agriculture-sous-lemprise-de-la-finance-et-des-multinationales, accessed on 5 September, 2015.

Comby, J.B. 2015. *La question climatique. Genèse et dépolitisation d'un problème public.* Paris: Raison d'Agir.

*Conflitti globali* 2007. Un mondo di controlli, 5, http://www.agenziax.it/wp-content/uploads/2013/03/conflitti-globali-5.pdf, accessed on 3 September, 2015.

Connor, K. M. and Davidson, J. R. T. 2003. Development of a New Resilience Scale: The Connor-Davidson Resilience Scale (CD-RISC). *Depression and Anxiety*, 18, 2, 76–82.

Cornish, D. B. and Clarke, R. V. (eds) 1986. *The Reasoning Criminal. Rational Choice Perspectives on Offending*. New York, NY: Springer-Verlag.

Cortelazzo, M. and Zolli, P. 1988. *Dizionario etimologico della lingua italiana*. Bologna: Zanichelli.
Cotterrell, R. 1997. The concept of legal culture, in *Comparing Legal Cultures*, edited by D. Nelken, Darmouth: Aldershot, 13–31.
Cozens, P. M., Pascoe, T. and Hillier, D. 2007. Critically Reviewing the Theory and Practice of Secured-by-Design for Residential New-build Housing in Britain, in *Burglary*, edited by R. Mawby. Aldershot: Ashgate, 345–361.
Crenshaw, K.W. 1991. Mapping the Margins: Intersectionality, Identity Politics, and Violence Against Women of Colour. *Stanford Law Review*, 43, 6, 1241–1299.
Crowe, T. D. 2000. *Crime Prevention through Environmental Design. Applications of Architectural Design and Space Management Concepts*. Woburn, MA: Butterworth–Heinemann.
Crozier, M. and Friedberg, E. 1997. *L'Acteur et le Système: les contraintes de l'action collective*. Paris: Seuil.
Cusson, M., Lemieux, F. and Dupont, B. (eds) 2007. *Traité de la sécurité intérieure*. Montréal: Presses de l'Université de Montréal.
Daase, C. 2010. Wandel der Sicherheitskultur, in *Aus Politik und Zeitgeschichte* 50, 13, 9–16. http://www.bpb.de/system/files/pdf/1EH2QT.pdf, accessed 03/09/2015
Daase, C. 2011. Der Wandel der Sicherheitskultur, in *Zivile Sicherheit*, edited by P. Zoche, S. Kaufmann and R. Haverkamp. Bielefeld: Transcript Verlag.
Dal Lago, A. and Palidda, S. (eds) 2010. *Conflict, Security and the Reshaping of Society. The Civilization of War*. Abingdon: Routledge.
Dalferth, S. 2008. Europäische Kriminalpolitik und die EU. Freiheit und Sicherheit in der Europäisierung, in *Globaler Terrorismus und Europa*, edited by P. Nitschke. Wiesbaden: VS Verlag für Sozialwissenschaften.
Davis, D. E. 2012. *Urban Resilience in Situation of Chronic Violence*. http://web.mit.edu/cis/urban_resilience_toolkit.pdf, accessed on 3 September, 2015.
Davis, M. 1990. *City of Quartz: Excavating the Future in Los Angeles*. New York: Verso.
Davis, M. 1998. *Ecology of Fear: Los Angeles and the Imagination of Disaster*. New York: Metropolitan Books.
De Biase, M. 2015. Fires of Pianura: The Antidump Struggle in the Western Outskirts of Naples. *Capitalism Nature Socialism*, 26, 1, 39–58.
De Giorgi, A. 2000. *Zero Tolleranza. Strategie e pratiche della società di controllo*. Roma: Derive Approdi.
De Marchi, B. 1991. La sociologia dei disastri, teorie ed esperienze, *Quaderni dell'ISIG*, 4, 12-43.
Dechenaud, D. 2010. Nouvelles menaces sur le droit pénal des affaires. Libres propos au sujet de l'avant-projet du futur Code de procédure pénale. *Droit Pénal*, 6 June.
Decker, O., Kieess, J. and Brähler, E. 2015. *Die stabilisierte Mitte – Rechtsextreme Einstellung in Deutschland 2014*. Leipzig: Universität Leipzig.

Deleuze, G. 1992. *Postscript on the Societies of Control*, Cambridge, MA: MIT Press, 3–7.

della Porta, D. 2004. *Comitati di cittadini e democrazia urbana.* Soveria Mannelli: Rubbettino.

della Porta, D. and Vannucci, A. 2012. *The Hidden Order of Corruption.* Farnham: Ashgate.

Delmas- Marty, M. 2004. *Le flou du droit*. Paris: Puf. (1st edition, 1986).

Delumeau, J. 1978. *La peur en Occident XIVe-XVIIIe siècles. Une cité assiégée.* Paris: Fayard

Delumeau, J. 1989. *Rassurer et protéger. Le sentiment de sécurité dans l'Occident d'autrefois*. Paris: Fayard.

Delumeau, J. 1990. Le sentiment d'insécurité dans l'histoire, *Les Cahiers de la Sécurité Intérieure*, 1, 19–26.

Demurger, A. 2002. *Jacques de Molay: le crépuscule des templiers*, Paris: Payot.

Demurger, A. 2005. *Les Templiers. Une chevalerie chrétienne au Moyen Âge.* Paris: Le Seuil.

Deveau, J. M. 2010. *Le retour de l'esclavage au XXIe siècle*. Paris : Karthala.

Di Federico, G., Fabri, M., Carnevali D. and Contini, D. 1994. *Organizzazione e gestione degli uffici giudiziari. Il caso di una Procura della Repubblica presso il tribunale*, Working Papers, IRSIG-CNR.

Di Luzio, G. 2003. *I fantasmi dell'Enichem*. Roma: Baldini Castaldi Dalai.

Di Palma, G. 2013. *The Modern State Subverted. Risk and the Deconstruction of Solidarity.* Colchester: Ecpr Press Essays.

Dillon, M. and Neal, A. N. 2009. *The Liberal Way of War. Killing to Make Life Live*. London: Routledge.

Dodson, F. and Hammar, S. P. (eds) 2006. *Asbestos: Risk Assessment, Epidemiology, and Health Effects*, Boca Raton, FL: Taylor & Francis Group.

Drabek, T. E. 1986. *Human System Responses to Disaster*. New York: Springer-Verlag

Droit R. P. 2004. *Foucault. Entretiens*. Paris: Odile Jacob.

Dubber, M. and Valverde, M. (eds) 2006. *The New Police Science: The Police Power in Domestic and International Governance.* Stanford, CA: Stanford University Press.

Duwe, G., Donnay, W. and Tewksbury, R. 2008. Does Residential Proximity Matter? A Geographic Analysis of Sex Offense Recidivism. *Criminal Justice and Behavior*, 35, 4, 484–504.

Dynes, R. R. 1970. *Organized Behavior in Disaster*. Lexington, MA: Lexington Books.

Ehrenreich B. and Hochschild R. A. 2002. *Global Woman. Nannies, Maids, and Sex Workers in the New Economy*. New York: Metropolitan Press.

Eisenberg, A. and Spinner, H.J. 2005. *Minorities within Minorities. Equality, Rights and Diversity*. Cambridge: Cambridge University Press.

Egeland, B., Carlson, E. and Sroufe, L. A. 1993. Resilience as Process. *Development and Psychopathology*, 5, 517–528.

Ekblom, P. 2012. *Design against Crime*. Boulder, CO: Lynne Rienner.

Engbersen, G. and Broeders, D. 2009. The State versus the Alien: Immigration Control and Strategies of Irregular Immigrants. *West European Politics*, 32, 5, 867–885.

EngineerSprey.2012.onF35.https://www.youtube.com/watch?v=BPqbN7OooRg, accessed on 3 September, 2015.

Esposito, R. 2002. *Immunitas. Protezione e negazione della vita.* Torino: Einaudi.

Fabri, M. 1997. Discrezionalità e modalità di azione del pubblico ministero nel procedimento penale, *Polis*, 11, 171–192.

Fanon, F. 2007. *I dannati della terra.* Torino: Einaudi.

Fassin, D. and Fassin, E. (eds) 2006. *De la question sociale à la question raciale? Représenter la société française*, Paris: La Découverte.

Fassin, D. 2009a. Another Politics of Life is Possible. *Theory Culture Society*, 26, 5, 44–60.

Fassin, D. 2009b. *Compassion and Repression: The Moral Economy of Immigration Policies in France.* http://www.sss.ias.edu/files/pdfs/Fassin/Compassion-and-repression.pdf, accessed on 3 September, 2015.

Fassin, D. 2011a. *Humanitarian Reason. A Moral History of the Present.* Berkeley, CA: University of California Press.

Fassin, D. 2011b. *La force de l'ordre, Une anthropologie de la police des quartiers.* Paris: Seuil (2013, *Enforcing Order. Ethnography of Urban Policing*, Cambridge: Polity Press).

Feeley, M. and Simon, J. 1994. Actuarial Justice: The Emerging New Criminal Law, in *The Futures of Criminology*, edited by D. Nelken, London: Oxford University Press, 173–201.

Feige, E.L. (dir.) 1989. *The Underground Economies, Tax Evasion and Information Distortion.* Cambridge : Cambridge University Press.

Fekete, A., Shen, X., Birkmann, J. and Bogardi, J. J. 2011. Umwelt als Gefährdung – Wahrheit und Wahrnehmung, in *Zivile Sicherheit. Gesellschaftliche Dimensionen gegenwärtiger Sicherheitspolitiken*, edited by P. Zoche, S. Kaufmann and R. Haverkamp. Bielefeld: Transcript Verlag, 91–109.

Felgentreff, C., Kuhlicke, C. and Westholt, F. 2012. *Naturereignisse und Sozialkatastrophen.* Berlin: Spektrum Akademischer Verlag.

Ferrajoli L. 2001 (and 2008). *Diritti fondamentali. Un dibattito teorico*, Rome-Bari: Laterza.

Ferrajoli. L. 2008. Il 'diritto penale del nemico': un'abdicazione della ragione, in *Legalità penale e crisi del diritto*, edited by Bernardi, A., Pastore, B. and Pugiotto, A. Milan: Giuffrè, 161–178.

Ferrari, L. 2010. *L'ascesa dell'individualismo economico.* Piacenza: Casa Editrice Vicolo del Pavone.

Ferrari, L. 2014. *Alle fonti del kafkiano. Lavoro e individualismo in Franz Kafka.* Piacenza: Casa Editrice Vicolo del Pavone.

Ferrari, L. and Veglio, O. (eds) 2006. *Donne e uomini nel mercato del lavoro atipico. La dimensione psicologica e di genere del lavoro precario e flessibile.* Roma: Franco Angeli.

Ferrari, L., Ferrari, L. and Venini, G. 2006. Crisi economiche estreme e psicologia: il caso dell'Argentina. *Ricerche di Psicologia*, 29, 4, 127–157.

Ferrari Bravo, L. 2007. Forma dello stato e sottosviluppo, in *Stato e sottosviluppo. Il caso del Mezzogiorno italiano*, edited by L. Ferrari Bravo and A. Serafini. Verona: Ombrecorte, 23–107.

Ficher, H. Collateral Damage, Crimes of War. http://www.crimesofwar.org/a-z-guide/392/, accessed on 3 September, 2015.

Fierro, E. 2008. In ostaggio dell'amianto. Isochimica di Avellino. *L'Unità*, 2 November.

Fischer, N. and Spire, A. (eds) 2009. L'Etat face aux illégalismes. *Politix,* 3, 87, 3.20

Fischoff, B et al. 1978. How Safe is Safe Enough: a Psychometric Study of Attitude Towards Technological Risks and Benefits. *Policy Sciences*, 9, 127–152.

Fletcher, D. and Sarkar, M. 2013. Psychological Resilience: A Review and Critique of Definitions. *European Psychologist*, 18,1, 12–23.

Flores, M. 2005. *Tutta la violenza di un secolo*. Milan: Feltrinelli

Florini, A. and Simmons, P. J. 1998. *The New Security Thinking: A Review of the North American Literature*. New York: Rockefeller Bros. Fund.

Fo, D. and Cazacu, F. 2015. *Un uomo bruciato vivo. Storia di Ion Cazacu*. Milan: Chiarelettere.

Fois, F. and Forino, G. 2014. The Self-built Ecovillage in L'Aquila, Italy: Community Resilience as a Grassroots Response to Environmental Shock. *Disasters*, 38, 4, 719–739.

Fortier, A. M. 2010. Proximity by Design? Affective Citizenship and the Management of Unease. *Citizenship Studies*, 14, 1, 17–30.

Foucault, M. 1971. Orders of Discourse. *Social Science Information* 10, 7–30 (orig. 1976; Gallimard 2001).

Foucault, M. 1977. *Cours du 7 janvier 1976*, in *Microfisica del potere: interventi politici*, edited by A. Fontana and P. Pasquino, Turin Einaudi, 163–177 (also in *Dits et écrit II. 1976–1988*, Paris: Gallimard, 2001).

Foucault, M. 1979. *Discipline and Punish: The Birth of the Prison*. New York: Random House.

Foucault, M. 1988a. *Technologies of the Self: A Seminar with Michel Foucault*. Amherst: The University of Massachusetts Press.

Foucault, M. 1988b. *La volontà di sapere. Storia della sessualità*, 1. Milano: Feltrinelli.

Foucault, M. 1994. Eterotopia, in *Eterotopia. Luoghi e non-luoghi metropolitan*, edited by M. Foucalt. Milano: Mimesis, 9–20.

Foucault, M. 1999. *Le pouvoir psychiatrique*, *Cours au collège de France. 1973-1974*, edited by J. Lagrange, F. Ewald and A. Fontana. Paris: Gallimard/ Seuil, Hautes Etudes.

Foucault, M. 2003. *Society Must Be Defended, Lectures at the Collège de France, 1975–1976*. London: Penguin Books.

Foucault, M. 2004a/2010. *Naissance de la biopolitique. Cours au Collège de France, 1978–1979*. Paris: Seuil/Gallimard. (*The birth of biopolitics: lectures at the Collège de France, 1978-1979*. New York: Picador).

Foucault, M. 2004b/2009. *Sécurité, territoire, population. Cours au College de France, 1977–1978*. Paris: Gallimard-Seuil (2009, *Security, Territory, Population (Lectures at the Collège De France 1977-1978)*, ed. by Michel Senellart. New York: Picador-Palgrave Macmillan.

Foucault, M. 2008. *Le gouvernement de soi et des autres*. Paris: Hautes Etudes/ Gallimard/Seuil.

Foucault, M. 2009. *Le gouvernement de soi et des autres – tome 2, Le courage de la vérité. Cours au Collège de France (1983–1984)*. Paris: Seuil-Gallimard.

Foucault, M. 2012. *Mal faire, dire vrai. Fonction de l'aveu en justice, Cours de Louvain 1981*, edited by F. Brion and B. Harcourt. Louvain: Presses Universitaires de Louvain (*Wrong-Doing, Truth-Telling. The Function of Avowal in Justice*, edited by F. Brion and B. E. Harcourt. Chicago: University of Chicago Press).

Fourier, P. 2014. The Neoliberal/Neurotic Citizen and Security as Discourse. *Critical Studies on Security*, 2, 3, 309–322.

Fradkin P. 2005. *The Great Earthquake and Firestorms of l906: How San Francisco Nearly Destroyed Itself*. Berkeley, CA: University of California Press.

Freire, P. 1970. *Pedagogy of the Oppressed*. New York: Continuum.

Fremlova, L., Georgescu, M., Hera, G., L.G. Marin and Miletic, G. 2014. *Barabaripen – Young Roma speak about discrimination*. http://enter.coe.int/roma/Roma-Youth-Action-Plan/Areas-of-action/Multiple-discrimination/Barabaripen-Young-Roma-speak-about-discrimination, accessed on 3 September, 2015.

Friedberg, E. 1996. *Power and Rules. The Organizational Dynamics of Collective Action*. London: Ashgate.

Fritzsche, A. F. 1986. *Wie sicher leben wir? Risikobeurteilung und Risikobewältigung in unserer Gesellschaft*. Köln: Verlag TÜV Rheinland.

Fuller J. F. C. 1945. *Armament and History: The Influence of Armament on History from the Dawn of Classical Warfare to the End of the Second World War*. London: C. Scribner's Sons.

Furedi, F. 1997. *Culture of Fear: Risk Taking and the Morality of Low Expectation*. London: Continuum International Publishing Group.

Furedi, F. 1999. *Courting Mistrust: The Hidden Growth of a Culture of Litigation in Britain*. London: Centre for Policy Studies.

Furedi, F. 2003. *Therapy Culture: Cultivating Vulnerability in an Uncertain Age*. London: Routledge.

Furedi, F. 2005. *Politics of Fear. Beyond Left and Right*. New York: Continuum International Publishing Group.

Gaissmaier, W. and Gigerenzer, G. 2012. 9/11, act II: A Fine-grained Analysis of Regional Variations in Traffic Fatalities in the Aftermath of the Terrorist Attacks. *Psychological Science*, 23, 12, 1449–1454.

Gallino, L. 2011. *Finanzcapitalismo. La civiltà del denaro in crisi.* Torino: Einaudi.
Ganor, M. and Ben-Lavy, Y. 2003. Community Resilience: Lessons Derived from Gilo under Fire. *Journal of Jewish Communal Service*, Winter/Spring, 105–108.
Garapon, A. 2001. *Bien juger. Essai sur le rituel judiciaire*. Paris: O. Jacob.
Garapon, A. 2012. *Lo stato minimo. Il neoliberalismo e la giustizia.* Milan: Raffaello Cortina (orig. 2010. *La raison du mondre Etat. Le neolliberalisme et la justice*. Paris: Odile Jacob).
Garapon, A. and Salas D. 1996. *La repubblica penale*. Macerata: Liberlibri.
Garland, D. (ed.) 2001. *Mass Imprisonment. Social Causes and Consequences.* London and New York: Sage Publications.
Garland, D. 2001. *The Culture of Control. Crime and Social Order in Contemporary Society*. Chicago, IL: The University of Chicago Press.
Garmezy, N. 1991. Resiliency and Vulnerability to Adverse Developmental Outcomes Associated with Poverty. *American Behavioural Scientist*, 34, 4, 416–430.
Garofalo, R. 1905/1887. *La Criminologie*. Paris: Félix Lacan, 1905.
Gerhold, L. 2011. *Dokumentation des Workshops „Kriminalität – alte und neue Herausforderungen für die Sicherheit'*. Berlin: Forschungsforum Öffentliche Sicherheit.
Ghelli, F. 2013. *Das Flüchtlingselend kommt Italiens Rüstungsindustrie gelegen.* http://www.zeit.de/politik/ausland/2013-10/italien-fluechtlinge-mare-nostrum, accessed on 3 September, 2015.
Giddens, A. 1999. *Runaway World. How Globalization is Reshaping our Lives.* London: Profile Books.
Giebel, D. 2012. *Integrierte Sicherheitskommunikation. Zur Herausbildung von Unsicherheitsbewältigungskompetenzen durch und in Sicherheitskommunikation.* Berlin. Münster: LIT.
Gigerenzer, G. 2004. Dread Risk, September 11, and Fatal Traffic Accidents. *Psychological Science*, 15, 4, 286–287.
Giordano, A. and Tarro, G. 2012. *Campania, Terra di veleni.* Napoli: Denarolibri.
Giroux, H. 2006. *Stormy Weather: Katrina and the Politics of Disposability.* Boulder, CO: Paradigm Publishers.
Giudice F., Roux M., MoulierBoutang Y. and Ruellan D. 1989. *Têtes de Turcs en France*. Paris: La Découverte.
Glassner, B. 1999. *The Culture of Fear. Why Americans are Afraid of the Wrong Things*. New York: Basic Books.
Godefroy, T. 2010. La délinquance économique et financière serait-elle en voie de disparition? *Après-demain*, 16, 31–34.
Godschalk, D. R. 2002. *Urban Hazard Mitigation: Creating Resilient Cities.* Paper for Presentation at the Urban Hazards Forum, John Jay College, City University of New York, January 22–24.
Godschaulk, D. R. 2003. Urban Hazard iMtigation: Creating Resilient Cities. *Natural Hazards Review*, 4, 3, 36–142.

Goffman, E. 1959. *The Presentation of Self in Everyday Life*, University of Edinburgh Social Sciences Research Centre. New York: Anchor Books Edition.

Goffman, E. 1974. *Frame Analysis. An Essay of the Organization of Experience*. London: Harper and Row.

Goold, B. Loader, I. and Thumala, A. 2010. Consuming Security? Tools for a Sociology of Security Consumption. *Theoretical Criminology*, 14, 1, 3–30.

Gordon, C. 1991. Governmental Rationality: An Introduction, in *The Foucault Effect: Studies in Governmentality*, edited by G. Burchell, C. Gordon and P. Miller. London: Harvester/Wheatsheaf, 1–51.

Gordon, J. E. 1978. *Structures*. Harmondsworth: Penguin Books

Graham, B. 2010. *Making Assessment Matter*. London: Continuum International Publishing Group.

Graham, S. and Wood, D. 2003. Digitizing Surveillance: Categorization, Space, Inequality. *Critical Social Policy*, 23, 2, 227–248.

Gramsci, A. 1975. *Gli intellettuali e l'organizzazione della cultura*. Turin: Editori Riuniti.

Greer, J. and Bruno K. 1996. *Greenwash: The Reality behind Corporate Environmentalism*. New York: Apex Press.

Griffiths, J. 1999. The Social Working of Anti-Discrimination Law, in *Non Discrimination Law: Comparative Perspectives*, edited by T. Loenen and P. R. Rodrigues. The Hague: Kluwer Law International, 313–332.

Griffiths, J. 2003. *The Social Working of Legal Rules*, 35, 48, 1-84, http://www.tandfonline.com/doi/abs/10.1080/07329113.2003.10756567?journalCode=rjlp20#.VehQzLztmko, accessed on 3 September, 2015.

Groenemeyer, A. 2001. Von der Sünde zum Risiko? – Bilder abweichenden Verhaltens und die Politik sozialer Probleme am Ende des Rehabilitationsideals. *Soziale Probleme*, 12, 1–2, 146–182.

Groenemeyer, A. 2010. Wege der Sicherheitsgesellschaft. Transformationen der Konstruktion und Regulierung von Unsicherheiten, in *Wege der Sicherheitsgesellschaft. Gesellschaftliche Transformationen der Konstruktion und Regulierung innerer Unsicherheiten*, edited by A. Groenemeyer. Wiesbaden: VS Verlag, 7–19.

Gu, Q. and Day, C. 2007. Teachers' Resilience: A Necessary Condition for Effectiveness. *Teaching and Teacher Education*, 23, 1302–1316.

Guadagnucci, L. 2015. *sTortura. Perché l'Italia non sa punire la tortura ed è incapace di una riforma democratica della forze di polizia*, Altreconomia. http://www.altreconomia.it/stortura/sTortura-1.pdf, accessed on 3 September, 2015.

Gunderson, L. 2000. Ecological Resilience – in Theory and Application. *Annual Review of Ecology and Systematics*, 31, 425–429.

Hackworth, J. 2007. *The Neoliberal City: Governance, Ideology, and Development in American Urbanism*. Ithaca, NY: Cornell University Press.

Hallenbeck, W. H. and Hesse, C. S. 1977. A Review of the Health Effects of Ingested Asbestos. *Reviews on Environmental Health*, 2, 3, 157–166.

Hamilton, C. 2010. *Requiem for a species. Why we resist the truth about a climate change*, London, Earthscan.

Hanak, G., Stehr, J. and Steinert, H. 1989. Ärgernisse *und Lebenskatastrophen. Über den alltäglichen Umgang mit Kriminalität*. Bielefeld: AJZ-Verlag.

Handmer, J. and Dovers, S. 1996. A Typology of Resilience: Rethinking Institutions for Sustainable Development. *Industrial and Environmental Crisis Quarterly*, 9, 4, 482–511.

Harcourt, B. 2007. *Against Prediction: Profiling, Policing and Punishing in an Actuarial Age*. Chicago, IL: The University of Chicago Press.

Hardt, M. and Negri, A. 2000. *Empire*. Cambridge, MA and London: Harvard University Press.

Harling, P. 2014. IS back in business, *Le Monde Diplomatique*. http://mondediplo.com/2014/09/04islamicstate, accessed on 3 September, 2015.

Harvey, D. 1991. *The Condition of Postmodernity: An Enquiry into the Origins of Cultural Change*. London: Wiley-Blackwell.

Harvey, D. 2000. Reinventing Geography. *New Left Review*, 4, 75–97.

Haverkamp, R. 2012. Gefühlte Sicherheiten und Sicherheitsgefährdungen – Barometer Sicherheit in Deutschland BaSiD, in *Auftrag der Deutschen Stiftung für Verbrechensverhütung und Straffälligenhilfe DVS*, edited by H.-J. Kerner and E. Marks. Hannover: Internetdokumentation des Deutschen Präventionstages. http://www.praeventionstag.de/html/GetDokumentation.cms?XID=1141, accessed on 3 September, 2015.

Haverkamp, R., Schwedler, A. and Wößner, G. 2012. Die elektronische Aufsicht von als gefährlich eingeschätzten Entlassenen. *Recht & Psychiatrie*, 30 1, 9–20.

Heers, J. 2008. *Jacques Cœur*. Paris: Perrin.

Heilmann, E. 2011. Video Surveillance and Security Policy in France: From Regulation to Widespread Acceptance. *Information Polity*, 16, 4, 369–377.

Heilmann, E. 2012. *Parlons sécurité en 30 questions*. Paris: La Documentation française.

Henderson, A. D. 2005. *Reconstructing Home: Gender, Disaster Relief, and Social Life after the San Francisco Earthquake and Fire*, 1906–1915. PhD thesis. Stanford, CA: Stanford University Press.

Herbert, S. and Brown, E. 2006. Conceptions of Space and Crime in the Punitive Neoliberal City. *Antipode*, 38, 4, 755–777.

Hester, S. and Eglin, P. 1992. *A Sociology of Crime*. London & New York: Routledge.

Hitthaler, A. 2011. Wieder ein Modewort – Resilienz. *PlanerIn*, 5, 43–44.

Hobbes, Th. 1881. *Leviathan*. Oxford: James Thornton.

Hobsbawm, E. J. 1994. *The Age of Extremes: The Short Twentieth Century, 1914–1991*. London: Michael Joseph.

Hobsbawm, E. 2007. *Globalization, Democracy and Terrorism*. London: Little, Brown

Holling, C. S. 1973. Resilience and Stability of Ecological Systems. *Annual Review of Ecology and Systematics*, 4, 1–23.

Hörnqvist, M. 2004. The Birth of Public Order Policy. *Race & Class*, 46, 1, 30–52.

Hugo, V. 2002/1864. *Océan, Oeuvres complètes*. Paris: Laffont.

Hummelsheim, D. and Oberwittler, D. 2014. *Festschrift für Baldo Blinkert*, edited by H.-J. Albrecht, H. Hoch and P. Zoche. Münster: LIT-Verlag.

Hutta, J. S. 2009. Geographies of Geborgenheit: Beyond Feelings of Safety and the Fear of Crime. *Environment & Planning*, 27, 251–273.

Huysmans, J. 2000. The European Union and the Securitization of Migration. *Journal of Common Market Studies*, 38, 5, 751–777.

Hytten, E. and Marchioni, M. 1970. *Industrializzazione senza sviluppo. Gela: una storia meridionale*. Milan: Franco Angeli.

Iavarone, I., Castellano, G,, Martinelli, W., Lerna, A., Suma, G. and Conversano, M. 2009. *Ecological and Human Biomonitoring in Taranto, an Italian Contaminated Site*. ISEE 21st Annual Conference. Poster Presentation. ISEE-0518. Dublin, Ireland, August 25–29.

Ignatieff, M. 1978. *A Just Measure of Pain: The Penitentiary in the Industrial Revolution, 1750–1850*. New York: Pantheon Books.

Innes, M. and Jones, V. 2006. *Neighbourhood Security and Urban Change. Risk, Resilience and Recovery*. York: Joseph Rowntree Foundation.

Institute for Economics and Peace. 2015. Peace and Corruption Report, http://www.visionofhumanity.org/#page/news/1162, accessed on 4 September, 2015.

Introna, L. D. and Wood, D. 2004. Picturing Algorithmic Surveillance: The Politics of Facial Recognition Systems. *Surveillance & Society*, 2, 2/3, 177–198.

IRPET 2013. Prato. Il ruolo economico della comunità cinese. Firenze: IRPET. http://www.irpet.it/index.php?page=pubblicazione&pubblicazione_id=487, accessed on 7 September, 2015.

IRPET 2015. Relazioni Locali e Transnazionali delle Imprese Cinesi di Prato e loro Contributo all'economia della Provincia, Firenze: IRPET. http://www.irpet.it/storage/pubblicazioneallegato/542_Volume%20Cina%20con%20copertina.pdf, accessed on 7 September, 2015.

Isbasoiu, G.-M. 2008. *Industrial Clusters and Regional Development. The Case of Timisoara and Montebelluna*. ICFAI University Press.

ISFOL, 2012. *Dimensioni e caratteristiche del lavoro sommerso/irregolare in agricoltura*. Rome: ISFOL

Jean, J. P. 2008. *Le système pénal*. Paris: La Découverte.

Jeffery, C. R. and Zahm, D. L. 1993. Crime Prevention through Environmental Design, Opportunity Theory, and Rational Choice Models, in *Routine Activity and Rational Choice*, edited by R. V. Clarke and M. Felson. New Brunswick, NJ: Transaction Publishers, 323–350.

Johnson, C. (ed.) 2011. *The Neoliberal Deluge: Hurricane Katrina, Late Capitalism, and the Remaking of New Orleans*. Minneapolis, MN: University of Minnesota Press.

Johnson, S. 2010. Prospective Crime Mapping, in *Encyclopedia of Victimology and Crime Prevention*, edited by B. Fisher and S. Lab. Thousand Oaks, CA: Sage Publictions, 685–689.

Johnson, S., Kaufmann, D. and Zoido-Lobaton, P. 1998. Regulatory Discretion and the Unofficial Economy. *American Economic Review*, 88, 2, 387.

Joseph, J. 2013. Resilience in the UK and French Security Strategy: An Anglo-Saxon Bias? *Politics*, 33, 4, 253–264.

Josling, T. and Tangermann, S. 2014. *Agriculture, Food and the TTIP: Possibilities and Pitfalls*. http://www.ceps.eu/system/files/SP%20No_99%20TTIP%20and%20Agriculture%20%281%29.pdf, accessed on 3 September, 2015.

Joxe, A. 1991. *Voyage aux sources de la guerre*. Paris: PUF.

Joxe, A. 2002. *Empire of Disorder*. Cambridge, MA: MIT/Semiotext(e)

Joxe, A. 2012. *Les guerres de l'empire global. Spéculations financières, guerres robotiques, résistance démocratique*. Paris: La Découverte

Joxe, A. 2015. Alain JOXE la Crise vers une euroguerre mondiale, comme d'habitude? http://blogs.mediapart.fr/blog/alain-joxe-debat-strategique/190515/alain-joxe-la-crise-vers-une-euroguerre-mondiale-comme-dhabitude, accessed on 5 September, 2015.

Kagan, E. and Jacobson, R. J. 1983. Lymphoid and Plasma Cell Malignancies: Asbestos-related Disorders of Long Latency. *American Journal of Clinical Patho*logy, 80, 1, 14–20.

Kaldor, M. and Martin, M. (eds) 2010. *The European Union and Human Security. External Interventions and Missions*. Abingdon: Routledge.

Kaldor, M., Martin, M. and Selchow, S. 2007. Human Security: A New Strategic Narrative for Europe. *International Affairs*, 83, 2, 273–288.

Kapuy, K. 2004. The Relevance of the Local Level for Human Security. *Human Security Perspectives Journal*, 1, 1, 1–6.

Kaufmann, F.-X. 2012. *Sicherheit als soziologisches und sozialpolitisches Problem*. Untersuchungen zu einer Wertidee hochdifferenzierter Gesellschaften. Unveränderter Nachdruck der 1973 im Ferdinand Enke Verlag, Stuttgart, erschienenen Auflage Münster.

Kaufmann, S. 2013. Die Stadt im Zeichen ziviler Sicherheit, in *Verfassungsstaatlichkeit im Wandel. Festschrift für Thomas Würtenberger zum 70. Geburtstag*, edited by D. Heckmann, R. P. Schenke and G. Sydow. Berlin: Duncker and Humblot, 1011–1027.

Kaufmann, S. and Blum, S. 2013. Vulnerabilität und Resilienz: Zum Wandern von Ideen in der Umwelt- und Sicherheitsdiskussion, in *Unberechenbare Umwelt. Zum Umgang mit Unsicherheit und Nicht-Wissen*, edited by R. von Detten, F. Faber and M. Bemmann. Wiesbaden: VS Verlag, 91–120.

Kimhi, S. and Shamai, M. 2004. Community Resilience and the Impact of Stress: Adult Response to Israel's Withdrawal from Lebanon. *Journal of Community Psychology*, 32, 439–451.

Kitsuse, J. I. and Cicourel, A. V. 1963. A Note on the Uses of Official Statistics. *Social Problems*, 11, 2, 131–139.

Klein, N. 2008. *The Shock Doctrine: The Rise of Disaster Capitalism*. London: Penguin.

Klein, R. J. T., Nicholls, R. J. and Thomalla, F. 2003. Resilience to Natural Hazards: How Useful is This Concept? *Environmental Hazards*, 5, 35–45.

Kloosterman, R. 2010. Matching Opportunities with Resources: a Framework for Analysing Migrant Entrepreneurship from a Mixed Embeddedness Perspective. *Entrepreneurship & Regional Development*, 22, 1, 25–45.

Kocze, A. 2009. *Missing Intersectionality. Race/Ethnicity, Gender, and Class in Current Research and Policies on Romani Women in Europe*. Budapest: CEU-Centre for Policy Studies.

Kreimer, A. M., Arnold, M. and Carlin, A. (eds) 2003. *Building Safer Cities: The Future of Disaster.* Washington, DC: The International Bank for Reconstruction and Development/The World Bank. http://www-wds.worldbank.org/servlet/WDSContentServer/WDSP/IB/2003/12/05/000012009_20031205154931/Rendered/PDF/272110PAPER0Building0safer0cities.pdf, accessed on 3 September, 2015.

Larson, E. V. 2007. *Misfortunes of War: Press and Public Reactions to Civilian Deaths in Wartime*. Santa Monica, CA: RAND Corporation.

Lascoumes, P. 1996. L'illégalisme, outil d'analyse – à propos de Surveiller et punir de Michel Foucault. *Sociétés et représentations*, 3, 78–84.

Lascoumes, P. 2013. Une tache sur le col blanc d'Edwin, *Champ pénal/Penal field*, X. http://champpenal.revues.org/8470, accessed on 3 September, 2015.

Lascoumes, P., Godefroy, T. and Favarel-Garrigues, G. 2009. *Les sentinelles de l'argent sale: les banques aux prises avec l'antiblanchiment*. Paris: La Découverte.

Lashaw, A. 2013. How Progressive Culture Resists Critique: The Impasse of NGO Studies. *Ethnography*, 14, 501–522.

Latour, B. 92015. Face à Gaïa. Huit Conférences sur le nouveau régime climatique. Paris: La Découverte.

Lecompere, T. 2014. Pour le droit universel à l'alimentation. http://blogs.mediapart.fr/blog/terezinha-lecompere/151014/pour-le-droit-universel-lalimentation, accessed on 5 September, 2015.

Le Goff, J. 1956. *Marchands et banquiers au Moyen Âge*, Paris: PUF (1979. The Usurer and Purgatory, in *The Dawn of Modern Banking*, ed. Center for Medieval and Renaissance Studies, New Haven: UCLA).

Legnaro, A. 1997. Konturen der Sicherheitsgesellschaft. Eine polemisch-futurologische Skizze. *Leviathan*, 25, 2, 271–284.

Lemieux, P. 2007. L'économie souterraine : causes, importance, options, *Les Cahiers de recherche de l'Institut économique de Montréal*. http://www.iedm.org/files/cdr_nov07_fr.pdf, accessed on 17 September, 2015.

Lentzos, F. and Rose, N. 2009. Governing Insecurity: Contingency Planning, Protection, Resilience. *Economy and Society*, 38, 2, 230–254.

Leogrande, A. 2008. *Uomini e caporali. Viaggio tra i nuovi schiavi nelle campagne del sud*. Milan : Mondadori.

Lippert, R. K. 2014. Neoliberalism, Police, and the Governance of Little Urban Things. Foucault Studies, 18, 49–65.

Lippert, R. and Stenson, K. 2010. Advancing Governmentality Studies: Lessons from Social Constructionism. *Theoretical Criminology*, 14, 4, 473–494.

Lippi, A. and Mrorisi, M. 2005. *Scienza dell'amministrazione*. Bologna: Il Mulino.

Loader, I. 1999. Consumer Culture and the Commodification of Policing and Security. *Sociology*, 33, 2, 373–392.

Lombroso, C. 1911. *Crimes: Its Causes and Remedies*. London: William Heinemann.

Longstaff, P. 2005. *Security, Resilience, and Communication in Unpredictable Environments Such as Terrorism, Natural Disasters, and Complex Technology.* Syracuse, NY: Author.

Lorenz, D. F. 2010. The Diversity of Resilience: Contributions from a Social Science Perspective, *Natural Hazards*, 67, 1, 7–4.

Luhmann, N. 1990. *Risiko und Gefahr*, St. Gallen.

Luhmann, N. 1993. *Risk. A Sociological Theory.* Berlin/New York: de Gruyter.

Luhmann, N. 1996. *Modern Society Shocked by its Risks*. Hong Kong: University of Hong Kong, Department of Sociology Occasional Papers, 17. http://hub.hku.hk/handle/10722/42552, accessed on 3 September, 2015.

Luthar, S. S., Cicchetti, D. and Becker, B. 2000. The Construct of Resilience: A Critical Evaluation and Guidelines for Future Work. *Child Development*, 71, 543–562.

Lyon, D. 1994. *Electronic Eye: The Rise of Surveillance Society* (1st edn). Minneapolis, MT: University of Minnesota Press.

Lyon, D. 2001. *Surveillance Society. Monitoring Everyday Life.* Buckingham: Open University Press.

Maccanico Y. 2011. England: At the Forefront in the Punishment of Minors, in *Racial Criminalization*, edited by Palidda, S. Farnham: Ashgate, 253–264.

Maguire, B. and Hagan, P. 2007. Disasters and Communities: Understanding Social Resilience. *The Australian Journal of Emergency Management,* 22, 2, 16–20.

Maltoni, C. et al. 1986. Mesoteliomi negli operai di officine di grandi riparazioni (OGR) delle Ferrovie dello Stato italiane, esposti ad asbesto. Primo resoconto. *Acta Oncologica*, 7, 159–186.

Maltoni, C. et al. 2002. The Mesotheliomas Related to Asbestos Used in Italian Railway: Report of 199 Cases. *European Journal of Oncology*, 1, 7, 51–55.

Maneri, M. 2010. Peacetime War Discourse: The Political Economy of Bellicose Metaphors, in *Conflict, Security and the Reshaping of Society. The civilization of war*, edited by A. Dal Lago and S. Palidda, London and New York: Routledge.

Maneri, M. 2011. Media Discourse on Immigration. Control Practices and the Language We Live by, in *Racial Criminalisation of Migrants in the 21st Century*, edited by S. Palidda, London: Ashgate, 77–94.

Maneri, M. 2013. Si fa presto a dire «sicurezza». Analisi di un oggetto culturale. *Etnografia e ricerca qualitativa*, 6, 2, 283–312.

Marcu, O. 2014. *Malizie di strada. Una ricerca azione con giovani rom romeni migranti.* Milano: Franco Angeli.

Martiniello, M. 1999. *Multicultural Policies and the State. A Comparison of Two European Societies*. London: Ashgate.

Marx, G. T. 1988. *Undercover. Police Surveillance in America.* Berkeley, CA, London: University of California Press.

Marx, G. T. 2002. What's New About the 'New Surveillance'? Classifying for Change and Continuity. *Surveillance & Society*, 1, 1, 9–29.

Marzbali, M. H., Abdullah, A., Razak, N. A. and Tilaki, M. J. M. 2011. A Review of the Effectiveness of Crime Prevention by Design Approaches towards Sustainable Development. *Journal of Sustainable Development*, 4, 1, 160–172.

Maslow, A. H. 1954. *Motivation and Personality*. New York: Harper.

Masten, A. S., Best, K. M. and Garmezy, N. 1990. Resilience and Development: Contributions from the Study of Children Who Overcome Adversity. *Development & Psychopathology*, 2, 425–444.

Mauss, M. 1950/1999. *Sociologie et anthropologie*, Quadrige. Paris: PUF.

Mbembe, A. 2003. Necropolitics. *Public Culture*, 15, 1, 11–40.

McGarry, B. 2014. Report: Pentagon to Buy Fewer F-35s in 2015. The F-35 is a Bad Deal. http://f35baddeal.com/2014/02/18/report-pentagon-to-buy-fewer-f-35s-in-2015/, accessed on 3 September, 2015.

Menichelli, F. 2013. The National Picture: The Reconfiguration of Sovereignty, the Normalization of Emergency and the Rise to Prominence of Urban Security in Italy. *European Journal of Criminology*, 12, 3, 263–276.

Mertens, D.M., 2009. *Transformative Research and Evaluation*. New York, London: The Guildford Press.

Merton, R. K. 1948. The Self-fulfilling Prophecy. *The Antioch Review*, 8, 2, 193–210.

Mini, F. 2015, CS intervista Fabio Mini. http://www.enzopennetta.it/2015/08/cs-intervista-fabio-mini/, accessed on 5 September, 2015.

Miraftab, F. 2004a. Public-Private Partnerships. The Trojan Horse of Neoliberal Development? *Journal of Planning Education and Research*, 24, 1, 89–101.

Miraftab, F. 2004b. Making neoliberal Governance: The Disempowering Work of Empowerment. *International Planning Studies*, 9, 4, 239–259.

Modood, T. 2007. *Multiculturalism: A Civic Idea*. Cambridge, MA: Polity.

Molotch, H. 1970. Oil in Santa Barbara and Power in America. *Sociological Inquiry*, 40, Winter, 131–144.

Monastersky, R. 2015 Anthropocene: The human age, *Nature*, http://www.nature.com/news/anthropocene-the-human-age-1.17085, accessed on 05/12/2015.

Montana, R. 2009. Pubblico ministero e pratiche di selezione del crimine. Cultura giuridica e rappresentazioni di senso comune: perché si può ancora sperare. *Antigone*, 2–3, 274–298.

Moore, Jason W. 2014. The theory of Capitalocene, http://www.jasonwmoore.com/About.html, accessed on 05/10/2015.

Morawska, E. 2003. Disciplinary Agendas and Analytic Strategies of Research on Immigrant Transnationalism: Challenges of Interdisciplinary Knowledge. *International Migration Review*, 37, 3, 611–640.

Morgan E. P. 1992. *The Sixties Experience: Hard Lessons about Moderm America*. Philadelphia, PA: Temple University Press.

Morice, A. and Potot, S. (dir.) 2010. *De l'ouvrier immigré au travailleur sans papier. Les étrangers dans la modernisation du salariat*. Karthala.

Mosconi, G. and Padovan, D. (eds) 2005. *La fabbrica dei delinquenti. Processo penale e meccanismi sociali di costruzione del condannato.* Torino: L'Harmattan Italia.

Moulier Boutang Y. 1998. *De l'esclavage au salariat: économie historique du salariat bridé.* Paris: PUF.

Mucchielli, L. (ed.) 2008. *La Frénésie sécuritaire: retour à l'ordre et nouveau contrôle social.* Paris: La Découverte.

Mudu, P. P. 2009. L'evasione del rischio: la produzione della dipendenza, in *Spazi e società a rischio*, edited by P. Saitta. Naples: Think Thanks.

Müller, B. 2010. Urban and Regional Resilience – A New Catchword or a Consistent Concept for Research and Practice? in *Urban Regional Resilience. How Do Cities and Regions Deal with Change?*, edited by B. Müller. Berlin: Springer-Verlag, 1–13.

Neufeld, M. 2004. Pitfalls of Emancipation and Discourses of Security: Reflections on Canada's 'Security with Human Face'. *International Relations*, 18, 1, 109–123.

Newburn, T. 2012. Riot Victims Showed Resilience But Too Many Were Failed by the State. *The Guardian*, 2 February.

Newman, E. 2010. Critical Human Security Studies. *Review of International Studies*, 36, 1, 77–94.

Newman, O. 1972. *Defensible Space. Crime Prevention through Urban Design.* New York: MacMillan.

Nimis, G. P. 2009. *Terre mobili. Dal Belice al Friuli, dall'Umbria all'Abbruzzo.* Rome: Donzelli.

Nocenzi, M. 2002. *Vivere con l'incertezza. Sociologia, politica e cultura del rischio ambientale nel caso dell'inquinamento elettromagnetico.* Milano: Franco Angeli.

Nocenzi, M. 2005. Environment and Value Challenge: Persistencies and Changes. *International Review of Sociology: Revue Internationale de Sociologie*, 15, 2, 291–292.

Norris, C. 2009. A Review of The Increased Use of CCTV and Video-Surveillance for Crime Prevention Purposes in Europe, European Parliament–Directorate General Internal Policies Policy Department C Citizens' Rights and Constitutional Affairs. http://www.statewatch.org/news/2009/apr/ep-study-norris-cctv-video-surveillance.pdf, accessed on 4 September, 2015.

Norris, F. H, Stevens, S. P, Pfefferbaum, B, Wyche, K. F. and Pfefferbaum, R. L. 2008. Community Resilience as a Metaphor, Theory, Set of Capacities, and Strategy for Disaster Readiness. *American Journal of Community Psychology,* 41, 1–2, 127–150.

O'Malley, P. 2010. *Crime and Risk.* London/Thousand Oaks, CA: Sage Publications.

Ong, A. 2006. *Neoliberalism as Exception: Mutations in Citizenship and Sovereignty*. Durham: Duke University Press.

Pajno, A. (ed.) ASTRID 2010. *La sicurezza urbana.* Santarcangelo di romagna: Maggioli Editore.

Palidda, S. (ed.) 1996. *Délit d'immigration. La construction sociale de la déviance et de la criminalité parmi les immigrés en Europe.* Bruxelles: COST-Communauté Européenne.

Palidda, S. (ed.) 2009. *Infortuni sul lavoro e malattie professionali. Cosa ne pensano i lavoratori.* Genova: Regione Liguria and INAIL Publishers, http://www.regione.liguria.it/archivi/archivio-allegati/doc_download/4679-infortuni-e-malattie-professionali-cosa-ne-pensano-i-lavoratori.html, accessed on 3 September, 2015.

Palidda, S. (ed.) 2011. *Racial Criminalization of Migrants in the XXI Century.* Farnham/Burlington, VA: Ashgate.

Palidda, S. and Sanfilippo, M. (eds) 2012. *Emigrazione e organizzazioni criminali.* Viterbo: Settecittà.

Palidda, S. 1992. L'anamorphose de l'Etat-Nation: le cas italien. *Cahiers Internationaux de Sociologie*, 39, 93, 269–298.

Palidda, S. 2000. *Polizia postmoderna.* Milan: Feltrinelli.

Palidda, S. 2007. Missions militaires italiennes à l'étranger: la prolifération des hybrides. *Cultures & Conflits*, 67, http://conflits.revues.org/3126, accessed on 6 September, 2015.

Palidda, S. 2008. *Mobilità umane.* Milan: Cortina, 2008.

Palidda, S. 2010. Revolution in Police Affairs, in *Conflict, Security and the Reshaping of Society: The Civilisation of War*, edited by A. Dal Lago and S. Palidda. London: Routledge, 118–128.

Palidda, S. 2013. Re-hybridizing the Legal and the Criminal in All Activities at the Local, National and Global Level: a 'Political Total Fact' in the 21st Century neoliberal Frame, in *Formal, Informal and Criminal Economy: An Outlook on Northern and Southern-Europe*, edited by P. Saitta, J. Shapland and A. Verhage. The Hague: Eleven International Publishing, 29–50.

Palidda, S. 2015a. Emigration, Immigration, Mobilités. Un Fait Politique Total, in *Mobilité(s). Actes du colloque de l'ACSALF 2014*, edited by M.-C. Haince and A. Tremblay, A. Montréal: Éditions de l'ACSALF, (forthcoming).

Palidda, S. 2015b. Italian Police Forces in the Neoliberal Turn. *European Journal of Policing Studies*, 3, 1, 52–78

Pasetto, R., Comba, P. and Pirastu, R. 2008. Lung Cancer Mortality in a Cohort of Workers in a Petrochemical Plant: Occupational or Residential Risk? *International Journal of Occupational and Environmental Risk*, 14, 2, 124–128.

Paton, D. and Johnston, D. 2001. Disasters and Communities: Vulnerability, Resilience, and Preparedness. *Disaster Prevention and Management*, 10, 270–277.

Pavarini, M. 2006. La neutralizzazione degli uomini inaffidabili. La nuova disciplina sulla recidiva ed altro ancora sulla guerra alle unpersonen. *Studi sulla questione criminale*, I, 2, 7–29.

Pease, K. and Farrell, G. 2011. Climate Change and Crime. *European Journal on Criminal Policy and Research*, 17, 2, 149–162.

Pech, T. and Padis, M. O. 2004. *Les multinationales du coeur. Les ONG, la politique et le marché.* Paris: Seuil.

Peck, J. 2001. *Workfare States,* Princeton, NJ/London/New York: Guilford Press.
Peck, J. and Tickell, A. 2002. Neoliberalizing Space. *Antipode*, 34, 3, 380–404.
Pelling, M. 2003. *The Vulnerability of Cities. Social Resilience and Natural Disaster*. London: Earthscan Publication.
Perelman, L. J. 2007. *Shifting Security Paradigms: Toward Resilience*, Critical Thinking Series. Arlington, VA: Critical Infrastructure Protection Program, George Mason University.
Perin, G. 2011. L'applicazione ai Rom e ai Sinti non cittadini delle norme sull'apolidia, sulla protezione internazionale e sulla condizione degli stranieri comunitari ed extracomunitari, in *La condizione giuridica di rom e sinti in Italia*, edited by P. Bonetti, A. Simoni and T. Vitale. Milano: Giuffrè, 363–414.
Perrotta, D. and Sacchetto, D. 2013. Les ouvriers agricoles étrangers dans l'Italie méridionale entre 'séclusion' et action collective'. *Hommes et Migrations*, 1, 57–66.
Persano, B. 2015. Sei mesi di condanna per un furto di 4 euro. http://genova.repubblica.it/cronaca/2015/02/12/news/sei_mesi_di_condanna_per_un_furto_di_4_euro-107161079/, accessed on 5 September, 2015.
Petrillo, A. (ed.) 2009. *Biopolitica di un rifiuto. Le rivolte anti-discarica a Napoli e in Campania*. Verona: Ombre Corte.
Petrillo, A. (ed.) 2015. *Il silenzio della polvere*. Milna: Mimesis.
Petrillo, A. 2003. *La città delle paure. Per un'archeologia dell'insicurezza urbana.* Castel di Serra: Elio Sellino Editore.
Petrillo A. 2011. Napoli globale: discorsi, territorio e potere nella 'città plebea', in *Città mediterranee e deriva liberista*, edited by S. Palidda, Messina: Mesogea, 45–70.
Petrillo A. 2014. 'Razze informali' al lavoro. Naturalizzazione della 'plebe' e postfordismo nella trasformazione del territorio napoletano. *Briganti o emigranti,* edited by Orizzonti meridiani, Verona: Ombrecorte, 137–150.
Pfefferbaum, B., Reissman, D., Pfefferbaum, R., Klomp, R., and Gurwitch, R. 2005. Building Resilience to Mass Trauma Events, in *Handbook on Injury and Violence Prevention Interventions,* edited by L. Doll, S. Bonzo, J. Mercy, and D. Sleet. New York: Kluwer Academic Publishers.
Piana, D. 2010. *I magistrati in Italia. Una professione al plural*. Rome: Carocci.
Piasere, L. 2004. *I rom d'Europa. Una storia moderna*. Bari-Rome: Laterza.
Pickering, S. 2005. Crimes of the State. The Persecution and Protection of Refugees. *Critical Criminology,* 13, 2, 141–163.
Pigliaru, A.. 1955. La vendetta barbaricina come orientamento giuridico, republished in Antonio Pigliaru, *Il banditismo in Sardegna* (2000). Nuoro: il Maestrale.
Pirastu, R., Zona, A., Ancona, C., Bruno C., Fano V., Fazzo L., Iavarone I., Minichilli F., Mitis F. and Pasetto R. (eds) 2011. Sentieri Project. Mortality Study of Residents in Italian Polluted Sites: Results. *Epidemiologia & Prevenzione*, 35, 5–6, 1–204.
Pitch, T. 2007. *La società della prevenzione*, Roma: Carocci.

Pizzorno, A. 1992. La corruzione nel sistema politico, in *Lo scambio occulto. Casi di corrruzione politica in Italia*, edited by D. della Porta. Bologna: Il Mulino, 13–74.

Pollner, M. 1987. *Mundane Reason: Reality in Everyday and Sociological Discourse*. Cambridge: Cambridge University Press.

Portaluri, M. 2006. Manfredonia Thirty Years Later. *Epidemiologia & Prevenzione*, 30, 4–5, 217–218.

Poyner, B. 1983. *Design Against Crime. Beyond Defensible Space*. London: Butterworths.

Prina, F. 2003. *Devianza e politiche di controllo. Scenari e tendenze nelle società contemporanee*. Roma: Carocci.

Purdy, J. 2015. The Triumph of Corruption. https://www.dissentmagazine.org/article/zephyr-teachout-review-triumph-of-corruption, accessed on 5 September, 2015.

Putnam, R. D., Leonardi, R. and Nanetti, R. Y. 1993. *Making Democracy Work: Civic Traditions in Modern Italy*. Princeton, NJ: Princeton University Press.

Quarantelli, E. (ed.) 1998. *What Is A Disaster?* London: Routledge.

Quassoli, F. 1999. Immigrazione uguale criminalità: rappresentazioni di senso comune e pratiche degli operatori del diritto. *Rassegna Italiana di Sociologia*, 40, 1, 43–76.

Quassoli, F. 2002. Il sapere dei magistrati: un approccio etnografico allo studio delle pratiche giudiziarie, in *Un certo sguardo*, edited by A. Dal Lago and R. De Biasi. Rome-Bari: Laterza, 196–217.

Quigley, B. 2010. Corporations Profit from Permanent War: Memorial Day 2010. *The Huffington Post*, 24 May. http://www.huffingtonpost.com/bill-quigley/corporations-profit-from_b_586896.html, accessed on 3 September, 2015.

Renn, O. 2010. Wer hat Angst vor Risiken? Wahrnehmung und Bewertung von Risiken in der Bevölkerung, in *Transparenz: multidisziplinäre Durchsichten durch Phänomene und Theorien des Undurchsichtigen*, edited by S. A. Jansen, E. Schröter and N. Stehr. Wiesbaden: Springer, 152–162.

Reyneri, E. 1988. L'innovazione produttiva nella rete delle relazioni sociali. *Stato e mercato*, 23, 147–163.

Richardson, J. E. and Colombo, M. 2014. Race and Immigration in Far- and Extreme-Right European Political Leaflets, in *Contemporary Studies in Critical Discourse Analysis*, edited by C. Hart and P. Cap. London: Bloomsbury, 522–542.

Rigouste, M. 2012. *La Domination policière. Une violence industrielle*. Paris: La Fabrique Editions.

Riolli, L. and Savicki, V. 2003. Information System Organizational Resilience. *Omega: The International Journal of Management Science*, 31, 3, 227–233.

Rivas, L.M. 2002. Invisible Labors: Caring for the Independent Person, in *Global Woman: Nannies, Maids and Sex Workers in the New Economy, edited by* B. Ehrenreich and A. R. Hochschild. *New York: Metropolitan Press.*

Roberts, A. 2010. Lives and Statistics: Are 90% of War Victims Civilians? *Survival,* 52, 3, 115–136.

Robert, Ph. 2009. Le politiche di prevenzione e di sicurezza in Europa. *Studi sulla questione criminale*, 4, 3, 85–108.

Robin C. 2004. *Fear: The History of a Political Idea.* Oxford: Oxford University Press.

Robine, J. M. Siu Lan Cheung, Le Roy, S., Van Oyen, H. and Herrmann, F. R. 2007. Report on excess mortality in Europe during summer 2003. EU Community Action Programme for Public Health, Grant Agreement 2005114. http://ec.europa.eu/health/ph_projects/2005/action1/docs/action1_2005_a2_15_en.pdf, accessed on 14 September, 2015.

Robins, S. 2002. At the Limits of Spatial Governmentality: A Message from the Tip of Africa. *Third World Quarterly*, 23, 4, 665–689.

Roché, S. 1993. *Le sentiment d'insécurité.* Paris: PUF.

Rogers, P. 2012. *Resilience & the City. Change, (Dis)Order and Disaster.* Farnham: Ashgate.

Rolinski, K. 1980. *Wohnhausarchitektur und kriminalitat.* Wiesbaden: Bundeskriminalamt (BKA-Forschungsreihe vol. 13).

Romano, O. 2006. Sviluppo, in *Lessico di biopolitica*, edited by R. Brandimarte, P. Chiantera-Stutte, P. Di Vittorio, O. Marzocca, O. Romano, A. Russo, A. Simone. Rome: Manifestolibri, 310–314.

Rondinone, A. (ed.) 2012. *Costruire il futuro: difendere l'agricoltura dalla cementificazione.* Rome: Ministero delle Politiche agricole, alimentari e forestali.

Rose, N. 1998. *Inventing Our Selves: Psychology, Power, and Personhood.* Cambridge: Cambridge University Press.

Rose, N. 2000. Community, Citizenship, and the Third Way. *American Behavioural Scientist*, 43, 9, 1395–1411.

Rose, N. and Miller, P. 2008. *Governing the Present: Administering Economic, Social and Personal Life*. Cambridge: Polity.

Rossi-Doria, M. 1982. *Scritti sul Mezzogiorno*. Torino: Einaudi.

Rostas, I. 2012. Roma Participation: From Manipulation to Citizen Control. ERRC, *Roma Rights, Challenges of Representation: Voice on Roma Politics, Power and Participation.* http://www.errc.org/article/roma-rights-2012-challenges-of-representation-voice-on-roma-politics-power-and-participation/4174/0, accessed on 3 September, 2015.

Rotter, J. B. 1966. *Generalized Expectancies for Internal Versus External Control of Reinforcement*. Washington, DC: American Psychological Association.

Rotter, J. B. 1990. Internal versus External Control of Reinforcement. A Case History of a Variable. *American Psychologist*, 45, 4, 489–493.

Roux, A. 2010. *La part d'ombre de la mondialisation, Paradis fiscaux, blanchiment et crime organisé: l'urgence de l'espace judiciaire européen*, Aix-en-Provence: Presse Universitaires d'Aix-en-Provence.

Ruffin, J. C. 2012. *Le grand Cœur*. Paris: Gallimard.

Ruggiero, V. 2013. *The Crimes of the Economy: A Criminological Analysis of Economic Thought.* London: Routledge.

Ruggiero, V. and Gounev, P. 2012. *Corruption and Organized Crime in Europe: Illegal Partneships.* London: Routledge.

Ruhne, R. 2003. Sicherheit ist nicht die Abwesenheit von Unsicherheit – Die soziale Konstruktion geschlechtsspezifischer Unsicherheiten im öffentlichen Raum, in *JahrbuchStadt/Region 2002: die sichere Stadt*, edited by N. Gestring, H. Glasauer, C. Hannemann, W. Petrowsky and J. Pohlan. Opladen: Leske – Budrich, 55–73.

Rusche, G. and Kirchheimer, O. 2009. *Punishment and Social Structure.* New Brunswick, NJ: Transaction Publishers.

Rutter, M. 1987. Psychosocial Resilience and Protective Mechanisms. *American Journal of Orthopsychiatry*, 57, 316–331.

Rutter, M. 1990. Competence under Stress: Risk and Protective Factors, in *Risk and Protective Factors in the Development of Psychopathology*, edited by J. Rolf, A. S. Masten, D. Cicchetti, K. H. Nuechterlin and S. Weintraub. New York, NY: Cambridge University Press, 181– 214.

Rutter, M. 1993. Resilience: Some Conceptual Considerations. *Journal of Adolescent Health*, 14, 626–631.

Sacchetto, D. and Andrijasevic, R. 2013. Integration verschiedener Arbeitsregime? Foxconn in Tschechien, *iSlave. Ausbeutung und Widerstand in Chinas Foxconn Fabriken, Mandelbaum*, edited by Pun Ngai u.a., Wien, 219–245.

Saitta, P. 2009. *Spazi e società a rischio.* Napoli: Think Thanks.

Saitta, P. 2013. *Quota zero. Messina dopo il terremoto: la ricostruzione infinita.* Roma: Donzelli.

Saitta, P., Shapland, J. and Verhage, A. 2013. *Formal, Informal and Criminal Economy: An Outlook on Northern and Southern-Europe.* The Hague: Eleven International Publishing.

Saletti Salza, C. 2010. *Dalla tutela al genocidio? Le adozione dei minori rom e sinti in Italia (1985–2005).* Roma: CISU.

Saletti Salza, C. 2012. Dal non intervento all'intervento indifferenziato. Riflessioni sulla tutela dei minori rom e sinti. *Minorigiustizia*, 2, 108–116.

Samara, T. R. 2010. Policing Development: Urban Renewal as neoliberal Security Strategy. *Urban Studies*, 47, 1, 197–214.

Sampson, E. 1977. Psychology and the American Ideal. *Journal of Personality and Social Psychology*, 35, 11, 767–782.

Sampson E. 1983. Identity Politics: Challenges to Psychology's Understanding. *American Psychologist*, 48, 12, 1219–1230.

Sampson, E. 1985. The Decentralization of Identity: Toward a Revised Concept of Personal and Social Order. *American Psychologist*, 40, 11, 1985, 1203–1211.

Sampson, E. 1988. The Debate on Individualism. *American Psychologist,* 43, 1, 15–22.

Sampson, R. J. 2012. *Great American City. Chicago and the Enduring Neighborhood Effect.* Chicago, IL: University of Chicago Press.

Sampson, R. J., Raudenbush, S. W. and Earls, F. 1997. Neighborhoods and Violent Crime: A Multilevel Study of Collective Efficacy. *Science*, 277, 5328, 918–924.

Santino, U. 2015, *Mafia and Antimafia: A History*. London: I.B. Tauris.

Sapelli, G. 1989. *L'Italia inafferrabile. Conflitti, sviluppo, dissociazione dagli anni Cinquanta a oggi*. Marsilio, Venezia.

Sarzotti, C. 2004. La cultura giuridica dei procuratori generali nelle relazioni inaugurali degli anni giudiziari, in *Cultura giuridica e attori della giustizia penale*, edited by A. Balloni, G. Mosconi and F. Prina, 229–254.

Sarzotti, C. 2007. *Processi di selezione del crimine. Procure della Repubblica e organizzazione giudiziaria*. Milano: Giuffrè.

Saviano, R. 2006. *Gomorrah: Italy's Other Mafia*. London: Pan.

Saville, G. and Cleveland, G. 2008. Second Generation CPTED: The Rise and Fall of Opportunity Theory, in *21st Century Security and CPTED*, edited by R. Atlas. Boca Raton, FL: CRC Press, Taylor & Francis Ltd, 79–90.

Savitch, H. V. 2008. *Cities in a Time of Terror. Space, Territory, and Local Resilience*. Armonk, NY: Taylor & Francis.

Schipper, L. and Pelling, M. 2006. Disaster Risk, Climate Change and International Development: Scope for, and Challenges to, Integration. *Disasters, Special Issue: Climate Change and Disasters*, 30, 1, 19–38, March.

Schneider. J. (ed) 1998. *Italy's 'Southern Question'. Orientalism in One Country*. Oxford and New York: Berg.

Schubert, H. and Veil, K. 2011. Kriminalprävention im Sozialraum. Explorative Validierung des ISAN-Präventionsmodells. *Monatsschrift für Kriminologie und Strafrechtsreform*, 94, 2, 83–101.

Sefira. 2015. Policy Address in View of the Vote of the National Emission Ceiling (Nec) Directive at the European Parliament, www.sefira-©-project.eu

Semprebon, M. 2011. Phone Centres and the Struggle for Public Space in Italy: Between Revanchist Policies and Practices of Resistance. *Journal of Urbanism*, 4, 3, 223–237.

Senellart, M. 2007. L'ennemi intérieur dans le discours de la défense sociale au XIXe siècle. *Erytheis*, 2. https://halshs.archives-ouvertes.fr/halshs-00368498, accessed on 3 September, 2015.

Senior, K. and Mazza, A. 2004. Italian 'Triangle of Death' Linked to Waste Crisis. *The Lancet Oncology*, 5, 9, 525–527.

Siegl, J. 2006. *Wenn Entscheidungen Sorgen bereiten – Untersuchungen zu Ausmaß und Inhalten von Worrying sowie zu dem Zusammenhang von Worrying und Decision-Making unter Berücksichtigung klinischer und nichtklinischer Gruppen*. Dissertation an der Otto-Friedrich-Universität Bamberg.

Sigona, N. 2003. How Can a Nomad be a Refugee? Kosovo Roma and Labelling Policy in Italy. *Sociology*, 37, 1, 69–79.

Sigona, N. 2008. *The 'Latest' Public Enemy: the Romanian Roma in Italy*. Report commissioned by OSCE/ODHIR. http://www.romsintimemory.it/assets/files/discriminazione/intercultura/guerra-ai-rom/III.3.d%20ALLEGATO%203%20OSCE_publicenemy.pdf, accessed on 3 September, 2015.

Sigona, N. and Nidhi, T. (eds) 2009. *Romani Politics in Contemporary Europe: Poverty, Ethnic Mobilisation and the Neoliberal Order.* Basingstoke: Palgrave Macmillan.

Simmel, G. 1999. *Etudes sur les formes de la socialisation*. Paris, PUF.

Simon, J. 2007. *Governing Through Crime. How the War on Crime Transformed American Democracy and Created a Culture of Fear*. New York: Oxford University Press.

Singelnstein, T. and Stolle, P. 2006. *Die Sicherheitsgesellschaft. Soziale Kontrolle im 21. Jahrhundert*. Wiesbaden: VS Verlag. https://www.socialnet.de/rezensionen/3656.php, accessed on 17 September, 2015.

Sironi, F. 2013. Licenza di tortura: ecco i volti delle vittime di Stato, *l'Espresso*, 25 November, http://espresso.repubblica.it/attualita/2013/11/25/news/i-volti delle-vittime-di-stato-1.142657, accessed on 3 September, 2015.

Sjöberg, L., Moen, B-E. and Rundmo, T. 2004. *Explaining Risk Perception. An Evaluation of the Psychometric Paradigm in Risk Perception Research*. Trondheim: Norwegian University of Science and Technology.

Smith, O. A. 1999. What is a Disaster? Anthropological Perspectives on a Persistent Question, in *The Angry Earth. Disaster in Anthropological Perspective*, edited by O. A. Smith and S. Hoffman. New York: Routledge.

Sobotka, E. and Veermersch, P. 2012. Governing Roma Inclusion: Can the EU be a Catalyst for Local Social Change? *Human Rights Quarterly*, 34, 3, 800–822.

Sonn, C. and Fisher, A. 1998. Sense of Community: Community Resilient Responses to Oppression and Change. *Journal of Community Psychology*, 26, 5, 457–472.

Sowby, F. D. 1965. Radiation and Other Risks. *Health Physics*, 11, 879–887.

Stallings R. A. 2006. On Sociological Theory and the Sociology of Disasters: Moving from Periphery to Center, for World Congress of Sociology, Durban, South Africa, 25 July. http://www-bcf.usc.edu/~rstallin/papers/presaddress.pdf, accessed on 4 September, 2015.

Starr, C. 1969. Social Benefit versus Technological Risk. *Science,* 165, 3899, 1232–1238.

Steffen, W., Grinevald, J., Crutzen, P. and McNeill, J. 2011. The Anthropocene: conceptual and historical perspectives, *Philosophical Transactions. Royal Society*. 369, 842–867

Stefanizzi, S. 2013. *Il teatro della sicurezza. Attori, pratiche e rappresentazioni.* Milano: ET Al Edizioni.

Stenson, K. and Sullivan, R. R. 2001. *Crime, Risk and Justice. The Politics of Crime Control in Liberal Democracies*. London: Willan Publishing.

Stummvoll, G. 2002. *CPTED. Kriminalprävention durch Gestaltung des öffentlichen Raumes*. Wien: Institut für höhere Studien.

Sudnow, D. N. 1965. Normal Crimes: Sociological Features of the Penal Code in a Public Defender Office. *Social Problems*, 12, 3, 255–276.

Sutcliffe, S. B. 2012. Cancer Control: Life and Death in an Unequal World. *Current Oncology*, 19, 1, http://www.current-oncology.com/index.php/oncology/article/view/994/822, accessed on 3 September, 2015.

Sutherland, E. 1945. Is White Collar Criminality a Crime? *American Sociological Review*, 1945, 10, 132–139 (in French: in *Les Cahiers de la sécurité intérieure*, 36).

Szasz, A. 1994. *Ecopopulism: Toxic Waste and the Movement for Environmental Justice*. Minneapolis, MT: University of Minnesota Press.

Tadjbakhsh, S. and Chenoy, A. 2007. *Human Security: Concepts and Implications*. London: Routledge.

Taleb, N. 2007. *The Black Swan. The Impact of the Highly Improbable*. New York: Random House.

Teachout, Z. 2014. *Corruption in America: From Benjamin Franklin's Snuff Box to Citizens United*. Cambridge, MA: Harvard University Press.

Teti, V. 2011 (1993). *La razza maledetta. Origini del pregiudizio antimeridionale*. Rome: Manifestolibri.

Thiers, A. 1858. *Histoire de law*. Paris: Editions Hetzel.

Thomas, W. I. and Thomas, D. S. 1928. *The Child in America: Behavior Problems and Programs*. New York: Knopf.

Tierney, K. 1999. Toward a Critical Sociology of Risk. *Sociological Forum*, 14, 2, 215–242.

Tierney, K. J. 2007. From the Margins to the Mainstream? Disaster Research at the Crossroads. *Annual Review of Sociology*, 33, 1, 503–525.

Tomovska, I. 2010. Poverty, Discrimination and the Roma: A Human Security Issue. *Human Security Perspectives Journal*, 7, 1, 63–82.

Tondini, M. 2007. Beyond the law of the enemy recovering from the failures of the global war on terrorism through law, in *Jura Gentium*. http://www.juragentium.org/topics/wlgo/cortona/en/tondini.htm, accessed on 17 September, 2015.

Torre S. 2004. *Alle origini della città contemporanea. Rendita fondiaria urbana e processi di accumulazione a Genova nel XIX secolo*. Catania: CUECM.

Torre. S. 2013. *Dominio, natura, democrazia*, Milan: Mimesis

Torrente, G. 2007. Le storie organizzative di due procure della repubblica tra obbligatorietà dell'azione penale e selezione del crimine, in *Processi di selezione del crimine*, edited by C. Sarzotti. Milan: Giuffrè, 226–356.

Ungar, S. 2009. Ignorance as an Under-Identified Social Problem. *British Journal of Sociology*, 59, 2, 301–326.

Vale, L. J. and Campanella, T. J. (eds) 2005. *The Resilient City: How Modern Cities Recover from Disaster*. Oxford: Oxford University Press.

Valverde, M. 2008a. Analyzing the Governance of Security: Jurisdiction and Scale. Police, Sovereignty, and Law. Foucaultian Reflections, *Behemoth*, 1, 1, 3–15.

Valverde, M. 2008b. Beyond Discipline and Punish: Foucault's Challenge to Criminology. *Carceral Notebooks*, 14, 4, 201–223.

van Baar, H. 2011. Europe's Romaphobia: Problematization, Securitization, Nomadization, Environment and Planning D. *Society & Space*, 29, 2, 203–212.

van Baar, H. 2012. Socio-economic Mobility and neoliberal Governmentality in Post-Socialist Europe: Activation and the Dehumanisation of the Roma. *Journal of Ethnic and Migration Studies*, 38, 8, 1289–1304.

Vassallo Paleologo, F. 2013. *Diritti sotto sequestro.* Rome: Aracne.

Vermeule A. 2007. Instrumentalisms. *Harvard Law Review*, 120, 2113–2132..

Vermeersch, P. 2012. Reframing the Roma: EU Initiatives and the Politics of Reinterpretation. *Journal of Ethnic and Migration Studies*, 38, 8, 1195–1212.

Vertovec, S. and Wessendorf, S. (eds) 2010. *The Multiculturalism Backlash: European Discourses, Policies and Practices.* London: Routledge.

Veyne, P. 2008. *Foucault. Sa pensée, sa personne*. Paris: Albin Michel.

Viganò, F. 2008. Diritto penale del nemico e diritti fondamentali, in *Legalità penale e crisi del diritto*, edited by A. Bernardi, B. Pastore and A. Pugiotto. Milan: Giuffrè, 107–148.

Violante L. 1995, Apologia dell'ordine pubblico. *Micromega*, 4, 95, 124–140.

Vitale, T. 2010. Rom e sinti in Italia: condizione sociale e linee di politica pubblica. *Osservatorio di politica Internazionale*, October. http://www.parlamento.it/documenti/repository/affariinternazionali/osservatorio/approfondimenti/Approfondimento_21_ISPI_RomSinti.pdf, accessed on 3 September, 2015.

Vollmer, B. 2011. Policy Discourses on Irregular Migration in the EU – 'Number Games' and 'Political Games'. *European Journal of Migration and Law*, 13, 3, 317–339.

Voss, M. 2010. Resilienz, Vulnerabilität und transdisziplinäre Katastrophenforschung, in *Jahrbuch für europäische Sicherheitspolitik 2009/2010*, edited by A. Siedschlag. Baden-Baden: Nomos, 67–84.

Wacquant, L. 1999a. Penal 'Common Sense' Comes to Europe – US Exports Zero Tolerance. *Le Monde Diplomatique*, April. http://mondediplo.com/1999/04/02zero, accessed on 3 September, 2015.

Wacquant, L. 1999b. Suitable Enemies: Foreigners and Immigrants in the Prisons of Europe. *Punishment and Society*, 31, 2, 13–38.

Wacquant, L. 2002. Toward a Dictatorship over the Poor? Notes on the Penalization of Poverty in Brazil, *Punishment and Society*, 52, 197–205.

Wacquant, L. 2008. *The Place of the Prison in the New Government of Poverty, in After the War on Crime.* New York: New York University Press.

Waever, O. 1995. Securitization and Desecuritization, in *On Security*, edited by R. D. Lipschutz. New York: Columbia Press, 46–86.

Walker, J. and Cooper, M. 2011. Genealogies of Resilience: From Systems Ecology to the Political Economy of Crisis Adaptation. *Security Dialogue*, 43, 143–160.

Walklate, S. 2011. Reframing Criminal Victimization: Finding a Place for Vulnerability and Resilience. *Theoretical Criminology*, 15, 2, 179–194.

Waller, M. 2001. Resilience in Ecosystemic Context: Evolution of the Concept. *American Journal of Orthopsychiatry*, 71, 290–297.

Wallraff, G. 1986. *Tête de turc*. Paris: La Découverte.

Waxweiler, R.J and *Robinson, R.*1983. Asbestos and non-Hodgkin's Lymphoma. *The Lancet*, 321, 8317, 189–190.

Weale, A. 1992. *The New Politics of Pollution.* Manchester: Manchester University Press.

Webb, G. R. 2002. Sociology, Disasters and Terrorism: Understanding Threat of the New Millennium. *Sociological Focus*, 35, 1, 87–95.

Weber, M. 2013/1960. *Sociologie de droit*. Paris: PUF.

Werner, E. and Smith, R. S. 1982. *Vulnerable but Invincible: A Longitudinal Study of Resilient Children and Youth*. New York: McGraw-Hill.

Werner, E. and Smith, R. 1992. *Overcoming the Odds: High Risk Children from Birth to Adulthood*. Ithaca, NY: Cornell University Press.

Whitehead, J. 2005. The Neoliberal State in Disaster Management. *Anthropology News*, 46, 9, 18–19.

Wildavsky, A. [1988] 2011. *Searching For Safety*. New Brunswick, NJ: Transaction Publishers.

Williams, R. 1961. *Culture and Society 1780–1950*. London: Chatto & Windus.

Willman, D. 2015. The Pentagon's $10-Billion Bet Gone Bad, ... Defense System Proves Unreliable, in Los Angeles Times, June 15, 14. http://graphics.latimes.com/missile-defense/, accessed on 5 September, 2015.

Würtenberger, T. and Tanneberger, T. 2009. Gesellschaftliche Voraussetzungen und Folgen der Technisierung von Sicherheit, in *Sicherheitsforschung – Chancen und Perspektiven*, edited by P. Winzer, E. Schnieder and F.-W Bach. Berlin: Springer Acatech, 221–240.

Zanier, M. L. 2009. *Tra il dire e il fare. Obbligatorietà dell'azione penale e comportamenti degli attori giuridici*. Macerata: EUM.

Zedner, L. 2003. Too Much Security? *International Journal of the Sociology of Law*, 31, 3, 155–184.

Zedner, L. 2007. Seeking Security by Eroding Rights: The Side-Stepping of Due Process, *in Security and Human Rights*, edited by L. Lazarus and B. Goold. Oxford: Hart, 257–276.

Zedner, L. 2009. *Security*. London/New York: Routledge.

Zedner, L. 2010. Security, the State and the Citizens: The Changing Architecture of Crime Control. *New Criminal Law Review*, 13, 2, 379–403.

Zedner, L. 2012. Risiko, Sicherheit und Terrorismus: Drei Konzepte auf der Suche nach einer akademischen Disziplin. *Kriminologisches Journal*, 10, 30–46.

Zincone, G. Penninx, R. and Borkert, M. (eds) 2011. *Migration Policymaking in Europe: The Dynamics of Actors and Contexts in Past and Present*. Amsterdam: Amsterdam University Press.

## International and National Institutions/Organizations

AIRC (Associazione Italiana per la Ricerca sul Cancro) 2015. Le Statistiche del Cancro. http://www.airc.it/cancro/cos-e/statistiche-tumori-italia/, accessed on 4 September, 2015.

AIRT (Associazione Italiana Registro Tumori) I numeri del cancro in Italia 2014. http://www.registri-tumori.it/PDF/AIOM2014/I_numeri_del_cancro_2014.pdf, accessed on 4 September, 2015.

Ance-Cresme 2012. Lo Stato del territorio italiano 2012. Insediamento e rischio sismico e idrogeologico. http://www.camera.it/temiap/temi16/CRESME_rischiosismico.pdf, accessed on 5 September, 2015.

Bundeskriminalamt (Federal Criminal Police Office) 2012. Police Crime Statistics 2011. Wiesbaden: Federal Republic of Germany, own publication. http://www.bka.de/nn_194552/EN/Publications/PoliceCrimeStatistics/policeCrimeStatistics__node.html?__nnn=true, accessed on 14 September, 2015 (quote PKS 2011).

Camera di Commercio di Avellino, Rapporto Irpinia 2013. http://www.av.camcom.gov.it/files/Statistica/rapporto-giornata-economia-2013-14-giu-2013%20_Ripristinato_.pdf. accessed on 16 September, 2015.

Commission nationale consultative des droits de l'homme. 2014. La lutte contre le racisme, l'antisémitisme et la xénophobie - Année 2013. http://www.cncdh.fr/fr/publications/rapport-racisme-antisemitisme-et-xenophobie-2013-banalisation-de-la-parole-raciste-et, accessed on 16 September, 2015.

CIVIC 2009. *Losing the People: The Costs and Consequences of Civilian Suffering in Afghanistan.* Washington, DC. http://civiliansinconflict.org/uploads/files/publications/losing-the-people_2009.pdf, accessed on 3 September, 2015.

Coe 2012. Lives lost in the Mediterranean Sea: who is responsible? http://assembly.coe.int/CommitteeDocs/2012/20120329_mig_RPT.EN.pdf, accessed on 3 September, 2015.

Council of Europe (CoE) 2012. *Lives Lost in the Mediterranean Sea: Who is Responsible?* http://assembly.coe.int/CommitteeDocs/2012/20120329_mig_RPT.EN.pdf, accessed on 3 September, 2015.

Deutsche Bundesregierung 2009. *Arbeit der EU-Grenzschutzagentur FRONTEX und Zurückschiebung von Asylbewerbern und Flüchtlingen aus Deutschland in andere EU-Staaten.* Berlin (Drucksache 16/12931).

ECRI - European Commission against Racism and Intolerance. 2014. Conclusions on the Implementation of the Recommendations In Respect of Italy Subject to Interim Follow-Up Adopted on 9 December, 2014 1. Published on 24 February, 2015. https://www.coe.int/t/dghl/monitoring/ecri/Country-by-country/Italy/ITA-IFU-IV-2015-004-ENG.pdf, accessed 17 September, 2015.

Environment Agency (EEA) 2011. *Technical Report No 15/2011.* http://www.eea.europa.eu/publications/cost-of-air-pollution, accessed on 3 September, 2015.

European Commission (EC) 2009a. *FP7 in Brief. How to Get Involved in the EU 7th Framework Programme for Research.* https://ec.europa.eu/research/fp7/understanding/fp7inbrief/what-is_en.html, accessed on 3 September, 2015.

European Commission (EC) 2009b. *Towards a More Secure Society and Increased Industrial Competitiveness. Security Research Projects under the 7th Framework Programme for Research.* Bruxelles: European Commission.

European Commission (EC) 2010a. *Horizon 2020.* http://ec.europa.eu/programmes/horizon2020/en/what-horizon-2020, accessed on 3 September, 2015.

European Commission. 2010b. The EU Internal Security Strategy in Action: Five steps towards a more secure Europe. http://ec.europa.eu/

commission_2010-2014/malmstrom/pdf/news/internal_security_strategy_in_action_en.pdf, accessed on 17 September, 2015.

European Commission (EC) 2010c. *Communication from the Commission to the European Parliament and the Council. The EU Internal Security Strategy in Action: Five Steps towards a More Secure Europe.* http://eur-lex.europa.eu/LexUriServ/LexUriServ.do?uri=COM:2010:0673:FIN:EN:PDF, accessed on 3 September, 2015.

European Commission (EC) 2011. *Communication on an EU Framework for National Roma Integration Strategies up to 2020.* http://ec.europa.eu/justice/policies/discrimination/docs/com_2011_173_en.pdf, accesed on 3 September, 2015.

European Commission (EC) 2012a. *Policy and Research in Security.* http://ec.europa.eu/dgs/home-affairs/financing/fundings/pdf/research-for-security/security_research_catalogue_2014_en.pdf, accessed on 3 September, 2015.

European Commission (EC) 2012b. *Discrimination in the EU in 2012 – Special Eurobarometer, No. 393.* http://ec.europa.eu/public_opinion/archives/ebs/ebs_393_en.pdf, accessed on 3 September, 2015.

European Commission (EC) 2013a. *Secure Societies – Protecting Freedom and Security of Europe and its Citizens.* http://ec.europa.eu/programmes/horizon2020/en/h2020-section/secure-societies-—-protecting-freedom-and-security-europe-and-its-citizens, accessed on 3 September, 2015.

European Commission (EC) 2013b. *Press Release: Smart Borders.* http://europa.eu/rapid/press-release_IP-13-162_en.htm, accessed on 3 September, 2015.

European Commission (EC) 2014a. *Frontex Joint Operation 'Triton' – Concerted Efforts to Manage Migration in the Central Mediterranean.* http://europa.eu/rapid/press-release_MEMO-14-566_it.htm, accessed on 3 September, 2015.

European Commission (EC) 2014b. *Shadow Economy and Undeclared Work.* http://ec.europa.eu/europe2020/pdf/themes/07_shadow_economy.pdf, accessed on 3 September, 2015.

European Commission (EC) 2015. Trade policy. http://ec.europa.eu/trade/policy/in-focus/ttip/documents-and-events/index_en.htm, accessed on 3 September, 2015.

European Commission against Racism and Intolerance. 2014. Conclusions on the Implementation of the Recommendations In Respect of Italy Subject to Interim Follow-Up Adopted on 9 December 2014 1. Published on 24 February 2015. https://www.coe.int/t/dghl/monitoring/ecri/Country-by-country/Italy/ITA-IFU-IV-2015-004-ENG.pdf, accessed on 17 September, 2015.

European Court of Human Rights, Judgement 23 February, 2012, Application no. 27765/09. http://ec.europa.eu/dgs/home-affairs/what-is-new/news/pdf/grandchamberjudgmenthirsijamaaandothersvitaly2302121_en.pdf, accessed on 17 September, 2015.

European Parliament 2009. *Note of Directorate General Internal Policies of the Union, Policy Department C Citizens' Rights and Constitutional Affairs. AReview of the Increased Use of Cctv and Video-Surveillance for Crime*

*Prevention Purposes in Europe*. http://www.statewatch.org/news/2009/apr/ep-study-norris-cctv-video-surveillance.pdf, accessed on 3 September, 2015.

European Parliament 2013. *German Social Dumping is Destroying Jobs and the European Socioeconomic Fabric* (written question). http://www.europarl.europa.eu/sides/getDoc.do?pubRef=-//EP//TEXT+WQ+E-2013-003157+0+DOC+XML+V0//EN, accessed on 3 September, 2015.

European Parliament 2014a. Parliamentary Questions, 29 January 2014: 'Push-backs' off the Coast of Greece (Farmakonisi) Resulting in Refugee Deaths.

European Parliament 2014b. *Report on the Proposal for a Regulation of the European Parliament and of the Council Establishing Rules for the Surveillance of the External Sea Borders*. http://www.europarl.europa.eu/sides/getDoc.do?pubRef=-//EP//TEXT+REPORT+A7-2013-0461+0+DOC+XML+V0//EN, accessed on 3 September, 2015.

European Parliament 2014c. Plenary Session, 16 April 2014: Migrants: Parliament Approves Search and Rescue Rules to Prevent Deaths at Sea.

European Roma Rights Centre (ERRC) 2000. *Campland: Racial Segregation of Roma in Italy.* http://www.errc.org/cms/upload/media/00/0F/m0000000F.pdf, accessed on 3 September, 2015.

European Roma Rights Centre (ERRC) 2011. *Shadow Report to the Committee on the Elimination of all Forms of Discrimination against Women on Italy under Article 18 of the Convention on the Elimination of all Forms of Discrimination against Women for its Consideration at the 49th Session 11 to 29 July 2011 concerning the Situation of Romani Women in Italy*.

European Roma And Travellers Forum & Romani Women Informal Platform 'Phenjalipe', 2014.

Making Early Marriage In Roma Communities A Global Concern. https://cs.coe.int/team20/cahrom/9th%20CAHROM%20Plenary%20meeting/Item%2009%20-%20ERTF%20and%20Phenjalipe%20paper%20Making_Early_Marriage_in_Roma_Community_a_global_concern_Final_Version_.pdf, accessed on 17 September, 2015.

European Society of Clinical Microbiology and Infectious Diseases (ESCMID) 2015. *Antibiotic Armageddon in UK and Europe by 2025*.

European Union Institute for Security Studies (EUISS) 2003. *A Secure Europe in a Better World. European Security Strategy*. http://www.iss.europa.eu/uploads/media/solanae.pdf, accessed on 3 September, 2015.

Eurostat 2014. *Large Increase to Almost 435 000 Asylum Applicants Registered in the EU28 in 2013*. http://ec.europa.eu/eurostat/documents/2995521/5181422/3-24032014-AP-EN.PDF/36a73587-7914-4a51-94a4-8e558a086fba, accessed on 3 September, 2015.

Frontex 2013. *About Frontex*. www.frontex.europa.eu, accessed on 3 September, 2015.

Frontex 2014. *Annual Risk Analysis 2014*. www.frontex.europa.eu/assets/Publications/Risk_Analysis/Annual_Risk_Analysis_2014.pdf, accessed on 3 September, 2015.

Government of Malta 2014. Il-PM f'laqgħa mal-President tan-Niġerja. www.inewsmalta.com/dart/20140403-il-pm-laqg-mal-president-tan-ni-erja, accessed on 3 September, 2015.

House of Lords – European Union Committee 2008. FRONTEX: the EU External Borders Agency. London: House of Lords. https://www.escmid.org/fileadmin/src/media/PDFs/2News_Discussions/Press_activities/ECCMID_2015/ESCMID_one_million_deaths_UK_consumer.pdf, accessed on 3 September, 2015.

IARC (International Agency for Research on Cancer), 2015. http://www.iarc.fr/, accessed on 4 September, 2015.

International Labour Organisation 2013. *Shadow Economy and Undeclared Work.* http://ec.europa.eu/europe2020/pdf/themes/07_shadow_economy.pdf, accessed on 3 September, 2015.

Istituto per gli studi sui servizi sociali onlus, 2013. Donne Rom condizione femminile, diritti umani e non discriminazione. http://www.istisss.it/wp-content/uploads/Donne-Rom-AAVV_IMPAGINATO.pdf, accessed on 17 September, 2015.

Istituto Superiore per la Protezione e la Ricerca Ambientale (ISPRA) 2013. *Annuario*. http://annuario.isprambiente.it/, accessed on 3 September, 2015.

Long, M., Weil, P., Braibant, G., Delvolvé, P. and Genevois, B. 2015. *Les grands arrêts de la jurisprudence administrative*. Paris: Dalloz.

Murphy, R. 2013. *Closing the European Tax Gap. A Report for Group of the Progressive Alliance of Socialists & Democrats in the European Parliament by Richard Murphy FCA Director*, Tax Research UK. http://europeansforfinancialreform.org/en/system/files/3842_en_richard_murphy_eu_tax_gap_en_120229.pdf, acccessed on 3 September, 2015.

National Office on Anti-Racial Discriminations. 2012. National Strategy for the Inclusion of the Roma, Sinti and Caminanti Communities – European Communication No.173/2011. http://ec.europa.eu/justice/discrimination/files/roma_italy_strategy_en.pdf, accessed on 17 September, 2015.

OECD 2011. *Global Forum on Transparency and Exchange of Information for Tax Purposes*. Paris: OECD.

Regione Emilia-Romagna-Il Difensore Civico. 2013. http://www.assemblea.emr.it/garanti/attivita-e-servizi/difensorecivico/documenti/0QuadernoVersoilsuperamentodeicampinomadi2.pdf, accessed on 17 September, 2015.

Resilience Alliance 2007. *A Research Prospectus for Urban Resilience. A Resilience Alliance Initiative for Transitioning Urban Systems towards Sustainable Futures.* Canberra.

Study Group on Europe's Security Capabilities, 2004. A Human Security Doctrine for Europe. The Barcelona Report of the Study Group on Europe's Security Capabilities. http://www.consilium.europa.eu/uedocs/cms_data/docs/pressdata/solana/040915capbar.pdf, accessed on 17 September, 2015.

The Stockholm Resilience Centre 2015. http://www.stockholmresilience.org/21/research/research-programmes/planetary-boundaries.html, accessed on 3 September, 2015.

UNAR 2012. http://www.unar.it/unar/portal/wp-content/uploads/2013/09/Relazione-2012.pdf, accessed on 3 September, 2015.
United Nations 2014. UN Daily News, 20 June 2014. www.un.org/News/dh/pdf/english/2014/20062014.pdf, accessed on 17 September, 2015.
United Nations Commission on Human Security 2003. *Human Security Now.* http://www.un.org/humansecurity/content/human-security-now, accessed on 3 September, 2015.
United Nations Development Programme (UNDP) 1994. *Human Development Report 1994.* New York and Oxford: Oxford University Press.
United Nations Educational, Scientific and Cultural Organization (UNESCO) 2007. *Promoting Human Security: Ethical, Normative and Educational Frameworks in Western Europe.* http://unesdoc.unesco.org/images/0015/001511/151144c.pdf accessed on 3 July 2013.
United Nations Educational, Scientific and Cultural Organization (UNESCO) 2009. *Human Security Approaches and Challenges.* http://unesdoc.unesco.org/images/0015/001593/159307e.pdf, acessed on 3 September, 2015.
United Nations High Commissioner for Refugees (UNHCR) 2001. *Rechtliche und praktische Aspekte der Rückstellung von Personen, die keinen internationalen Schutz benötigen.* www.unhcr.de/fileadmin/rechtsinfos/fluechtlingsrecht/1_international/1_8_rueckfuehrung/FR_int_rueck-HCR_Rueckueberstellung.pdf, accessed on 3 September, 2015.
United Nations High Commissioner for Refugees (UNHCR) 2014a. Statement on Boat Incident off Greece Coast. www.unhcr.org/52df83d49.html, accessed on 3 September, 2015.
United Nations High Commissioner for Refugees (UNHCR) 2014b. War's Human Cost – UNHCR Global Trends 2013. www.unhcr.org/5399a14f9.html, accessed on 3 September, 2015.
United Nations Interregional Crime and Justice Research Institute UNICRI and MIT Senseable City Lab 2011. *New Energy for Urban Security. Improving Urban Security through Green Environmental Design.* Turin.
United Nations News Centre Online 2013a. *Nearly 93,000 People Killed in 'Vicious' Syria Conflict – UN Human Rights Chief.* http://www.un.org/apps/news/story.asp?NewsID=45162&Cr=syria&Cr1=#.Ud_OFazTKSI, accessed on 3 September, 2015.
United Nations News Centre Online 2013b. *One in Three Europeans Potentially Exposed to Asbestos, New UN Study.* http://www.un.org/apps/news/story.asp?NewsID=50739#.VeiYlLztmko, accessed on 3 September, 2015.
United Nations Trust Fund for Human Security (UNTFHS) 2009. *Human Security in Theory and Practice.* https://docs.unocha.org/sites/dms/HSU/Publications%20and%20Products/Human%20Security%20Tools/Human%20Security%20in%20Theory%20and%20Practice%20English.pdf, accessed on 3 September, 2015.
United States. Dept. of Homeland Security. Office of Inspector General. 2006. *A performance review of FEMA's disaster management activities in response*

*to Hurricane Katrina,* Dept. of Homeland Security, Office of Inspections and Special Reviews.

US Department of Defense. 2006. A Performance Review of FEMA's Disaster Management Activities in Response to Hurricane Katrina.

World Bank 2010. *The Costs of Roma Exclusion.* http://siteresources.worldbank.org/EXTROMA/Resources/Economic_Costs_Roma_Exclusion_Note_Final.pdf, accessed on 3 September, 2015.

World Health Organisation (WHO) 2013. *Asbestos: Use, Bans and Disease Burden in Europe*. http://www.who.int/bulletin/volumes/92/11/13-132118.pdf, accessed on 3 September, 2015.

World Health Organisation (WHO) 2015. *Asbestos: Elimination of Asbestos-related Diseases*. http://www.who.int/mediacentre/factsheets/fs343/en/, accessed on 3 September, 2015.

## Centres of Research or Information, Associations, NGOs

350.org & Attac France. 2015. Crime Climatique Stop ! Paris: Seuil

Amnesty International: http://www.amnesty.eu/.

Amnesty International 2010. *The Wrong Answer: Italy's 'Nomad Plan' Violates the Housing Rights of Roma in Rome.* http://www.amnesty.ch/de/themen/armut-und-menschenrechte/slums-zwangsraumungen/dok/2010/zwangsraeumung-roma/italy-the-wrong-answer-italys-nomad-plan-violates-the-housing-rights-of-roma-in-rome, accessed on 3 September, 2015.

Amnesty International 2012. *Italy. Briefing to the UN Committee on the Elimination of Racial Discrimination.* 80th session, February 2012. https://www.amnesty.org/en/documents/EUR30/001/2012/en/, accessed on 17 September, 2015.

Amnesty International 2013a. *Frontier Europe – Human Rights Abuses on Greece's Border with Turkey*. London: Amnesty International.

Amnesty International 2013b. *Libya: Refugees, asylum-seekers and Migrants Held Indefinitely in Deplorable Conditions*. http://www.amnestyusa.org/news/press-releases/refugees-asylum-seekers-and-migrants-held-indefinitely-in-deplorable-conditions-in-libya, accessed on 3 September, 2015.

ASGI: http://www.asgi.it.

Associazione 21 Luglio 2013. *Mia Madre era ROM. Le adozioni dei minori rom in emergenza abitativa nella Regione Lazio.* http://www.21luglio.org/wp-content/uploads/2013/10/Rapporto-Mia-madre-era-rom_Associazione-21-luglio2.pdf, accessed on 17 September, 2015.

Associazione 21 Luglio 2014. In Rome, the "Nomad Camp system" is worth 24 million euros. http://www.21luglio.org/en/nomad-camps-inc-in-rome-the-camp-system-is-worth-24-million-euros, accessed on 3 September, 2015.

Associazione Vittime Uranio Impoverito. http://inchiestauranio.blogspot.it/, accessed on 17 September, 2015.

Calabresi, C. 2013. Società Nazionale Operatori della Prevenzione. http://www.snop.it/attachments/article/274/Tessile%20e%20infortuni.pdf, accessed on 03/09/2015.

Consiglio Italiano per i Rifugiati 2013. *Survey on the Phenomenon of Statelessness Among Roma Communities Living in Italy*. http://www.cir-onlus.org/images/pdf/In%20the%20sun_apolidia_CIR_rapporto%20finale_italiano.pdf, accessed on 3 September, 2015.

Human Rights Watch 2002. *Fatally Flawed: Cluster Bombs and Their Use by the United States in Afghanistan*, 18 December, G1407. http://www.unhcr.org/refworld/docid/3f4f594b7.html, accessed on 3 September, 2015.

Human Rights Watch 2012. *Italy, Leave 'Nomad Emergency' in the Past*. http://www.hrw.org/news/2012/05/11/italy-leave-nomad-emergency-past, accessed on 3 September, 2015.

IlariaAlpi. http://en.wikipedia.org/wiki/Ilaria_Alpi, accessed on 15 September, 2015.

Immigrazione Onlus 2015. Drogarsi per lavorare la neo-schiavitù dei braccianti sikh alle porte di Roma. http://www.inmigrazione.it/UserFiles/File/Documents/87_2014%20-%20DOPARSI%20PER%20LAVORARE%20COME%20SCHIAVI.pdf, accessed on 8 June, 2015.

Industrial Disasters: http://en.wikipedia.org/wiki/List_of_industrial_disasters.

International Conference "How to think the Anthropocene? Anthropologists, philosophers and sociologists face to climate change", http://www.fondationecolo.org/l-anthropocene; http://www.fondationecolo.org/l-anthropocene/resumes, accessed on 10/11/2015

*International Journal of Mass Emergencies and Disasters*: www.ijmed.org/.

ISIS Explained 2014. http://www.infoplease.com/news/2014/isis-explained.html#ixzz3YJmdwEWn, accessed on 3 September, 2015.

Lunaria. 2013. *An Inhuman Costs*. http://www.lunaria.org/wp-content/uploads/2013/07/Inhuman_cost.pdf, accessed on 3 September, 2015.

Maccanico, Y. 2012. TESAT Report Shows Decrease in Terrorist Activity in 2011 but National Police Forces See a Continuing Threat. *Statewatch*, 22, 2/3, http://www.statewatch.org/analyses/no-216-tesat-terrorist-activity.pdf, accessed on 3 September, 2015.

Meltingpot web immigrants: http://www.meltingpot.org.

Migrants at Sea 2011. Libya Seizes Two Italian Fishing Boats in Renewed Dispute Over Libyan Territorial Waters. http://migrantsatsea.org/2011/11/, accessed on 17 September, 2015.

Migrants at Sea 2012. *Italian Appeals Court Acquits 2 Tunisian Fishing Boat Captains Who Rescued Migrants in 2007*. http://migrantsatsea.wordpress.com/2011/09/29/italian-appeals-court-acquits-2-tunisian-fishing-boat-captains-who-rescued-migrants-in-2007, accessed on 3 September, 2015.

Mucchielli website: http://www.laurent-mucchielli.org/.

Naga web Migrants and Roma people: http://www.naga.it.

Osservazione 2015. I miei vicini (rom) ed io, il presidente dellaToscana. http://www.osservazione.org/?lg=it&id=2_76&prn&prn, accessed on 17 September, 2015.

Sciurba, A. and Carnemolla, D. 2013. Due volte sfruttate. Le donne rumene nella fascia trasformata del ragusano. http://www.meltingpot.org/Due-volte-sfruttate-Le-donne-rumene-nella-fascia.html#.VNdsu2SG83I, accessed on 3 September, 2013.

SIPRI 2015. *SIPRI Yearbook 2015*. http://www.sipri.org/yearbook/2015, accessed on 5 September, 2015.

Statewatch 2007. Velsen Treaty 2007. http://www.statewatch.org/news/2007/oct/eu-gendarmerie-treaty-sept-2007.pdf, accessed on 3 September, 2015.

Vassallo Paleologo F. 2008. Accordo Libia-Italia sulla pelle dei migrant. Un'analisi del recente accordo sottoscritto a Bengasi da Silvio Berlusconi e Gheddafi. http://www.peacelink.it/ac/a/27117.html, accessed on 3 September, 2015.

Vassallo Paleologo, F. 2012. Nuove e vecchie intese tra Italia e Libia: continua l'embargo della stampa. Aumentano le vittime. http://old.asgi.it/home_asgi.php%3Fn=2131&l=it.html, accessed on 3 September, 2015.

Vassallo Paleologo, F. 2015. Blog, http://dirittiefrontiere.blogspot.it/, accessed on 3 September, 2015.

Workers of The World Journal. 2014. https://workersoftheworldjournal.files.wordpress.com/2014/10/workers-of-the-world_-international-journal-on-strikes-and-social-volume-1-number-5.pdf, accessed on 17 September, 2015.

## Newspapers, TV, Videos

ABC News Nightline 2012. *Apple Chinese Factory Foxconn Nightline*. http://www.youtube.com/watch?v=h1_XAuJ8qCc, accessed on 3 September, 2015.

Adragna, P., Bagnariol, G., Monaco, L. and Nencioni, P. 2013. *L'italia del lavoro nero*. http://inchieste.repubblica.it/it/repubblica/rep-it/2013/04/30/news/lavoro_nero-57761041/, accessed on 8 June, 2015.

Asud'Europa 2015. I nuovi schiavi. http://www.piolatorre.it/public/art/i-nuovi-schiavi-in-agricoltura-arrivano-dall-africa-330, accessed on 8 June, 2015.

Bilton, R. 2014. Apple 'failing to protect Chinese factory workers'. http://www.bbc.com/news/business-30532463, accessed on 17 September, 2015.

Bonifacio, M., Cangemi, A., Di Foggia, C. and Paudice, C. 2013. MUOS, l'antenna americana. http://inchieste.repubblica.it/it/repubblica/rep-it/2013/03/30/news/niscemi-55621210/?inchiesta=%2Fit%2Frepubblica%2Frep-it%2F2013%2F03%2F29%2Fnews%2Fmuos_l_antenna_americana-55619791%2F, accessed on 17 September, 2015.

Bordenet, C. 2014. *Bilan de la délinquance: la valse des stratégies de communication*. http://www.lemonde.fr/societe/article/2014/01/23/bilan-de-la-delinquance-la-valse-des-strategies-de-communication_4353620_3224.html, accessed on 3 September, 2015.

Castaldo, A. 2015. Il racconto della bracciante rumena schiavizzata, dans Corriere. http://video.corriere.it/mi-diceva-sei-schiava-l-incubo-erika-sfruttata-serre-costretta-quattro-aborti/4de492be-0e78-11e5-89f7-3e9b1062ea42?refresh_ce-cp, accessed on 3 September, 2013.

Corriere 2013. http://www.corriere.it/economia/13_dicembre_01/denunciati-5-mila-dipendenti-pubblici-ecco-chi-sono-nuovi-falsi-poveri-04a008b6-5a58-11e3-97bf-d821047c7ece.shtml, accessed on 3 September, 2015.

Corriere.it 2007. Amato: Facciamo come Giuliani Tolleranza zero nelle nostre città.http://archiviostorico.corriere.it/2007/agosto/17/Amato_Facciamo_come_Giuliani_Tolleranza_co_9_070817057.shtml, accessed on 3 September, 2015.

Corriere 2013. http://www.corriere.it/inchieste/reportime/societa/denis-verdini-confessa-ho-preso-soldi-nero/62573a58-39bd-11e3-893b-774bbdeb5039.shtml, accessed on 3 September, 2015.

CorriereTv 2015. *Insidie TTIP*. http://video.corriere.it/tutte-insidie-ttip-l-accordo-scambi-commerciali-usa-ed-europa/119fa006-cd6d-11e4-a39d-eedcf01ca586, accessed on 3 September, 2015.

Crispino, A. 2014. Così si muore nelle celle zero italiane. Dai pestaggi ai suicidi sospetti. Le foto incredibili, *Corriere della Sera.* http://www.corriere.it/inchieste/cosi-si-muore-celle-zero-italiane/94cd432c-8eb0-11e3-afb4-50ae7364e5b3.shtml, accessed on 3 September, 2015.

Dentini, F. 2009. https://ilsecoloxxi.wordpress.com/tag/scidone/, accessed on 3 September, 2015.

Diamanti, I. 2008. Se anche l'Italia di sinistra adesso respinge lo straniero, *La Repubblica,* 9 June, http://www.repubblica.it/2008/06/sezioni/cronaca/sicurezza-politica-7/mappe-demos-incubo/mappe-demos-incubo.html, accessed on 3 September, 2015.

Foschini, G. and Pleuteri, L. 2006. Quei 119 spariti dalla Polonia e adesso scomparsi in Puglia. http://www.repubblica.it/2006/09/sezioni/cronaca/scomparsi-polonia/scomparsi-polonia/scomparsi-polonia.html, accessed on 3 September, 2015.

Glawogger, M. 2005. *Workingman's Death,* movie.

German News Agency (dpa) 2014. Mehr als 140 000 Tote in Syrien. In: Frankfurter Allgemeine. Zeitung, 15 February, 2014. http://www.faz.net/aktuell/politik/ausland/naher-osten/friedensverhandlungen-wieder-gescheitert-mehr-als-140-000-tote-in-syrien-12803713.html, accessed on 17 September, 2015.

Italiani in Romania . http://www.italianiromania.com/ and http://www.expat-blog.com/it/nazionalita/italiano/in/europa/romania/timisoara/., accessed on 8 June, 2015.

Knaebel, R. 2012. *Libertés publiques 2013: comment l'Europe se prépare à espionner ses citoyens.* http://www.bastamag.net/auteur201.html, accessed on 3 September, 2015.

*La Nazione* 2014. *Cinesi, Milone attacca Rossi 'Se ne accorge soltanto ora'.* http://www.lanazione.it/prato/cronaca/2014/01/23/1014732-cinesi_milone_attacca_rossi.shtml, accessed on 3 September, 2015.

Leogrande, A. 2009. I desaparecidos polacchi nei campi di pomodoro del Tavoliere delle Puglie, 28 January, 2009. http://www.unita.it/italia/i-desaparecidos-polacchi-nei-campi-di-pomodoro-del-tavoliere-delle-puglie-1.26660, accessed on 8 June, 2015.

Lynn, G. and Davey E. 2015. Security guard fraud exposed by undercover researchers, BBC News, London, 23 March. http://www.bbc.com/news/uk-england-london-31876590, accessed on 18 September, 2015.

Médiapart 2014. *Tumers et silence autour de l'étang de Berre*. http://www.dailymotion.com/video/x1cp4qy_tumeurs-et-silences-autour-de-l-etang-de-berre_news, accessed on 3 September, 2015.

Médiapart 2015. http://www.mediapart.fr/journal/international/280415/le-secret-des-affaires-un-intense-lobbying-dans-les-couloirs-de-bruxelles, accessed on 17 September, 2015.

NOTAV. https://strugglesinitaly.wordpress.com/reappropriation/en-the-no-tav-struggle-in-the-susa-valley/, accessed on 3 September, 2015.

Orange, M. 2015. Trade Secrets, Intense Lobbying in the Corridors of Brussels. Médiapart. http://www.mediapart.fr/journal/international/280415/le-secret-des-affaires-un-intense-lobbying-dans-les-couloirs-de-bruxelles, accessed on 3 September, 2015.

Osservazione 2009. *EU Values: The Roma Migration Challenge. Italy Report,* D. Storia. http://www.osservazione.org/it/3_6/eu-values-the-roma-migration-challenge.htm, accessed on 3 September, 2015.

Palidda, S. 2012. Etica delle Polizie e spirito liberista. http://www.poliziaedemocrazia.it/live/index.php?domain=ricerca&action=articolo&idArticolo=2738, accessed on 4 September, 2015.

Panorama 2014. *La strage di cinesi a Prato ha colpe ben precise*. http://news.panorama.it/cronaca/strage-cinesi-prato-lavoro-nero, accessed on 3 September, 2015.

Peled. 2005. *China Blu.*

Presa Diretta RAI. 2009. *Caccia agli Zingari.* http://www.presadiretta.rai.it/dl/portali/site/puntata/ContentItem-f5076c88-0197-4235-b705-70e0bfd4c25a.html, accessed on 17 September, 2015.

Presa diretta RAI 2013. *Lavori in corso. Great works*. http://www.rai.it/dl/portali/site/puntata/ContentItem-70b803ce-184d-447a-83b7-f075a2d0346b.html, accessed on 3 September, 2015.

Presa Diretta RAI 2014a. *Rana Plaza, a un anno dal crollo*, http://www.presadiretta.rai.it/dl/portali/site/articolo/ContentItem-cd95c7a2-c2f0-4dda-97a1-35ab3969afd6.html, accessed on 17 September, 2015 .

Presa Diretta RAI 2014b. *Made in Italy*. http://www.rai.tv/dl/RaiTV/programmi/media/ContentItem-32d3b3be-421e-4164-af7b-d75aac92e530.html, accessed on 3 September, 2015.

Presa Diretta RAI 2015. *Il business dei rifugiati*. http://www.presadiretta.rai.it/dl/portali/site/puntata/ContentItem-3ce00810-9147-4dd3-a97b-b780845251b8.html, accessed on 3 September, 2015.

Quarrata news. 2012. http://quarratanews.blogspot.it/2012/09/banche-tempesta-nella-bcc-area-pratese.html, accessed on 7 September, 2015.

Report RAI 1997–2015. http://www.report.rai.it/dl/Report/Page-8200206a-0877-46c1-ab57-c774db2c6d38.html, accessed on 3 September, 2015.

Report RAI 2015. Report Cult: Nutrire il pianeta, http://www.report.rai.it/dl/Report/puntata/ContentItem-6218ddf2-6771-46ae-8a38-bbd715430e9b.html, accessed 4 September, 2015.

Rai Report 2014. *Transnistria.* http://www.rai.tv/dl/RaiTV/programmi/media/ContentItem-f166f9d7-bb8f-4b14-89e2-e3343a8e0273.html, accessed on 3 September, 2015.

RAI Blu Notte. 2012. Amianto: Le Morti Silenziose, 13 February. http://www.blunotte.rai.it/dl/portali/site/puntata/ContentItem-99ba6c31-c0b0-4a0f-9cd4-e28a5c0a2933.html, accessed on 17 September, 2015.

RAI TV 2009. *Viaggio nell'Italia dei veleni. Cinquasette luoghi tossici.* http://www.rai.tv/dl/RaiTV/programmi/media/ContentItem-49fe1052-f31e-406b-aee7-d520585f2551.html, accessed on 3 September, 2015.

RAI TV 2010. *Lo stipendio del capo della polizia italiana: 621.253,75 euro annui.* http://www.tg1.rai.it/dl/tg1/2010/articoli/ContentItem-8fbd9aa9-da93-49f9-8b2c-e50d213d6b66.html?p=0, accessed on 3 September, 2015.

*Rai3 Presadiretta 2009. Caccia agli Zingari.* http://www.presadiretta.rai.it/dl/portali/site/puntata/ContentItem-f5076c88-0197-4235-b705-70e0bfd4c25a.html, accessed on 3 September, 2015.

Rashbaum, W. K. 2010. Retired Officers Raise Questions on Crime Data. http://www.nytimes.com/2010/02/07/nyregion/07crime.html, accessed on 3 September, 2015.

Redazione online 2007. Livorno, incendio in campo rom: morti 4 bimbi. *Corriere della sera.* http://www.corriere.it/Primo_Piano/Cronache/2007/08_Agosto/11/incendio_campo_rom_livorno.shtml, accessed on 13 November, 2013.

Redazione online 2010. Case ai rom, la Procura indaga su discriminazione razziale. *Corriere della sera.* http://milano.corriere.it/milano/notizie/cronaca/10_dicembre_22/case-rom-procura-apre-fascicolo-discriminazione-razziale-comune-181128010796.shtml, accessed on 3 September, 2015.

Redazione online 2011. Roma: incendio in campo rom, quattro bambini morti. *Corriere della sera.* http://www.corriere.it/cronache/11_febbraio_06/roma-incendio-baracca_5c213c98-3230-11e0-a054-00144f486ba6.shtml, accessed on 3 September, 2015.

Ristretti web PRISON. http://www.ristretti.it/areestudio/territorio/opera/documenti/politiche/baratta.htm, accessed on 3 September, 2015.

Scroccaro, A. and Sivieri, C. 2009. Timiș Oara e l'imprenditoria della Calzatura Veneta. Dal Distretto dello Sportsystem di Montebelluna a Treviș Oara. https://unorzoperfavore.files.wordpress.com/2013/02/fne-upd-071-090.pdf, accessed on 9 June, 2015.

Statewatch: http://www.statewatch.org.

*The Guardian* 2014. http://www.theguardian.com/environment/gallery/2014/feb/27/agbogbloshie-worlds-largest-e-waste-dump-in-pictures, accessed on 03/09/2015

The Riots Communities and Victims Panel 2012. *After the Riots. The Final Report of the Riots Communities and Victims Panel*. London: The Riots Communities and Victims Panel.

Tourancheau, P. 2014. *Sécurité: le trafic de stats démonté.* http://www.liberation.fr/societe/2014/03/03/securite-le-trafic-de-stats-demonte_984292, accessed on 3 September, 2015

Toxic Europe. 2011. https://www.irpi.eu/it-toxic-europe/, accessed on 17 September, 2015.

*Una montagna di balle* 2009. https://vimeo.com/6000381, accessed on 17 September, 2015.

Various. 2011. *Santa's Workshop – Inside China's Slave Labour Toy Factories*. http://www.youtube.com/watch?v=yF8jUDzz5bE, accessed on 3 September, 2015.

# Index

For Product Safety Concerns and Information please contact our EU
representative GPSR@taylorandfrancis.com
Taylor & Francis Verlag GmbH, Kaufingerstraße 24, 80331 München, Germany

www.ingramcontent.com/pod-product-compliance
Ingram Content Group UK Ltd.
Pitfield, Milton Keynes, MK11 3LW, UK
UKHW010956180926
475701UK00004B/390

* 9 7 8 0 3 6 7 5 9 6 8 3 5 *